The Laws of Cool

The Laws of Cool

Knowledge Work and the Culture of Information

Alan Liu

The University of Chicago Press
Chicago and London

Alan Liu is professor of English at the University of California, Santa Barbara. He is the author of *Wordsworth: The Sense of History* (1989) and essays on eighteenth- and nineteenth-century literature, literary theory, cultural studies, and information culture. He is also the creator of *Voice of the Shuttle,* a Web site for humanities research, and the director of Transcriptions: Literary History and the Culture of Information, a project funded by the National Endowment for the Humanities.

The University of Chicago Press gratefully acknowledges the generous support of the John Simon Guggenheim Memorial Foundation toward the publication of this book.

The University of Chicago Press, Chicago 60637
The University of Chicago Press, Ltd., London

Printed in the United States of America
12 11 10 09 08 07 06 05 04 1 2 3 4 5
ISBN: 0-226-48698-2 (cloth)
ISBN: 0-226-48699-0 (paper)

Library of Congress Cataloging-in-Publication Data

Liu, Alan, 1953–
The laws of cool : knowledge work and the culture of information / Alan Liu.
p. cm.
Includes bibliographical references and index.
ISBN: 0-226-48698-2 (cloth : alk. paper)—ISBN 0-226-48699-0 (pbk. : alk. paper)
1. Information society. 2. Knowledge workers. 3. Humanities—Social aspects. 4. Education, Higher—Aims and objectives. 5. Internet—Social aspects. 6. Digital media. 7. Literature and technology. 8. Art and technology. 9. Popular culture—History—20th century. 10. Work—Social aspects. I. Title: Knowledge work and the culture of information. II. Title.

HM851.L56 2004
303.48′33—dc22

2003024504

♾ The paper used in this publication meets the minimum requirements of the American National Standard for Information Sciences—Permanence of Paper for Printed Library Materials, ANSI Z39.48-1992.

For Patricia Fumerton

Contents

Acknowledgments

Since I can here reproduce only a "lossy" image[1] of the full network of those who have contributed to this book, let me bring to the fore just those to whom I am most consciously in debt, hoping that faulty human memory (supplemented by additional acknowledgments in my notes) will not drop too many bits.

I want first to thank—and honor—Helen Tartar, who is a humanities editor for the ages. She saw my previous book into press with remarkable, intimate attention to matters both intellectual and practical; supported the writing of this book with equal acumen; and was only prevented from seeing this book join the inspired literature and theory lists she built at Stanford University Press because of a restructuring and downsizing at the press that set her aside. It is with a piercing sense of pain and irony that I am forced in my acknowledgment of Helen also to acknowledge the power of the post-industrial forces that are major themes in my book about the fate of the humanities in the age of knowledge work. Thanks to Helen for her courage, loyalty, rigor, and ferocious intellectual curiosity. I remember a crackling fireplace we sat by when she visited me in Bethany, Connecticut, during a blackout caused by an ice storm. In the morning the trees were covered in a brilliant sheen of ice—every twig and leaf. Fire

[1] A term in digital imaging for the irretrievable loss of information in some compressed image formats.

and ice: these stay with me as emblems of the clarity of Helen's editorial vision.

Over the years, the following individuals have made suggestions, read my manuscript (in whole or part), and offered other important support: Jay Clayton, Robert Essick, Jonathan Freedman, Jennifer Jones, Katherine Hayles, Richard Helgerson, Matthew Kirschenbaum, Laura Mandell, Jerome McGann, J. Hillis Miller, David Palumbo-Liu, Mark Poster, Rita Raley, David Simpson, Randolf Starn, and William Warner. The complexity of their assistance—ranging from detailed suggestions to overall perspective—is belied by any simple list. A special thanks to Rita Raley, who has worked closely with me on issues related to new media and its implications. I have benefited greatly from her suggestions, insights, and advance scouting of works and trends in digital literature and art.

To this list—or, better, network—I wish to superimpose (with some overlap) two others. One is the group of colleagues and students I worked with closely at University of California, Santa Barbara, on the Transcriptions Project; its curricular offshoot, the Literature and Culture of Information Specialization; and such other digital projects as the *Voice of the Shuttle*, the *Romantic Chronology*, and the UCSB English Department Web site. These include Robert Adlington, Charles Bazerman, Carolyn Brehm, Sharon Doetsch, Jeremy Douglass, Laurie Ellinghausen, Eric Feay, Andrea Fontenot, Michael Frangos, Elizabeth Freudenthal, Tassie Gniady, Robert Hamm, Jennifer Hellwarth, Chris Hoffpauir, Sheila Hwang, Zia Isola, Jennifer Jones, Gisela Kommerell, Sarah McLemore, Christopher Newfield, Carol Pasternack, Michael Perry, Rita Raley, Christopher Schedler, Jeanne Scheper, Diana Solomon, Carl Stahmer, Melissa Stevenson, Jennifer Stoy, Anna Viele, William Warner, Eric Weitzel, Vincent Willoughby, and Jeen Yu. Administrative support from Christina Nelson, Laura Baldwin, and Lyn Thompson, along with technical support from Brian Reynolds and Karen Whitney, were invaluable. Thanks for all this group's help in building the "new media" community in my English department that has nurtured my ideas about the relation of the humanities and arts to information culture.

Another network of people from whom I have gained much knowledge, and just plain enjoyment, consists of the artists, writers, programmers, computer scientists, and scholars in other disciplines or institutions with whom I have collaborated or whom I have consulted about new media, information culture, and digital technology. These include especially Amr El Abbadi, Bruce Bimber, Anna Everett, Andrew Flanagin, Janet Head, George Legrady, Marjorie Luesebrink, Robert Nideffer, and Victoria Vesna, as well as the officers of the Electronic Literature Organization and that organization's working group for the preservation, archiving, and dissemination of electronic literature (PAD). Thanks to this group for widening my horizons.

Research fellowships from the John Simon Guggenheim Memorial Foundation, the National Endowment for the Humanities, and the University of California Office of the President have helped create the time to educate myself in the relevant fields and to write. Thanks also to the NEH for the institutional grant that allowed me to start the Transcriptions Project, and to the College of Letters and Science, the Division of Humanities and Fine Arts, the Interdisciplinary Humanities Center, and the Department of English at my home institution, the University of California, Santa Barbara, for substantial complementary project funding. David Marshall, dean of Humanities and Fine Arts at UCSB, has been instrumental in supporting my work and that of others on my campus interested in digital media and culture. A special thanks to Alan Thomas at University of Chicago Press for believing in my long manuscript; to Randy Petilos, Leslie Keros, and others at the Press who saw the manuscript so smartly into production, and to my superb copy editor and indexer, Lys Ann Shore.

A version of the preface to part 1 appeared previously as "Knowledge in the Age of Knowledge Work" in *Profession* (1999) and is reprinted by permission of the copyright owner, Modern Language Association of America. Portions of my introduction, chapters 1 and 5, the preface to part 3, and appendix C appeared in earlier form as "The Future Literary: Literature and the Culture of Information," in *Time and the Literary*, edited by Karen Newman, Jay Clayton, and Marianne Hirsch (New York: Routledge, 2002). Portions of my introduction and chapter 1 also appeared in a version abridged by Randolph Starn under the title "The Downsizing of Knowledge: Knowledge Work and Literary History" in *Knowledge Work, Literary History, and the Future of Literary Studies*, edited by Christina M. Gillis (Berkeley: Doreen B. Townsend Center/Regents of the University of California, 1998).

This book is dedicated to Patricia Fumerton, who, with our daughter, Lian, makes up my closest, most important network. The intellectual interest that Paddy shares with me in the question of history and culture, the tolerance she has shown for the long hours of digital activity I have had to devote to this book, her own exploration of digital technology for the Early Modern Center she directs at UC Santa Barbara, and the fine interweaving of our life that is the true texture behind my "voice of the shuttle" and other works: these remind me how precious is human life—far more robust and fragile, both, than any virtual or "posthuman" life we now know. In this, my nearest network, we may not be cool, but we are warm.

Introduction

Literature and Creative Destruction

This book is a study of the cultural life of information or, more broadly, of contemporary "knowledge work." The specific question I prepare for—but that I can turn to only speculatively here and in my concluding chapter—concerns the role of literature in that cultural life. What is the future of literature and literary study when all culture is increasingly the culture of information and when even literary scholars subordinate literature to an apparent clone of information—cultural context? And a related question: what is the future in general of the humanities and arts when the former seems destined only for what information industries call "content" and the latter for "multimedia entertainment"?

To be honest, my concern is not really with works of literature as such, which from the viewpoint of general society have effectively lost their category distinction on the gradient that blurs textuality and information, imagination and entertainment, authors and celebrities, and publishers and conglomerates. My concern, more crucially, is with the underlying sense of *the literary,* which is even now searching for a new idiom and role. After all, whether we should elegize or celebrate "the death of literature" (as in Harold Bloom's *Western Canon* or John Beverley's *Against Literature*) is now beside the point.[1] Literature as traditionally understood no longer survives as an autonomous force or, put in the cultural-critical terms of the current academy, as a force positioned by larger forces in the guise of autonomy. Since the high point of its avowed self-possession (roughly from the eighteenth through the nineteenth century), literature has merged with mass-market, media, educational, political, and other institutions that reallocate, repackage, and otherwise "repurpose" its assets. Such churning of literary capital has only accelerated in the information age as major institutions compete to appropriate that capital under the spotlight of media coverage (e.g., in the canon wars, which pitted political pundits against academics).

But all that is done, and we need harbor no false romanticism about the literature that was. Whatever one thinks of cultural criticism, it has been brutally effective in demonstrating that the churning of literary capital has always characterized literature. Literature could not have been part of the life of culture otherwise. What is of interest now is the distinctive form of that churning in relation to the general economic and social churning that Joseph A. Schumpeter, in his classic phrase about capitalism, called "creative destruction." A "perennial gale of creative destruction," Schumpeter wrote in 1942, "incessantly revolutionizes the economic structure *from within*, incessantly destroying the old one, incessantly creating a new one."[2] The real competition, Schumpeter said, is not the normal furor over prices, quality, and sales effort, but "competition from the new commodity, the new technology, the new source of supply, the new type of organization . . . [competition] which strikes not at the margins of the profits and the outputs of the existing firms but at their foundations and their very lives."[3] Recent scholars of business and economic history take such furious creativity to be simply postindustrial business as usual. In "post-capitalist society," Peter F. Drucker says, "creative destruction" is "innovation," compelling the "systematic abandonment of the established, the customary, the familiar, the comfortable."[4] The "spirit of informationalism," Manuel Castells adds, "is the culture of 'creative destruction' accelerated to the speed of the optoelectronic circuits that process its signals."[5] Meanwhile, in the thriving print and television journalism of business, Schumpeter's dictum has become cliché (his other work and especially his prediction of the eventual demise of capitalism conveniently forgotten). A special double issue of *Business Week* on "The 21st Century Corporation" in 2000 takes it for granted that "knowledge-based products and networks can quickly disappear in a burst of Schumpeterian creative destruction. So corporations must innovate rapidly and continuously."[6]

The vital task for both literature and literary study in the age of advanced creative destruction, I believe, is to inquire into the aesthetic value—let us simply call it *the literary*—once managed by "creative" literature but now busily seeking new management amid the ceaseless creation and re-creation of the forms, styles, media, and institutions of postindustrial knowledge work. In the regime of systematic innovation, is the very notion of the literary doomed to extinction even if—or, rather, especially if—it begins to venture "creatively" into the province of knowledge work, if it dares to imagine a literature of the database, spreadsheet, report, and Web page? After all, next to the great institutional documents of our times heralding "innovation" in their very logos—the legions of "dot com" company prospectuses, Web sites, advertisements, and so on—what could literature be but a minor act of creativity, like a screensaver? This is one way to read the powerful, repeated dirge that John Guillory, Alvin Kernan, J. Hillis Miller, and others have sounded over (in Guillory's words, in *Cultural Capital*) the "perceived decline in the cultural significance of literature itself, the perceived marginality of literary culture to the modern social order."[7] In such wakes, there is a note of mourning that seems excessive until we realize that what is being mourned is not so much literature as the "literary culture" that is the very possibility of literature. Or if literariness is to rise from the dead—to entertain Guillory's concluding prophetic elegy, his surprising surmise of a redemptive "aestheticism unbound"—where could it go and what does it yet have to do?[8] What is the future of the literary when the true aestheticism unbound of knowledge work—as seen on innumerable Web pages—is "cool"? Cool is the techno-informatic vanishing point of contemporary aesthetics, psychology, morality, politics, spirituality, and everything. No more beauty, sublimity, tragedy, grace, or evil: only cool or not cool.

Or rather, since the fate of the literary is an abstraction unless we also address the fate of literary people, the operative question becomes: what is the relation between the now predominantly academic and other knowledge workers (even "creative writers") who manage literary value in "cultural context" and the broader realm of professional, managerial, and technical knowledge workers who manage information value in "systems"?* What do the well-read who once held power in the name of the aesthetic still have to teach the well-informed who now hold power under the cover of cool?[9]

As may be discerned in the conflicted way I have so far invoked cultural criticism, these questions are interwoven for me with questions about what role that method, and the contemporary humanities education it represents, might play in the emergence of the future literary. I was one of the academic intellectuals in the 1980s and 1990s who staked their careers on joining literary studies at the hip to cultural studies.[10] But now it is time to reflect on the legitimacy of cultural criticism in ways that shift earlier controversies

* On the term *knowledge worker*, see appendix A, "Taxonomy of Knowledge Work."

about cultural materialism, the New Historicism, multiculturalism, and so on to new ground.

It might be said, with Kafkaesque irony: I went to sleep one day a cultural critic and woke the next metamorphosed into a data processor. It is not just that cultural context and information have come to approximate each other in their gross anatomy (each requiring the same kind of gathering, collating, and filtering work); it is that now even the fine structure seems to mate. We can extend to informationalism Arif Dirlik's general point about postindustrialism in his *Postcolonial Aura* (especially the chapter "The Postmodernization of Production and Its Organization").[11] In a convergence so massive as to be all but indiscernible in normal academic practice, advanced literary study has since the 1970s evolved from structuralism through deconstruction to cultural/multicultural criticism, so as to swing into conjunction with an information society that meanwhile evolved in parallel from logocentric corporations and broadcast empires to the postindustrial equivalents of cultural diversity—flexible-team corporations and distributed information networks. To put it rudely, in other words, perhaps the academic controversies of the past two decades were not really about supplanting the author or canon with the deconstructive intertext or cultural context. Perhaps such controversies were really about recruiting professional interpreters for an impending mental merger with the software-telecom-cable-Hollywood conglomerates now promising that ultimate intertext or context, high-bandwidth information.

After all, any cultural critic who today uses a personal computer to write "files" about literature is from the first incorporated within an information culture closest to hand in the operating system itself.[12] One might say that the well-known epilogue to Stephen Greenblatt's paradigm of New Historicist criticism, *Renaissance Self-Fashioning*, needs to be updated. To reshoot for the 2000s the scene in which Greenblatt reads Clifford Geertz's *Interpretation of Cultures* while sitting on a plane with a man whose son is grievously ill would require that our camera pull back to reveal the device that all the corporate intelligentsia up and down the aisle have open instead of a book—a laptop or handheld computer.[13] Greenblatt romancing Geertz (and, it must be said, myself romancing Wordsworth's "sense of history" in the 1980s) is as expert at opening archives of cultural memory as the managerial/professional/technical intelligentsia are at neuromancing databases and spreadsheets. Cultural-critical experts, in other words, read in a manner originally schooled by the technical rigor of formalism.[14] At the same time the corporate intelligentsia processes with an equal technical ritual schooled in the burgeoning corporate learning industry—Dana U., Disney U., Motorola U., Solectron U., and so on.[15] Even the technical jargon seems congruent. Such "politically correct" academic antifoundationalisms as *différance* and *difference* are matched by the mytho-Japanese antifoundationalisms of the new corporate correctness: "continuous improvement *(kaizen)*, just-in-time delivery, total quality, statistical process control, and 'design for' manufacture and assembly."[16]

The one consistent difference is that cultural criticism is fundamentally historical. I mean by this more than the obvious fact that most humanities fields are now *ipso facto* historical (in the noncontroversial sense in which "literature department" is synonymous with "literary history department"). I mean also that even as cultural criticism has rejected older modes of literary or intellectual history, it has not repudiated the necessity of historical consciousness. It has instead proposed rude ways to examine the seeming obviousness of such necessity. Cultural criticism wants to know why historical consciousness became the core of humanities education from the Enlightenment on. What was such consciousness for, and whom did it serve?

Put in the past tense, such questions concern what Jean-François Lyotard has called the "metanarratives" of progressive humanity and speculative reason that academic historicism once sustained but that now, from the viewpoint of cultural critics, seem just so many empty postures.[17] But it is the *present* tense of these questions—the sense that they bear on a gigantic "now" inclusive of the Enlightenment and the nineteenth and twentieth centuries *together*—that cultural criticism has found most compelling. That now is modernity. In the broadest sense, the underlying historical concern of cultural criticism has been modernization, the centuries-long "progress" of rationalization, routinization, institutionalization, organization building, and empire building (with their attendant political, market, and media effects) engineered by post-Enlightenment industrial societies.[18] Cultural criticism is the critique by disjunction of such progress. Its characteristic practice has been to bring pressure to bear on the apparently seamless necessity of modernization by foregrounding the otherness of early modern, subcultural, multicultural, postmodern, and other alternative historical paradigms. Queen Elizabeth I's unchanging portrait-face and the singer Madonna's ever-changing face (the early modern and postmodern, respectively) thus come to look alike in such criticism because both hold up a mirror of historical difference to modern understandings of the relation between cultural display and institutionalized power (the queen's face constituting a power of display unconfirmed by modern police or military apparatuses, the singer presiding over a "society of spectacle" also empty of rational reality).[19] There are risks in such a notion of history, including the tendency to flatten out all historical difference into the single, politically inert "difference" that

Dirlik has criticized. But there is also much to gain from such a notion, according to which history is not dead to modernity, but other to modernity.[20]

By contrast, the world of "just-in-time" knowledge work—of astonishingly rapid yet finely calibrated turnovers in supplies, inventory, documents, mission statements, and finally people—waits upon the death, or layoff, of no one, certainly not that of the son of the man in the plane. As arbitrageurs of "creative destruction" might say: he's history. History—including the history of modernization itself (now identified with smokestack industries and ossified organizations)—is obsolescence.[21]

Of course, what I called the merger between academic humanities "research" (the very term is symptomatic) and corporate, government, media, medical, and military knowledge work has developed over time—whether we date such convergence from the period circa 1900 when U.S. universities first modernized under the influence of corporate capitalism, from 1900–1930 when the academic and white-collar sectors grew in tandem, or from the boom after World War II when the relation between the academic sciences and the military-industrial-government complex claimed attention.[22] Lyotard's *Postmodern Condition* is one well-known recent critique of the resulting convergence upon "performative" knowledge or what Bill Readings, extending Lyotard's case in *The University in Ruins*, ironically calls corporatist "excellent" knowledge.[23] Excellent knowledge, Readings writes, is a posthistorical knowledge that has dispensed with the progressive narratives of Humanity and Reason that once afforded the university its mission, in favor of mere specifications of technical and informational systematicity.[24]

But only with the new millennium are the conditions in place for humanists to grasp the full implications of such convergence upon "excellent knowledge"—above all, to recognize that the defining issue for a field like literary studies really is its position with regard to information and, borne on the carrier wave of information, the juggernaut of postindustrial knowledge work. This is because the combined ideological and material build-up to the year 2000—to its installation in social consciousness as an "event"—focused the problem with sudden clarity. Ideologically, "2000" unlocked an end-of-history enthusiasm that theorizes knowledge work as *millennial* knowledge—that is, as knowledge that is antihistorical (anti-obsolescent) on principle.[25] The centrality of the challenge to academic knowledge thus stands starkly revealed: knowledge work is not just indifferent to humanistic knowledge, it opposes it on principle. The material instantiations of "2000" have been just as bracing. Among all the technological, political, and economic infrastructures put in place to install the new world order and new economy, just one may be mentioned as epitomizing the whole: networked information technology. Networked IT crossed a threshold of scale in the mid-1990s beyond which—as evidenced most spectacularly by the World Wide Web—competing models of knowledge work, once rooted semi-autonomously in academic, business, media, health-industry, government, and other sectors, suddenly seemed to fuse into a single, parsimonious continuum—so-called "worldwide"—able to afford just one global understanding of understanding.[26]

Nor, it should be added, has the build-up to "2000" been merely a remote abstraction for academics. As indicated by controversies in the late 1990s over plans to "partner" the information systems of major U.S. universities with technology corporations, to restructure other universities according to the philosophy (declared by one university president) of "pretending" to be "a corporation," to gear still others for the technological future by eliminating whole suites of liberal arts programs, or, in New Zealand, to reorganize the higher education system into strictly "accountable" corporate units, the academy is increasingly being told by administrators and legislators to attend to business, or else.[27] Commentary by humanities scholars such as Bill Readings, J. Hillis Miller, Paul Lauter, Wesley Shumar, Jeffrey Williams, and (in the United Kingdom) Kevin Robins and Frank Webster thus began appearing in the 1990s to express concern over the corporatization of the university.[28] And meanwhile the ballooning numbers of temporary, part-time, nonladder, and other itinerant educators and graduate students have been even more pointed in their concern.[29]

Wherever the academy looks in the new millennium, it sees the prospect of a world given over to one knowledge—a single, dominant mode of knowledge associated with the information economy and apparently destined to make all other knowledges, especially all historical knowledges, obsolete. Knowledge work harnessed to information technology will now be the sum of all worthwhile knowledge—except, of course, for the knowledge of all the alternative historical modes of knowledge that undergird, overlap with, or—like a shadow world, a shadow web—challenge the conditions of possibility of the millennial new Enlightenment.

If cultural criticism is to be legitimate, I speculate, then together with the creative arts, it must metamorphose not so much into Kafka's insect as into a different kind of "bug." I have in mind the slow, sprawling, yet ever so graceful "Kuang" computer virus at the end of William Gibson's *Neuromancer*, which can break the densest ICE (intrusion countermeasures electronics) of corporate databases because it transforms its own substance into that of the database, draws nearer and nearer until there seems to be no difference, and then at last injects the one powerful difference it has treasured at its core.[30] My highest ambition for cultural criticism and the creative arts is that they can in tandem become "ethical hackers" of knowledge work—a problematic role in the information world but one whose general cultural paradigm needs to be explored.* Many intellectuals and artists will become so like the icy "New Class" of knowledge workers that there will be no difference; they will be subsumed wholly within their New Economy roles as symbolic analysts, consultants, and designers. But some, in league with everyday hackers in the technical, managerial, professional, and clerical mainstream of knowledge work itself, may break through the ice to help launch the future literary. For it is the future literary—or whatever the peculiarly edgy blend of aesthetics and critique once known as the literary (and its sister arts) will be named—that can serve as witness to the other side of creative destruction: not the boundless "creation" that has powered the market rallies of the New Economy, but the equally ceaseless destruction that produces historical difference. This is why it now makes sense to think of cultural criticism and the creative arts as having come into special conjunction. Where once the job of literature and the arts was creativity, now, in an age of total innovation, I think it must be history. That is to say, it must be a special, dark kind of history. The creative arts as cultural criticism (and vice versa) must be the history not of things created—the great, auratic artifacts treasured by a conservative or curatorial history—but of things destroyed in the name of creation.

Nor, we may further conjecture, will it be sufficient merely to *witness* the history of things destroyed in the name of creation. Reviewing the ways in which the avant-garde and subcultural arts of the twentieth century have already influenced the arts of information at the new millennium (as exemplified in the writings and projects of such collectives as the Critical Art Ensemble and Electronic Disturbance Theater), it is difficult to avoid the conclusion that the most ambitious art will henceforth "make history" by itself performing acts of destruction—or at least of blockage and trespass—in a certain manner, against certain targets.[31] Whether it is expressed as appropriation, sampling, defacement, or hacking, there will be nothing more cool—to use the term of the nascent, everyday aesthetics of knowledge work—than committing acts of destruction against what is most valued in knowledge work—the content, form, or control of information. Instantaneous, simultaneous, and on-demand information is the engine of the post-industrial "now" submitting history to creative destruction, and it is the destruction of this eternal "now" or self-evident presence of information, therefore, that will have the most critical and aesthetic potential. Strong art

* See appendix C on ethical hacking.

will be about the "destruction of destruction" or, put another way, the recognition of the destructiveness in creation.

The lesser practices of such an aesthetic will be acts of delay, displacement, oblique representation, and stylization—all that information "cool" already unconsciously practices to impede or parody the self-evident force of information. The stylistic repertory from which the future literary may emerge, therefore, is not new. But what *may* be "new" among the writers, artists, programmers, designers, critics, scholars, and others who push workaday cool to an extreme is the rejection of the aesthetic ideology of critical innovation ("make it new") in favor of an ideology of critical destruction. Such an ideology of art, I argue, will be intended to reorganize the residual avant-garde, subcultural, and counter-cultural elements in cubicle culture (psychedelic screensavers and so on) into a cool that achieves what may be called an "ethos of the unknown." The ethos of the unknown is a zone where those who live and work nowhere but inside the system of contemporary knowledge can paradoxically, and with more than the normal (and normalizing) irony of cool, seem to stand outside it.[32]

Teaching the difference between such an ideology of critical destruction and terrorism—the subtle difference between dark historicism and even darker inversions of creative destruction—will be the special concern of both humanities educators and writers or artists in the future.*

But before we can glimpse the future literary, we need to make a serious attempt to grasp contemporary knowledge work and its information culture. What is knowledge work? How does information work sustain it? And how might the *culture* of such information—self-named "cool"—challenge knowledge work to open a space, as yet culturally sterile (coopted, jejune, anarchistic, terroristic), for a more humane hack of contemporary knowledge?

In parts 1 through 3 of this book, I offer a historical sketch of knowledge work and a theoretical frame for investigating its culture of cool. In part 4, I follow up with an argument about the role of humanities education and the arts in the world of cool. This latter argument turns on the general character of historical and aesthetic knowledge in the information age, and will bring me just to the verge—but no further—of discussing literature and art in particular, both in their similarity and in their difference. There is much "more," of course, as the button at the bottom of any Internet search-

* This introduction, and much of this book, was written before the events of September 11, 2001. The issue of "terrorism" is conceptually integral to some of the "destructively creative" art I discuss in chapter 11 (certain instances of which were imputed to be "terrorist" in principle well before the new millennium). I have chosen not to alter or inflect the issue in hindsight.

engine page promises, and the brief prolegomenon on the future literary that closes the book will only mark the path of research and practice yet to come. That is the path along which literary knowledge might yet converge with the other kinds of aesthetic knowledge upon cool. For there *is* a ghost in the machine. It is called culture, and all the hope of the future literary is to be what Sir Philip Sidney called the companion of the camps of that culture—diviner, soldier, critic, and hacker all at once.[33]

PART I

The New Enlightenment

We are all aware how work both emboldens us and strangles our soul life in the very same instant. It reveals how much we can do as part of a larger body, literally a corpus, a *corporation,* and how much the wellsprings of our creativity are stopped at the source by the pressures of that same smothering organization.

. . . We stand to gain a marvelous involvement in our labors, but must relinquish a belief that the world owes us a place on a divinely ordained career ladder. We learn that we do have a place in the world, but that it is constantly shape-shifting, like the weather and the seasons, into something at once new and beautiful, tantalizing and terrible.

Confronted with the difficulty and drama of work, we look into our lives as we look into deep water. We kneel, as if by the side of a pool, seeing in one moment not only the fleeting and gossamer reflection of our own face, clouded and disturbed by every passing breath and the lives of all the innumerable creatures that live in its waters, but the hidden depths below, beyond our sight, sustaining and holding everything we comprehend.

David Whyte, *The Heart Aroused: Poetry and the Preservation of the Soul in Corporate America* (1994)

Preface

"Unnice Work"

Knowledge Work and the Academy

> Even this warning did not prepare Robyn for the shock of the foundry. . . . Her first instinct was to cover her ears, but she soon realised that it was not going to get any quieter, and let her hands fall to her sides. The floor was covered with a black substance that looked like soot, but grated under the soles of her boots like sand. The air reeked with a sulphurous, resinous smell, and a fine drizzle of black dust fell on their heads from the roof. Here and there the open doors of furnaces glowed a dangerous red, and in the far corner of the building what looked like a stream of molten lava trickled down a curved channel from roof to floor. . . . Everywhere there was indescribable mess, dirt, disorder
>
> David Lodge, *Nice Work* (1988)

Was David Lodge's 1988 novel simply behind the times when it challenged its heroine, Robyn Penrose, temporary lecturer in English literature, to confront the sooty business managed by its hero, Vic Wilcox, product of a Midlands technical college? Was this the utmost challenge that Lodge could imagine for the contemporary academic sensibility: to come to grips with the realism of "smokestack" industrialism as it has appalled fiction since the nineteenth-century industrial novel (Lodge's elaborate allusion) through at least D. H. Lawrence's *Sons and Lovers*? If so, then we can adequately attribute Lodge's comedy to the slow, sly romance he builds between the academy and industry (and their protagonists)—to his deft dance of opposites that at last issues, if not in a classically comic wedding, then at least in the fleeting copulation of two faculties of expertise divorced since Victorian sages presided over the "idea of a university."

Or should we allow Lodge's minor prophets of the new world order—Robyn's investment banker brother; his financial exchange dealer consort; or, perhaps most tellingly, the "CNC," or computer-numerically-controlled manufacturing machine in Vic's factory—to shift the comedy into an altogether different register of satire? Robyn's brother says cheekily while on holiday from financial London: "Companies like [Vic's] are batting on a losing wicket. . . . the future for our economy is in service industries, and perhaps some hi-tech engineering." Vic says somberly, as he and Robyn stare across a Perspex pane at the CNC machine's inhumanly "violent, yet controlled" motions, "One day . . . there will be lightless factories full of machines like that. . . . Once you've built a fully computerised factory, you can take out the lights, shut the door and leave it to make engines or vacuum cleaners or whatever, all on its own in the dark." "O brave new world," Robyn responds.[1]

To glimpse even peripherally such a brave new world order is to recognize that Lodge's last, best joke—so cruel that only his furiously contrived happy ending can salve the bite of the satire—is the obsolescence of the entire, tired opposition between the academy and industry. "Shadows" of each other, as the novel calls them, Robyn and Vic both inhabit a twilight order on the other side of the Perspex—or more fittingly, computer screen—from true *post*-industrial night. That night, which in its own eyes seems the dawning of a new enlightenment, is knowledge work, and the information work that is its medium. Knowledge work is the *Aufhebung* of both academic "knowledge" *and* industrial "work."

"We grew up in the Industrial Age," Thomas A. Stewart writes in *Intellectual Capital: The New Wealth of Organizations* (1997):

> It is gone, supplanted by the Information Age. The economic world we are leaving was one whose main sources of wealth were physical. The things we bought and sold were, well, *things;* you could touch them, smell them, kick their tires, slam their doors and hear a satisfying thud. . . .
>
> In this new era, wealth is the product of knowledge. Knowledge and information—not just scientific knowledge, but news, advice, entertainment, communication, service—have become the economy's primary raw materials and its most important products. Knowledge is what we buy and sell. You can't smell it or touch it. . . . The capital assets that are needed to create wealth today are not land, not physical labor, not machine tools and factories: They are, instead, knowledge assets.[2]

The clarion call of the new millennium is clear: *Let the academies have pure ideas. Let the Third World (represented in Lodge's novel by the swarthy, immigrant underclass who serve Vic's factory) have pure matter work. You, the New Class destined to inherit the earth (or at least cubicle), you who are endowed with the inalienable right to process a spreadsheet, database, or report: have you counted your knowledge assets today?*

But this is too facile a caricature of the age of knowledge work. Just as Lodge's academic romance can be read in different tones, so too can our contemporary romances of knowledge work. I refer to the immensely influential and best-selling works of fiction-blended-with-realism—let us loosely call them "novels"—by the Victorian sages of our time: management "gurus" (Stewart among them). The mold, as John Micklethwait and Adrian Wooldridge note in their survey of the genre, was set in 1982 by Tom Peters and Robert Waterman, Jr.'s best-selling *In Search of Excellence: Lessons from America's Best-Run Companies.*[3] The genre came into its own in the early 1990s, with the appearance of works of wide impact such as these:

Michael Hammer and James Champy, *Reengineering the Corporation: A Manifesto for Business Revolution*
Joseph H. Boyett and Henry P. Conn, *Workplace 2000: The Revolution Reshaping American Business*
Robert M. Tomasko, *Downsizing: Reshaping the Corporation for the Future*
William H. Davidow and Michael S. Malone, *The Virtual Corporation: Structuring and Revitalizing the Corporation for the 21st Century*
Peter M. Senge, *The Fifth Discipline: The Art and Practice of the Learning Organization*
Don Tapscott, *The Digital Economy: Promise and Peril in the Age of Networked Intelligence*
Tom Peters, *Liberation Management: Necessary Disorganization for the Nanosecond Nineties*
Jon R. Katzenbach and Douglas K. Smith, *The Wisdom of Teams: Creating the High-Performance Organization*
Peter F. Drucker, *Post-Capitalist Society*

"These books," Armand Mattelart comments, "which enjoyed a transnational readership far broader than just business executives, provided a medium for the followers of the new business doctrine," "a veritable cult of enterprise, bordering on the religious."[4] And this is not even to mention the new journalism of business that, everywhere we look—in newspapers, magazines, cable business channels, and ordinary TV news—amplifies the dominant fictional realism of our times by rehearsing the mantra of right-sizing, just-in-time delivery, flat-structuring, disintermediation, flexibility, teamwork, lifelong learning, diversity management, and (that ultimate arbiter of collective fiction) shareholder value—all instantiated in networked information technology, or "IT."

Inverting my questions about Lodge, we may pose the following puzzles for the postindustrial business imagination. On the one hand, is the new business literature simply ahead of (rather than, as with Lodge, behind) the times when it promises an age of business that is all information processing? Is the new business literature, in other words, what the word *virtual* may really mean—posthistorical? "When someone asks us for a quick definition of business reengineering," Hammer and Champy declare in their *Reengineering the Corporation*, "we say it means 'starting over.' It *doesn't* mean tinkering with what already exists or making incremental changes that leave basic structures intact." Peter F. Drucker, the dean of U.S. management theory, sums it up: "Innovation," he says, "means, first, the systematic sloughing off of yesterday."[5] Read virtually or posthistorically in this way, the business bestsellers are utopian prophecies of what Michael Dertouzos calls *What Will Be,* and Bill Gates refers to as *The Road Ahead* (to cite two titles from the affiliated genre of information-technology prophecy).[6]

On the other hand, is the new business literature so dystopian (pessimistic about the future, where Lodge's minor prophets are optimistic) that its real subject is the impassibility of history? Witness, for example, the rhetorical dependence of the genre not just on broad denunciations of traditional ways of living and working, but also on long catalogs of specific historical "obstacles." "So, if managements want companies that are lean, nimble, flexible, responsive, competitive, innovative, efficient, customer-focused, and profitable," Hammer and Champy ask, "why are so many American companies bloated, clumsy, rigid, sluggish, noncompetitive, uncreative, inefficient, disdainful of customer needs, and losing money?" The answer is history: "Inflexibility, unresponsiveness, the absence of customer focus, an obsession with activity rather than result, bureaucratic paralysis, lack of innovation, high overhead—these are the legacies of one hundred years of American industrial leadership."[7] Similarly, we can take the measure of how Davidow and Malone's *The Virtual Corporation* excoriates old ways of doing things from the following excerpt, in which history is purely a process of decay:

> Here in the United States, [the] sense of distortion and confusion, mixed with considerable fear, has become an uncomfortable part of our daily lives.

> Everywhere there is a disquieting sense of decay—in government, within boardrooms, on shop floors. . . .
>
> One by one our industries are losing competitiveness and market share to industries of other nations. Our government seems more concerned with lifelong job security for politicians and spending money it doesn't have than in enhancing the economic prosperity of the citizenry. Our manufacturing sector often insults consumers with shoddy products and workers with unearned executive compensation—and then blames its woes on foreign competition. By the same token, workers are frequently unmotivated and selfish compared with their foreign counterparts; and consumers have in the past replaced good sense and security with almost pathological acquisitiveness.
>
> Meanwhile, our major cities, once the jewels of our culture, have become violent, ungovernable places perpetually teetering on bankruptcy.[8]

Education, of course, comes in for special attention as the very pharaoh of decadent old ways. A plague on education, Boyett and Conn thus say, or, to be exact (citing the business spokesmen they quote in their chapter on education), schools are a "national disaster," "a third world within our own country," "the American dream turned nightmare," "the greatest threat to our national security," and so on.[9]

To emphasize such harsh, corrosive, often satiric denunciation, we recognize, is to see that the new business literature walks the dark side of the street (the "road ahead") of prophecy. From this perspective, works in this genre are fire-and-brimstone jeremiads damning sinners in the hands of an angry global competition. Gurus are not seers of the new millennium. They are witnesses to a damned history that is everywhere and nowhere, present in every manifest obstacle imposed by the past, yet profoundly unknowable in the discourse of knowledge work except as monstrous other. History is an imaginary Third World—a reservation for peoples who remain historical—couched within the First World itself. It is the other of the future.

To reflect on the relation between knowledge work and the academy today, in short, is to discover a profound ambivalence. I raise questions of stance or tone about Lodge's novel and the new literature of business not to suggest that such questions can be firmly decided. Rather, the considerable vitality of both the novel and the business discourse I cite depends precisely on their undecidability. Comedy or satire, prophecy or jeremiad: the underlying contradictions glossed by these modes are structural. As such, they are best approached in the spirit not of decision but of suspense. We are, after all, on the scene of the abiding suspense of the contemporary middle class, which is even more structurally contradictory than the original white-collar class of the twentieth century. To be a white-collar or salaried worker in the 1950s, for example, was to stake the entirety of one's authority not on the self-owned property, business, goods, or money of the predecessor entrepreneurial classes of the nineteenth century, but on an existentially anxious property of "knowledge" that had to be re-earned from scratch by one's children.[10] Thus was laid the foundation for the overdetermined relation between business and education. But to be a professional-managerial-technical worker now is to stake one's authority on an even more precarious knowledge that has to be re-earned with every new technological change, business cycle, or downsizing in one's own life. Thus is laid the foundationless suspense, the perpetual anxiety, of "lifelong learning."

Understanding knowledge (and information) work from the point of view of the academy, therefore—and especially understanding the contemporary relation between knowledge work and the academy—means exploring a complexly ambivalent zone of antipathy, cross-purposes, and also, at times, unexpected sympathy. To do so requires the opposite of the decisiveness heard so preemptively—to cite an important example—in the "Final Report of the MLA Committee on Professional Employment" (1998), a white paper commissioned by the Modern Language Association to suggest ways to ameliorate the "job crisis" in the humanities. However acute, consistent, or substantial its recommendations may be (or not), the "Final Report" is clearly fighting the last war when it legitimates itself in its preface:

> Lawmakers call for greater productivity on campus, and advisers trained in business management counsel various forms of "downsizing." In numerous instances, indeed, formal commissions, college presidents, boards of trustees, and the media have pressed for a new efficiency in higher education based on corporate models in which students are defined as "clients" or even "products" and academic institutions are regarded as sites of production. But *of course* the object of business corporations is to make a profit, while the object of institutions of higher education is to acquire and disseminate knowledge as well as, most important, to develop in students sophisticated intellectual strategies they will use for the rest of their lives, in and out of the workplace.[11]

The Maginot Line here, of course, is the "of course" that refuses at all costs to acknowledge the complexity of the contemporary overlap between the work of education and the knowledge work of business. What the perspective represented by the "Final Report" thus refuses to see is the seriousness of the challenge to academic knowledge posed by a knowledge work that has been redefined—in the mode of Fritz Machlup's and Marc Uri Porat's magisterially inclusive accounts of knowledge industries—as both a practical *and* intellectual pretender to the academic throne.[12] In the academy, educators have long been accustomed to accommodating the *practical* logic of business, as personified comfortably and endogamously in those whom they love to hate—administrators.[13] To follow the lead of both the "Final Report" (in its discussion of "careers outside the academy") and the Presidential Forum at the 1998 Modern Language Association convention, the humanities are just now thinking about going exogamous—that is, asking business nicely for work as part of a general enterprise of "going public."[14] But what has been generally missing is any engagement with the full *intellectual* force of business in its new persona as knowledge work.

My meaning is illustrated by a single passage from one of the most influential and widely cited gospels of the new knowledge work, Peter Senge's *Fifth Discipline: The Art and Practice of the Learning Organization* (1990). This is how Senge defines his fifth, climactic discipline of business change, "Metanoia—A Shift of Mind":

> When you ask people about what it is like being part of a great team, what is most striking is the meaningfulness of the experience. People talk about being part of something larger than themselves, of being connected, of being generative. It becomes quite clear that, for many, their experiences as part of truly great teams stand out as singular periods of life lived to the fullest. Some spend the rest of their lives looking for ways to recapture that spirit.
>
> The most accurate word in Western culture to describe what happens in a learning organization is one that hasn't had much currency for the past several hundred years. It is a word we have used in our work with organizations for some ten years, but we always caution them, and ourselves, to use it sparingly in public. The word is "metanoia" and it means a shift of mind. The word has a rich history. For the Greeks, it meant a fundamental shift or change, or more literally transcendence ("*meta*"—above or beyond, as in "metaphysics") of mind ("noia," from the root "*nous,*" of mind). In the early (Gnostic) Christian tradition, it took on a special meaning of awakening shared intuition and direct knowing of the highest, of God. "Metanoia" was probably the key term of such early Christians as John the Baptist. In the Catholic corpus the word metanoia was eventually translated as "repent."

> To grasp the meaning of "metanoia" is to grasp the deeper meaning of "learning."[15]

For the humanities scholar, there is inexpressible irony in the thought that the single most influential contemporary visionary of the One Life and Imagination (as the Romantic poets called it) should be a management guru.[16] Senge offers a whole scholarship of, and about, learning *bypassing* academic historical knowledge in favor of a fantastic pastiche of classical, Christian, and (the real school of his work) New Age lore. It would be easy for the professional scholar to demystify Senge's lore (etymological speculation on the meaning of words, after all, has been one of the preferred gambits of deconstruction). But this would miss the point, which is that the academy can no longer claim supreme jurisdiction over knowledge. The narrowness of much contemporary academic research is perhaps precisely what allows Senge to claim high-court jurisdiction in the breadth and daring of his intellectual-cum-practical will to know what it might mean to "know."

We are now in a position to understand why a serious study of knowledge work from the perspective of the academy is necessary. Scholars are themselves knowledge workers in a complete sense: they are intellectuals, but they are also middle managers responsible for an endless series of programs, committees, performance reports, and so on.[17] More important, their entire mission is the education of students who, however diverse their backgrounds, are destined for service in the great, contemporary prosperity corps of knowledge work. (Even students who aim for more idealistic, Peace-Corps-style service in government, nonprofit, or low-profit sectors are likely to experience the increasing colonization of all work by the principles and technologies of corporate knowledge work.) Only if scholars now think about business as an intellectual *and* practical partner in knowledge work, therefore, can the critical issues in the relation of the academy to business be joined. Asking business for nice work need not mean selling out, but only if the contemporary academy engages business in a full act of critique in which it both gives and takes. Such reciprocal critique cannot even be initiated unless it is elevated to the proper level, where scholars first assume that the academy and business have a common stake in the work of knowledge and, second, ask, "What *then* is the difference?" What is the postindustrial, and not nineteenth-century, difference between the academy and the "learning organization"?[18]

Since the mid-twentieth century, we may reflect, the U.S. academy has increasingly understood its business to be the education of "all"—or at least as many of the "all" as a relatively liberal notion of the white-collar middle class (and its more recent New Class techno-managerial-professional overlords) can accommodate. But now knowledge work has called the academy's bluff. Here is a partial listing of the areas of knowledge production that Machlup included in his 1962 survey:

> Education (at home, on the job, in church, in the armed services, elementary and secondary, higher)
> Research and development (basic, applied)
> Media of communication (printing and publishing, photography and phonography, stage and cinema, broadcasting, advertising and public relations, telephone, telegraph, and postal service)
> Information machines (instruments for measurement, observation, and control; office information machines; electronic computers)
> Information services (legal, engineering and architectural, accounting and auditing, medical, financial, wholesale trade)

Now that knowledge—in its training, exercise, and possession—really is presumed ideologically, if not in fact, to be the business of "all," and especially of business, how will be the academy adapt to its diminished role as one among many providers in a potentially rich and diverse—but also potentially impoverished and culturally uniform—ecology of knowledge? In the "knowledge economy," education occurs across a whole lifetime in an unprecedented variety of social sectors, institutions, and media: not just schools, community colleges, and universities, but also businesses, broadcast media, the Internet, even the manuals or "tutorials" that accompany software applications. Education, in other words, is now a decentralized field where no one institution individually corners the market and where we encounter a dizzying dispersion of the kinds and scales of learning—all the way from educational programs leading to degrees to CNN "factoids" leading only to the next commercial.[19] What voice will brave, dear Temporary Lecturer Robyn Penrose have in the cacophonous new world of knowledge work? In broader terms, how can society create the most inclusive, flexible, and intelligently interrelated mix of educational options to take care of all its citizens hungry to "know"?[20]

What is the idea of knowledge work? What is its relation to the knowledge of the humanities in the contemporary academy? And how does focusing specifically on "information work"—on its technologies, techniques, and, ultimately, culture—help us understand that relation?

Chapter 1

The Idea of Knowledge Work

To understand knowledge work from the perspective of the humanities, let us start by reviewing in a single frame of analysis three explanations of the concept that arose independently and largely in ignorance of each other. Two are academic approaches characteristic of the humanities in their now prevailing cultural critical personality. The third is the neo-corporate business thesis that seems destined to buy out the others. Where there was "identity group" and "cultural class," there will now be only that elementary unit of corporate knowledge work, the team.

Subject Work

Recall, to begin with, that since about 1980 the dominant, if unwitting, explanation of knowledge work in the humanities, especially in literature departments, has been the cultural criticism of identity and subject. Sketched very broadly, the paradigm of 1980s-style cultural criticism was as follows.[1]

The paradigm started with the assumption that cultural value—or, put negatively, discrimination—is determined by social structure. Specifically, value is "constructed" by a structure whose implicit or explicit patterning after some hard-core segment of society (e.g., economic structure, patriarchal family structure) made it seem a unitary regime of social "containment" no matter what the evidence of inner "subversion."[2] An example of such a formulation is Foucault's social "discursive formation" as it became identified in common academic usage with the penitentiaries or bedlams that Foucault made into such strange attractors. Discursive formations were poststructuralist in their constitutive inner contradiction or scandal, but they were also palpably structuralist in their characteristically all-of-a-piece historical behavior. (In the standard narrative: first there was one discursive regime or "episteme" that hung all together, then another . . .)[3] Of course, social constructionism that hung all together in this way ran the risk of being intolerably reductive and totalizing. Therefore (and here we glimpse the pertinence of the problem of knowledge work), cultural criticism in the 1980s found it crucial to introduce within social determination at least the *thought* of indeterminacy, by foregrounding the mediating role of "ideological state apparatuses," "representations," "mentalities," and other dream-states of *imaginary* constructionism supervised by such "relatively autonomous" institutions of identity formation as the church, school, or media.[4] To restage the well-known Althusserian scene in which a policeman hails someone walking on the street ("Hey, you there!"): no wires are necessary for society to jerk one around because one's body is already wired that way by one's own head, by one's inculcated sense of who one is (e.g., a responsible citizen).[5]

My anatomization here of a body subordinated to a head is indicative. Though much cultural criticism argued that ideology manifests itself first of all in bodily constructions of gender, ethnicity, and race, the body as register of ideology has always been an emanation of the real point of ideology critique—figuratively, the head.[6] The ultimate aim of 1980s-style cultural criticism, in other words, was the point-singularity or black hole of the imaginary named "the subject," for which identity groups (gender, ethnicity, race) were the social event horizon.[7] Knowledge work was a subject or identity work as vast as all culture. It was everywhere.

More complexly, subject work was both an intense localism rooted in the "here and now of my life story" and a strangely generalized localism ("the personal is the political"). Like a Hollywood movie, it played *locally everywhere*—and thus, to advert to the well-known impasse of social or political "agency" in cultural criticism, also apparently nowhere in particular (thus necessitating the supplementary notion of the Foucaultian "specific intellectual" to locate sectors where knowledge work could be seen to concentrate in active fashion, as in Hollywood itself in my analogy).[8]

The exact personality of such generalized localism may be demonstrated by comparison with the (alleged) essentialism of nationalist and identity politics—uncanny doppelgängers that steadily lost credibility in the academy even as they flourished in other arenas throughout the 1980s. We can grasp the underlying similarity between these two otherwise antithetical stances by recalling their classical precedent. In classical and neoclassical philosophy, the paradox of local yet also general identity had been assuaged, if not solved, by premising a universal "nature" embracing all. (For Sir Joshua Reynolds, for example, "the perfect state of nature" guarantees that just "as there is one general form, which . . . belongs to the human kind at large, so in each of these classes [the specific types of human kind] there is one common idea and central form, which is the abstract of the various individual forms belonging to that class.")[9] Right-wing nationalism updated classicism by supposing instead that general-local identity was grounded within a *nativist* version of nature: "the American character," which was both particular (even regionalist) and somehow universal ("when in the course of human events . . ."). Essentialist identity politics then arose precisely to challenge the melting pot of nationalist nature through "groupisms" that imposed a different standard of nature. Decried by the cultural right as tribalism, yet also comprehensible as a return to a purer, city-state classicism, groupism on this model closed the contradiction between the local and the general—"me and my group," on the one hand, and universal human rights, on the other—on the basis of a hybrid cultural/biological "nature" (the essential nature of "woman," "African American," and so forth).[10]

Caught in the contest between nationalist and identity group essentialisms, constructivist cultural criticism in the 1980s sympathized fitfully with the latter. But really, it threw out both as twin horns of the same dilemma of chauvinism, leaving apparently no ground at all to take a stand upon.[11] All the grounds of human "nature" were removed from identity, and the relation between "me and my group" and the world was thus radically destabilized. While the relation did not become any more or less fundamentally contradictory, it became more dynamic, unfixed. Cultural criticism sometimes stressed the particularity of "my (group's) life story" and sometimes the generality that identity is universal.[12]

Now we can understand the less noted role of *cyber*cultural criticism as it emerged in the mid-1980s to mid-1990s among communications scholars and sociologists of cyberspace as well as such literary or cultural critics of digital technology as J. David Bolter *(Turing's Man)*, Michael Heim *(Electric Language)*, George P. Landow *(Hypertext)*, Mark Poster *(The Mode of Information, Second Media Age)*, Howard Rheingold *(The Virtual Community)*, Allucquère Rosanne Stone *(The War of Desire and Technology at the Close of the Mechanical Age)*, and Sherry Turkle *(Life on the Screen)* (in the company of others like Donna Haraway, who revised identity in light not just of information culture but of technoculture generally). As has become even clearer in subsequent cyber- and new-media theory (for example, the many essay collections that have appeared with *culture* in their titles), cybercultural criticism was the flank movement of cultural criticism that pivoted the whole problem of the missing ground of identity toward the information front and, on that front, found a concrete way to think about the radical instability of constructed identity in terms well suited for the new millennium.[13]

It did so by unfolding the "locally everywhere" subject as the "virtual" subject. In the tradition of Marshall McLuhan writing about "typographic man" versus the "extensions of man," the object of study in much cybercultural criticism was presumed from the first to be the subject—that is, the sensorium, cognition, affectivity, or identity of cyber-citizens and cyber-authors and -readers compared to their predecessors. Perfectly expressive is Mark Poster's concern that "super-panoptic" databases will construct subjects as "dispersed identities," or the pronounced anxiety of both advocates and foes of new media over the "erosion" of the textual "subject" (e.g., George Landow or Michael Heim on the authorial self, or Sven Birkerts in "Paging the Self: Privacies of Reading").[14] But as heard precisely in such epithets as *dispersed* or *eroded,* the assumption was that the virtual subject is centered not on any essential ground (and certainly not any natural or national ground) but instead on empty ground. This empty ground is the "network."[15] Identity in this thesis is schizophrenically both local and general, both locked in solitude before a "personal computer" and (like the users Turkle studies "cycling," "morphing," or "slipping" between "multiple" virtual personae) driven to distribute itself compulsively over the wires everywhere.[16] In between the local and general is the postnatural and postnational network (more accurately, Inter-network) that is the great contemporary construction. What is the ontological, social, political, economic, and other status of this construction? What, in other words, *is* networked identity when—following the original military specifications for the Internet (then the ARPAnet)—such identity is so dynamic and flexible that it can route around interruptions in system integrity as big as a nuclear strike? Borrowing from poststructuralist theory, cybercultural criticism provided an essentially negative answer—what such identity was not. It was not centered identity, stable identity, and so on.

By way of illustration we can recall all the fictional "shape shifters" or "morphs" who suddenly appeared in mass media in the early 1990s: computer-morphed singers in music videos; the morph villain from the future in *Terminator II*; the morph security expert on *Star Trek: Deep Space Nine* who periodically lost his shape and had to sleep in a bucket; the networked, all-assimilating Borg on *Star Trek: The Next Generation*, who were really a collective morph; or even the comic morph represented by Jim Carrey's bizarrely flexible bank account manager in *The Mask*.[17] Cybercultural criticism wanted to know who the morph was who occupied all those security-expert, networked, managerial, and other flexible information jobs.

(New) Class Work

A second explanation of knowledge work also arose in the academy, but from the direction of sociology rather than literary cultural studies. This understanding, while originally of 1970s vintage, crossed disciplinary boundaries into general intellectual prominence in the early to mid-1990s just in time to tell us who the morph was: the white-collar professional/managerial/technical class, or "New Class." I refer to New Class critique together with the related "cultural capital" approach to class experience—a set of revisionary assumptions and methods about social stratification that came powerfully to bear on literary studies in 1993 in John Guillory's *Cultural Capital* (and that had previously made inroads in cyber- and technocultural studies in Andrew Ross's *No Respect: Intellectuals and Popular Culture* and Myron C. Tuman's *Word Perfect: Literacy in the Computer Age*).[18]

For convenience, I will treat the broader and narrower aspects of such critique in succession: first, the general sociology of culture-based class distinctions proposed by Pierre Bourdieu; then, the more specific sociology of the New Class. Both levels of the critique can be said to be constitutionally amorphous. To borrow a term that Loïc J. D. Wacquant applies to Bourdieu, their logic is fuzzy.[19]

In the Bourdieu variant, which is the major influence on Guillory's book, social distinction is distributed unequally on the basis of a fuzzy logic "general" economy of capital (simultaneously economic, cultural, social, and symbolic).[20] This economy is organized not so much by total social structure or system as by a more shapeless version of structure, social "fields." Defined by occupational, professional, institutional, and/or geographical boundaries, social fields resemble the cultural critical notion (some would say caricature) of structure because they are determinative.[21] But fields are unlike structure because they have no necessary coherence, either internally or in external relation to some social totality patterned after a master field. "Every field," Bourdieu writes, "constitutes a potentially open space of play whose boundaries are *dynamic borders* which are the stake of struggles within the field itself."[22] So fuzzy are such fields that Bourdieu's characteristic tables and diagrams of multivariant analyses are really not a method but a world-view: the fields creative of social reality behave like quantum shells that are knowable only statistically and diffusely.[23]

Consequently, the issue of the totalistic determination of *a* subject by *a* structure is from the first a nonissue for Bourdieu. Determination may act in certain fields in certain ways and in specific combinations with other fields (and with historical change) to give particular people distinction. But there is nothing certain or even interesting to say about *overall* social determination as such, other than that it *is* statistically "overall." The entire epicycle of "imaginary" (in)determination that 1980s-style cultural criticism added to its Ptolemaic system to resist totalizing determination is thus moot. Not only does Bourdieu have no use for the imaginary, he actively prosecutes it in favor of his thesis that fields affect human experience through the opposite of the body as register of ideology—the body *without* ideology. In his characteristic idiom, social fields are inhabited "organically" and "durably" by "incorporated" "bodily practice," "sense," and "disposition" to create a gigantic end-run around the head called *habitus*. Habitus is the body-resident instantiation of practical beliefs, or *un*imaginary ideology, that Bourdieu calls *doxa*: "the relationship of immediate adherence that is established in practice between a *habitus* and the field to which it is attuned, the preverbal taking-for-granted of the world that flows from practical sense."[24] (As we will see later, however, the question of the imaginary in such analysis returns when the field in question is information.)

Now we can bring the Bourdieu paradigm to its head, which is not the figurative head at all but something much fuzzier. What is the identity of habitus? Not subjectivity at all, but a different kind of body (rather than head) politic. "To speak of habitus," Bourdieu argues, "is to assert that the individual, and even the personal, the subjective, is social, collective. Habitus is a socialized subjectivity."[25] Habitus confers the identity that throughout Bourdieu's work goes by the name of "class habitus" or *cultural class*—the fuzzy, lifestyle version of economic class.[26] Like subjectivity, cultural class is as vast as all culture because class work is locally everywhere. But the class concept is designed from the ground up to stratify the local by distinct levels and sectors—dominant, petit bourgeois, and working class, with their component occupations—so as to scale up to the general system of class relations with articulated precision (in contrast to the more or less uniform "difference" celebrated by subject critique).[27] As instanced in the educators Bourdieu studies in *Distinction* and *Homo Academicus*, therefore, class work may occur everywhere, but knowledge work is a sector within that universality positioned exactly with reference to other sectors. Rather than being a free-floating imaginary performed by anyone and everyone who has a subject, it is a discrete set of practices performed by a designated class rooted in a particular if widespread habitus of educational, linguistic, occupational, political, musical, gustatory, and other commonplaces.

Yet however divergent its emphases, the thesis of cultural class also ultimately converges with that of the subject. Given that cultural capital is expressly *not* just material capital (and habitus therefore not just the body but also practice, sense, and disposition), cultural class is clearly not essentialist class. Instead, class for Bourdieu is open to the same style of instability we saw in subject critique. Consider, for example, the oscillation between (in my notation) Bourdieu's emphasis on [a] the local individual and [b] the general system in the following set of passages from *An Invitation to Reflexive Sociology*, where the oscillation has an energy and abruptness that exceed the carefully balanced pendulum swings by which Bourdieu sometimes consciously practices hermeneutics ("a sort of hermeneutic circle . . . an endless to and fro movement in the research process that is quite lengthy and arduous"):[28]

> The notion of field reminds us that the true object of social science is not [a] the individual, even though one cannot construct a field if not through individuals, since the information necessary for statistical analysis is generally attached to individuals or institutions. [b] It is the field which is primary. . . . [a] This does not imply that individuals are mere "illusions," that they do not exist: [b] they exist as *agents*—and not as biological individuals, actors, or subjects—who are socially constituted as active and acting in the field.

> [a] Habitus is not the fate that some people read into it. Being the product of history, it is an *open system of dispositions* that is constantly subjected to experiences, and therefore constantly affected by them in a way that either reinforces or modifies its structures. It is durable but not eternal! [b] Having said this, I must immediately add that there is a probability, inscribed in the social destiny associated with definite social conditions, that experiences will confirm habitus, because most people are statistically bound to encounter circumstances that tend to agree with those that originally fashioned their habitus.
>
> [a] In truth, the problem of the genesis of the socialized biological individual, [b] of the social conditions of formation and acquisition of the generative preference structures that constitute habitus as the social embodied, is an extremely complex question.

> This study [one "recently launched on the experience of 'social suffering'"] is premised on the idea that [a] *the most personal is* [b] *the most impersonal,* that many of [a] the most intimate dramas, the deepest malaises, the most singular suffering that women and men can experience find their roots in [b] the objective contradictions, constraints and double binds inscribed in the structures of the labor and housing markets, in the merciless sanctions of the school system, or in mechanisms of economic and social inheritance. . . . Armed with full knowledge of [a] the individual's social trajectory and life-context, we proceed by means of very lengthy, highly interactive, in-depth interviews aimed at helping interviewees discover and state [b] the hidden principle of their extreme tragedies or ordinary misfortunes; and at allowing them to rid themselves of this external reality that inhabits and haunts them, possesses them from the inside, and dispossesses them of initiative in their own existence in the manner of the monster in *Alien.*[29]

The abruptly self-interrupting rhythm of such a locution as "It is durable but not eternal! Having said this, I must immediately add . . ." may be likened to Bourdieu's image of that brilliantly realized morph subject, the creature in *Alien.* Individual identity is breached from within by a principle of general sociality that—as in the exoskeletal hideousness of the creature bursting from its victim's chest cavity—is nothing other than pure exteriority. Local being bonds to general being, in other words, on the grounds of a common essence (we are all food for the monster) that is an *alien* essence.

More could be said about how the lack of natural essence destabilizes the local/general relation in Bourdieu. Take the instability of national essence in his work, for example. Premised on class rather than nation, identity for Bourdieu is geopolitically unpredictable. On the one hand, he is always careful to root his analyses of social distinction firmly in a particular nation, France. His conclusions, he thus qualifies, are local, not general. On the other hand, there is an unsettling tendency in his writings (and certainly in his reception) to reach across national borders toward general class experience. "This book will strike the reader as 'very French,'" he begins in the preface to the English-language edition of *Distinction,* only to continue: "But I believe it is possible to enter into the singularity of an object without renouncing the ambition of drawing out universal proportions." There may thus be "structural invariants" and "equivalent institutions" bridging France and America.[30]

It would be stretching it to compare Bourdieu at this point so fully to subject critique that the logical outcome would again be cybercultural criticism (as best instanced here, perhaps, by the Bourdieu Forum list on the Internet, which in the past has made the nationality of class an explicit topic). Still, we can spot the basic congruence of Bourdieu's class thesis and the network thesis in a passage like the following: "a field may be defined as a network, or a configuration, of objective relations between positions. . . . The social cosmos is made up of a number of such relatively autonomous social microcosms, i.e., spaces of objective relations that are the site of a logic and a necessity that are *specific and irreducible* to those that regulate other fields."[31] Identity is locally specific and irreducible. But it is also generally comparable to "other fields" because it exists on the plane of abstract, empty continuity that is the "network" of "social space and its transformations" (so titled in a chapter of *Distinction*).[32] At once postnatural (nonmaterial) and postnational, such social space is Bourdieu's network. As he says at one point in *An Invitation to Reflexive Sociology*, "I have analyzed the peculiarity of cultural capital, which we should in fact call *informational capital* to give the notion its full generality, and which itself exists in three forms, embodied, objectified, and institutionalized. Social capital is the sum of the resources, actual or virtual, that accrue to an individual or a group by virtue of possessing a durable network of more or less institutionalized relationships of mutual acquaintance and recognition."[33] It is almost as if Boudieu were equating "network" specifically with the Internet.

That so much of Bourdieu's work focuses on intellectuals indicates the degree to which his sociology dovetails with New Class critique.[34] New Class critique began when class theorists observed that the emphasis of old-style Marxism on binaristic struggle between capitalists and workers had been fatally embarrassed by a massive extrusion of the middle class in the twentieth century—knowledge workers, who had come to bulk so large and influential as a group that they seemed to crowd out all others.[35] Defined most narrowly, the New Class refers to our current elite or ruling class: the professionals, managers, technical intelligentsia, intellectuals, semi-autonomous credentialed employees, and others who constitute the upper and middle crust of the "new middle class." Or rather, since the term *new middle class* was coined to refer to the salaried middle class that emerged early in the twentieth century from the older, entrepreneurial and small-farm middle classes, we can say that the New Class is the second-generation or *new* "new middle class." Defined more broadly, the New Class overlaps complexly and sometimes undecidably with what may be called its "trailing edge": clerical and other clean-collar service workers who identify (or are identified) with the code of professionalism but who do not enjoy the full perquisites of that status. Both the higher and lower strata of the class have a stake in "knowledge work"; whether we should stress their differentiation or commonality depends on the context of discussion.[36]

The main thesis about the New Class, especially in its Marxist or leftist variants (most famously, Alvin Gouldner's *Future of Intellectuals and the Rise of the New Class*), is remarkably similar to Bourdieu's fuzzy logic in relation to subject critique. Again, the premise is that cultural value stems from a general economy of cultural or "human capital" ("moneyed capital is . . . a special case of capital"). Again, this economy is thought to be organized not so much by totalizing social structure as by the evidentiary basis of New Class critique—occupational fields.[37] Such fields are so far from being total that there is no consensus about the nature of the structural whole they piece together. As epitomized in Val Burris's comparative chart of several New Class theories, for example, views of the modern social whole are contradictory (fig. 1.1). Crucially, there is marked disagreement over where to draw the line between socially higher and lower occupations, which results in the divergence between the narrow and broad definitions of the New Class I referred to above and also touches upon the key structural contradiction underlying the whole discussion. On the one hand, salaried knowledge workers are in the camp of capitalist owners because they supervise wage workers—to the extent that expertise-based control has actually replaced ownership-based control in most major enterprises. On the other hand (a circumstance only partially mitigated in high-tech firms by the distribution of stock options to nonexecutives), they are largely excluded from actual ownership and are therefore themselves ultimately vulnerable to exploitation.[38] Owning a capital of pure knowledge that "cannot be hoarded against hard times" and whose reproduction requires education and self-discipline "visited, in each generation, upon the young as they were upon the parents," as Barbara Ehrenreich puts it, knowledge workers are "an insecure and deeply anxious" elite.[39] No wonder that in some circumstances and at some times, knowledge workers act like owners, while at other times they behave like proletariat. And so, too, with their politics, which has been of particular interest to the left: today knowledge workers are left, tomorrow right-leaning.[40]

Consequently, the cultural critical hypothesis of the imaginary is hardly needed to mediate structural determination. When indeterminacy *is* the structural condition, the thought of indeterminacy is redundant. The extent to which ideology critique effectively drops out of the equation in New Class theory is thus striking. Mental work is the crucial fact, but such work clearly has nothing to do with the work of the imaginary, even when (as in

<table>
<tr><th>Detailed Class Fractions</th><th>Poulantzas's Classes</th><th>Mills's Classes</th><th>Ehrenreichs' Classes</th><th>Carchedi's Classes</th><th>Wright's Classes</th></tr>
<tr><td>Managers and Supervisors</td><td rowspan="4">New Petty Bourgeoisie</td><td rowspan="3">New Middle Class</td><td rowspan="2">Professional-Managerial Class</td><td>New Middle Class</td><td>Managers and Supervisors</td></tr>
<tr><td>Professional and Technical Workers</td><td rowspan="4">Proletariat</td><td>Semi-auton. (credentialled) Employees</td></tr>
<tr><td>Routine Mental Workers</td><td rowspan="3">Proletariat</td><td rowspan="3">Proletariat</td></tr>
<tr><td>Unproductive Manual Workers</td><td rowspan="2">Proletariat</td></tr>
<tr><td>Productive Manual Workers</td><td>Proletariat</td></tr>
</table>

Figure 1.1 Burris's table of alternative models of class divisions among salaried workers
From Val Burris, "Class Structure and Political Ideology," *Insurgent Sociologist* 14, no. 2 (Summer 1987): 33.

Gouldner's thesis) it carries the potential for vanguardist resistance to determination. Mental work remains foremost an occupational phenomenon whose significance lies more in the demographics of *work* than in the uniqueness of *mental* work . The New Class notion of mental work is thus broadly aligned with Bourdieu's antimentalist habitus (though here again, recentering the discussion on information work will require that mentality be reconsidered). In Gouldner's influential explanation, for example, knowledge work is primarily a linguistic habit. It is rooted in a shared "culture of critical discourse" that originates in schooling to inculcate habits of analytical distance, judgment through rationality (as opposed to authority), technical mastery, professional autonomy, and other ultimately class practices.[41] Mentalism is a way of life.

To bring the argument to its head is therefore again to skip the head for a different order of identity, named in "New Class" itself: a revisionary notion of "class." New Class critique, we may say, is a method that has been dragged protesting from economically grounded class analysis to what seems to some of its practitioners the Never-Never Land of a new class concept—not just a concept of the New Class but a new concept of class. Depending on which theorist one reads, the class of knowledge workers is just an assemblage, a "weakly formed class," or, in Erik Olin Wright's important formulation, a "contradictory class."[42] It is the class of morphs, or amorphous class. Yet the whole point is that such amorphousness is concretely fractionable and locatable, sector by sector. Knowledge work may be as vast as all culture because it is everywhere. But instead of being "locally everywhere," as in subject critique, now it is sociologically everywhere, in a manner to be counted in such places as the central office, the branch office, the line, and so forth.

Yet however different its emphases from those of cultural criticism, New Class critique is also like Bourdieu's approach in finally converging with the thesis of the constructed subject. It does so because the void essence that underlies its "weakly formed" class is susceptible to the same unstable relation between local and general identity we previously witnessed. Here the postnational aspect of the problem is most salient. On the one hand, Gouldner throughout localizes the New Class by nation (the New Class of the United States, USSR, Japan, France, and so on).[43] On the other hand, his overall goal is to elide the "nation" concept altogether to generalize upon what he calls in the subtitle of his book the "international class contest of the modern era." The New Class is thus a phenomenon of "all countries that have in the twentieth century become part of the emerging world socioeconomic order," he says on page one. Sometimes, then, the New Class seems very local (a matter of particular occupations in particular cultures), and sometimes it appears hyperbolically global. In this and other ways, the New Class is permanently "under construction."

Ultimately, we might reflect, New Class critique matches the dimensions of any of modernity's great, universalizing theories of civilization—but with a postmodern, localizing twist.[44] Civilization on this model means the tidal ebb of unconscious work and the emergence of a global condition that the Enlightenment had once called reason; Freud, repression; Weber, instrumental rationality; and Foucault, discipline or power/knowledge. New Class theory—in a dark *ricorso* to the Enlightenment that incorporates all the more sinister modern theories en route—simply accounts it "knowledge." Thus Gouldner's oft-cited evaluation of the New Class perfectly captures the mixed bright and dark, Enlightenment and twentieth-century hue I indicate: the New Class, he judges, is an embryonic "universal class" that, while "profoundly flawed," is also "the most progressive force in modern society and . . . a center of whatever human emancipation is possible."[45] Yet the universal class is local and limited, too: it is "elitist and self-seeking and uses its special knowledge to advance its own interests and power."[46]

New Class critique originated before the era of server-client and personal computing that made the network (as we will see in more detail in chapter 4) the avatar of unstable local/global identity. But again, as with Bourdieu, we may notice that the "universal class" that is "elitist and self-seeking [in its use of] special knowledge" is identical to the virtual subject studied by cybercultural criticism. The New Class is the shapeshifter seen in the deliberately underdrawn characters of Scott Adams's *Dilbert* comics, where weakly individuated office workers (characterized more by occupational stereotype than personality) occupy a common, generic space of information technology. When the identity group becomes "class" and ideology becomes "habitus" or "the culture of critical discourse," then subjectivity morphs into the networked collective subjection that may be called, in every sense of the word, *corporate*.

Teamwork

Finally, then, we must consider a third major explanation of knowledge work that brings the approaches of subject critique and cultural class critique into their sharpest contemporary focus. It is this model above all that raises our present idea of knowledge to the pitch of millennial consciousness. I refer to the new or postindustrial corporatism, which argues (in what Christopher Newfield calls the deliberately "prophetic" tones of "the business future") that both the constructed, dispersed subject and the weakly formed, contradictory class described by subject critique and New Class critique, respectively, have been surpassed by a single kind of amorphous knowledge entity: the "networked," "web," "flat," "fishnet," "cluster," "relational," "virtual," "crazy," "boundaryless," "democratic," "teamwork," or "learning" corporation.[47] Among these terms symptomatic of what Dirlik calls "postmodern" corporatism, I will emphasize *teamwork*, because neo-corporatism puts flexible worker teams at the base of all its network structures, and *learning*, because teamwork is presumed to be all about—and only about—knowledge.[48]

The event that calls forth the new corporatism, of course, is the massive change in the status of knowledge work that has occurred since the 1970s and 1980s. In that earlier era, itself emergent from what Theodore Roszak calls the personalism of 1960s counterculture, not just academic subject critique and New Class critique but the whole popular media myth of the "me" generation arose on the assumption that a universal knowledge identity or class was coming into being.[49] Given the seeming *fait accompli* of a consciousness- or expertise-based culture (opposites in Roszak's analysis, siblings from my perspective here), knowledge's emancipatory potential seemed to be the issue. Could consciousness be liberating? Could expertise be vanguardist? But now a great, mind-numbing retrenchment has come. That retrenchment is restructuring or downsizing, the U.S.-led (but increasingly global) systemic reorganization of knowledge work that—together with such corollaries as total quality management, just-in-time production, outsourcing, flex-timing, and, above all, information technology (IT)—has been overtaking all our major public and private sectors (corporate, government, military, academic, media, medical, and so on). First, the 1981–82 U.S. recession gave rise to a burst of enthusiasm for IT and "demassing layoffs" in the industrial equipment and commodity-oriented manufacturing sectors. Then the recession of the early 1990s prompted a gold rush toward ever more IT and increasingly pervasive, if also more surgical, "reengineering," "flattening," "optimizing," "rightsizing," "decruiting," "disintermediation," "unassignation," and "proactive outplacement" layoffs.[50] This continued through the "jobless recovery" of the mid-1990s to the enigma of job creation along with downsizing in the late 1990s and early 2000s.*

The rationale for all the wirings and firings lagged behind events briefly, but then arrived so fulsomely and confidently across all the major registers of social experience (whether the mass media or specialized sector discourses) that in Bourdieu's terms it might well be called our postindustrial doxa or foundational belief. It would be tempting to play this explanation on multiple tracks. For example, we could stay with academic theory by reading such severely revisionary accounts of New Class critique as Stanley Aronowitz and William DiFazio's *Jobless Future* (1994) or Nicholas Garnham's "Media and Narratives of the Intellectual" (1995)—both of which are acutely attuned to the twin phenomena of contemporary IT and downsizing. We could branch out to academic administration by attending to the rationales for decreased staffing and increased IT in major university systems during the 1990s (often issued under the credo of "corporatization").[51] From there, it would be just a short jump to the controversies generated by consolidations in such other IT-intensive sectors as the military, health industry, and media.[52] Clearly relevant as well in the mid-1990s was the credo of smaller government in the Republican-dominated 104th U.S. Congress, accompanied by heavy-handed legislation of IT and telecommunications, on the one hand, and intended purges of smart people (whether intellectuals or bureaucrats), on the other.

But I believe we must focus on the corporate sector, whose present *conceptual* and not just practical influence most academics have barely begun to recognize even amid the clamor over the corporatization of the university. As Michel Vilette puts it, "The field of management has contaminated all segments of society and is perceived as a universal cultural model."[53] The corporate sector is where the new paradigm of IT-enabled knowledge *as* knowledge (and not just as the bottom-line mentality academics customarily dismiss) has been fashioned for wholesale application to other realms—

* See appendix B for a chronology of downsizing.

even to the extreme, for instance, of such unabashedly corporatist imitations as the Continuous Quality Improvement (CQI) and Just in Time (JIT) initiatives in higher education.[54] I will therefore choose as my object of study precisely the kind of business literature I cited previously. In their sum, the prolific business bestsellers of the 1990s and after—by Hammer and Champy, Boyett and Conn, Tomasko, Davidow and Malone, Senge, Tapscott, Peters, Katzenbach and Smith, Drucker, Stewart, and others—are not just revisionary in their view of business and society but treat revisionism itself as the organizing principle of business and society. As Robert Johansen and Rob Swigart put it in their *Upsizing the Individual in the Downsized Organization* (1994), this is the revolution of the "*re*-word preachers": "restructuring," "reinvention," "redesign," "reengineering," and "recession."[55]

Of course, books on management theory (the work of the "witch doctors," as Micklethwait and Wooldridge dub them) do not by themselves tell the whole gospel of the new business—even when filled with case histories, adopted by companies as official or semi-official organizational literature (put on every desk like Gideon Bibles of the cubicle), and used to authorize the most massive, far-reaching changes in the lives of institutions and individuals.[56] This is because they represent only the orthodoxy of the new corporate doctrine. In its broader lineaments, however, postindustrialism is indeed a doxa, which like any such belief—as we learn from Guillory's extension of Bourdieu's analysis—is a three-body problem of orthodoxy, heterodoxy, and (as Guillory terms the New Critics' reaction to modern industrial rationality) "paradoxy." We can adapt such a three-body theory of doxa for our purposes by first detailing the problem in Bourdieu's and Guillory's terms and then significantly updating those terms.

Doxa, we heard Bourdieu say, is "the relationship of immediate adherence that is established in practice between a *habitus* and the field to which it is attuned, the pre-verbal taking-for-granted of the world that flows from practical sense." Or a bit more fully: "In the extreme case, that is to say, when there is a quasi-perfect correspondence between the objective order and the subjective principles of organization (as in ancient societies) the natural and social world appears as self-evident. This experience we shall call *doxa*, so as to distinguish it from an orthodox or heterodox belief implying awareness and recognition of the possibility of different or antagonistic beliefs."[57] Doxa is thus belief that is not just undeclared but—unlike the controversies of orthodoxy and heterodoxy—never even seems to require declaration because it simply *is* the uncontroversial state of things or, more accurately, the underlying eternal or natural condition that makes the current state of things possible. This is credence so deep, as Bourdieu likes to say, that it is "incorporated" in bodily habit. By contrast, orthodoxy and heterodoxy urge changes in habit. They point in one direction or another to some supra- or subculture's version of truth (e.g., the assertive cultures of the clergy, business gurus, or youth).

Guillory adds the following complication to adapt the argument to modern societies (where doxa "is condemned . . . to complex relations with both orthodoxy and heterodoxy") and especially to the early and mid-twentieth-century milieu of the New Criticism (the topic of his most immediately relevant chapter). Doxa in itself may be undeclarable, he says, but it may be "gestured" toward in modern times by the apparent elision of both orthodoxy and heterodoxy that the New Critics called paradox. "Paradox names the very condition by which the poem [in a New Critical reading] does not *name* the truth to which it nevertheless gestures," Guillory observes. Or again: "The condition of paradox is precisely the fact that a certain truth *(doxa)* stands alongside *(para)* the poem itself."[58] The paradox that surpasses understanding (which the New Critics also called "irony," "ambiguity," and so forth) is thus itself neither ortho/heterodox opinion nor doxa, neither a polemical view of the truth nor truth direct. Rather, to follow Guillory's explication of "gesture," it is a position formed in the wake of orthodoxy and heterodoxy (the literary criticisms of the T. S. Eliot and F. R. Leavis years, the *I'll Take My Stand* politics of Southern Agrarianism) by redirecting controversial opinion into a now apparently undecidable intuition of the truth analogous to—but as if displaced anamorphically from—the doxa at the center of all truth.[59] It is as if paradox says, "I cannot point to the truth on the right (orthodoxy) or the left (heterodoxy), but the very fact that I cannot point to the truth is *like* the truth, which never points but simply is ["ontologically," John Crowe Ransom said in his *New Criticism*]."

Yet, we must quickly add, precisely because New Critical paradox stood in the wake of orthodoxy and heterodoxy, its gesture could not be free of tendentiousness. The New Critics could not abide the prevailing doxa of modern times: the industrial-scientific rationality that the Southern Agrarians (from whom the New Critics emerged) called Northernism. What the New Critics' cherished paradox gestured toward, then, was not the truth but a slant vision of truth according to which the very action of gesturing was as dogmatic as any attempt by orthodoxy or heterodoxy to point to polemical versions of truth. Specifically, gesturing was opposed to making a point in rational (logical, scientific) propositions.[60] As Guillory thus summarizes the New Critical position: "The teacher or interpreter of the poem can only point to the truth which must not be spoken, but the very unspokenness of that truth elevates it to a status vastly greater than that of scientific truth, which always falls to the level of mere fact."[61] Like orthodoxy and heterodoxy, paradoxy displaced the central doxa toward a supra- or subcultural viewpoint (first southern, later academic), but unlike orthodoxy and heterodoxy, it was able to use its apparent undecidability or lack of a point to mimic centrality. The cult of paradox thus became a strangely orthogonal double or "recusant" of doxa, to adapt Guillory's concluding characterization of the New Criticism.[62] It was a shadow doxa.

To adapt the logic of doxology for our purposes now requires two additional steps. The first is to update the paradigmatic scene of doxa from Bourdieu's "ancient societies" directly to modern industrialism (rather than doing so at one remove via the New Critics). This requires looking beyond the New Critical poetic analysis of modernity to parallel analyses by early twentieth-century sociologists, "critical theorists," and others concerned with the broad underpinning of industrial society in rationality.[63] The most convenient way to do so is to update the two poles of the recognizably Cartesian dilemma fused within Bourdieu's notion of doxa: the objective "world," on the one hand, and "practical sense" or "subjective principles of organization," on the other. For objective world in an industrial context, we may say "technology." And for practical sense, we may say "technique" (the repertory of modern rationalized procedures, routines, protocols, standards, codes). The doxa of industrialism, then, is the unquestioned *alignment* of technology and technique under the assumption that the fundamental value is efficiency or productivity. The governance of such alignment is management; and the institution of that governance is the modernization of Bourdieu's "incorporation"—the corporation. This is the overall doxological ensemble that Herbert Marcuse spoke of in the 1940s as "technological rationality" (also "technological truth," "technics," "the apparatus"), and which Jacques Ellul called in the 1950s the total life of "technique."[64] Of course, a paradox within technological rationality gestured as much toward alternative rationalities as did the New Criticism. That paradox was leisure, which craved ease of entertainment as desperately as the New Critics craved difficulty of interpretation. But unprotected by the recusancy of the academy, leisure was effectively harnessed to the system through the regularized division of life into weekday work and weekend leisure, so that leisure not only alternated with work but complemented it as consumption to production.[65]

The second step needed to adapt the logic of doxology to our context is to shift the scene from modern industrialism to postindustrialism. At this point, paradox becomes central in a manner not readily off-loaded to the academic margins or time-shifted to the weekend. Unlike industrialism, postindustrialism is a technological rationality that integrates in its core—

within the technology and techniques of *information*—a paradox whose critical potential cannot be off-loaded, time-shifted, or differentiated as consumption because it is central to the processes of information technology–enabled production.

Preliminary to my thought here is Castells's crucial insight into the distinction between postindustrialism as a mode of production and informationalism as what he calls a "mode of development."[66] While the collapse of any viable, competing Soviet or statist mode of production now makes it seem that postindustrial capitalism and information technology are simply dual aspects of a single event, Castells argues, modes of production and modes of development are logically separate. "Capitalist restructuring" and "the rise of informationalism" are distinct, he says, "and their interaction can only be understood if we separate them analytically."[67] Put another way, the co-evolution of advanced capitalism and informationalism proceeds not by necessity but by historical contingency:

> What truly matters for social processes and forms making the living flesh of societies is the actual interaction between modes of production and modes of development, enacted and fought for by social actors, in unpredictable ways, within the constraining framework of past history and current conditions of technological and economic development. Thus, the world, and societies, would have been very different if Gorbachev had succeeded in his own *perestroyka*, a target that was politically difficult, but not out of reach. Or if the Asian Pacific had not been able to blend its traditional business networking form of economic organization with the tools provided by information technology.[68]

Keeping in mind this analytical distinction between postindustrialism and informationalism, we can see that the technologies and techniques of IT are not necessarily fused to the doxa of postindustrialism (restructuring, reengineering, lean production, and all the rest of the new management dicta). As a mode of development, IT also generates what amounts to a semi-autonomous doxa (a *belief* in information or in technology), akin to the faith that a rural laborer in revolutionary France, about to burn the house of the nearest aristocrat, might have expressed by picking up a pitchfork and saying, "*This* is what I believe." The challenge posed by dumb, contrarian uses of agrarian technology to production during the French Revolution is potentially the same challenge posed by the much "smarter" technologies and techniques of IT to postindustrial capitalism during the information revolution. Shrouded in glossy plastic and glass, softened with beige colors or occasionally dramatized in matte black, touched with the magic of winking LED lights, and, above all, plugged through the slenderest of wires to a network that is what Steven Johnson calls our contemporary understanding of infinity, the computer that sits on our desk is not just a tool.[69] It is a shrine of belief whose inner doxological structure, when we look at it closely, is seen to coincide not necessarily but only contingently with the new business plan for the future. To use an expression that recalls clearly the lineage of pitchfork, pike, and hammer and sickle in the background, the computer is both a tool and a "hack" of advanced capitalism.

A full statement of the inner doxological structure of information will need to wait for a social history of information technology (which I sketch in Part II of this book), but it will be useful here to anticipate some main points. First, if we were to define the semi-autonomous doxa of information strictly on the premise that information is something we *consume,* like news or entertainment, then we might say that our current fundamental belief is that all useful knowledge (what Bourdieu calls "practical sense") simply coincides with the "objective" world as that world comes to us mediated through information technologies whose real-time, interactive quality neutralizes the sense of mediation and so constitutes the *meaning* of objectivity. (The sense of being "live" on today's news broadcasts, for example, is conveyed by such unedited, real-time effects as the percussion of wind upon a journalist's microphone—a phenomenon that has no natural counterpart and thus amounts to a *technology* of natural reality. "Real-time" or "live"—no matter how mediated—simply *is* the same as "real" or "true.") Information consumed without concern for technological mediation, we may thus say, is our contemporary habitus. It is the habitual information environment in which "subjective principles of organization" (as Bourdieu puts it) are deeply in-formed by a world defined as technology-object. If there is also a strong sense of dissonance in such habitus (the conviction that IT overwhelms us with a "data smog" of mere facts that don't make sense), then such dissonance is subsumed within what could be called the general consonance of dissonance (the "expected" meaninglessness of the contemporary world).[70] We might think by comparison of "common sense" in pre-industrial societies. Common sense may have appeared only in the contradictory maxims, proverbs, clichés, and other incoherent bric-a-brac of practical knowledge (e.g., "look before you leap," but "he who hesitates is lost"). Yet any dissonance bowed to an overall faith in consonance named Nature or God.[71]

But, secondly, any such characterization of information as a habit of consuming objectivity would be incomplete if we did not immediately move beyond the perspective of consumption to the greatest habit of contemporary postindustrial life—work. What is missing thus far in my rendering of the doxa of information is the viewpoint of production rather than consumption, the viewpoint that makes information the hallmark of knowledge *work*.[72] Adding the vantage point of production changes the feel of information considerably. In this light, there is less faith that the latter-day Nature or God—the corporation—can overcome the fundamental dissonance of the information technology it manages. While the technological mediation of information is often largely camouflaged in the sphere of consumption, the technological *management* of information has certainly not been so in the sphere of production. From the viewpoint of production, the doxa of information may be defined as follows (using phenomenological vocabulary drawn from Shoshana Zuboff's interviews with early business computer workers who said their computers let them "see it all").[73] Information is at once an emancipatory, quasi-transcendental "vision" of the total system of technological rationality (like opening a file directory on a company computer and suddenly seeing a list of all related files) and a frustrating sense that one is seeing only disassociated pieces of the big picture (mere facts and data) because the principle of *managed* information aggressively denies access to that total vision. Information systems are designed to enable real-time, interactive connectivity; synopticism is their premise and goal. Yet, paradoxically, such systems are also designed to resist connectivity in the name of what James R. Beniger, in his expansively conceived *Control Revolution: Technological and Economic Origins of the Information Society*, simply terms "control" and what the workaday corporate world now calls "security."[74] If the network is our contemporary intuition of infinity, then its boundlessness is matched by an equally infinite, equally unreal hunger for security—indeed, for what amounts to a metaphysics of security (paradoxically secure yet connected) compensating symbolically for all the other vulnerabilities of life in the post–Cold War era of global connections: immune systems, national borders, jobs, and so on. This is the fundamental meaning of the password that is the first information demanded of a user when logging onto a system. As the sign of the metaphysics of security, the password is a piece of pure information that is ideally unrecognizable in *any* real-world context or pattern of information (e.g., "mKiJ84sY"). While users wish their passwords could be habitual to the way they work and live, actually—as anyone knows who has ever forgotten a password or been forced to change passwords on systems with automatic expiration rules—passwords are never habitual unless they contravene security. Passwords are oaths to a transcendental security.

While the doxa of information may be central to the doxa of postindustrialism, then, there is a curious doubling within such centrality that makes information (at once synoptic and quarantined, visionary and blind, connected and secure, managerial and managed, transcendental and paranoid, free and enslaved, decentralized and centralized) also the *paradox* of postindustrialism.[75] Such an information paradox is our latter-day, secular recusancy—rusticated not just to the academy (though it is there, too) but to a simultaneously empowered and resentful culture of knowledge named "cool" within the center of knowledge work. Such paradox, as I foreshadowed in discussing Bourdieu and New Class theory, restores the relevance of the specifically mental character of information work. While mentality as such may now be a category mistake when thinking about knowledge work as *work*, the very irrelevance (and sometimes irreverence) of such mentality—an excess of knowledge about the connectivity of information that the *management* of information deems a "hack"—grants the knowledge worker the mental distance needed to "gesture" beyond the doxa of postindustrialism to the semi-autonomous para-doxa of cool.

By reserving room for such paradox in what might otherwise seem either too dogmatic a rehearsal of business orthodoxy or, criticizing it from a humanities viewpoint, too reductive a heterodoxy, we can now recount the basic thesis of the management bestsellers. On the one hand, it has become orthodox that in the age of global competition knowledge work is essential—so much so that such work has caused a foundational shift in the "being" of business organizations.[76] Where once matter was the essence (in the industrial age a corporation not only processed matter but *was* its material factories, inventory, and people), now matter work is for the developing world. If the U.S.-led West is to stay ahead of the rest of the world, so goes the thesis, *post*industrial corporations must de-essentialize themselves until they are nothing but information processing or (since durables and consumables must still be produced) at least matter that can be made to act like information processing—that is, plants, goods, and people endowed with the quick-turnaround responsiveness, flexibility, and ultimate eraseability of bits.[77]

This leads to the basic logic of downsizing. Picture the postindustrial corporation as an optical fiber or superconducting wire designed for resistance-free flow of information—an alignment of technology and technique so perfect that it is free from the friction of matter. The ideal organization is one that has stripped out all intermediary levels of equipment, inventory, processes, organizational units, and people so that the information necessary to produce, for example, a new car flows laterally between customers, sales, design, plant, and suppliers with the speed of light. Everyone and everything is part of an information network whose basic units are flexible team workers incessantly communicating with each other and with the larger organization. So, too, any vertical hierarchy remaining in the flattened organization is greased for information flow: the ideal CEO communicates with operations directly or through severely pared down middle management layers. The image of an optical fiber or superconducting wire is not far from the literal truth. Enabling the mania for information flow from the mid-1980s on (and achieving dominance in the networked 1990s) was a symptomatic form of corporate IT: local area networks, groupware, intranets, collaborative project schedulers, conferencing programs, and other such team communication products. "Hey, look," IBM's Lotus division declared in a 1996 advertisement for its SmartSuite programs: "Another software ad promising to get everyone working together in *complete harmony*. Be still, my heart." The suite included TeamReview and TeamConsolidate features. Similarly, Microsoft pitched its Project program under the slogans "Graphical views let teammates visualize where the project is going" and "Keep everyone moving on the same track with open lines of communication."[78] Information flow is everything.

And so, too, according to the orthodoxy, knowledge workers shall be everything. The information flow thus envisioned can function only if the network is operated at all its nodes by super-informed people able to subsume the roles of laid-off middle managers. Every teamworker in the new regime, therefore, is trained and equipped for smart work. According to what Katzenbach and Smith call "the wisdom of teams," there will be frequent team brainstorming meetings, perpetual retraining motivated by "pay for knowledge," and constant dissemination of company-wide philosophy and performance data. "The newest and lowest-level employee," Joseph Boyett and Henry Conn say in *Workplace 2000,* "will be expected to know more about the company that employs him or her than many middle managers and most supervisors knew . . . in the 1970s and 1980s."[79] At the top of the hierarchy, the CEO must also be personally super-informed. Reappropriating many of the middle management and even clerical techniques that, as Zuboff observes, were spun off from the executive position during earlier stages of office automation, he or she becomes something like an ultra-clerk able to make fluent use of company IT (e-mail, spreadsheets, databases, charting programs, project schedulers) to gain instantaneous apprehension of what scores of underlings once would have taken weeks to synthesize.[80]

The entire organization, in sum, will live or die by the current *idée fixe* of lifelong learning. As Boyett and Conn note, "Continuous learning will become commonplace to create a more flexible work force, provide employees with the skills necessary to take advantage of rapidly changing technology, and prepare employees for new jobs inside or outside the company when their old jobs are replaced by technology or eliminated due to changes in customer demands."[81] Not surprisingly, then, the jeremiad of neo-corporatism in book after book concerns the failure of traditional education to prepare the new force of smart workers. Instead, we learn, the new corporation must shoulder this responsibility and become, in Peter Senge's now famed phrase, the "learning organization." This is "where new and expansive patterns of thinking are nurtured, where collective aspiration is set free, and where people are continually learning how to learn together."[82] To academic ears accustomed to the rhetoric of commencement ceremonies, this may sound like wholesale appropriation of the educational mission, and that is exactly what it is. Tom Peters opens a section of his *Liberation Management* by redefining Stanford University as the creation of a great "information organizer" (Leland Stanford). That stroke accomplished, the conclusion follows with pitiless efficiency: "Organizations are pure information processing machines—nothing less, nothing more."[83] With similar efficiency, Don Tapscott's section entitled "Learning Is Shifting Away from the Formal Schools and Universities" enumerates the astounding growth of "formal budgeted employee education" and then makes an aggressive comparison: the numbers, he says, represent "the equivalent of almost a quarter of a million additional full-time college students—thirteen new Harvards. This is more growth in just one year than the enrollment growth in all the new conventional college campuses built in the United States between 1960 and 1990."[84] No longer *veritas* or *lux et veritas* on university seals, in short, but the brilliant insignia of the new corporate enlightenment: "Intel," "Sun," "Lucent," and so forth.

So much for the new orthodoxy. But the new corporatism can also be viewed from a considerable heterodox angle (as is often the case in mainstream media coverage of downsizings). The heterodox view is that as knowledge work gains in value, knowledge workers are systematically devalued.[85] In the corporatist prophecy, that is, the millennium is when someone must be damned for the corporation to be saved. The simplest case concerns the damned middle managers, who when laid off in the millions are by and large irrevocably shunted off the track of traditional job security and career growth into a wholly different school of hard knocks—one of nervous corporate job-hopping and/or high-risk independent business, purely lateral career movement, permanent re- or deskilling, and long-term salary and benefits reduction.[86] More complex is the case of knowledge workers who remain behind in a firm as "survivors." Greater information work assigned to fewer workers, of course, is in one sense vastly empowering (to use the precious human resource term).[87] Instead of laborers chained to piecemeal tasks, multi-competent work teams are now supposed to oversee projects holistically with perspective on total company strategy. And instead of bureaucratic managers overseeing their small piece of turf, the managers who remain ostensibly have greatly increased "spans of control," more flexibility in their new roles as "facilitators" or "coaches" of work teams, and more cross-field expertise gained on "management teams."[88] But clearly also, eating of the apple of postindustrial knowledge is not without pain. On the line, as Boyett and Conn put it, "tremendous responsibilities will be shifted to workers and their peers. . . . This new and expanded role for employees will exert enormous pressures on employees and companies alike to invest in education and retraining."[89] As spelled out by the refrain in the business books, "some" will not adapt: that is, not all workers will be willing or able to commit to lifelong learning, ever quicker just-in-time production, riskier evaluation and pay schemes pegged to team- and company-wide performance, and, in general, what has been called management by stress. In middle management, similarly, increased spans of control and flexibility are accompanied by the classic symptoms of corporate demassing: multi-tasked overwork, anxiety over nontraditional management roles (what exactly is a facilitator?), worry about late career retraining, broken faith in the firm, and so on.[90]

Yet, despite critique in the press and elsewhere, and even during the tight labor market of the late 1990s (boom times when many firms struggled to hire or retain scarce skilled workers while *simultaneously* downsizing), the new corporatism has not seriously had to acknowledge heterodoxy about restructuring as anything more than a public relations problem.[91] And it has certainly not had to accommodate the possibility, as academic cultural criticism might put it, that the knowledge workers it employs could be meaningfully contestatory. Neo-corporatism assumes fundamentally that total determination is a done deed, enforced by global competition, and that dissent among employed individuals, groups, or classes is largely irrelevant. More accurately, it would be too facile to say that the new corporatism never accommodates any critique. Criticism from individuals within the restructured corporate system takes the visible form of messages posted by disaffected employees to company e-mail systems or online public forums, memos leaked to Web sites such as InternalMemos.com (affiliated with the FuckedCompany.com site), legal actions for distresses ranging from wrongful termination to stress injuries, internal hacking of company IT, and even "whistleblowing" complaints to public authorities.[92] Yet such critique tends to be normalized either as routine or exceptional "disgruntlement." The salient fact is that the most significant internal critiques of the corporation, even if ultimately motivated by the wish to mollify individuals and thus to fend off legal or political scrutiny, are those whose policy implications emerge on a different level than that of the individual. Representative are the social policy issues that William Davidow and Michael Malone pose in their last chapter when they ask if the "virtual corporation" can be "virtuous," or the recommendation by Herman Maynard, Jr. and Susan Mehrtens that "fourth wave" corporations must "make the intellectual shift from wanting to beat the competition to wanting to serve the world."[93] Equally symptomatic is the increasing formal attention of corporations to "business ethics" and "business and society" agendas.[94] Insofar as the new corporatism accommodates a critique of restructuring, it bypasses the individual and projects the problem onto the abstract plane of the *whole corporate entity*.

Such a focus on the overall corporation goes to the heart of the matter. At this point in our review of explanations of knowledge work, we can ask: What is the ultimate point of the new corporatism? The point, which will strike humanities scholars with force in the wake of subject and cultural class critique, is precisely *neither* the subject nor class. Rather, it is the "dispersal," "weakness," or "contradiction" of both (as characterized by academic theoreticians), extrapolated to become the total negation of both—that is, the whole corporate entity. The new corporatism, in other words, is at once the logical extreme and the annihilation of identity and New Class critique. It is the subsumption of identity and class. Here we might turn to two of the diagrams that neo-corporatist literature frequently uses to visualize its new world order: Don Tapscott's fivefold scheme of corporate identity (from the "Effective Individual" up to the global "Internetworked Business") and Jessica Lipnack and Jeffrey Stamps's similar scheme (from the "Small Group" up to the "Economic Megagroup") (figs. 1.2, 1.3). In this corporatist order, we recognize, the fundamental move is made when identity is from the first swallowed alive by the cult of the team (and thereafter, as Susan Albers Mohrman and colleagues call it in their definitive work on the topic, the higher forms of "team-based organization").[95]

What is a team? The positive answers offered by business are many and complex. In the terms I have developed here, the team is the unit of ephemeral identity that most flexibly fuses technologies and techniques into skill sets (called "innovation," "creativity," or "resourcefulness") adapted to the changefulness of the global economy. But in our current context the negative answer is the most compelling: by definition a team is *not* an identity group, and it is assuredly *not* a class formation.

Yet we must rethink the notion of negation if we are to avoid relapsing into once robust but now increasingly hapless modes of describing social "hegemony" and class "oppression" in the oppositional (or dialectical) terms of "them" versus "us" or big "company" versus little "worker." What is

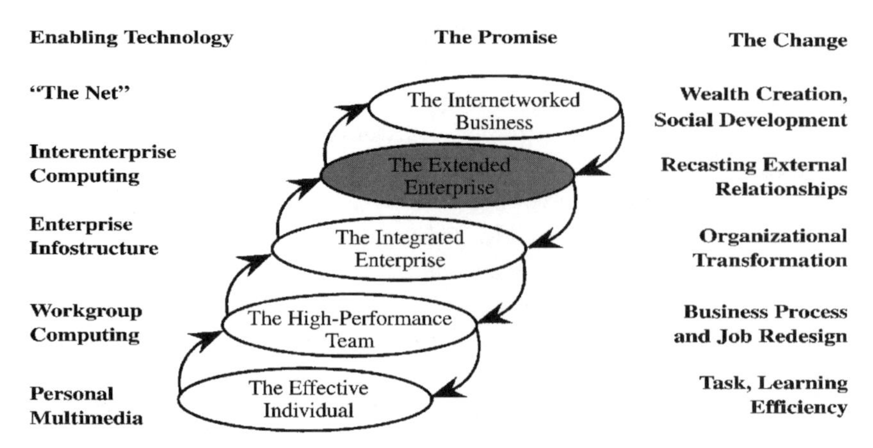

Figure 1.2 Tapscott's "Extended Enterprise"
From Don Tapscott, *The Digital Economy: Promise and Peril in the Age of Networked Intelligence* (New York: McGraw-Hill, 1996), p. 85. Copyright 1996 by The McGraw-Hill Companies, Inc.

distinctively new about contemporary corporate "identity management," as Thomas B. Lifson calls it, is that it does not so much master, marginalize, discriminate against, or exclude group and class identities as *take them over,* complete with their contestatory force (thus harnessing "diversity" to "competitiveness"), through that most powerful manner of postindustrial negation—simulation.[96] The team is designed to "simulate away" identity groups and class by incorporating them.

To understand how this is possible, we must examine the antihistoricism of business prophecy. As should already be clear, this antihistoricism is so profound—yet often also so automatic and slick—that it might be likened to a shrink-wrapped, one-minute version of Alvin Toffler's *Third Wave* bundled together with Francis Fukuyama's *End of History and the Last Man.* On the menu bar of each of the business bestsellers, as it were, there is a big button marked "Delete History." For example, consider the precision model of historical obsolescence offered by Maynard and Mehrtens's *Fourth Wave*: "The First Wave of change, the agricultural revolution, has essentially ended and will not be of concern here. The Second Wave, coincidental with industrialization, has covered much of the Earth and continues to spread, while a new, postindustrial Third Wave is gathering force in the modern industrial nations. We see a Fourth Wave following close upon the Third."[97] The possi-

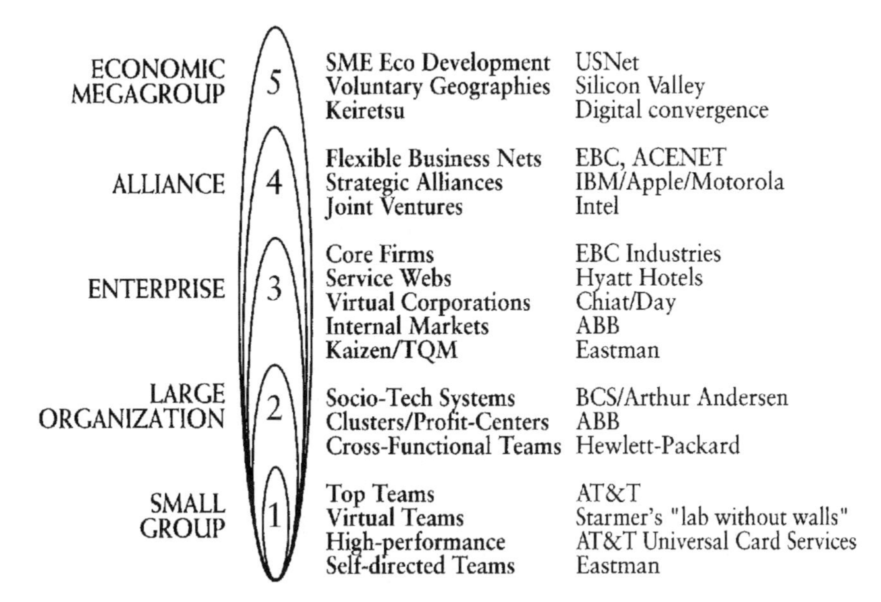

Figure 1.3 Lipnack and Stamps's "Teamnets along the Organizational Scale"
Reprinted, with permission, from *The Age of the Network: Organizing Principles for the 21st Century* by Jessica Lipnack and Jeffrey Stamps, p. 98. Copyright © 1994 by Jessica Lipnack and Jeffrey Stamps.

bility that the agricultural revolution may just now be reaching broad expanses of the globe in complex collaboration with the industrial revolution (tractors in Africa, for example) is moot: in this starkly unlayered view of history, the world is a diskette that can be reformatted any number of times. Change, we heard previously, means "starting over," "the systematic sloughing off of yesterday." Hammer and Champy add: change must be "fundamental," "radical," "dramatic."[98]

We might put the case in largest view as follows. Recall the succession of universalizing theories of civilization I earlier recited under the names of reason, repression, instrumental rationality, knowledge/power, and knowledge work. Contemporary management theory (descendant of the "managerial demiurge" C. Wright Mills identified in his classic study of the white-collar middle class) is the latest claimant in this Enlightenment genealogy.[99] As perhaps the most unmediated of antihistoricisms since the French Revolution (which presumed to throw out the relevant past to start over again in French Revolutionary Year One), management theory reveals with stark clarity that its universalism, progressivism, and rationalism resolve at base into a neo-Enlightenment theory of history. This theory (in a severe truncation of Fukuyama's thesis) holds that the only significant history is the end of history, or Reason. That end may be imagined at the teleological close to history, but more ordinarily it is implanted as a sort of temporary telos in every point of significance in history. History in this view (and even in its counter-Enlightenment critique, as in Foucault) is finally all about "epochal," "revolutionary," or "epistemic" moments of change, when history effectively starts anew—when historicity itself, in other words, disappears from or, in the modern critique, becomes repressed in the apparatuses of Reason. Our greatest contemporary holdover from the Enlightenment is thus the belief that civilization is managed history. Whether one champions or critiques it, civilization is the faith that history can be rationally managed. Insofar as history from another point of view is precisely that which is *un*manageable and *ir*rational (revolutionary mobs, the unconscious, Unreason, etc.), it must finally be managed through the overthrow/repression of history itself. ("Management is the organ of institutions," Peter Drucker writes, "the organ that converts a mob into an organization, and human efforts into performance.")[100]

That much contemporary management theory in the Tom Peters mold appears to dance on the Bastille in its ludic celebration of "chaos" management and other forms of organizational irrationality is not an exception to the rule but the crowning instance of the institutional recapture of disruptive history (as heard of old in such formulae as "The king is dead, long live the king!" or its revolutionary variant, "The king is dead, long live the people!"). Irrationality in all its terrible energy is now to be harnessed *within* the organization as the means of strong, rational management—especially as personified in the form of a charismatic leader able to rationalize otherwise irrational total changes of "organizational culture."[101] (Thus occurs one of the outstanding contradictions of recent management theory: its simultaneous worship of decentralization and, often in a separate chapter, of executive "leadership.") Postindustrial management is the *ne plus ultra* of the restructuring revolution initiated by the French Revolution with all its radical political, economic, geographical, calendrical, military, educational, and other reengineerings.

Now we can understand why those immediate predecessors to academic subject critique and cultural class critique—so-called "essentialist" identity politics and traditional Marxism, respectively—seem so distinctly unenlightened at present. These perspectives were predicated above all on remembering the historical struggles of peoples and classes, to the point that the process responsible for "dispersed" identity (as subject and New Class critique calls it) was experienced historically as the saga of diaspora.[102] By extension, we also understand what the postindustrial corporation must do to manage identity groups and classes so thoroughly (in a constructionism that is the extreme of subject and cultural class critique) that it elides their last vestiges. It has to hide their history. Just as the French Revolution attempted to suppress the historical identity of estates and regions to ensure that everyone, high and low, Parisian and provincial, would be fraternal, so now the new corporation does the same for subject groups and classes.

But there is a difference. The French Revolution substituted for historical identity a representation of national identity that was, first of all, what recent scholars of the period have called "theatrical" representation. The new representative politics of the nation-state, in other words, was theater at the same time that it was political reality, and one could only be fraternal if one "performed" the role, complete with improvised clothes and speech forms, of *citoyen*.[103] In contrast, postindustrial corporations substitute for historical identity another style of unreal identity that makes even the posthistorical or "eternally modern" nation-state obsolete (and implies instead the globalism that is now the necessary complement to any progressive nationalism). In the postindustrial version of the local/general problem, multinational corporations base themselves on the local *and* the global, the intra- *and* the international, to the exclusion of the nation concept in the middle, which now appears merely as a regulatory obstacle or trade barrier.[104] Such corporations create a *post*representational identity that is not so much theater as what postmodern theorists call simulation (e.g., Baudrillard's *Simulations*). Simulation is representation backed up by no reality—or, what is the same, by "mass" realities too simultaneously local and global for representations of nation-state vintage to capture. On the side of consumption, for example, the pattern was set early on by cable TV channels that reached out from local markets (the city of Atlanta, for example) to blanket the world. On the side of production, the cellular unit of such corporate simulated identity is the team. At once local and global, here at home and there in Japan, teams are a fraternity that bypasses both the essential nation and, arising to contest the melting-pot identity of such nation, essential group and class formations.

To say that the team simulates away identity, then, is to say that it deletes the historical identity of the folk's habitus, or (as the Birmingham cultural studies group termed it) customary corporation, and then preempts the recapture of that identity by the great political, economic, media, and other institutions of the nation-state era that had arisen specifically to manage that recapture—that is, to modernize and bureaucratize the folk.[105] Identity is recaptured instead for a new, "flexible" bureaucracy—the team corporation—designed to forge no recognizable national, group, or class identification at all ("our country/group/class: love it or leave it"), but instead simulated identification with the team, simulated identification with all the higher organizational forms of the team corporation depicted in figures 1.2 and 1.3 (enterprise, alliance, megagroup), perhaps even simulated identification with global competition itself ("global competition: love it or leave it").

Consider, therefore, the recent fate of identity groups and classes in the corporation. In regard to groups, the most instructive instance is U.S. diversity management. Diversity management theory arose in several works of 1991—most influentially, R. Roosevelt Thomas, Jr.'s *Beyond Race and Gender,* David Jamieson and Julie O'Mara's *Managing Workforce 2000,* and Marilyn Loden and Judy B. Rosener's *Workforce America!* It has since swollen into a whole movement of studies, guidebooks, video series, training programs, and corporate initiatives.[106] Prompted by the highly influential forecast of *future* demographic diversity contained in the 1987 Hudson Institute report *Workforce 2000,* the movement explicitly discounts diversity efforts driven by statutes redressing past discrimination. Instead, it makes diversity all about present business self-interest.[107] To cite the subtitle of Lee Gardenswartz and Anita Rowe's *Diverse Teams at Work,* it is about "capitalizing on the power of diversity." Or as Roosevelt Thomas puts it: "Burning the increasingly diverse human resource fuel" and unleashing the "power that all the various groups in our national work force have to offer" is "an opportunity for competitive advantage."[108] "Workforce 2000," in other words, means that topics such as "Slavery Pre-1861," "Railroads 1860s," "Ellis Island 1892–1954," or "California 1990s" are not to the point—except insofar as their power may be drilled out of the historical ground, clarified of their dinosaur identities, and pumped into a corporate engine that had run on gas but now must guzzle any diesel, methane, or perhaps even gasohol fuel. However, we cannot be content with mere bumper-sticker labels for the history that is thus rejected—part of the "Workforce 2000" mindset itself—if we are to grasp the exact manner in which diversity management expends the past to compete in the present. What is the relevant history that is distilled out of the picture?

The broadly historical intent of the question will be served if we answer it this way: identity groups as we now know them originally arose on the industrial age "line" or "shift" populated in different ways by emigrants from the Old World to the New, the South to the North, and the country to the city.[109] Previously, the identities of many of these migrants had been rooted in agrarian age work formations on the order of the yeoman "household" (the extended family with or without live-in servants).[110] Together with its clan, village, parish, estate, and other envelopments, such household culture was saturated with the kind of blood, kin, religious, gender, language, *paisan*, and other solidarities that would later be called "race," "ethnicity," and so on. But crucially, such solidarities were not in themselves racial, ethnic, gender, or other constructs in our sense. Constructs of this modern sort only became experientially real in dislocation from their original location—that is, in the diaspora that transplanted the household onto the line/shift as a fragmented social system needing to be reconsolidated in a new set of cultural formations in competition with those of other diasporic groups.[111] "Experientially real" in my phrasing here means that, while dislocation *de*-realized known cultural reality, it was compensated for by an abiding sense of history, a sense that the process of reality-turning-unreal (historical change) was a deeper reality cognate with identity. "Everything we once knew is gone and we are as strangers in this land," history says, "and *therefore* we are who we are."

The working subject, in other words, knew that it was a subject only insofar as it was part of a "people" or "tribe" who, in a daily reenactment of their diaspora, all went on the line/shift together. This is why the Fordist-Taylorist line in its heyday was not just dehumanizing, no matter how much it demanded that the lived reality of human technique mold itself to inhuman technology. Exile from one's humanity simultaneously created the sense of a whole working-class neighborhood away from neighborhood, a community with so much potential for solidarity that it could have its own after-hours hangouts, charities, bowling leagues, youth subcultures, and so on (though cultural solidarity, we note, was often effectively quarantined in "leisure").[112]

We now know how diversity management must proceed. Increasingly, it concentrates on team building or, in the now standard terminology, creating "diversity teams."[113] For the team concept is the perfect way to deinstall not so much the actual line/shift (U.S. auto makers, for instance, did not adopt the Swedish innovation of station-based assembly) as the *relation* between the worker and the line.[114] Join the team: leave your identity group and history behind to enlist in a "small group culture" so semi-autonomous that it meets, talks, works, and plays together and even has the celebrated power to "stop the line." Once team culture is in place, then all the rest of downsizing culture follows: smart work, flexible competence, flat management, and so on. The general principle, in other words, is to couple diversity to the total restructuring effort. Thus witness Thomas's restructuring of the "empowerment" concept in *Beyond Race and Gender*:

> Another word for the process of tapping employees' full potential is "empowerment". . . . In fact, a managing diversity capability is implicit in several innovations already in process in progressive organizations. Some corporations, for example, are moving to "push decision making down." Others are implementing "total quality" initiatives. Still others have downsized their work forces in search of greater efficiency and productivity. All of these initiatives, however they differ, have one aspect in common: Their success depends on the ability to empower the total work force.[115]

So, too, Anthony Patrick Carnevale and Susan Carol Stone assert that diversity corresponds point by point to "total quality management," "reengineering," and other major restructuring imperatives. The real work of achieving diversity, it is now believed, lies in achieving a proper team and corporate culture in the first place.[116]

The real work, then, is *capitalizing* the concept of culture so that the pure business culture that remains at the end of the process becomes definitive of all culture. Such capitalization may be read everywhere in the implementation agendas that diversity management issues for diversity "inventories," "networking," mentoring, or training exercises. I take as illustration the section titled "Archaeology 101: Creating a Team Culture" in Rafael Gonzalez and Tamara Payne's essay "Teamwork and Diversity." "Team culture," Gonzalez and Payne list, "is a structure of experience that gives individuals":

> [1.] A sense of who people are
> [2.] A sense of belonging
> [3.] A sense of behavior and an understanding of what they should be doing
> [4.] A set of problem-solving tools for daily coping in a particular environment
> [5.] The capacity and mechanisms for transmitting coping skills and knowledge[117]

We can parse this definition of culture into two main components, as follows. One is customary culture or culture as a way of *being*: "who people are" and "belonging" (1, 2). The other is a way of *doing* and, in particular, of doing business: the functionalism (3) and technological rationality (4, 5) whose specifically postindustrial form is the frictionless alignment of technology and technique called "smart work." Building a team culture means adapting the notions of customary culture and technological rationality so that they slot easily into each other—ways of being into ways of doing business. Put differently, both customary and business cultures must jack into a common model of capital specifically wired for postindustrialism. To adopt a governing metaphor whose relevance will become clear later, the particular cultural platform no longer matters if all equipment for being and doing links up through standard protocols to the same router.

Starting with nos. 1 and 2 above, we see that customary culture is jacked into capital in a single bold move: it is put in present tense ("who people *are*"). History is thus cut out of the circuit so cleanly that we might almost miss its absence for lack of obvious torn edges. Yet this one move makes possible the entire procedure of diversity management, which we might imagine on the basis of our governing metaphor, as follows. Picture history as an original "file" on a cultural "server" that must be transmitted to the corporate server by TCP/IP protocol (the Internet file transmission protocol according to which information is broken into discrete packets, routed or switched semi-autonomously over the Net, and then reassembled at the other end). Now take a single, instantaneous cross-section of the network carrying the transmission. The result is an impression of multiple strands and nodes, each holding discrete, self-contained packets that are not only oblivious to other packets from their file but may be intermixed with other files. Just so, the standard procedure of diversity management is to approach culture as disassociated traits (cultural packets) that do not appear to cohere logically because their historical "roots" (an older metaphor for networked identity) lie concealed.[118] In Gonzalez and Payne's own metaphor, therefore, diversity management is an archaeology limited to a single layer in a dig: it is the archaeology of the present. Here, for example, is what "Archaeology 101" uncovers in "team members' cultural norms": greetings; dress and appearance; breaks, mealtimes, food; timeliness, time needs; relationships with co-workers; definition of family and roles of family members; primary values; beliefs; celebrations.[119] Thus "dress and appearance" are discovered lying in one spot; "breaks, mealtimes, food" in another; relics of "beliefs" and "celebrations" in a third. A gigantic amplification of such archaeology occurs in Gardenswartz and Rowe's *Managing Diversity* and *Diverse Teams at Work*. Packed with worksheets for profiling workers as aggregates of space sense, communication habits, dress and appearance, eating customs, time consciousness, and so on, such books demonstrate that, despite the best holistic intentions, diversity management always seems to come down just to managing piecemeal the way people say hello, shake hands, eat, and keep their appointments.

From the accountant's rather than the archaeologist's viewpoint, that is the whole goal of the exercise. The goal is to download human resources into the corporation less as a stockpile of raw psychology (the "empowerment," "sensitivity," and so on that human resource theory had nurtured) than as capital. Or, rather, "stockpile" with its just-in-case philosophy is here exactly the wrong notion of capital. The goal is instead to make human resources conform to the nervous, jumpy, constantly assembled and reassembled, just-in-time capital of postindustrialism. As Avery Gordon comments in her insightful critique of diversity management, "to the extent that the cultural reorganization of the corporation fastens on the element of variability, it is consistent with the metaphysics of capital-logic: mobility and flexibility. The hypermobility of capital and the transcoding of flexibility into a new form of social regulation find a consonance in the notion of corporate culture."[120] Why should the new corporations accommodate diverse cultural identity, in other words? Not just because it is moral, legal, or even psychologically effective to do so, but because when cultural identity is managed as traits that are modular, flexible, and just-in-time, it doesn't matter which cultures from Workforce 2000 enter the mix: it is all grist for the lean production mill. We might translate affirmative action for the new millennium as follows: *We affirm the right of workers to carry around only as much of their inventory of customs as needed, to adapt their customs* [now called "skills"] *as flexibly as possible to the final arbiter of custom, the customer, and then to turn around on a dime to adopt entirely new configurations of customs as needed* [called "changing one's culture"].

Such a post-ethical and -statutory approach to affirmative action, of course, might still ring hollow to anyone who expects capital to be justified even minimally on customary cultural grounds. Therefore, we must now turn to nos. 3–5 in Gonzalez and Payne's definition to observe that, even as diversity management retools the notion of custom to suit corporate culture, so it reciprocally alters the idea of corporate culture to accommodate custom. Or rather—and this is its real power—it inflates corporate culture until it subsumes culture at large and becomes self-legitimating.

We will need to attend more fully to the theory of corporate culture in Part II. For the moment, we may note that the inflation I indicate is implicit in Gonzalez and Payne's prefatory statement that "when you create a work community . . . you're actually developing a culture," or again "work communities are a microcosm of life."[121] Such inflation is also evident in other works of diversity management that explicitly define culture as all technological-rational—for example, Harris and Moran's *Managing Cultural Differences:* "culture is a distinctly human capacity for adapting to circumstances and transmitting this coping skill and knowledge to subsequent generations" and "in essence, human beings create culture . . . as an adaptation to their physical or biological environment."[122] Crucial in this regard is the way diversity routinely disappears into its purely technical lookalike, "diversity of talent and expertise." For example, Davidow and Malone in their section "The New Breed" document the need to hire from an increasingly multicultural workforce and then, in the next section, "Teamwork," immediately transpose diversity into this purely "talent"-oriented notion: "the empowerment of employees, combined with the cross-disciplinary nature of virtual products, will demand a perpetual mixing and matching of individuals with unique skills. These individuals, as their talents fit, will coalesce around a particular task, and when that task is completed will again separate to reform in a new configuration around the next task. The effect will be something like atoms temporarily joining together to form molecules, then breaking up to form a whole new set of bonds."[123] Diversity in this mold is not multicultural because it is not *any* cultural in a customary sense. It is constructed not so much out of men and women as from culture-bare "atoms" (a common trope).[124] These are "configured" around totally ad hoc tasks, and assisted in their chemistry by totally ad hoc social supports (the "parties, hoopla, and celebrations" that Katzenbach and Smith and others say are vital to high performance teams).[125] The team thus disintegrates all bonds of race, ethnicity, gender, and so on to create just-in-time or on-the-fly cultures diversified by "skills" and "talents" oriented toward maximizing results.[126] Such capitalization of customary culture is legitimate because *all* culture is merely a modular capital of techniques, skills, and talents.

The appropriateness of subordinating archaeology, atoms, and so on to our governing technological-rational metaphor of jacks, networks, and TCP/IP protocol can now be explained. Davidow and Malone's "atoms," we note, are subordinate to "bits" in their overall depiction of the virtual corporation as all about information. Similarly, Gardenswartz and Rowe's master trope turns out to be "cultural programming": "All of us are programmed by cultural 'software' that determines our behaviour and attitudes."[127] The image that best describes culture as understood by diversity management is indeed a network through which culture circulates in packet-traits of instantaneous, historyless (called on the Internet "stateless") information. The great *allegory*—and not just medium—of postindustrial capital, in other words, is digital. According to the allegory, just as all good capital is as uniform yet flexible as a bit, so, too, all good culture is as uniform in its multiculturalism as Silicon Valley technoculture. Not enough color? Just use the managerial equivalent of a graphics program like Photoshop to change the "palette" of traits, apply a "filter," and instantly repixelate the image. A million different images of culture result, yet—and this is why such images are finally simulations of culture—all the differences turn out to be part of the same culture of information management. Not accidentally, we notice, the final consequence of the "diversity" = "diversity of talent" thesis is that diversity inevitably expands to include everyone. All differences are technical, after all. Here is Roosevelt Thomas, for example, on who gets to count as diverse:

> Diversity includes everyone; it is not something that is defined by race or gender. It extends to age, personal and corporate background, education, function, and personality. It includes lifestyle, sexual preference, geographic origin, tenure with the organization, exempt or nonexempt status, and management or nonmanagement. It also shows up clearly with companies involved in acquisitions and mergers. In this expanded context, white males are as diverse as their colleagues.[128]

Lewis Brown Griggs opens the door even wider:

> I believe *diversity* should be defined in the broadest possible way. Not only does *diversity* include differences in age, race, gender, physical ability, sexual orientation, religion, socioeconomic class, education, region of origin, language, and so forth but also differences in life experience, position in the family, personality, job function, rank within a hierarchy, and other such characteristics that go into forming an individual's perspective. Within an organization, diversity encompasses every individual difference that affects a task or relationship.[129]

Diversity of this sort, as Gardenswartz and Rowe put it in the title of one of their sections, sees "Each Individual as a Culturally Diverse Entity."[130]

In the end, however one may admire the practical achievement of diversity management (its success in adapting diversity to business self-interest must account in large part for the tenacity of corporate diversity initiatives during the political backlash against affirmative action in the United States beginning in the mid-1990s), it is accurate to say that the team culture and corporate culture it molds are only a simulated culture with a simulated diversity, which might be called a "monoculture of diversity."[131] In an unusually self-reflective section titled "Team Building: Is the Whole Idea Culturally Biased?" Gardenswartz and Rowe stop to consider an objection raised by a minority worker hostile to teams. Their discussion is worth quoting at length:

> When a colleague of ours was called to facilitate intergroup problem solving between an Anglo and a Latino group, the leader of the Latino group didn't want to participate. He saw team building as one more Western invention

that he didn't want or need. Was he right? The answer to this question is both yes and no. His perception that team building is a culturally biased intervention is accurate in three significant ways:

> 1. It is a linear, American intervention, designed to "fix" dysfunctional work groups or teams. It is reflective of the mainstream American culture's need to problem-solve whatever ails it and, in so doing, make it better.
> 2. The directness of the team-building method reflects the directness of American mainstream culture. . . .
> 3. Team-building priorities differ. Mainstream America values both task accomplishment and satisfying relationships, but good relationships are not viewed as an end in themselves. . . . In most of the world, relationships have intrinsic merit. . . .

. . . In spite of these examples of team building, American-style, a part of the answer to the question . . . is still no, in the sense that every culture wants work groups to function productively, profitably, collaboratively, and harmoniously. The goal of having productive work teams is universal. It is the process of how to achieve this goal that has many cultural variations.[132]

What is most worth remarking here is the way the whole issue of multiculturalism—as opposed to universal culture—has become undecidable. All cultures are different, and yet to the extent that all cultures want to achieve productivity and profitability, they will end up being quintessentially American. Mainstream American culture, in other words, gets to be both a distinct cultural identity (Yankee practicality and problem solving) and something universal. It sets the pattern for an undecidable multi-/monoculture unlike any previously known: a simulated culture so big with ethnic, racial, and gender identity groups, yet neutral with respect to those groups, that it is not just "beyond race and gender" but *beyond culture*. In Harris and Moran's words, "team culture" testifies to "a universal microculture of work." Or, as they prophesy with chilling matter-of-factness: "a unique global culture with some common characteristics may be emerging."[133]

So, too, the case is abundantly clear in the context of class. In examining diversity management, we have already encountered notions of class complexly interwoven with ethnicity, race, gender, and language. That Thomas's and Griggs's expansive definitions of diversity can simply list "management or nonmanagement" and "socioeconomic class" among other categories of identity is one evidence. Another is the chapter "Hierarchy and Class" in

Katherine Esty, Richard Griffin, and Marcie Schorr Hirsch's *Workplace Diversity*, or the chapter on diversity and unions in Carnevale and Stone's *American Mosaic* (which reminds us that not just management but also blue-collar organizations began by excluding minorities).[134] My own earlier argument regarding the transition from line or shift work to teamwork links the notions of group and class because the line was the locale not just of ethnicity or gender but of working-class identity. The subsequent drift of the team concept to the white-collar tiers then does not so much dilute the pertinence of specific class experience as steep it in the complexity of overall U.S. class history—in particular, the history by which the executive sphere successively ejected from itself routine functions that became niches for ever more *déclassé* lower management and clerical populations (among whom, to factor in the extra complexity of generational history, were included the descendants of blue-collar diasporic workers for whom degraded clerical or management occupations were a step up).[135]

As complex as the relation between class and identity groups may be, however, we can make good sense of it here—at least for the purpose of explaining one part of the relevant class history—by resolving that relation into a fundamental similarity between class and identity groups. Both are founded on identity by *classification*. Classificatory identity, as it may be called, is the sense that one's identity, whether manifested in social group or class, ultimately rests on a difficult equilibrium between being assigned a position in a hierarchy and actively staking out such a position ("*my* place on the line," "*my* shift," "*my* job description"). The history that makes me part of "my group," in other words, is cognate with what makes me part of "my class," because both are at base histories of negotiating the fulcrum point "mine" in a world where the real leverage is corporate.

The best history to study in this regard is that of the unions. I refer in particular to the long struggle of U.S. organized labor in the twentieth century to win work rules and job classifications that—in what is only apparently a contradiction—expressed the solidarity of the working class through a perpetual contest of status differentiation within the class. The success of this struggle may be measured in the now familiar tangle of workplace regulations specifying myriad job classifications (welder, fitter, assembler, set-up man, and so on), decreeing which classifications can perform exactly which jobs, and laying down the law of seniority governing the "bumping," or hierarchical reassignment, of workers.[136] In regard to the solidarity of the class as a whole, however, this struggle also drew a firm line between working class and management in the aftermath of the infamous "straw boss" period between 1910 and 1930—when the U.S. auto industry, for example, raised the proportion of foremen to line workers by 50 percent and allowed foremen to form cadres of "working leaders" to manage insidiously from within the ranks.[137] The internal and external faces of the classification struggle—of differentiation from others in the same class and of differentiation from other classes—were homologous. Bumping and other such practices of internal status contest could be brutal, but they reveal that workers identified with each other as a class *contra* management only insofar as they could stake out their position on the line or shift according to openly understood, fair rules that applied class-wide without the toadyism, favoritism, and arbitrariness endemic to the straw boss system.

Given this history, it should be clear that the impact of the new team concept that U.S. factories cloned from the paradigm-setting Toyota/GM NUMMI plant in Fremont, California (which began operations in 1984), was devastating. To hear management tell it, the team concept was good on all counts. It empowered workers democratically in the new spirit of corporate correctness, which was then also cutting back on such perks as separate cafeterias for separate ranks.[138] And it did so in strict conformity with the total restructuring effort to "flatten" everything for reasons of the bottom line. By definition, after all, the team concept declassifies or, as it is called, "broadbands" workers. Only thus can high-performance teams be assembled from cross-disciplinary and cross-rank skills and talents.[139] At once a gesture toward egalitarianism and a legitimation of severe reductions in the number of job classifications, the team concept seemed to corporations a win-win deal.

But from the perspective of its union detractors, just as obviously, the broad band team concept could seem a losing proposition precisely because it deleted the entire apparatus of classification earned through class struggle by flattening everyone to the status of all-purpose, anonymous worker in an ant hive. A key work to read in this regard is a book co-authored by two former auto workers: Mike Parker and Jane Slaughter's *Choosing Sides: Unions and the Team Concept* (1988). Throughout this simultaneously factual/statistical and scathingly ironic study, there runs a deep current of anger against declassification.[140] First, Parker and Slaughter serve up a nice parody of the management vision of teamwork:

> *Team members are multiskilled or cross-trained so they can help each other out and do different jobs as the situation requires. The teams handle many of the foreman's previous functions. . . . Foremen become "advisers" or group leaders for several teams.*
>
> *Dignity and respect are the new bywords. Management respects workers' knowledge, while workers come to understand the real problems faced by the company.*

> *Barriers crumble: foremen take off their neckties; time clocks are eliminated; workers and management share the same cafeterias, parking facilities and bathrooms.*[141]

Then, with relentless energy, they prosecute this vision on the grounds that "drastic reduction of classifications" and "interchangeability" of workers are about nothing other than dismantling the hard-won classification, work-rule, and seniority regulations that once protected workers from arbitrary treatment and straw boss infiltration. Nor is the effect merely symbolic or trivial. According to time-motion studies of just-in-time assembly lines, Parker and Slaughter assert, the result of the team concept is a regime of "management by stress" that puts more physical and mental pressure on workers than even the original Fordism-Taylorism could manage.[142]

The ironic effect of such reverse labor history, as labor historian Nelson Lichtenstein puts it in an essay included in Parker and Slaughter's book, is that "the corporations' recent efforts to combine and eliminate job classifications represent an effort to restore to lower level management [i.e., foremen and straw bosses] much of the shop authority they were forced to relinquish when the UAW won bargaining rights in the late 1930s."[143] That is, teams of "flexible," diversely talented workers presided over by a "team leader" or "facilitator" seem *to reproduce exactly* the earlier exploitative conditions that fomented organized labor in the first place. Now that the team concept tried out on blue-collar culture has been transposed to managerial and professional levels, the same reverse history applies even to white-collar cultures. Here, the deinstallation of assembly lines pegged to classified occupations has its analog in the deinstallation of "linear process flows" pegged to professional identities. Where once work flow proceeded in distinct, ordered stages of design, engineering, manufacture, distribution, sales, and so on, such that the underlying group identities of workers came to be overlaid by discrete professional identities, now "concurrent engineering," "design for manufacture," and "point of sale" intervention have scrambled everything up.[144] Join the team, and leave behind both one's ancestral identities and the carefully nurtured professional identities that once allowed second- or third-generation descendants of diaspora to gentrify their ancestry.

The result for both the working and white-collar middle classes is that class solidarity has transmuted into the simulation of solidarity that Charles C. Heckscher terms—in a phrase that erases itself even as we speak it—the New Unionism:

> As craft gave way to industrial unionism, so industrial unionism will, I believe, be replaced by *associational* unionism—a form more appropriate to

> rapid economic change, flexible systems of management, and shifting employee loyalties. . . . Concretely, it can be glimpsed in efforts by many corporations and an increasing number of unions to establish direct worker participation on the shopfloor; in the expansion of joint committees at different levels of decision making; in the growing willingness to reduce the emphasis on fixed work rules and contractual uniformity; and in the growing assertiveness of many associations of white-collar and semiprofessional employees in response to their lack of effective voice.[145]

As Boyett and Conn summarize, Heckscher "forsees a kind of 'associational unionism' that encompasses a greater number of employees, not just traditional 'workers,' and is more decentralized and flexible. He notes that the American tradition and labor law recognizes—even insists upon—a clear distinction between workers (labor) and management. . . . Yet, [he] argues, in the new organization 'the lines between management and workers are blurred.'"[146] "Associational unionism," we recognize, means simulating away class-based unionism. Associational unionism is how blue-collars are assimilated to the New Class (accepting, for example, not just new work rules but new paradigms of collaboration with management).[147] Then because the New Class, in its internal confusion of manager and laborer identities, is itself out of joint with traditional class logic, in the end it is how blue-collars and white-collars alike are assimilated to the class of those who have *no* class. The assemblage or "weakly formed" class of the New Class, in other words, proves to have been from the first a patsy for the new corporatism because it was always really only a nascent team. We recognize the assemblage class of those with no class by their characteristic lumpen-managerial dress in the most enlightened Silicon Valley corporations: the dressed-down, tieless, casual look epitomized in the jeans and open-necked shirt that Marc Andreessen, co-founder of Netscape Communications Corp., wore on the cover of *Time* magazine in 1996.[148] Looking at this portrait of Andreessen, one sees blue-collar, managerial, and capitalist-owner classes all rolled up in one.

Whether we turn our gaze upon identity groups or class in the new corporatism, the "same" look of simulated identity thus stares out at us—the same composite, repixelated, and endlessly mutable monoculture of diversity backed up by no more history than a daily backup of the hard drive. This is a homogeneity of culture that far exceeds in its impact the consumer culture regularly targeted by both academic and mainstream critics of mass media and mass marketing. This is the homogeneity of *producer* culture.[149]

One more step in the argument, and we will be done. It is on the transnational scale that we find the clearest confirmation that suppressing history allows neo-corporatism to simulate away cultural identity in favor of a monoculture of diversity. In the theater of the multinationals, monocultural diversity is called "global competition."[150] No testimony is more compelling in this regard than Joel Kotkin's *Tribes: How Race, Religion, and Identity Determine Success in the New Global Economy* (1992), a work whose hybrid status as an academic and business book allows us to see both the massive fact of history behind the new global economy and the consequences of the subtraction of that history.

On the one hand, Kotkin's book is properly described as scholarly because it offers one of the few researched, historical studies of the rise of the new corporatism to make it onto store bookshelves alongside bestselling works of business literature. Indeed, Kotkin's entire philosophy of business is a philosophy of history. *Tribes* traces the long history of five peoples—first the Jews and then the Anglo-Americans, Japanese, Chinese, and subcontinental Indians—to show that a particular evolution of ethnicity has positioned, or is positioning, these history-laden "tribes" as the standard bearers of today's networked, global enterprises. All these peoples, according to Kotkin, share the following characteristics:

1. A strong ethnic identity and sense of mutual dependence that helps the group adjust to changes in the global economic and political order without losing its essential unity.
2. A global network based on mutual trust that allows the tribe to function collectively beyond the confines of national or regional borders.
3. A passion for technical and other knowledge from all possible sources, combined with an essential open-mindedness that fosters rapid cultural and scientific development critical for success in the late-twentieth-century world economy.[151]

We recognize here the basic components of Gonzalez and Payne's definition of team culture (a sense of "who people are" and "belonging" combined with technological rationality). And yet, although Gonzalez and Payne's idea of culture lines up virtually point by point with Kotkin's, there is no real comparison, for Kotkin's basic position throughout is that all of the above—a strong sense of ethnic identity, of belonging even amid global dispersion, and of the value of smart work and lifelong learning—inhere in a deep sense of cultural history founded on the experience of diaspora (a perhaps overly broad concept in Kotkin that includes the emigration not just of the dispossessed but of colonizers and even Japanese managers, who rotate out to foreign countries in a "diaspora by design"). Fleshing out his account with details gathered from the centuries, Kotkin allows us to understand in a profound way why a certain kind of ethnicity is still central to the new corporatism. Clearly, for example, global competition in the 1980s and early 1990s was from the U.S. viewpoint really competition against the "Japanese," who were less a nation than the ethnic and racial way of life that reasserted itself precisely in the extinction of sovereign nation immediately after World War II.

Yet Kotkin's book balances its orientation toward the past with a progressivist, future-directed argument, which aligns it with the millennial business bestsellers. Kotkin dissolves the notion of ethnicity even as he gives historical evidence for it. Thus, while the Jews receive the place of ancestral honor in his book (the first case study after the introduction), the logical center of his argument lies in the succeeding chapter on the British, whom he tethers to the Americans to create a general category of the Anglo-Americans.[152] Here he draws on the ideas of R. H. Tawney, Max Weber, and others to argue that the business acumen accompanying the British on their great adventure of imperialism was originally a particular kind of ethnicity consisting in equal parts of religious and commercial fervor. Imperialism and global business expansion were the "Calvinist diaspora."[153] But the unique quality of British and Yankee ethnicity was that it deployed itself in ways that ultimately transcended the history of blood ties and religion (though not language) until it simply *became* world business culture:

> Unlike other empires, whose legacies follow racial bloodlines, bureaucratic systems or great religious concepts, the practical and commercial character of the British diaspora brought in its wake something more subtle, yet arguably more essential to the operations of modern society. In a host of critical fields—from accounting and advertising to culture, science, and, finally, the operations of government—the Anglo-Saxons created standards not just for their own race, or for their colonies, but also for the entire modern world. Even the very word *international* sprang from the writings of British eighteenth-century philosopher Jeremy Bentham.[154]

In a particularly interesting discussion, for example, Kotkin notes that British accounting techniques reflected an essentially ethnic mindset (accounting was "quasi-religious" for the English and Scots), which from the nineteenth century on became the global mindset. As it were, Anglo-Saxon-Celtic Calvinism (supplemented by the commercialism of other dissenting sects) powered a diaspora of accounting standards and British bookkeepers that gave modern technological rationality its very "account" of rational value, monetary or otherwise. The magnification of ethnicity this implies is particularly clear in the following passage, from Kotkin's interview with an Indian partner in the historically crucial Price Waterhouse firm. In this passage, British accounting (and the entire British way of doing business) appears at once as culturally neutral and as an exact replacement for subcontinental ethnicity: "I am more committed to Price Waterhouse than to my wife and family. It's in my blood. I think it's a great organization. Here if I am good, I rise. If I'm poor, I fail. I don't need a godfather. And if my son joins, he gets to be partner only if he performs to standards. That's not the way things usually work in India." In the entirety of its institutions, instruments, methods, ethics, and language, in other words, corporate globalism as we now know it *is* ethnic. "Even today," Kotkin says, "when mastery in many of the industrial arts has passed to Asia, British- and American-descended firms continue to establish the basic standards in *how* business is conducted and managed worldwide."[155]

From this logical center of his book, everything else follows, for Kotkin's basic understanding of the evolution of world business toward the new millennium is that such recent or emergent global business tribes as the Japanese, Chinese (and Taiwanese), and Indians are "new Calvinists." While driven by unique histories and ethnicities of their own, they either chose or were forced (e.g., by the outcome of World War II) to accept as their channel of expression the global ethnicity of the Anglo-Americans—even to the point of adopting English, now the idiom of technological rationality, as the world business language.[156] Each saga that Kotkin tells, therefore, is a variation on the story of how, for their own purposes, with their own accents, "new Calvinists" and "new Americans" learned to inhabit the business techniques and technologies of the diasporic WASPs. As Kotkin summarizes, marking an exact pivot point between past-oriented historicity and "twenty-first-century" prophecy:

> Just as Roman systems of law and government became the foundation of the major European nation-states and a broad range of Chinese traditions shaped the development of the Asian tributaries of the *t'ien hsia*, or "All-under heaven" empire, so the Anglo-American systems have been incorporated by the new players in the evolving global economic system.
>
> In the twenty-first century, we are likely to see the further development of this multiracial world order running along British-American tracks of market capitalism, political pluralism and cultural diversity.[157]

And so the juggernaut of history keeps rolling on its tracks inexorably toward the new millennium. There, we are to enter a postindustrial promised land or golden mountain where (and this is the *ne plus ultra* of the simulation I indicate) ethnicity and class will simply *be* global business culture and vice versa, where ethnicity and the new corporatism are indistinguishable. Thus in Kotkin's pages (and even more so in other business writings that unreflectively refer to *keiretsu, chaebol,* and other Japanese or Korean terms for business networks and conglomerates) it is at last radically ambiguous whether the Japanese are organized *ethnically* into extended fealty/family groups governed by the ethos of *ie* or *corporately* into the tightly knit, mega-industry groups called *keiretsu*. So, too, it is radically ambiguous whether clannish ethnicity or corporate identity is the agenda of such organizations as "rotating credit associations among Japanese and Chinese, *kye* among Koreans and the *susus* of West Indians."[158] (Attesting further to this ambiguity is the undecidable Western discourse of "corruption"/"inefficiency," at once Calvinist moral and technological rational, often used to analyze such formations.) Bloodless and featureless the great international businesses may seem, but the more complex reality is that they are compounded from both ethnic diaspora and enterprise networking in a manner now so globally standard that the line between "tribe" and "corporation" is inscrutable.

Harris and Moran write in their book on managing multinational cultural differences:

> The transnational corporation that moves beyond the culture of a single country and operates comfortably in the multicultures of many nations obviously will develop a unique microculture of its own. Its organization model and environment will reflect the synergy of the diverse macrocultures in which it functions, as well as the varying managerial approaches to business, government, and people. Thus, far-flung business activities require a new organizational culture that is able to accommodate itself to cross-cultural realities.[159]

The monoculture of diversity that is team culture at the local level converges with the monoculture of diversity that is corporate culture at the global level. Thus does the new corporation take over all identity positions, saturating the field of identity from local to universal so as to leave customary identity groups and classes no room to stand except in the corporation. Such is the sameness of difference in a world without history. Such is what Hammer and Champy, in the subtitle to *Reengineering the Corporation,* call "A Manifesto for Business Revolution." A specter is haunting Europe, Marx and Engels said. But their vision of a specter was based on a historical vision of class experience. The historyless specter now haunting the United States and much of the world besides is one that may *seem* to have the lumpen-ethnic traits of Japanese, Chinese, Koreans, Indians, and so on, but in reality washes out both ethnicity and class to focus on technological/technical structures and processes. After all, the neo-corporatist perception of global competition characteristically makes the history of foreign peoples invisible (subordinating it to the concept of "customs" or "traits") so as to obscure that those peoples may have their own equivalents of internal identity groups and class formations. Global competition thus always resolves into the challenge to workers to be as minimalist in identity—that is, to be as self-sacrificing, to work as hard, to save as much—as zero-people, or so they seem, in the Far East or elsewhere.[160]

It is as if one were to take a magnet and drag it across the face of a diskette: knowledge as we now understand it is the function of a civilization that believes it is headed toward what cyberpunk writers like William Gibson *(Neuromancer)* and Neal Stephenson *(The Diamond Age)* vividly imagine to be corporate "arcologies," "claves," and "phyles" appropriating all that remains of historical groups, classes, and nations. Erase the history file; make room for the people of the phyles.

"X"

The merger of the new corporatism with academic subject critique and cultural class critique, I believe, puts in place the last piece we need to characterize our contemporary understanding of knowledge work. In a purely academic idiom, we might equate the kind of merger I indicate with *sous rature* in literary theory. As in some Shelleyan "Triumph of Life" (as read by Paul de Man), our contemporary understanding of understanding arises when the subject is at once erased and simulated (Shelley: "masked") by the New Class, and then both subject and class are in turn masked by the new corporation. Such is the real-life deconstruction of knowledge that is the Triumph of the Corporation. Indeed, perhaps that is how Shelley's late Enlightenment poem, with its unfinished vision of multitudes driven before the chariot of Life, would have had to be completed: upon the clarification that the ultimate multitude, and life, is corporate. Capping the sequence of Enlightenment and post-Enlightenment theories of civilization I earlier cited,* the new corporatism is an end-of-history vision of civilization that throws into the dustbin of history not only all its predecessors but also the very notion of relevant

* See page 34 above.

history. History conceived as erasure—the disfiguration of one age by another—itself operates under the sign of erasure.

Now, perhaps, we can understand the true meaning of the emancipation proclamation of the information age first uttered in 1984 by Stewart Brand, publisher of *The Whole Earth Catalog*: "information wants to be free."[161] In the era of the Whole Earth Corporation, "information wants to be free" is ultimately how we are no longer allowed to say "we" want to be free. "We," the subject and class of information culture, come fully to know our world only in the blinding moment of illumination when the world network routes around our knowledge—that is, the *us* in our knowledge that Fukuyama (in the other half of his thesis) terms "the struggle for recognition" and Castells (in the second volume of his *Information Age* trilogy) calls "the power of identity."[162] We do not even need the hyperbole of cyberpunk science fiction, with its unerring instinct for the mutilation of subjects (e.g., the silicon-punctured bodies and flat-lined subjectivities of Gibson's *Neuromancer*), to grasp the intensity of our loss—nor the uncanny double of that intensity, the blurred anomie of it all. "X" marks the spot where the whole generation of incipient knowledge workers in the United States succeeding the baby boomers—the generation caught in the "pipeline" from education to the corporation—has been deleted from the network.[163] Indeed, we may speculate that the purely generational identity of "Gen X" (and now "Gen Y" after them) looms large at this moment precisely because it is an empty solidarity reflecting—as if in cyberpunk "mirrorshades"—the hollow form of the corporate world's own generational identity as "workforce 2000." "We" are no more than this transient moment when we have nothing more in common—as Jean-Luc Nancy might say in his *Inoperative Community*—than our finitude, our extinction, our "death."[164]

My tone is once more elegiac. But that is because the fatal problem with which I began has been resumed at another level. The "death of literature," we now know, cannot be understood except against the background of the death of knowledge in the information age—that is, the dying of "our" knowledge into a paradigm of knowledge work that grants us virtual freedom only by freeing us from all those things once thought to give freedom its point. Like refugees of consciousness embarked on the diaspora of the new millennium, we are given the opportunity to be free of our identity, settled home, peoples, security, everything. Or at least such would be the most heterodox critique of the doxa—the total identification of culture with technology and technique—posited by the new corporatism. Representative of such heterodoxy is Jeremy Rifkin's *End of Work*, which argues elegiacally that perhaps the only thing "we" have left is a sense of worthlessness internalizing the end of history that he calls the "end of work":

> The death of the global labor force is being internalized by millions of workers who experience their own individual deaths, daily, at the hands of profit-driven employers and a disinterested government. They are the ones who are waiting for pink slips, being forced to work part-time at reduced pay, or being pushed onto the welfare rolls. With each new indignity, their confidence and self-esteem suffer another blow. They become expendable, then irrelevant, and finally invisible in the new high-tech world of global commerce and trade.[165]

What is bleakest about this vision is the utter poverty of collective identity in Rifkin's analysis—the way in which he is interested only in "individual deaths" of "confidence and self-esteem" (merely the inverse of economic individualism) because even the possibility of a group or class experience of loss is unthinkable.

So what is a knowledge worker—in school, looking for work, at work, or laid off from work—to do? How to think and feel and live amid such loss of identity and history so as to make reparation, to seek what one of our earliest poets of knowledge work, William Wordsworth, called "abundant recompence"?[166] And how to do so, moreover, without being nostalgic for foreclosed group and class identities in a manner that would inauthentically mime the great fundamentalist, nationalist, and ethnic reactionisms of peoples of the world *excluded* from "knowledge" (as Castells so eloquently studies)? After all, the situation of someone whose identity is excluded precisely through being included in the pipeline of knowledge work is different. Even if someone in this position is heterodox enough in relation to "team culture" to mourn the culture of identity groups and classes, he or she is increasingly powerless to believe that such identity formations can any longer meaningfully organize the experience of work except insofar as they are accommodated within the new corporation as simulacra of themselves (i.e., as a "diversity" of "skills"). Just as literature is dead while the value of the literary may yet survive to seek new forms, as I have suggested, so groups and classes among knowledge workers have now been neutralized where they most count—in the culture of production—even as the need for the identity function they once supplied continues unabated.

Or rather, since the technological rationality of a phrase like *identity function* is symptomatic of the evacuation of knowledge worker identity, perhaps we would do better to think of the hunger for identity that survives within the pipeline as a craving for a restorative ethos preliminary to any particular identity or function—an ethos, in particular, able to withstand the otherwise relentless ethos of postindustrialism. For make no doubt: the right life that can now make the knowledge worker not just informed but well informed must be a whole *ethos* (Greek, "custom, character") able not just to live with, but to make a life of the great succession of *ethoi* that have been cumulatively recruited to industrialism and postindustrialism to achieve the present spirit of technological rationality: Weberian "Protestant ethic," Schumpeterian "creative destruction," Japanese *Wa* or Chinese neo-Confucianism (as pressed into service to explain the then blossoming "Asian economic miracle"), and now Castells's "spirit of informationalism." "What is, then, this *'ethical foundation of the network enterprise'* this *'spirit of informationalism'*?" Castells asks, and answers (I quote more fully a passage touched upon earlier):

> It is a culture, indeed, but a culture of the ephemeral, a culture of each strategic decision, a patchwork of experiences and interests, rather than a charter of rights and obligations. It is a *multifaceted, virtual culture,* as in the visual experiences created by computers in cyberspace by rearranging reality. It is not a fantasy, it is a material force because it informs, and enforces, powerful economic decisions at every moment in the life of the network. . . . The "spirit of informationalism" is the culture of "creative destruction" accelerated to the speed of the optoelectronic circuits that process its signals. Schumpeter meets Weber in the cyberspace of the network enterprise.[167]

We can go right to the heart of the paradox of contemporary identity by inverting Castells's question in this way: not "what is [the] ethical foundation of the network enterprise" or "spirit of informationalism," but what ethical foundation enables identities to live an *un*-networked and *counter*-informational fantasy *within* the spirit of informationalism? What room might there be for a counter-ethos within the dominant ethos of informationalism that spends its days and nights locked within the cubicles of postindustrialism but also "gestures" all the while that it is something else—something like what Raymond Williams called a "structure of feeling" that, if it can no longer be identity or class in customary ways, can nevertheless reconstitute the *basis* for a renewed folk identity? Ethos, after all, is not in itself identity but the inchoate coming-to-be or basis of identity; it is identity at the point of emergence from collective, undifferentiated doxa.* As such, it necessarily participates in the paradoxical logic of doxa outlined earlier. Even the dominant postindustrial ethos is constantly shadowed by internally displaced ethoi that, if not revolutionary or even aggressively subversive, nevertheless have an unsettling neither/nor status. If there is to be what I called a right life of information that can make the knowledge worker not just informed but, in a fundamental sense, *well* informed, such rightness must lie precisely in a paradoxical ethos or shadow doxa flourishing within the Valley of Death of contemporary knowledge—one that from the viewpoint of the new corporation both does its job and seems to gesture toward some contrarian reserve of knowledge of the sort that William R. Paulson and J. Hillis Miller, in their works on the role of literature in the information age, call respectively a "noise of culture" and a "black hole."[168]

It is upon such a contrarian "unknown ethos" or "ethos of the unknown" secreted within knowledge work, I believe, that the humanities and arts (and my special concern, literary study) must now come to bear or not at all. For the humanities and arts, this is where the contest for "humanity" now lies: to educate the ethos of the unknown that broods within knowledge work so that it is not also the same as an ethos of unknowing, of resenting the fated life of knowledge work so much that one could "care less" for knowledge. After all, even the mass ethics visible in such TV sitcoms originating in the 1990s as *Friends* (where an ensemble cast playing the part of Gen Xers at once enacts and repairs the concept of "team") entertains an abiding inquiry into what being well in the age of the well informed really means. Why not the humanities and arts, too, informed as they are by a history (e.g., literary history) able to testify to so many alternative experiences of identity, community, knowledge, action, work, feeling, and reparation? Why not the humanities and arts, too, especially since in the academy they are already in the pipeline alongside—but profoundly separated from—the paradoxical contemporary ethos of disenchanted knowledge whose name we have only to utter to realize how difficult is the task of reparation: *cool?*

* *Ethos* in this book is a flexible term that, depending on context, expands to be as capacious as "Zeitgeist" and contracts to be as specific as "ethic" or "aesthetic." A fuller gloss would triangulate the term somewhere in the midst of Williams's "structure of feeling," Bourdieu's "habitus," Habermas's "lifeworld," and Gouldner's "culture of critical discourse." These concepts vary in their emphasis on the visceral versus rationalist, collective versus individual, and preverbal versus articulate poles of social experience (and are incommensurable in other ways as well). But it is precisely such variance that I mean *ethos* to designate because the concept is best understood dynamically. *Ethos* locates the *process* of emergence (and submergence) where identity at once cleaves from and to inchoate social experience. Cool is thus an ethos, not an identity.

Part II

Ice Ages

First, then, the men must be brought to see that the new system changes their employers from antagonists to friends who are working as hard as possible side by side with them, all pushing in the same direction.

Frederick Winslow Taylor, *Shop Management* (1903)

Windows never would have gained popularity and reached critical mass without the benefits of innovative, user-friendly technologies developed by our Office team.

Bill Gates, “The Case for Microsoft: Why Windows and Microsoft Office Should Stay under One Roof” (15 May 2000)

Preface

"We Work Here, but We're Cool"

Amid the coldest and richest of our contemporary seas—awash in the bright, quick data streams and great knowledge surges—lies the continent of cool.

The sea is the sea of information in which we variously surf, navigate, explore, and drown (the standard metaphors).[1] We do so in so many ways and with such enthusiasm that it is perhaps no longer possible even to attempt the kind of survey of the "information sector" or "knowledge industry" that economic statisticians such as Fritz Machlup and Marc Uri Porat performed so sweepingly some decades ago.[2] Information today would require a Borges—a Library of Babel or Funes the Memorious—to keep an account. In any event, it will require in this book that we gradually scale back the topic from information at large to the epitome of contemporary information—the Internet and (especially) the World Wide Web.[3]

We know that the Web—and the combination of networking and GUI (graphical user interface) technologies it encompasses—is the "user friendly" face of information. But the friendship of the Web, and everything it represents in the long history of work leading up to current knowledge work, is also strangely cold. It is from this coldness—remoteness, distantiation, impersonality—that *cool* emerges as the cultural dominant of our time. Strip away all the colorful metaphors of information seas, webs, highways, portals, windows, and the rest (like picture calendars tacked to the wall), and what comes to view is only the stark cubicle of the knowledge worker. Yet precisely in this cold space of nonidentity, cool appears as the *cultural* face—perhaps not the best or truest face, but the interface by which it knows itself—of knowledge work.

What makes knowledge workers clutching a console in a cubicle think they are as cool as the jazz musicians, black British or African American youths, white greasers, and other such subcultures exiled from knowledge work during the early to mid-twentieth-century "birth of the cool"? Or, to refer to a different phylum in the genealogy of cool, what makes such workers feel as secretly "beat" or "hip" as the countercultures of the 1950s and 1960s that borrowed subcultural cool precisely to drop out of the knowledge work for which they were destined (school, business, the "military-industrial complex")?[4] Finally, how do all the phyles of cool mentioned so far—subcultural, countercultural, informational—relate to contemporary, mainstream consumer cool?[5] Certainly, the technology that has been the necessary buzz of everything really cool (e.g., hot rods, reggae "sound systems," electric guitars, designer drugs, high-speed processors) plugs into the whole cargo cult of industrial age consumer leisure (stereos, DVD players, special-effects movies). How do cool graphics on the Web, then, differ from cool computer-animated dinosaurs in a mass-market movie like *Jurassic Park*?

Or, to ask my questions in a specifically postindustrial rather than industrial context, what is cool in an age when *producer culture* dominates everyday life so thoroughly that it renegotiates the whole relationship of production to consumption, work to play? After the "big split" of work from leisure in the first half of the twentieth century (as C. Wright Mills termed it), consumer culture became the darling of both academic and journalistic critique.[6] It stood for the numbing totality of mainstream culture, and also for the limiting horizon of possibility—at once the originary stylistic resource and final *embourgeoisement*—of the subcultures and countercultures that protested mainstream culture.[7] But today it is producer culture that governs work life and home life alike in the name of a ubiquitous new regime of knowledge: not just company-mandated "lifelong learning," but also the "home office," "telecommuting," "edutainment," "investment clubs," and, as Paul A. Strassmann has observed, the thousand and one other routine jobs that have been shifted to, or created ex nihilo within, leisure time.[8] Even athletic training, tennis, sailing, and other recreations, Witold Rybczynski notes, have become occupations of skill and technique offering the chance "to work at recreation."[9] As documented in Juliet B. Schor's *Overworked American: The Unexpected Decline of Leisure* and Arlie Russell Hochschild's *Time Bind: When Work Becomes Home and Home Becomes Work,* therefore, the rise of "smart" work coincided with trends that not only added the equivalent of an extra month of work per year from the late 1960s to the late 1980s (and subtracted almost 40 percent of leisure time), but ultimately blurred the sense of a boundary between job and home.[10]

Increasingly, knowledge work has no true recreational outside. Cool therefore arises *inside* the regime of knowledge work as what might be called an *intra*culture rather than a subculture or counterculture. Cool is an attitude or pose from within the belly of the beast, an effort to make one's very mode of inhabiting a cubicle express what in the 1960s would have been an "alternative lifestyle" but now in the postindustrial 2000s is an alternative *workstyle. We work here, but we're cool. Out of all the technologies and techniques that rule our days and nights, we create a style of work that is "us" in this place that knows us only as part of our team. Forget this cubicle; just look at this cool Web page.*

At once wise to it all and profoundly ignorant, simultaneously arrogant and vulnerable, and—the specific paradox that is the key to our puzzle here—both strangely resistant to and enthralled by the dominating information of postindustrial life, cool is the shadow ethos of knowledge work. It is the "unknowing," or unproductive knowledge, within knowledge work by which those in the pipeline from the academy to the corporation "gesture" toward an identity recompensing them for work in the age of identity management. Whether watching cool graphics on the Web or cool dinosaurs in that Spielberg film allegorizing the fate of knowledge workers in the age of global competition (where the real drama occurs in the out-of-control computer control room behind the leisure theme park), knowledge workers are never far from the cubicle, where only the style of their work lets them dream they are more than they "know."[11]

Can we dream of anything more humane or less existentially lonely than driving information highways or surfing data seas? Can we do so without immediately fantasizing Jurassic, cold-blooded monsters (not Tyrannosaurus Rex and raptors, but CEOs and shareholders) that embody the worst nightmare of any age of global competition living in denial of history—*pre*history?

To evolve an answer, it will be useful to review the history of contemporary knowledge work—specifically, the history of how knowledge workers grew so cold they had to be cool. Cold lies at the heart of the paradoxically bright yet noir alienation of cool ("hiding in the light," Dick Hebdige calls it)—in the way, for example, that a crisp pair of sunglasses worn indoors is very cool.[12]

Cold work originated in alienation as Marx understood it. But what is most relevant in our context is the specifically twentieth-century history of alienation, which I unfold in three ages of ice named after the leading work paradigms of their times: *automating, informating,* and *networking.* Automating was dominant from the late nineteenth century through the Cold War (I concentrate on the 1920s through 1950s). Its great sociological achievement was the regimentation not just of blue-collar workers, but also of the salaried "new middle class" ancestral to knowledge workers. What Shoshana Zuboff calls "informating" then took the lead from the late 1950s through the 1970s, when mainframe computing altered the character of automation to redefine white-collar workers (and to some extent blue-collars) as information workers. Networking, together with the personal computer, emerged after about 1982 to escort white-collars into the new millennium of knowledge work.

Chapter 2

Automating

To understand the fate of alienation in the first half of the twentieth century, it will be useful to review the "Estranged Labour" fragment of Karl Marx's *Economic and Philosophic Manuscripts of 1844,* where he defined alienation. I use the term *review* in something like its dramatic sense, for Marx's definition was itself dramatic in its insistence upon staging alienation as a struggle between, or within, identities. Drawing on a Romantic metaphysics of identity, Marx first characterized alienation as the estrangement of Object from Subject: "The object which labour produces—labour's product —confronts it as *something alien,* as a *power independent* of the producer. The product of labour is labour which has been embodied in an object."[1] From this foundational disorder in identity, Marx derived the following sequence of further intra- and intersubjective alienations: estrangement from "the labour process" or "*producing activity* itself" (accompanied by the "loss of [the worker's] self"), estrangement from the "species-being," and estrangement of "man from man" (3:274–77).[2] Thus Marx's initial metaphysics of alienation at last implied a politics of alienation, of incipient struggle between classes of "man."

Alternatively, since the "Estranged Labour" piece was not yet a manifesto, we might better call it Marx's *tragedy* rather than politics of alienation: not struggle but agon. Waxing openly dramatic after theorizing the alienation of man from man, Marx stages the following soliloquy:

> Let us now see, further, how the concept of estranged, alienated labour must express and present itself in real life.
>
> If the product of labour is alien to me, if it confronts me as an alien power, to whom, then, does it belong?
>
> If my own activity does not belong to me, if it is an alien, a coerced activity, to whom, then, does it belong?
>
> To a being *other* than myself.
>
> Who is this being?
>
> The *gods*? (3:278)

Alienation, we recognize, is Marx's version of the classical gods, or fate, and "Estranged Labour" is his dramatization of fate as agon (man versus self) as well as antagonism (man versus man). For who are the gods in Marx's firmly nontranscendental universe? The gods possessing the worker "can only be *man* himself": "the capitalist (or whatever one chooses to call the master of labour)" (3:278, 279).

To change the scene to twentieth-century automation, we need then only take all the drama out of alienation. Of course, antagonism between labor and capitalists continued. We have merely to read Frederick Winslow Taylor's account of the pivotal scene in his early professional life—his "bitter," "mean," "continuous struggle" with machinists at the Midvale Steel Company where he was promoted to foreman in the 1880s—to witness the kind of chronic antagonism that erupted in organized form in the pitched battles of the Wobblies era or of the Ford Hunger Marchers and the UAW in the 1930s.[3] Alienation of "man from man," after all, was heard in slang from 1918 on as the mixed deference/resentment of workers for "the Man" *(OED)*. But to cite the opening words of Mills's *White Collar*, the true tragedy of labor in the twentieth century was that another hero—or, rather, unhero—"slipped quietly into modern society" to upstage the loud fight of laborers versus capitalists. I refer to the character that Mills, ironically conflating laboring "man" and "the Man," dubbed "Little Man": the white-collar middle class.[4] Growing at a rate over six times that of wage-workers, to reach approximately 25 percent of the U.S. workforce by 1940, this new middle class of salaried workers outstripped the old middle class of self-employed farmers and small entrepreneurs to exert an ever greater dominance over national lifestyle.[5] White-collars were the "new cast of actors" that Mills (like such other chroniclers of the 1950s as William H. Whyte, Jr., and David Riesman) heralded in ironically antidramatic terms as "the hero as victim, the small creature who is acted upon but who does not act, who works along unnoticed in somebody's office or store, never talking loud, never talking back, never taking a stand. . . . [he is] the generalized Little Man."[6]

Little Man (and even more diminutive Clerical Woman) were by midcentury the living advertisement for the commodity fetish and anomie effects that twentieth-century Marxist alienation theory seized upon. As Mills put it with clear reference to such theory, these little heroes inhabiting "the enormous file" and "the great salesroom" were "estranged from community and society in a context of distrust and manipulation; alienated from work . . . expropriated of individual rationality, and politically apathetic."[7] Yet as also attested by Mills's ongoing struggle to adapt alienation theory to such other intellectual currents as Weberian sociology and American pragmatism (paralleling the effort to rethink alienation in the work of mid-century authors more closely identified with Western Marxism), the very terms of his mock-dramatic analysis conflating the heroic with the pedestrian seemed forced.[8] This was because by Marx's standards nothing about white-collars felt particularly tragic. White-collars were an incompletely tragic class. Like some Oedipus of the suburb who never actually gets around to killing his father or marrying his mother because he has to work late at the office, white-collars posed a Sphinx's riddle for alienation theory: what is estranged from its "labour process" and "species-being" (the middle stages in Marx's fourfold derivation) but *not* from its "product" and "other men" (the initial and terminal stages)?

Alienation from labor process (Marx: "the alienation of activity, the activity of alienation . . . alienation, in the activity of labour itself" [3:274]) still fit because the signature of modern automation was not just that it forced work technique to adapt to technological process, but that it did so in ways that rapidly spilled across class lines from the factory to the office. Taylor's "scientific management" (the rationalization of procedures, routines, protocols, standards, and codes, inaugural of modern work technique) almost immediately spread from the shop floor to the office through the influence of William Henry Leffingwell's *Scientific Office Management* (1917) and similar initiatives.[9] Both the factory and the office, therefore, witnessed the modern changeover to assembly-line technologies and routinized labor techniques.[10] According to the now familiar paradigm of the assembly-line or office slave (in the latter case especially among the increasingly female clerical corps), workers gave up the rhythms of agrarian or craft labor for the routine of deskilled tasks pegged to the movement of the conveyor or message belt.[11] The tempo of work changed (now being measured in seconds rather than seasons or harvests). Every procedure became accountable at an unprecedented level of detail. And the autonomy, variation, or, in a word, *other* values vested in traditional labor processes were proscribed as officially other, alienated. In the terms of Marx's section on labor process, work became "external" to identity, and the worker "denied" rather than "affirmed" himself, did "not feel content but unhappy," and did "not develop freely his physical and mental energy but mortifie[d] his body and ruin[ed] his mind" (3:274).[12]

Similarly, alienation from the human "species-being" could also be applied to the twentieth-century milieu, though only if considerable effort is made to update Marx's Romantic (and in modern Marxist theory, largely disregarded) terms of organic naturalism. Borrowing from Ferdinand Tönnies, Max Weber, or other turn-of-the-century sociologists who linked alienation theory to modern "rationalization" theory, we could translate Marx roughly as follows.[13] Rather than remaining organically, holistically "natural," white-collar labor in the twentieth century was rationalized by the unnatural *Gesellschaft* and bureaucracy that arose to manage the new alignment of technology and technique. White-collar labor thus spawned a postorganic species-being of work: the "bureaucrat" and "organization man" whose status no longer rested on nature or even on what might be called the primary displacement of nature into property, but instead on the secondary displacement of property into "profession." A profession is an occupation that is unnatural because it is wholly technical, and unpropertied because its technique serves a production technology owned by someone else.

But rather than continue translating Marx's theory of alienation into twentieth-century terms, let us break off with the realization that in resorting to rationalization sociology, we are already anticipating the wrenching modernization that later twentieth-century alienation theory had to undergo, because the all-important initial and concluding steps in Marx's theory—alienation from the "product" and "man from man"—were sufficiently askew from twentieth-century experience that any attempt at direct translation risks missing the point.

First, white-collar labor could not easily be said to be alienated from its product because its defining feature—as indicated by the defensiveness of Leffingwell's question "Is Office Work Nonproductive?" in his *Textbook of Office Management*—was that it had no clear object-product at all.[14] Instead, it produced "service": the legion of organizing, communicating, publicizing, accounting, and other facilitating or mediating tasks whose combined demographic and economic significance gradually eclipsed "product" in the twentieth century. "As a proportion of the labor force," Mills observes, "fewer individuals manipulate *things*, more handle *people* and *symbols*. . . . The one thing [white-collars] do not do is live by making things; rather, they live off the social machineries that organize and coordinate the people who do make things."[15]

Second, white-collar labor did not culminate in alienation of "man from man" in a manner akin to the face-off of laborer and capitalist. Anticipating the New Class, white-collars both supervised or coordinated labor (thus representing ownership) and were crucially debarred from actual ownership (thus resembling wage-labor). This is why, in attempting to translate Marx's species-being argument, it becomes necessary to speak of the professionalism of white-collars in a way that bypasses the notion of property. Chimerical creatures who owned no productive property yet represented proprietors, the class of managers, professionals, supervisors and (at the trailing edge of the class) bookkeepers and clerks interposed what Mills calls a semi-autonomous "new pyramid" within "the old pyramid" of proletarians and capitalists.[16] White-collars were incommensurable with the binaristic agon scripted by Marx. Their identity was neither that of alienated labor nor that of capitalist gods. Instead, siring the pedigree of the "contradictory" New Class, they were mongrels of identity. They were the class of *un*tragic labor and *un*demonic godhood, of merely routine submission and petty bossiness.

Thus arose the necessity for all the twentieth-century revisionist theories of alienation that transformed Marxist theory to account for the new middle class, or, more broadly, to account for social history in its entirety as seen from the perspective of that class. There is no secret to the nature of these theories. They consist of the genealogy of Weberian sociologies, humanist and structural Marxisms, structural anthropologies, Foucaultian historicisms, media studies, and so on that arose to describe—in lieu of relations between persons—the nonrelation of persons to institutional structures, machinelike organizations, and rationalized techniques. Such theories now make it difficult to identify a specifically Marxist alienation theory at all (or at least a theory reminiscent of the early Marx manuscripts from which alienation theory took its inspiration).[17] It is all "cultural criticism" and "subject critique"—a mode of grasping alienation that refuses to dramatize it in terms of essential identity, and instead views it only in terms of constructed identity. Alienation becomes a structural function of "modes of production," "bureaucracies," "institutions," "systems," "apparatuses" (as in Althusser's "ideological apparatuses"), "technologies," "media," "discourses," and other such macro-agencies that construct identity but are fundamentally incommensurable with the identity concept. Mills put it this way: "The world market, of which Marx spoke as the alien power over men, has in many areas been replaced by the bureaucratized enterprise. . . . The more and the harder the white-collar man works, the more he builds up the enterprise outside himself, which is, as we have seen, duly made a fetish and thus indirectly justified." Marx's commodity fetish became an afterthought of modern bureaucracy fetish.[18]

An exemplary case of such alienation theory in our context is Robert Blauner's *Alienation and Freedom*, whose appearance in 1964 sparked an intense debate in Marxist circles. Was industrial machinery, as Harry Braverman's subsequent *Labor and Monopoly Capital* argued, inherently alienating? Or, as Blauner suggests, did industrialization evolve through different kinds of technology systems (with their accompanying techniques or "skills") until it actually makes workers feel "free"?[19] The disputed answer to this puzzle is less important for us than the enormity of its undergirding assumptions: capitalism is a drama not of workers versus capitalists but of bureaucratic systems of technology/technique. "It was Max Weber in his classic analysis of bureaucracy," Blauner notes in discussing his assumptions, "who expanded Marx's concept of the industrial worker's separation from the means of production to all modern large-scale organizations."[20] Such was alienation in a century dominated not by the blue-collar printers, textile workers, chemical workers, and others who are the ostensible subjects of Blauner's book, but ultimately by the white-collar organization man. In a telling confession in his preface, Blauner writes: "Possibly it would be more appropriate to study the problem of freedom and self-expression in work among white-collar and professional people, whose higher education has awakened the aspirations for fulfillment and creativity that often lie dormant among the mass of less-educated manual workers. Here I plead the continuity of intellectual tradition and my own personal history to defend my choice. The concept of alienation, in its classical form, was an attempt to explain the changes in the nature of manual work brought about by the industrial revolution."[21] For Blauner, the white-collar middle class was the absent presence that "classical" alienation theory could not fully see.

We now come to the crux of this review of alienation theory. As indicated by my complementary review of subject critique in chapter 1, the point is not that the new, massively authoritative explanations of modern technological and technical structures could simply expunge the problem of subjective identity. Rather, modern theory opened a rupture between the structural and subjective dimensions of alienation that Marx had fused.[22] The problem, therefore, became how to think the "relation of nonrelation" between modern social structures, on the one hand, and what Mills called the total "psychology," "experience," and "tone" of working life, on the other.[23] Or, put in later cultural critical terms: how are we to think the relation between the condition of modernity and cultural sensibility, mentality, episteme, *habitus*, lifeworld, structure of feeling, or, in my summary concept, ethos?[24] Of these overlapping terms, Raymond Williams's "structure of feeling" is perhaps most specifically suggestive in our context. His definition is worth quoting at length:

> The term is difficult, but "feeling" is chosen to emphasize a distinction from more formal concepts of "world-view" or "ideology." It is not only that we must go beyond formally held and systematic beliefs, though of course we have always to include them. It is that we are concerned with meanings and values as they are actively lived and felt, and the relations between these and formal or systematic beliefs are in practice variable (including historically variable), over a range from formal assent with private dissent to the more nuanced interaction between selected and interpreted beliefs and acted and justified experiences. . . . We are talking about characteristic elements of impulse, restraint, and tone; specifically affective elements of consciousness and relationships: not feeling against thought, but thought as felt and feeling as thought. . . . We are then defining these elements as a "structure": as a set, with specific internal relations, at once interlocking and in tension.[25]

What was the structure of the feeling of alienation, we may ask, in an age whose dominant white-collar identity seemed to have no feeling for alienation comparable to agon? whose most prevalent feeling, indeed, was a detached cynicism for the technological/technical structures ("the System," "the Establishment") that neutralized feeling by furnishing its very pattern as pure structure ("a set, with specific internal relations, at once interlocking and in tension")? By comparison with Marx's agon "man from man," after all, modern hatred for the System seems a thin, bloodless hatred—a mere technical diagram of hatred.

Here we arrive upon "cold," which names the structure of nonfeeling that converts this enigma into its own answer. If the twentieth-century ethos of alienation did not feel agon, that must be because such ethos consisted in the inability to feel agon. If nothing about the white-collar middle class "felt particularly tragic," as I earlier suggested, that must be because its real tragedy was to be the class that by definition could not *have* feelings, at least not at work. White-collars were the class of those without "productive" feelings. Anaesthesia—or, rather, an "at once interlocking and in tension" structure of oscillation between anaesthesia at the weekday core of productive life and hyperaesthesia during weekend leisure—*was* the white-collar structure of feeling. Such was the paradox of feeling inherent in industrialism, whose weekend acts of unproductive irrationality might have criticized technological rationality, except that consumption bought so fully into the whole system of modern production (a pattern that a chapter of Schor's *Overworked American* titles "The Insidious Cycle of Work-and-Spend").

The evidence for such a (non)feeling of alienation is of broad venue, since it potentially includes all the corporate, government, medical, journalistic, private, anecdotal, and other records that remain to indicate how people felt or were supposed to feel the rationalized structures of their existence. To reduce the scale of the problem, therefore, I will track the underlying nine-tenths of the iceberg of the modern ethos by looking just to its coldest tip: "feeling" in its most obvious sense of "emotional feeling."[26]

I take my lead here from such sociologists and social historians of "constructionist emotionology" as Arlie Russell Hochschild, Peter N. Stearns, Robert I. Sutton, John Van Maanen, Gideon Kunda, Jürgen Gerhards, Anat Rafaeli, Amy S. Wharton, Cas Wouters, and Abram de Swaan.[27] Though various in their topics and methods, these researchers allow us to piece together a composite history of early to late twentieth-century affect in the United States (and Northern Europe) that—by contrast to such landmark studies of earlier affectivity as Lawrence Stone's *Family, Sex, and Marriage in England, 1500–1800*—is notable for its emphasis on institutional rather than domestic or personal life. Not the family, in other words, but the great impersonal organizations of modernity—above all, the workplace—have set the tone of modern emotional experience. It is no accident, as Peter Stearns notes, that later editions of Dr. Spock's book on child care drew "from the areas of management and sales" for examples of the "work goals for which children were being prepared," nor, as Hochschild observes, that "what a person does at work may bear an uncanny resemblance to the 'job description' of being the child of such a worker at home."[28] In what is much more than a figure of speech in Hochschild's *Managed Heart: Commercialization of Human Feeling*, Stearns's chapter "Reprise: The New Principles of Emotional Management," or de Swaan's and Wouters's essays on "emotional and relational management" and "emotion management," twentieth-century emotional life dominated by the middle class was all about "management." It was about managing the allowable range and intensity of productive affect, displacing excess affect into indirectly productive acts of consumption, and thus establishing the modern paradox of deadpan professionalism and binge leisure that Stearns, in the title of his book on the subject, calls *American Cool: Constructing a Twentieth Century Emotional Style*.[29]

Concentrating on the first half of the twentieth century, but also glancing at older periods to deepen our historical understanding, we can recount the genesis of modern managed emotion as follows.[30] In previous centuries, the responsibility for managing the emotions belonged primarily to what might be called an extended personal sphere consisting of the individual amid his or her family, extended family, friends, village, parish, local market or trade circle, and so forth.[31] Customary sociality, in other words, swathed individuals from joyous birth to grievous death in what Hochschild calls "feeling rules." Feeling rules regulate the relation "between 'what I do feel' and 'what I should feel.'" Should one feel joy or sadness, for example, when apprenticed out to another family? Should one receive one's new husband openheartedly or with reserve? And what does one do—meaning, whom does one listen to—when in the most pivotal passages of life "do feel" and "should feel" (not to mention "do feel" and "should display") may not match?[32] The answers to this complex dynamic were to be found in the home, village, parish, and so on, along with the usually limited set of texts and pictures that such settings afforded as homilies.

But with the rise of "affective individualism" and the "companionate marriage" (as Lawrence Stone describes the nuclear family ethos), a subtle change occurred in the extended personal sphere. On the one hand, emotional management increasingly concentrated in the immediate family and shrank away from the wider customary milieu. It was father, mother, and child (the triangle internalized by Freud in the self) who managed the emotions among them in their fierce, tight tangle. Wider society also imposed feeling rules through an increasing array of conduct books, child-rearing manuals, youth literature, and scientific discourse. Yet the very volume of such media reveals that custom was now mediated rather than immediate.

On the other hand, regardless of the nuclear family's dominance in personal emotional life, the total significance of the personal diminished in relation to a different control of affect. The distinctively modern moment in the history of the emotions arrived when the entire province of the private/familial was marginalized by the great modern, impersonal organizations that stepped into the breach of customary sociality to commandeer the rules of feeling—to such an extent that one researcher has suggested we now conceive of corporate organizations as themselves the entities that have emotions.[33] Modern military organizations in the era of the Great War, modern big government, modern medicine, and modern education have all had famous roles to play in the rationalization and training of the emotions. (How many feelings, for example, are now acceptable in any of these institutions?)[34] The most important such institution in the modern management of feeling was business, which from the late nineteenth century led the way in developing what James Beniger calls the great, bureaucratic structures of "generalized control" (at once economic, technological, and social) that supervised lives for as much as sixty hours each week.[35] Wider sociality did not become less important in the age of the nuclear family, but it now exercised its control through organizational entities much more distant from the family form than the village, parish, and so on. However important it was for a child to learn to manage anger in the presence of father or mother, it was now even more important that father and mother learn to manage their anger, resentment, joy, lust, distraction, or boredom in a workplace cut off from the farm or town that had been the customary, ambient field of emotion.[36]

What was the particular lesson of the heart taught by the modern workplace? In the peremptory tone of Frederick Taylor during his fight with the machinists at the steelworks, we might at first say that the lesson was a mean one. (Relating an encounter with the workmen: "I do not want to take the next step, . . . but I warn you if I have to take this step it will be a durned mean one.")[37] The lesson was, sternly: *Thou shalt not know joy or sadness at work. Neither celebration nor protest nor mourning is on the clock.* But such a formulation emphasizing the sternness of the disciplinarian—and thus inevitably a lord-servant dialectic returning us to Marx's agon—is finally misleading. The real meanness of the modern workplace was that it refused to offer a stage for meanness at all. The whole point of "scientific management" as Taylor invented it after his fight was to subtract from the workplace any antagonism of "man from man" by absorbing antagonism into a relationship with something one could safely hate (or mourn, or love) with no practical effect at all: the technological/technical *system.*

Taylorism was not just the first rationalized system of labor management. It was also the first rationalized system of *emotional* labor management. It was the predecessor to what Hochschild calls commercialized "emotional labor": "the management of feeling to create a publicly observable facial and bodily display . . . [that can be] sold for a wage." Taylorism was what we might call the "zero degree" of such management of feeling. In semiotics, zero degree describes a "significantly absent" sound or sign whose omission "functions by itself as a signifier." There are thus words like *mana* in some cultures, as Claude Lévi-Strauss has observed, that have no specific or fixed content other than pure "meaningfulness" (i.e., *not* meaninglessness).[38] Taylorism was the zero degree of emotional labor management because, to adapt Hochschild's definition, it was the management of feeling to create a publicly observable facial and bodily display of *no* feeling. It was a feeling for nonfeeling.

The crucial step that Taylor made after his fight, therefore, was to recognize that his real antagonist had been not the workmen at all but an obsolete *system* of shop management: "In this contest, after my first fighting blood which was stirred up through strenuous opposition had subsided, I did not have any bitterness against any particular man or men. My anger and hard feelings were stirred up against the system." This recognition inspired the time-motion and other systematic studies of work that eventually led to his four cardinal mandates of scientific management:

1. "Develop a science for each element of a man's work."
2. "Scientifically select and then train, teach, and develop the workman."
3. "Cooperate with the men so as to insure all of the work [is] done in accordance with the principles of the science which has been developed."
4. Ensure that "there is an almost equal division of the work and the responsibility between the management and the workmen."[39]

The final mandate here, which Taylor took utmost care to explain throughout his career, meant that managers must "share" in the workmen's labor (later commentators would say "extract" the autonomous skills of that labor) by taking responsibility for all the technique—the planning, routing, timing, positioning, and so on—that workmen had once improvised for themselves or learned from custom. In practice, this required restructuring the role of the traditional gang-boss or shop foreman by appointing multiple "functional foremen" (e.g., the "order of work and route clerk," the "instruction card clerks," the "time and cost clerk," and the "shop disciplinarian"). Such foremen ancestral to a modern management staff could oversee labor in unprecedented detail during every moment of the workday.[40] They did so through technical rules, procedures, and "instruction cards" that depersonalized the presence of the gang-boss or shop foreman. This meant that not they but an invisible, silent, faceless monitor—the indifferent and mediated personality of the system itself—now dogged the worker's every step.

Taylor himself ascribed a persona to such dogged management that would be fateful for subsequent industrial history. He called it "friend." According to a consistent logical fallacy (of the excluded middle) in all his writings from *Shop Management* (1903) through his testimony before the Special House Committee (1912), antagonism between workmen and foremen gives way to a scientifically managed system of work that, because it is not recognizable as opposition in any customary sense, must therefore be friendship.

"First, then," he says, "the men must be brought to see that the new system changes their employers from antagonists to friends who are working as hard as possible side by side with them, all pushing in the same direction." *Friends, friendly,* and *friendship* are staples in Taylor's discourse, sometimes appearing so often as to seem a tic.[41] Similarly, there is his discourse of "harmony": "when [workmen and managers] realize that it is utterly impossible for either one to be successful without the intimate, brotherly cooperation of the other, the friction, the disagreements, and quarrels are reduced to a minimum. So I think that scientific management can be justly and truthfully characterized as management in which harmony is the rule rather than discord." Friendship means harmony or *alignment* with the system ("all pushing in the same direction"). It is a purely systemic coefficient of the fit between labor technology and technique, analogous to rate of flow, efficiency, and so on; as such, it ultimately characterizes the state of the system itself rather than any persons in it. As Taylor said on the last day of his congressional testimony, "Of all the devices in the world they [labor unions] ought to look upon scientific management as the best friend that they have."[42] Not any particular managers but "scientific management" itself is the friend. Such is the genesis of what can be called modern *systemic* or *technical* friendship.

Systemic or technical friendship—the ideology that enforces the adjustment of technique to technology—has been the primary instrument of emotional labor management in the twentieth century. There is not a single major change in industrial paradigm from automation through more recent networking that has not been sanctioned through a resocialization of work under the name of systemic friendship or, as it would later be reconfigured, "corporate culture." We will come to the explicit, positive content of such friendship when we turn to later corporate cultures. For the moment, let us consider the negative content of technical friendship under the management of what I called zero-degree emotional labor. In the theory of Taylor and the practice, most notably, of Henry Ford, such friendship was originally a wholly privative concept. It *proscribed* the entire gamut of expressive behaviors associated with customary work.

We can take up the case of blue-collar work first, for which the Ford factories were the great proving ground of the new friendship.[43] Whatever the debated truth of Henry Ford's self-styled image as the benefactor of the Five Dollar Day and the "workingman's best friend," it is clear that his bold, labor-friendly initiatives—including the famous five-dollar day, eight-hour shift, and five-day week—were not just offered as proverbial carrots. As noted by such critics of the "Ford Legend" as Keith Sward, they came along with a big stick of labor management that fattened profits by forcibly harmonizing the hearts and minds of workers with the faster, more cramped, and more brutal conditions of line speed-up. Boosting pay, for example, allowed Ford counter-intuitively to boost profits as well, by giving him the cover he needed to speed up the line, be more selective in hiring, and so on. So, too, cutting the day from nine to eight hours allowed him to run three shifts over twenty-four hours.[44]

In this regard, the famously personal zeal with which Ford promoted hard work, frugality, teetotaling, compulsory gardening, and so on among his workers is superficial. What matters is that, while he was still more of a feudal lord than most later chairmen of the board or CEOs, he was not just *il padrone*. Rather, he stood atop a nascent layer of bureaucratic *management* dedicated to administering the carrots and sticks of friendship. The five-dollar/eight-hour day announced in 1914, for instance, was carefully supervised by the Ford Sociological Department so that full pay went only to workers whose family arrangements, lifestyle, and moral conduct met company standards. Those standards bore directly on the emotional life of workers because they forcibly curtailed all customary agrarian, craft, or ethnic traditions of celebrating, carousing, and mourning that exceeded Ford's exacting Protestant ethic. Within a week of the new five-dollar day, for example, the company penalized "several hundred of its foreign-born workers who had taken a day off on their own to observe a holiday of the Greek Orthodox Church." Other actions in 1917 targeted workers for "Polish wedding, drunk . . . domestic trouble . . . crap game while on duty." Nor was it only on-the-job emotional conduct that mattered. The home visit corps of the Sociological Department (numbering 150 officers by 1919) routinely fanned out to search homes for workmen who spent their evenings "unwisely" or indulged in practices of "unwholesome living." In 1916 some 30 percent of Ford employees were disqualified from the full five-dollar wage for one reason or another.[45]

Even more infamous was the "Terror" at the Detroit Rouge Plant instigated by the Ford Service security corps under Harry Bennett. Notorious for the thugs it recruited, the corps used strong-arm tactics and an informant system to enforce an elaborate, detailed system of work rules effectively banishing all affective behavior. As Sward notes:

> Any number of idiosyncrasies bore witness to the stifling effect of regimentation. . . . Chatting or fraternizing with work-mates during the lunch hour was taboo. . . . It was then the rule during the noon-day spell to see a Ford employee squat on the floor, glum and uncommunicative, munching

> his food in almost complete isolation. The locked-in manner of the Ford worker supposedly at ease was remarked by Raymond J. Daniell who observed in the *New York Times* on October 31, 1937, "The visitor is struck by the restraint among the workers; even in moments of idleness, men stand apart from one another."[46]

So rigid was the system that "Fordization of the face" was the rule "inasmuch as humming, whistling or even smiling on the job were, in the judgment of Ford Service, evidence of soldiering [deliberate resistance to speed-up] or insubordination." One employee, for example, was fired in 1940 for "smiling" and "laughing with the other fellows." The only note of dissonance was the so-called "Ford whisper" through which workers communicated in covert undertones.[47]

Ultimately, Ford ushered into modern industrial culture a complete system of emotional labor management that disallowed workers any "productive" emotion at all—any emotion, that is, legitimately vested in work in the way, for example, harvest and work songs had once invested cutting the stalks, hauling them in, or separating the chaff with gusto. Ford minced no words on this score in his autobiography: "I pity the poor fellow who is so soft and flabby that he must always have 'an atmosphere of good feeling' around him before he can do his work. . . . Unless they obtain enough mental and moral hardiness to lift them out of their soft reliance on 'feeling' they are failures."[48] Such intolerance was no different in kind (though perhaps different in degree) from that of Ford's fellow industrialists. As Ray Batchelor notes, business and social groups of the time acting through the Detroit Urban League and related organizations tried to give the correct modern American temper to immigrants, ex-peasants, black sharecroppers, and anyone else who did not know how to harmonize with the system. Each of these latter groups

> brought with them traditions, festivals and social habits which to the industrialists were unfamiliar, threatening and incompatible with the increasing discipline they demanded of their labour force. If a Pole married, the celebrations might last two or three days. . . . Under the auspices of the Urban League, various social and reforming groups were brought together to "Americanize" Detroit's ethnic populations. The League sought, for example, to "educate" the "loud noisy type of Negroes unused to city ways" by publishing books of advice [that included such prescriptions as] "don't be late for work; don't wear flashy clothes; don't use loud or vulgar language."[49]

Don't was the word for feeling. As Peter Stearns and Shoshana Zuboff have extended the catalogue of industrial age commandments: Don't engage in "spitting, lack of cleanliness, sloppy posture, brazen stares, drumming of fingers," or "being dirty, . . . throwing water, seducing females, being drunk, arriving late, or not showing up at all."[50]

Into the twentieth-century emotional hopper went undisciplined agricultural workers, independent craftsmen, hot-tempered foremen, emotionally "spontaneous" immigrants and minorities, and excessively "sentimental" or, worse, angry women with all their personal, familial, ethnic, gender, and other such customary rules of emotion management. In went their anger, resentment, unruliness, distraction, chattiness, and all the other deadly sins of the new business moralism. In went the coordination of work life with the major events of social life: the birth or wedding occasioning an impromptu holiday, the Sunday carousing that spilled over into Monday sickouts, the entire intricate weave of personal and group interactions/antagonisms that gave a tone and feeling to customary experience.

Out the other end of the hopper came the only legitimate personality of labor in the Taylor/Ford era: the so-called "robot" or "automaton," whose Fordized face expressed at most a "Ford whisper." Work at the Ford plants under the speed-up regimen of the five-dollar-day era, one ex-worker recalled, was "a form of hell on earth that turns men into driven robots."[51] Echoing the same trope, the chairman of the House committee examining Taylor asked, "If the workman has to obey instructions implicitly as to how the work should be done, would he not thereby simply become an automaton?"[52]

But we have not yet come to the real robots. Only when we look from blue-collar workers to their superiors do we recognize the true automaton of the century destined to fulfill the mandates of systemic harmony and friendship. A tremendous irony attends this part of our story. While the emotional labor management practiced on blue-collar workers was explicit and brutal, the ability to make workers actually hew the company line was never certain. The real line or shift, as I suggested in chapter 1, developed its own social solidarity and, often, political agency. As attested by the constant threat of strikes, quitting, and "soldiering," there was just a hair's difference between the Ford-faced worker on the line and the sullen or angry worker on the picket line. The irony is that the most definitive, lasting success of the new emotional labor management was not the blue-collar worker at all, but rather, as Peter Stearns describes his topic in *American Cool*, "primarily . . . the middle class of business people and professional families."[53] Emotional labor management affected most deeply the class that evolved from Taylorist "functional foremen" and Fordist managers to *administer* emotional labor management: the white-collar middle class. White-collars internalized emotional labor management—the friendship of systemic alignment between technology and technique—in their own identity. To borrow William Blake's phrase, they became what they beheld.

The following passage in Leffingwell and Robinson's *Textbook of Office Management* is emblematic:

> *Disturbing factors* must be known and guarded against. A very common one is noise, not the noise of a boiler shop, but the constant sound of batteries of typewriters, adding machines, the babel of loud talking, and so forth, noises that would not be noticeable on a busy street but may become distracting in an otherwise quiet office. Excitement of any kind is a condition that should be guarded against. The efficiency of an office, for example, can be utterly destroyed, for the time being, by such a common accident as the fainting of a girl.[54]

Leffingwell and Robinson shift their attention from "disturbances" in noise level, lighting, and the other objective factors that Leffingwell spent so much of his career analyzing to the fainting girl as a figure of nervous, *subjective* "excitement."[55] They turn to feeling management.

Enacted in this passage is the crucial transition that Taylorist scientific management underwent when it migrated to the white-collar environment. Presided over in the 1910s through 1950s not just by Leffingwell but by proliferating experts in the new fields of industrial psychology (most notably, Elton Mayo), salesman training (e.g., Dale Carnegie), personnel counseling, employee testing, and so on, white-collar Taylorism gave birth to that ultimate system of emotional labor management in the twentieth century—"human relations" or "human resources" management. Such management went one better than the strong-arm tactics of the Ford Sociological Department or Ford Service in suppressing feeling. Rather than implementing obvious incentives backed up by repressive apparatuses, it enforced a norm of low affect through cultural means designed to secure not just the "harmonization" but the *identification* of the manager, professional, supervisor, and (as much as possible) clerical workers with the technological/technical system. The human resources movement only superficially contradicted the stern paternalism of Taylorism-Fordism with a sort of Dr. Spock approach to nurturing workers. More realistically, as Beniger argues, it was the next logical stage in the evolution of bureaucratic "generalized control." Human relations, in Mayo's words, was "a new method of human control."[56]

Elton Mayo himself began his career in the 1920s in Taylorist fashion by researching time and fatigue factors. But as a result of the now famous Hawthorne experiments at the Western Electric Company in Chicago, which showed that social factors—even the simple fact of being interviewed by researchers—could boost work performance, he increasingly turned his attention to human relations.[57] This meant that he had to address workplace emotions. As Peter Stearns notes, Mayo "was soon drawn into a concern about anger as he discovered a level of 'irritability' among workers that he could not ignore."[58] More broadly, he studied not just anger but also "monotony," "pessimism," "melancholy," "anxieties," and all other "morale" issues affecting what he termed (in his version of systemic "harmony") the "balance" or "inner equilibrium" of workers.[59] The emphasis on inner equilibrium was especially relevant to the white-collar middle class because it inculcated a standard of *internal* emotional labor management. Mayo made a paradigm of one factory where low absenteeism could be attributed to foremen who had been "taught the very great importance of three elementary rules or methods of approach to human problems . . . 1. Be patient. 2. Listen. 3. Avoid emotional upsets." He shows similar concern for the internalized emotional equilibrium of white-collar workers when he notes that it is "the administrator himself in these days [who] is frequently a victim of the emotional doubt or opposition." And near the end of a chapter in *Social Problems of an Industrial Civilization* (1945) titled "'Patriotism Is Not Enough; We Must Have No Hatred or Bitterness towards Anyone,'" he concludes: "Modern civilization is greatly in need of a new type of administrator who can, metaphorically speaking, stand outside the situation he is studying. The administrator of the future must be able to understand the human-social facts for what they actually are, unfettered by his own emotion or prejudice."[60]

Ultimately, Peter Stearns notes, counseling and personnel handbooks in the Mayo era multiplied "to instruct foremen, middle managers, and office personnel about the importance of keeping their own tempers and avoiding provoking. . . . 'Bullying begets bullying' was the new cry as a growing amount of attention turned to the anger habits of the middle and lower-middle class itself."[61] The result, as William Whyte said about human relations doctrine in *The Organization Man* (1956), was a system of management that did not push the white-collar worker around or even argue with him. Instead, in perfect synchronization with systemic balance and inner equilibrium, it "adjusted" him. "They will adjust him," Whyte wrote. "Through the scientific application of human relations, these neutralist technicians will guide him into satisfying solidarity with the group so skillfully and unobtrusively that he will scarcely realize how the benefaction has been accomplished."[62] "Adjustment," indeed, became in industrial psychology and human relations parlance the soul of the low-affect "mature" worker. For example, in a chapter titled "Employee Adjustment" of his textbook *Psychology of Industrial Behavior* (1955), Henry Clay Smith quotes a psychiatrist who defines human happiness as systemic adjustment: "the adjustment of human beings to the world and to each other with a maximum of effectiveness and happiness." Smith then internalizes systemic adjustment as a problem of inner adjustment: "For most men, the critical obstacles are within themselves." Finally, he confers upon the worker whose internal system matches the outer system the title of "mature personality." The mature personality is one who "focuses his attention upon realistic goals rather than upon his anxieties" so as to be "free" in his emotional life. But such freedom of the emotions is in effect privative. It is freedom *from* anxiety (including "executive anxiety"), aggression, regression, and so on.[63] "Impersonal, but friendly" is how another human relations book of the era puts it. Still another such work of the era instructs: "it is of the utmost importance that the foreman remain cool."[64]

Into the twentieth-century *white-collar* emotional hopper, then, went "personnel" preselected and predisposed to harmonize in friendly fashion with the corporate system. By definition, *personnel* meant not just people but—as Whyte observes at length—specifically people hired after taking standardized personnel tests (such as the widely used Bernreuter Personality Inventory or the Worthington Personal History test).[65] Typical of these tests were questions such as the following:

> "Does it annoy you to be interrupted in the middle of your work?"
> "Indicate whether you agree, disagree, or are uncertain: . . . The sex act is repulsive."
> "A woman walks up abruptly and demands to be waited upon before you. What would you do? a) Do nothing b) Push the woman to one side c) Give her a piece of your mind."

Equally typical was how answers to such questions were resolved into "contentment indexes," "irritability indexes," and so forth, which at once selected for and insidiously shaped a perfect breed of low-affect worker.[66]

Out the other end of the hopper came the great achievement of management in the human relations era: the self-regulating white-collar middle class, which henceforth knew to differentiate itself from the lower classes as the embodiment—standardized and well-adjusted—of the ethos of automation. The difference between higher and lower social standing, it came to be known, was not just white collar and clean hands versus blue collar and grime. The difference lay in emotional identity. "It was revealing that efforts to curtail anger at work, initially directed at blue-collar workers on the factory floor," Peter Stearns writes, "shifted focus toward the middle and lower-middle class, which was expected . . . to internalize the necessary restraints. Here was implicit recognition of some continued social gaps in emotional standards."[67] The lower classes were "unruly," "tardy," "undependable," "surly," "lazy," and so on. As a mid-century sociology book observes, the life of the lower-class person "is drunkenness, momentary hedonism, sexual license, violence and street scenes." By contrast, the white-collar middle class, and even clerical workers bound to the class by a complex mix of aspiration and imposition, were supposed to be self-controlled—in a word, professional. "The superego . . . tends to be stronger in the middle class," a 1957 sociology book says.[68] Whether one was a higher manager, middle manager, professional in the specific sense (e.g., lawyer, engineer), office supervisor, or even secretary, the rules were the same: *one* family picture on the desk (at most two), the *occasional* cartoon or satire tacked on the wall, a *few* office parties per year on well-defined occasions. These and other now-familiar markers of affect in the white-collar office define an emotional landscape as pure and clean as a desert.

The desert is a trope I introduce to track the mythopoetic landscape of work that will finally depict the workplace as a digital "highway" and "sea." Upon that desert arose the deeply paradoxical, yet also profoundly normalized, structure of feeling I have called weekday anaesthesia and weekend hyperaesthesia. Peter Stearns observes that just when "American cool" arrived as the mainstream emotional style, popular culture began, antithetically, to use the term *hot* for depersonalized sexuality and commodity goods. White-collar middle-class workers were cold in their subject life at work and hot solely in their object life of compensatory leisure. So it is, as Mills memorably put it, that "they must be serious and steady about something that does not mean anything to them, and moreover during the best hours of their day, the best hours of their life. Leisure time thus comes to mean an unserious freedom from the authoritarian seriousness of the job." So it is, in other words (as Witold Rybczynski notes in *Waiting for the Weekend*) that there arose the "problem of leisure" debated in sociological and journalistic essays from the 1920s on.[69]

Cold and hot, serious and unserious: the full history of this paradox of feeling enslaved to the "insidious cycle of work-and-spend" cannot be explored here as it evolved from the 1920s through the 1950s. But one snapshot from the end of this span will be useful. By the 1950s and early 1960s, we remember, Hollywood knew precisely what to make of the oscillation between cold and hot feeling. It created for fantasy a whole breed of characters able to bring that oscillation to a perfect point of convergence: steely-eyed heroes of the Western, gangster, noir, war, biker, and other film genres who had ice in their veins, who were so "cool" they were "hot," and who thus imaged even in the high-affect mode of popular leisure the low affect of work life. As Bruce Robbins memorably allegorized at the opening of his *Secular Vocations: Intellectuals, Professionalism, Culture,* no one could be more coldly professional than the technically proficient, mercenary "experts" played by Lee Marvin, Robert Ryan, Woody Strode, and Burt Lancaster in Richard Brooks's 1966 Western *The Professionals.*[70] The distinguishing trait of these Good, Bad, and Ugly experts, we may say (referring to another cinematic saga of the desert), is not just that they were all businessmen out for a few dollars more, but that they were "pros" as cold as machines. Behind the hot glare of all the Western deserts and urban jungles of film fantasy was only the fluorescent light of the office, where technique matched technology as perfectly as crosshairs laid steadily on a target.

We are now at a momentous point in our argument. When we contemplate the cultures of modern production and consumption in their convergence rather than fetishistic isolation, we witness the birth of twentieth-century cool proper. How could the paradox of cold/hot feeling be slaved so neatly to the production/consumption cycle that the implicit critical force of such paradox—the bleak, wry, or existential irony of cool—dispersed entirely into what Mills called unseriousness (such that critique became a purely ironic habit or style)? In a variation on the "safety valve" theory of social containment, I propose the following scenario. Cold production vented excess feeling into hot consumption, but the culture of consumption stopped short of registering any serious challenge to the culture of production because of a crucial act of symbolic distancing. The object of consumption became identified with "outsiders" to the entire system of work/leisure, and in contrast those who merely went to a hot film or bought a hot record still felt themselves to be comfortably "insiders." White-collars, in other words, displaced the very experience of alienation onto outsiders who could do the heavy lifting of being alienated for them.[71]

I refer here to the basic engine of cultural cool: the consumption by middle-class workers of forms of entertainment, journalism, and dress influenced by that part of culture excluded by definition from normal work—*subculture.*[72] Through commodified forms of leisure safely tethered to the system of production, white-collars directly or through their children (the emerging, mid-century "youth market") increasingly came into vicarious contact with the really cool cultures of the first half of the twentieth century. First, black music culture, glimpsed in ever more attenuated forms in jazz clubs, nightclubs, musicals, recordings, and films.[73] Second, the urban, gangster, teen, or otherwise "sexy" side of working-class culture (glimpsed, for example, in the image of Marlon Brando in a tee shirt). In a word, *Elvis.* Or, rather, we should not succumb to the easy essentialism that so often deems subculture the "really cool" and middle-class culture merely the appropriation of cool.[74] There *was* appropriation, of course, and with it all the fastidious distinctions (like a white picket fence around a suburban yard) that domesticated cool, until in the 1950s the label *delinquency* was widely used to quarantine teenagers who walked too close to the wild side of cool.[75] But middle-class appropriation was rooted in an overall process that may be called "reciprocal appropriation" or, as Chuck Kleinhans puts it in his "Cultural Appropriation and Subcultural Expression," "dual appropriation." I mean by this that the realms of subculture and the mainstream each appropriated the resources of the *other* to create their differing stances—their "style" or "lifestyle"—toward the workaday life of technological rationality. Cool was a displacement, circumvention, or "work-around" for a life dominated by the culture of work, and the "around" part of that formulation could only be discovered if subculture and mainstream each wandered out of bounds onto the turf of the other.[76]

Subculture appropriated from the mainstream the paradox of hot versus cold. *Hot* appeared in its perceived expressiveness, social or sexual fluency, violence, ethnicity, "lower class" behavior, and so on. But what is less commonly recognized—and what makes subculture truly an appropriation—was that it also affected a version of *cold* in a complex enactment of technological rationality that may be called mock- or camouflage-technology.[77] *Camo-tech,* to coin a phrase, was the construction of a bodily and social pose that perfectly expressed the adjustment of technique to technology—but for *unproductive* purposes. It was the cool jazz musician blowing his guts into the intricate machine beauty of a saxophone, the cool gangster religiously oiling and assembling his Thompson machine gun (as performed in the topos of "arming scenes" in genre films), or the cool, tee-shirted "hood" sprawled under a hot rod performing miracles with a set of metric spanners. Whatever was "really cool," in other words, was on a different level also truly routinized and automated—like working in a machine shop, only without the shop. There is no clearer example of such cool (though drawn from English subculture of the 1960s and 1970s) than Dick Hebdige's description of the reggae "sound-system":

> In clubs like the Four Aces, in the Seven Sisters Road, North London, an exclusively black audience would "stare down" Babylon, carried along on a thunderous bass-line, transported on 1000 watts. Power was at home here—just beyond the finger tips. It hung on the air—invisible, electric—channelled through a battery of home-made speakers. . . . The music itself was virtually exiled from the airwaves. It could live only in and through the cumbersome network of cabinets and wires, valves and microphones which make up the "system."[78]

Camo-tech, in short, was the hijacking of the system of technological rationality by what Hebdige and the Birmingham cultural studies group conceived to be the soul of subculture: its counter-system of "style." Though style is most often studied in terms of consumption culture, in our context it must be understood to fuse consumer fashion with producer sensibility to create the ubiquitous feeling that in every instance—in the way one dresses, walks, talks, or drives—there is an exact adjustment to be made between technology and technique. All the most influential subcultural styles of the twentieth century, we may thus say, were variants of techno-style. They were how subculture ritualized the sense that the particular adjustment defining one's identity—whether one revved or did not rev a car engine at a light, whether one wore loose or tight pants, or whether one spoke normal English or slang—was secondary to a fundamental adjustment between technology and technique. There was only one cool way to grease one's hair, just as there was only one right way to oil a drill press. Style was the delinquency, but also the mimicry, of Taylorism.

Meanwhile, the mainstream reappropriated subculture's appropriation of technological rationality. From one point of view, the middle class glamorized subculture (or at least its photogenic aspects) for having found a workaround for routinization that challenged the entire system of production/consumption from the outside. Subculture, after all, was by definition the part of society chronically excluded from meaningful work. It did not have a resume or career. It had style instead, the defamiliarization of normal lifestyle (like a greaser taking a perfectly ordinary comb and wearing it in his hair).[79] Mainstream culture was gratified to enact such defamiliarization during its nights and weekends of leisure, grasping after a fleeting sense of identification with what lay beyond the pale of the system.[80] For just a moment in the hot rush, cold chills, and otherwise contradictory cold/hot experience of deep leisure—leisure so profoundly bound up with identity that it is akin to "deep play" in Clifford Geertz's study of the status drama enacted in a Balinese cockfight—white-collars got to play the part of someone who could see the entire structure of modern cold and hot from the outside.[81] Thus the *drama* in alienation, it seems, returns.

But not really. From another point of view, the deep play of white-collar culture was never a matter of fully identifying with subcultural actors excluded from the life of work. Instead, the middle class identified primarily with the empty *position* rather than *identity* of the outsider. It identified less with the actor or even action of the outsider's challenge to the system, in other words, than with the empty stage or site of the exotic on which that challenge was enacted. This is why, for example, its most iconic, "cool" film stars had inscrutable faces signifying either nothing or the blankness of the middle class itself at work: such faces mark the place of the Outsider, but there is no one actually *there* in that place. Or, to put it in a way that has everything to do with the mythopoetic landscapes of work leading from filmic deserts to screensaver-like digital highways and seas: the mainstream identified with the pure *milieu, ambience,* or *texture* of challenge, with style emptied of agents and agency to become a world sufficient unto itself. The real hero of the Westerns, gangster films, detective novels, and so on of leisure was the style, texture, or "feeling" of a moral-geological wasteland so autonomous that its ambience could simply be lifted whole from one noir fantasy and transplanted to another. If subculture had style, then what the mainstream had was a style of style or ambience of style, according to which technique seemed to challenge the life of technique while actually it just confirmed the same old routine recurring eight hours a day, five days a week.

Thus did cool first come into its own as the ethos of modern alienation: a whole attitude or character of low affect that at once harmonized with the system of technological rationality and disengaged itself from that system by identifying just enough with outsiders to live a depersonalized fantasy of the Outside as pure style or decor. The system became in imagination just a "scene" from which one distanced oneself through regular indulgences in outlaw scenes. *I work here* [amid all these machines, protocols, procedures, codes], *but I'm cool.*

Yet all the time, we know, "youth culture" was growing stronger. Constituting a population of resident outsiders *within* the middle class, mainstream youth increasingly bought up, and bought into, the postures of subculture. The next stage in the evolution of cultural cool would come when such resident outsiders fell in love with the general notion of subculture ("the people"), transformed leisure into a lifestyle apparently transcending consumption (in such radicalized forms as the intake of drugs), and took to the streets to protest the whole system of technological rationality. A new "feeling" was in the air—most famously declared in the slogan "Make love, not war." However, the motto could just as well have been (as in that archetypal movie about the youth of knowledge work, *The Graduate*) "make love, not plastic" (or insurance policies, contracts, ads, and so on).

But before we can grasp this next episode of *cool* in our terms, we need to advance the underlying story of *cold*.

Chapter 3

Informating

As if telling an epic, we can best approach the next stage of our story *in medias res* by visiting two work locales of the early 1980s depicted in Shoshana Zuboff's *In the Age of the Smart Machine* (1988). Here is Zuboff's sketch of one work site paradigmatic of a new mode of blue-collar work:

> In 1981 a central control room was constructed in the bleach plant [of the Piney Wood pulp mill]. A science fiction writer's fantasy, it is a gleaming glass bubble that seems to have erupted like a mushroom in the dark, moist, toxic atmosphere of the plant. The control room reflects a new technological era . . . in which microprocessor-based sensors linked to computers allow remote monitoring and control of the key process variables. . . . Inside the control room, the air is filtered and hums with the sound of the air-conditioning unit built into the wall between the control room and a small snack area. Workers sit on orthopedically designed swivel chairs covered with a royal blue fabric, facing video display terminals. . . . The terminals each face toward the front of the room—a windowed wall that opens onto the bleach plant. The steel beams, metal tanks, and maze of thick pipes visible through those windows appear to be a world away in a perpetual twilight of steam and fumes, like a city street on a misty night, silent and dimly lit.[1]

And here is Zuboff's matching description of the new white-collar office of the time:

> For the millions of people who work in or have occasion to visit corporate facilities, the on-line office has become a familiar, even a clichéd, image. There are the clusters of desks, usually separated by partitions, and each is home to a video display terminal. The office may be softly lit; some even have a smattering of potted palms or rubber plants. A quick glance can give the impression that the women and men who work in these offices have been mesmerized by the green or amber glow of their video screens, as they spend the better part of each day with attention fixed on luminous electronic numbers and letters.
>
> In 1982 I began a series of visits to one such [newly computerized] office. . . . The rows of desks and paper clutter had given way to clean surfaces and desktop terminals. The work stations were separated by tall partitions, which created a cubicle effect around the work space of each clerk.[2]

Zuboff takes such care to dramatize these two scenes—almost as if she were setting the stage for a play—because in each case she is about to introduce an anecdote that momentarily revives the agonism we saw in Marx's theory of alienation. But I will temporarily suspend us in the immaculate bubble or cubicle space she depicts, prior to the unfolding of the story. As also documented by such other historians and sociologists of computing as James H. Bair, Tora K. Bikson, Martin Campbell-Kelly and William Aspray, Manuel Castells, Dennis Chamot, Roslyn L. Feldberg and Evelyn Nakano Glenn, Barbara Garson, Joan Greenbaum, and Robert E. Kraut, this bubble or cubicle represents the moment in the late 1970s and early 1980s when—at the "cross-over point in which an hour of computer time became cheaper than an hour of a person's time"—mainframe computing achieved its zenith as the new dominant industrial paradigm.[3]

Mainframe computers (under which rubric I also include minicomputers) entered the workplace in a series of overlapping stages from the 1950s on.[4] After many of the largest companies first adopted computers in their routine work in the 1950s and 1960s, computerization broadened across the range of industries and firms, until by the mid-1970s all major industrial sectors were reporting expenses for data processing.[5] First to be computerized "were complex numerical tasks such as actuarial analysis, mortality studies, and valuation of reserves"; next, "large volume, standardized calculations that had previously involved large numbers of clerks" (e.g., billing); then, centralized recordkeeping in the form of databases; and finally, in the

1970s, word processing.[6] By the early 1980s, as shown by the field research of Bikson, 64 percent to 81 percent of workers in every U.S. white-collar rank from the secretarial up to the professional-managerial were using computers, and even 36 percent of executives were spending time at the keyboard.[7]

The impact of computerization was even greater than the numbers alone registered because the change was indeed paradigmatic: it altered the overall pattern or configuration of work. Mainframe computing inaugurated not just sweeping changes in infrastructure and work routine, but the possibility of a whole new mentality of work.[8] Of course, every major new industrial paradigm may be said to be complemented by a new mindset, but computing was unique for the degree to which its mentality was bound to the nature of its underlying technology. We can think by contrast of the Industrial Revolution. While the steam, internal combustion, and electric engines "caused" a new production mentality (men as "dynamos" or "robots"), they did not literally induce such a mentality.[9] Any deduction of a hard and fast connection between the instrumental and mental bases of the Industrial Revolution, therefore, is to some degree open to the charge (as heard in the "technological determinism" debate) that it is just a figure, a reification.[10] In any case, the primary goal for industrialists was to make the lines run faster; mentality was a secondary or spillover effect that could always be adjusted after the fact in the heavy-handed manner of the Ford Sociological Department. But the engines that drove the Information Revolution were different because a fundamental parity between their instrumental and mental bases made it impossible to separate "primary" from "secondary," and "causal" from "figurative." The original intent of IT was to extend the automation paradigm by using information to act on matter (or in the office, paper) with additional speed, efficiency, and precision. But there arose from within this primary action, and bound inextricably to it, an unplanned, secondary mental action that from the perspective of business was as disturbing as it was promising. To use Zuboff's neologism, work was not just automated, it was "informated."

Zuboff coined this term to mean that computers generated an inescapably thick wrapping of second-order information (information acting on information) *around* the primary interface of information acting on matter where automation occurred—a wrapping so thick and interesting that it called into question the distinction between primary work and secondary mentality:

> The devices that automate by translating information into action also register data about those automated activities, thus generating new streams of information. For example, computer-based, numerically controlling machine

> tools or microprocessor-based sensing devices not only apply programmed instructions to equipment but also convert the current state of equipment, product, or process into data. . . . The same systems that make it possible to automate office transactions also create a vast overview of an organization's operations, with many levels of data coordinated and accessible for a variety of analytical efforts. . . . Information technology . . . introduces an additional dimension of reflexivity: it makes its contribution to the product, but it also reflects back on its activities and on the system of activities to which it is related. Information technology not only produces action but also produces a voice that symbolically renders events, objects, and processes so that they become visible, knowable, and shareable in a new way.[11]

Alongside information-for-automation, in other words, there arose a parallel universe of information-for-information that manifested itself not just as more information but as a mental construct of that information: an "overview," "visible" rendering, "voice," or "reflection." This construct was secondary and figurative relative to the "real" work at hand (as witnessed in Zuboff's decidedly figurative vocabulary of vision and voice). But secondary and figurative overlapped undecidably with primary and real. The mental construct that Zuboff identifies provided an awareness of the organized *whole* of work in which any work at hand was merely a part ("a vast overview of an organization's operations, with many levels of data coordinated and accessible"). Such holism or syncretism was fundamental rather than extraneous to computerization.

The essence of computerization, we may say, is not speed, flexibility, and comprehensiveness in themselves but what these qualities contribute to: the dynamic assemblage of separated pieces of information in an interlinked contextual field that can be grasped whole at the point of action, the rapid and flexible amassing of information in a synoptic frame within which the *systematicity* of technological rationality comes into view even as one is engaged in practical action. Enter a sale in a database table, for example, and instantly update inventory, customer base, and other so-called "relational" or interlinked tables in the total system. Or, to take the case of the spreadsheet (slightly anachronistic in the context of mainframes because it was the "killer app" of the subsequent personal computer age), alter the numbers in one cell and watch changes ripple outward through other cells to bring everything into visible alignment.[12] Necessarily, therefore, even the most automatic, fragmented, and Taylorized routines of technological rationality became with computerization potentially aware of what has been called "organizational cosmopolitanism." Computerization was a way to sense the overall systematicity of work.[13]

A mentality of overview was thus latent in the nature of the new technology. Indeed, "overview" and other such figures of vision are particularly apt in Zuboff's prose because they express the tight coupling of the instrumental and the mental—the system and the sense of system—in computerization. How did computerization allow routine work to become "aware of" or "sense" the systematicity of work, as I put it? The key sense, it turns out, was "vision," whose phenomenology is ubiquitous not only in Zuboff's discourse but in that of the many workers and managers she interviewed about the new IT. One manager reported, for example, "We'll be able to see what's happening. Not only will we have numbers, but we'll be able to see the dynamics for yesterday, today, and tomorrow. Using the projection capability, you can see immediately the impact on earnings or the portfolio. We'll be able to see the business through the terminal." Other managers stressed "overview":

> The new technology makes you look at the whole. Tasks become more comprehensive as a result. You need to know where to look for what you need and how to get it. You need to see patterns in relation to the whole.

> With the data-base environment, there is one information system for all to see. Tasks become more comprehensive. You can see the whole, not just the part. People will need a broader skill base to take more of a helicopter view.[14]

Significantly, IT-enabled workers themselves began to sound like managers with "vision" and "overview." An operator in a pulp mill observed: "The more I learn theoretically, the more I can see in the information. Raw data turns into information with my knowledge. I find that you have to be able to know more in order to do more. It is your understanding of the process that guides you." A line-level office supervisor added:

> The best part about having this new system is knowing what is in the unit and being able to feel like I have control over the work. That is one of my responsibilities, but I never felt like I had that control before. We were constantly chewed out by management—"You should have done this or that." If I had known what was going on, if I had had a clear picture of it, I might have been able to do the right thing. Now that I can see the total functioning of the office, I feel more ownership towards all of the units, not just my own. . . . I am not just a record keeper, but I can really use my brain.[15]

Similarly, other observers of computing and their interviewees speak of computers as providing a "window on the corporation," an "electronic window on the entire business world," "*visions* of constellations of activities," and "whole system overviews, integrative pictures, and near-holistic cognitive maps."[16]

Such language shows, we may say, that the "vision thing" that the first George Bush would soon prophesy for the "new world order" was farcical only when held to the standard of the high old way of transcendental vision. But transcendence is the wrong context. Bush's vision thing was profoundly white-collar, a mystification of bureaucratic and managerial vision. From the white-collar viewpoint, the only meaningful transcendence—the real new world order—was a vision thing equivalent to the *system thing*. Vision meant gaining a crystal-clear sense of the total system surrounding one's workaday world by means of reflective distance from that world (what Zuboff calls the "textualizing" or abstracting effect of computerization).[17] But rather than being transcendental in an older sense, such reflective distance was hard-wired or programmed into the system itself—particularly into the ability of the new machinery to let users "see" what they were doing in a wider context. By the time of the personal computer, IT would exercise its mandate for quickly and flexibly assembling information in synoptic frames of view by creating *actual* frames of view: graphical user interfaces that permitted elaborate sequences of technical procedures to be adjusted to equally complex series of technological actions simply by dropping the icon for one over the other. The ultimate vision thing, in other words, would soon after the mainframe age be the computer itself, the literal machine. Computers were the vision box that had the potential to make transcendence—or, in corporate-speak, supervision and oversight—routine. When the system is god, even as routine an act as opening a file list on a companywide system—let alone logging onto a global network—is big with revelation.

Informating thus meant building into automation the capacity metaphorically (and soon literally) to see the systemic whole of technological rationality—to glimpse not just individual files but the entirety of what C. Wright Mills named by synecdoche the "Enormous File" of the white-collar world.[18] At least in principle, therefore, computerization installed a whole culture of symbolic analysis and critical reflection in everyday work antithetical to the earlier industrial culture of deskilling and routinization. As attested by the quasi-managerial statements of lower level employees, moreover, such a culture was available not just to higher-ups but—again in principle—to anyone in the white-collar pyramid. It was as if the whole etiology of white-collar work, as Zuboff calls it, were suddenly reversed. Originally, Zuboff reminds us, an owner would likely not only manage his own company but also keep his own books. As the twentieth century progressed, however, the executive function successively extruded its routine functions into a new middle management cadre and later (at lower levels) an expanded clerical corps. By contrast, computerization seemed to elevate routine workers back to the level of the executive—to restore to middle managers the vision of the top executive and to clerks the purview of middle management. As one manager told Zuboff, "We are 'killing' the clerks as we know them now. The new clerk must be trained to make decisions, to deal with the information relevant to that function. It will mean a need for more educated people." And the clerks echoed: "We need a more global view so we can solve our problems when we have them. . . . The most important thing is to know my job in the context of all the tasks of the bank. How else can I exercise any judgment?"[19] Supervision and oversight, it seemed, were about to be distributed through the ranks.

The potential payoff of such distributed supervision was vast. As Zuboff says in a utopian tone, anticipating the postindustrial business prophecies we read in part 1 above:

> Imagine this scenario: Organizational leaders recognize the new forms of skill and knowledge needed to truly exploit the potential of an intelligent technology. They direct their resources toward creating a work force that can exercise critical judgment as it manages the surrounding machine systems. . . . This marks the beginning of new forms of mastery and provides an opportunity to imbue jobs with more comprehensive meaning. A new array of work tasks offer unprecedented opportunities for a wide range of employees to add value to products and services.[20]

Imagine, that is, an airline or bank, such as the pseudonymous "Global Bank Brazil" studied by Zuboff, that leverages its information systems into entirely new kinds of products designed to distribute the responsibility for data gathering and decision making throughout the firm—computerized airline reservation systems, complex new financial instruments, and so on.[21]

But such an imagination was only half the picture of computing in the mainframe age. There is a reason why the clearest examples of the visionary power of computing I offered above derive anachronistically from the last two decades of the twentieth century, when personal computers, graphical user interfaces, and networks came into their own to give a different view of the "visionary." As the matching, dystopian side of Zuboff's book emphasizes, business in the mainframe era repressed as often as it recognized the informating power of computers. Distributed supervision was repressed as a risk that had to be brought under the control of central oversight. After all, most major firms in this era still boasted massively tiered organizational trees mirroring the industrial age pattern of top-down management. Acknowledgment of this fact makes Zuboff's book—whose second and third parts include an extended critique of the "dominion of the smart machine"—the definitive social history of *mainframe* computing. The defining feature of the mainframe era was precisely that it insisted that IT conform to organizational patterns optimized for the earlier automation paradigm. As Zuboff puts it, technology was "treated narrowly in its automating function" to perpetuate the "logic of the industrial machine that, over the course of [the twentieth] century, has made it possible to rationalize work." "Like scientific management," therefore, "computer-based automation provide[d] a means for the managerial hierarchy to reproduce itself."[22] Or, as Beniger sums up, computerization in the mainframe era was the logical extrapolation of the apparatuses of "generalized control" that originally fostered the great bureaucratic organizations of industrial society.[23]

In particular, mainframe computing held information work hostage to industrial age standards of both hierarchical control and accountability. Hierarchical control in this context meant defending against the perceived threat of distributed supervision ("managers perceive workers who have information as a threat," one manager tells Zuboff) by implementing computerized versions of Taylorism: extracting the knowledge of workers, codifying or programming it, and assigning control of the resulting data structure to management information services (MIS) departments reporting to upper management. Computers were the gigantic successor to Taylor's "instruction cards" programmed by "functional foremen" in the "planning room."[24] The effect, Barbara Garson observes in her *Electronic Sweatshop* (1988), was "to centralize control and move decision making higher up in the organization." In complementary fashion, accountability meant compliance with such control and being subservient to ever stricter norms of productivity, efficiency, defect rate, and so on—no simple matter when, unlike lifting bricks or shuffling paper, much of the new work could not be meaningfully measured while in process. Who knew what work was or was not being done by a worker staring at a screen, ranging through the company database, or, even more worrisome, frantically typing in one of the early corporate bulletin board systems?[25]

Management had to know. So blinders were put on the "vision thing" to channel computerized work through new control and accounting mechanisms. Here the physical architecture of the computerized office is revealing.

As documented by Joan Greenbaum, who acquired an inside view of computing in the 1960s and 1970s as a programmer, analyst, project manager, and consultant, the layout of the computerized office increasingly converged on variations of the earlier, mid-century theme of "office as factory." Clerical pools outfitted with dumb terminals or word processors sat in windowless back-office interiors, while routine programmers and system operators (as opposed to higher level system designers) were similarly expected "to stay in the 'machine room' tending the computer."[26] Garson's experience as a data entry clerk in 1981 tallies with that of Greenbaum: "I entered the Office of the Future through a door that led into a windowless basement where dozens of women sat spaced apart, keying with three fingers of one hand."[27] The distinctive architecture of the office-as-factory achieved full expression in the 1980s, just in time to epitomize the dominance of the mainframe. This architecture was the partitioned cubicle, which physically inhibited gossiping, staring into space, looking around, and any other means of gaining contextual "overview."[28] With the cubicle, it was no longer even necessary to set aside inflexible, central spaces for clerical pools; each worker (now including those at a higher level) could be a modular pool of one.

How to control the vision thing, then? The Leffingwellian solution was obvious: wall it off.[29] Or rather, because the computer within the walls was the real outlet of vision, cubicles were just a supplementary control. Their ultimate task was not so much to prevent all interaction as to ensure that any interaction went through an even more effective control architecture housed in the machinery itself—"information architecture." In this regard, Mills was prescient when at mid-century he diagnosed the overall trend in office mechanization: "The new machines, especially the more complex and costly ones, require central control of offices previously scattered throughout the enterprise. This centralization . . . prompts more new divisions of labor."[30] The mainframing of the office fulfilled Mills's scenario to a Taylorist "T." By securing data on a single, highly expensive machine (or fleet of machines), quarantining programming from users, differentiating levels of access, implementing "vertical" software applications that regimented the relations between tasks and between work units, preferring "form based" input or "turnkey" programs that mandated a tightrope walk of fragmented procedures, and in general coding for routinized, deskilled "idiots" (as in the much touted desire of the time for "idiot proof" applications), mainframe computing acquired its reputation as the ultimate instrument of bureaucratic control—the last, best time-motion stopwatch. The mainframe took the "overview" implicit in informating and locked it in a black box.[31]

Witness, therefore, the following litany of complaints about the fragmenting, isolating, routinizing, and deskilling effects of computerization among Zuboff's interviewees:

> [A benefits analyst:] Now, once you hit that ENTER button, there is no way to check it, no way to stop it. It's gone and that's scary. Sometimes you hit the buttons, and then it stares you in the face for ten seconds and you suddenly say, "Oh no, what did I do?" but it's too late.
>
> [Another benefits analyst:] The computer system is supposed to know all the limitations, which is great because I no longer know them. I used to, but now I don't know half the things I used to. I feel that I have lost it—the computer knows more. I am pushing buttons. I'm not on top of things as I used to be.
>
> [A transfer assistant:] I have no decision making on that computer. It's been programmed to do this, and this, and this, and we are programmed to do the same thing. I don't want to be programmed. It does things automatically, and if you feel it's wrong, you really have no choice but to let it go that way on this formatted screen.
>
> [A bank manager:] You don't need to think, because the machine makes the calculations and performs the control. People don't understand the meaning of their jobs. It's a void. A clerk goes in and out of a task, but whoever made the system is the only one who knows what is in the box.
>
> [A manager:] There is a great opportunity for misinterpretation of data when everyone can see what is happening but their narrow perspective means that they can't tell why it is happening. Most people have a one-sided, functionally oriented sense when they look at the data. It gets worse in that the technology lets you look down that data tunnel at lightning speed—then the tunnel turns into a dot. You end up with one number, one reason, and you react to it.[32]

Just as vision was opening up ("everyone can see what is happening"), it disappeared down the data tunnel to the vanishing point.[33] As Zuboff sums up, "While the informating power of the technology resulted in a more comprehensive textualization of office work, it did not lead to an increase in the intellectual content of clerical tasks. Instead, these tasks embodied the uneasy requirement of sustained concentration and procedural reasoning without offering substantive content to naturally anchor the clerk's attention.

This is because, in both cases, managers and designers chose to emphasize the automating rather than the informating capacity of the new technology."[34]

Nor was control the only advantage of the black box philosophy of computing. Accountability was its other forte. Cubicle architecture was itself a step in this direction. Walls may not have ears, but thin, low-walled cubicle partitions do. Designed to isolate workers but also to leak privacy, cubicles assured that their occupants remained aware of a constant ambience of passive watchfulness. In principle, one could at any moment be monitored by a roaming supervisor. But it was information architecture that turned the art of background monitoring into method. To start with, monitoring functions were coded into application software itself, in such a way that, for example, "if/then" or "case switch" routines in a database input form watched over keystrokes and rejected anomalous entries ("the entry on line 25 is not a valid date; please reenter"). More remotely, watch guard programs running alongside applications monitored the speed or quantity of work through such mind-numbingly literal means as counting keystrokes per hour.[35] And the grand overseers of all such lower level watch guard functions were systemwide monitoring programs of the sort documented in Zuboff's chapter "The Information Panopticon."[36] Zuboff's exemplary case is the Overview System installed at one of the firms she studied. Designed to "read and record key instrument values throughout the plant (more than 2,500 pieces of data) every five seconds," the system exactly fulfilled Foucault's paradigm of the disciplinary panopticon. A manager tells Zuboff:

> We have disciplined and terminated people based on information from the Overview System. It provided information on incidents which showed that the individual user was not performing to basic knowledge requirements. By recording what happens to all of the instrumentation in a given part of the process on a five-second basis, we can see exactly what was done, what should have been done, and what was not done. Since you know the people that were there, you know what they did or did not perform. This becomes independently verifiable because the system knows it.[37]

Central supervision thus captured the "visionary" power of computing; Overview supervised "overview."[38] "The system knows it."

Mainframe computing, then, was the continuation of automation by other means. But now automation contained at its heart a crucial paradox—vision versus supervision—created in the incommensurability between the new informated technologies and old automated organizations. This paradox fostered not just organizational and procedural conflicts hampering the productivity of the new tools but also much felt tension in the workplace. When the transfer assistant quoted above complains, for example, "I don't want to be programmed. . . . if you feel it's wrong, you really have no choice," she clearly "feels" like saying to someone, "*You're* wrong." Or again, consider the tensions in human relations simmering within this comment by a telecommunications crew foreman: "I can't see my men in the field now, because I look through the computer. If someone has a problem, I can't work it out with him like I used to. Does he need some time off? The computer doesn't know his problems. It doesn't want to know if his kid just passed away or if his wife has problems. Maybe I need to lay off him for a while and then later I'll know I can count on him."[39]

How to neutralize or at least normalize such tense feelings inherent in what I have called the paradox of computing? How to do so, moreover, when the nature of information work made it increasingly undesirable for industry to rely on the safety valve that had once vented work frustrations into leisure? On the old assembly line, the contradiction that workers also thought (or daydreamed) for themselves was only marginally important so long as they were being properly supervised. Any tensions arising from the contradiction between unsupervised mentality and supervised work could be massaged away on weeknights or weekends in front of the TV by means of the industrial age predecessor of "vision"—fantasy. While fantasmatic leisure was crucial to the system of production because it created the conditions for the complementary system of consumption, it had no direct bearing on the routines of productivity. But in the information workplace, the equivalent contradiction between vision and supervision lay at the center of the new technologies and techniques of work. Any reliance on leisure to bleed off the felt tensions that resulted would thus miss the point that there was a core issue of productivity at stake: how best to use the new machinery to allow *both* unsupervised mentality (e.g., workers using the informated aspect of computing to fantasize "what if" revisions of a business model) and supervised work (workers merely plugging in data) to be productive. The great challenge for business in this era, in other words, was not just to palliate the tensions of mainframe computing work (thus leaving untouched the underlying structural contradiction between vision and supervision) but to capture those tensions *productively*—to feed them back into the system in order to optimize it, contradiction and all, for work. If there was to be a safety valve, it had to vent *inside.*

To understand how business created such a Klein bottle safety valve to "solve" the paradox of information, we must now return to the story of emotion at work. What I called the "felt tensions" stemming from the collision between vision and supervision were at the time framed within an even more general paradox of feeling in the workplace. Only by normalizing this general paradox of feeling could business develop its eventual strategy for meta-managing the paradox of information—that is, for managing the contradictions caused by its own management of information work. Making the general paradox of feeling normal, as we will see, was the means for making the underlying paradox of information (and its felt tensions) optimal.

What did it *feel* like to work in a corporation in the era dominated by mainframes? On the one hand, it felt cold. Workers in back offices and cubicles, certainly, remained in the same subject position as before—robots alienated not so much "man from man" as man from system. ("The system knows it," the manager says, as if the system were a person. "We are programmed," the transfer assistant says, as if she were the machine.) Just as the warm sociality of customary work had previously been chilled by the Leffingwellian redesign of the office, so now any chance to use the innate connectivity of information technology to foster new forms of social connectivity (e.g., wired forms of "gossip") was quickly barred. To start with, early computer work simply proscribed many forms of social expression and interaction. There was no way, for example, that database users could attach personal notes or commentary to a "form" or "field," except through that guerilla subversion of formatted information introduced in 1974—Post-It notes.[40] And the experiments that were attempted to accommodate individual or collective social expression often fell victim to heavyhanded supervision. Definitive on this score is Zuboff's account of a computer conferencing system tried at one drug firm in the late 1970s and early 1980s. Called DIALOG, the wildly popular system allowed employees to create discussion forums both for sharing of knowledge and for social purposes: "DIALOG gives me a better awareness of the company as a whole. Now I know it's not just me, myself, and I in a small group—I'm in a whole company." "It's like being tapped into the grapevine." But once managers and supervisors began to log on, the uncontrolled, free-ranging discussions quickly became controversial. To management, the breakdown in hierarchical control, normal circuits of communication, and professional tone encouraged by the system was "like a disease that must be walled off and contained." Or, as one sympathetic administrator ruefully reflected, "Management has not been able to appreciate how the electronic network improved morale. The company hierarchy does not have a very good understanding of the social and emotional needs of people to communicate." DIALOG was shut down, and a more tightly controlled system was put in its place in 1983.[41]

As literalized by the air conditioning in the "gleaming glass bubble" we saw Zuboff visit at the Piney Wood bleach plant, then, the feeling enforced by mainframe computing was indeed cold. And in the white-collar cubicles that are my main concern here, the feeling was perhaps coldest of all. Blue-collar workers in the factories Garson researched still "talked, joked, cursed and even yodeled when the supervisor was out of sight," but data entry clerks in the white-collar zone "rarely stopped to talk or stretch."[42] After all, they lived under the unblinking gaze of machines that far outmatched any mere corps of Taylorist "functional foremen" (humans, finally) in the consistency with which they ensured that work was ruled by a *lack* of feeling internalizing the worker's inability to relate to the system. It was not one's boss or manager who forbade laughing, weeping, cursing, shouting, or celebrating at work; it was a blank cubicle wall that simply shut off social interaction and, within that cubicle, database forms that accepted not a jot of humor, not a single expletive. It was the "system" or "computer," one felt (and only at a further remove one's boss), that treated one like an idiot. Moreover, such mind-numbing routine now stretched in various degrees all the way up the white-collar ladder to the higher ranks whose members had previously left the typing to others but now themselves had to set hands to keyboard to learn how to be idiots.[43] Supervisors were no longer foremen coddling, cheering, and tongue-lashing their "gang"; now they were managers with "spans of control" overseen through an Overview System sailing as high and serene—and effectively outside personal control—as some military spy satellite in space.

Just one especially poignant illustration of "cubicle cold" will punctuate the case. At one point in her research, Zuboff asked a number of women benefits analysts to draw themselves before and after computerization. The resulting crude stick drawings made a mockery of the vision thing promised with ever greater graphical proficiency by computing. Yet for that very reason they were all the more compelling. "Before" appears as a hubbub of workers so literally warm in their customary ethos that a bright sun hangs in the background of two of the sketches—almost as if office work could recuperate the spirit of the ancestral agrarian fields where grain and not data was the harvest. "After," by contrast, is cold, waste, and solitary. It is a Plato's cave of workers bereft of sunlight, wearing blinders, literally chained to their desks, and forbidden any interaction except through the cool luminescence—that other sun—of the database displayed on their screens.[44] Perhaps the best caption for these drawings might be a general observation about computing that Zuboff makes elsewhere in her book: "Absorption, immediacy, and organic responsiveness are superseded by distance, coolness, and remoteness."[45]

But now for the paradox of feeling in the mainframe age. The back office and cubicle were not the entirety of the office. There was the whole other hemisphere of the front office, salesroom, public relations department, and so on, which was expanding at the time to constitute an ever vaster "boundary spanning" zone linking firms to customers, clients, suppliers, trade groups, government agencies, courts, and media.[46] Over this latter hemisphere, it seemed, the sun shone so fully that work didn't just grow warm. Emotionally, it overheated.

The rise of information work cannot be understood in isolation from the larger economic phenomenon to which it was intimately linked (and which by the late 1980s accounted for 84 percent of the total U.S. investment in IT): the rise of service work.[47] In the United States, the ratio of service jobs to industry jobs rose between 1920 and 1970 from 1.1:1 to 2.0:1, and then *accelerated* in growth in the next twenty years—precisely in the heyday of computing—to 3.0:1.[48] A 1990 study estimated that "services account for approximately three-fourths of the gross [U.S.] national product and nine out of every ten jobs the economy creates." Projections for 1990–2005, based on data from the U.S. Bureau of Labor Statistics, extrapolate a similar curve.[49] As Amy S. Wharton points out, much of this growth occurred not only in the service sector as commonly defined ("the finance, insurance, and real estate industries; wholesale and retail trade; transportation, utilities, and communications; business, professional and personal services; and government") but also throughout business, wherever agents were needed to negotiate the widening public sphere of commerce. Indeed, it is more accurate to speak of the "service elements" in business than of the "service sector."[50]

In retrospect, the economic rationale for the parallel booms in information work and service work is clear. Neither was cause or effect of the other, because both were part of a single development. In an age when communications media increasingly supplanted physical transportation in carrying the bulk of exchanges between firms and their public, suppliers, contractors, and regulatory agencies, business necessarily had to be conducted in a compound idiom of information and service we might call *informed service*. That is, information—its timely collection, storage, processing, management, and delivery, whether conducted by a firm itself or outsourced to specialized accounting, legal, public relations, courier, and other service firms—did not simply facilitate service: it now *was* the ultimate service. For a firm to make effective contact across its boundary with other sectors or organizations, armies of salesmen were no longer adequate or, in many cases, even possible. Instead, firms had to work the telephone, the TV, the newspaper, and so on, which meant cumulatively that they had to transmit an ever larger proportion of both their image and products as information. Competitive advantage was to be gained in assigning larger numbers of workers—in a long chain of production stretching from the back office to the so-called front line—to the simultaneous creation and "impression management" of information.[51] In this light, both information workers keying data and front-line workers dealing with the public must be seen to form a continuum of service through which information flowed (a convergence in office work implicitly recognized by the U.S. Bureau of Labor Statistics when it recategorized all white-collar workers as "information workers" in the 1980s).[52] The back office crunched the numbers, input the data, and processed the text, but it did so as part of an overall work process whose product was service. Reciprocally, front-line workers looked the public in the eye, but increasingly did so with their other eye on a computer screen (e.g., at an airline ticket counter). Positioned near a dumb terminal at the outer perimeter of a firm's infosphere, in other words, front-line workers were the public inter-"face" of back-office computerization. They were living dumb terminals.

But precisely because the economic rationale for the convergence of information work and service work is so clear, the *emotional* rationale is puzzling. Information work, though a form of service, was supposed to be cold, efficient service, while service work—as in the case of the ticket agent who looks *up* from a computer screen to tell the customer how many window seats are available—was supposed to be warm ("service with a smile"). To take the full measure of this warmth, and thus of the severity of the overall paradox in workplace feeling, I draw once more on the emotion researchers I cited in chapter 2, this time with special attention to Arlie Russell Hochschild.

Mills had earlier observed about the rise of service work: "When white-collar people get jobs, they sell not only their time and energy but their personalities as well. They sell by the week or month their smiles and their kindly gestures." Inspired by Mills on the "personality market," and herself subsequently inspiring a whole branch of socioeconomic research, Hochschild set out in her *Managed Heart* (1983) to take a close look at the nature of commercialized emotional warmth.[53] The case she builds around her central empirical example, flight attendants, is compelling in both its breadth and its depth.[54] A major consequence of the rise of service work, Hochschild argues broadly, is that by 1970 (the date of the last census available to her) an unprecedented 38 percent of U.S. workers, and 55 percent of all women workers, held jobs that placed them in direct contact with the public.[55] Some were in manual service or low-level retail sales, but the majority were in the middle-class domain, including not only white-collar jobs at all levels (lawyer, personnel director, public relations or publicity writer, salesman, bank teller, clerical supervisor, claims adjuster or examiner, receptionist, secretary, and so on) but also what might be called "clean collar" pseudo-professional positions, such as flight attendants.[56]

What all these jobs required was some element of "emotional labor" governed by increasingly plentiful and rigorous feeling rules coded into hiring, training, promotion, disciplinary, and other personnel practices, together with their attendant discourses. Women interviewing for flight attendant jobs, for example, learned from the 1979–80 *Airline Guide to Stewardess and Steward Careers* to show a "modest but friendly smile," "be generally alert, attentive, not overly aggressive, but not reticent either," and maintain moderate eye contact demonstrating "sincerity and confidence." Trainees were then subjected to an even more exhaustive regimen of rules: "Really work on your smiles," "Your smile is your biggest asset—use it," and so on. As Anat Rafaeli and Robert I. Sutton later put it in studying similar emotion work industries, "friendliness was operationalized."[57] Such operationalization also affected positions less directly involved in front-line service. Studying a high-tech engineering firm, for example, John Van Maanen and Gideon Kunda noted the ubiquity of internal documents, such as "management reports, memos, summaries of popular books, . . . reprints of . . . articles dealing with [the firm]; slide presentations and talks, . . . and videorecordings of top managers," designed to boost the particular mix of gung-ho aggressiveness and "friendly" sociality of the ideal company worker.[58] Such behavioral scripts were reinforced in workshop sessions that prompted employees to volunteer the answers "friendly" and "amicable," when asked, "What are the characteristics of people at [this firm]?"[59]

The overall result was that the service worker became indentured to the diametrical opposite of what I earlier called zero-degree, or privative, emotional labor management. In every sense, emotional labor was *positive*. Whether one was a flight attendant, secretary, salesman, or public relations director, one engaged to some degree in "voice-to-voice or face-to-face delivery of service" that was "socially engineered and thoroughly organized from the top" to display all shades of bright, perky, and cheerful.[60] Positive rather than merely systemic or technical "friendship" became the order of the day.

But Hochschild's argument for depth is even more compelling. Her key insight is that although the control of emotional labor may have been instituted from the top, there was a disturbing way in which the line between external and internal emotional governance—between the emotion one was told to "display" and the emotion one "really felt"—tended to blur. Control was also self-control: it went all the way down into identity. And it was from this obscure overlap between "displayed" and "real" feeling that emotion work reaped all its profit (surplus emotional value, it might be called), as well as sowed all its potential hurt. In particular, what Hochschild notices about the myriad feeling rules controlling a worker's public persona is how they systematically hybridized the "displayed" and "felt" by instrumentalizing the latter to produce the former. Drawing an analogy to deep acting method in theater, Hochschild observes that the more programmatic kinds of emotion training commonly asked workers to produce a desired emotional display by "acting out" scenarios that mentally transported the workplace into domestic or other interior zones of personal identity. Treat a hostile airline passenger, for example, as if he were a child in your household. Or again, act "as if the cabin is your home," or "as if this unruly passenger has a traumatic past."[61] The same strategy applied in industries where the service work was less obviously customer-related. In the high-tech firm studied by Van Maanen and Kunda, internal company documents scripted display rules as *character* rules validated by analogy to domestic or personal life: "Do not harbor grudges. . . . Form the habit of considering the feelings of others," "WE ARE ALL ONE FAMILY."[62]

In short, as Hochschild puts it, the external control of emotional labor required an interiorization of control that was no less than a personal "transmutation." When a self was transmuted in this way, displayed and felt emotion came together in a strained consonance that held in check—if barely—the risk of "emotive dissonance":

> Display is what is sold, but over the long run display comes to assume a certain relation to feeling. As enlightened management realizes, a separation of display and feeling is hard to keep up over long periods. A principle of *emotive dissonance*, analogous to the principle of cognitive dissonance, is at work. Maintaining a difference between feeling and feigning over the long run leads to strain. We try to reduce this strain by pulling the two closer together either by changing what we feel or by changing what we feign. When display is required by the job, it is usually feeling that has to change; and when conditions estrange us from our face, they sometimes estrange us from feeling as well.[63]

Yet, as Hochschild's remark about self-estrangement suggests, and as her examples of bitter, cynical, "gone robot," or otherwise burned-out flight attendants reinforce, emotive dissonance was never very far under the surface. As she diagnoses in a section called "Emotional Labor and the Redefined Self": "When a flight attendant feels that her smile is 'not an indication of how she really feels,' or when she feels that her deep or surface acting is not meaningful, it is a sign that she is straining to disguise the failure of a more general transmutation."[64]

We remember that in the earlier era of systemic friendship the white-collar worker was expected to "harmonize" or "adjust" by internalizing technological rationality in a neutral personality identical with professionalism and efficiency. Now that service was part of the job description, *positive* friendship had to be internalized as well. Indeed, when internalization became "transmutation," and "harmony" or "adjustment" became "consonance," then there was no longer any difference between systemic and personal friendliness. Workers were no longer people; they were friendship systems. While looking down at the computer monitor, therefore, an airline ticket agent may exhibit "Fordization of the face," but the moment he or she looks up and asks, "Can I help you?" the sun comes out, the day warms, and the smile blossoms so sincerely that no one on the scene—not the worker, not the customer, not the company—even feels the need to reconcile the contradiction between the "Fordized face" of an instant ago and the radiance now. Both are simply part of the same system of friendly and informed service. Alienation is everywhere, but it is no longer the Marxist drama of antagonism; it is instead a drama ("deep act") of friendship. One may be feeling glum or doomed, but there is no escape. Friendliness, like fate, is everywhere.

The emotional rationale that explains the convergence of information and service work now begins to come clear. The distance between the cold work of the back-office information worker and the warm work of the front-line service staff was simply the organizational-level version of the distance between the two faces of the individual front-line worker sketched above. If it made sense that the cold and warm sides of the individual service employee were really consonant within a single system of friendship, then it could also make sense that the cold and warm divisions of the company as a whole were consonant within a similarly total, unitary system of friendship. What kind of sense did it make? The answer is dramatic sense—that is, pretense—in the manner of deep acting. The answer, in other words, was a *fantasy* that was not just integrated within the system but constitutive of the integrity of that system. Where deep acting or fantasy enforced the consonance of contradictory emotions in the individual worker, a *corporate-wide* version of such deep acting enforced the similar consonance of back-office information work and front-line service work. The name that business gave this total, unitary fantasy of friendship was "corporate culture," which came into vogue in the early 1980s through such business bestsellers as Tom Peters and Robert Waterman's *In Search of Excellence,* Terrence E. Deal and Allan A. Kennedy's *Corporate Cultures,* and William G. Ouchi's *Theory Z* (together with coverage of the corporate culture concept in business and general journalism) precisely at the peak of the dominance of the mainframe paradigm.[65]

All professional and efficient service, and thus the very identity of contemporary corporations, would henceforth be performed in the name of corporate culture. From now on, corporate culture caught up both the encubicled data processer and the front-office people processer in the same, collective rehearsal of a single culture of informed service. Nor need we look far for the methods of "acting" that rehearsed workers in their corporate culture: the many company retreats, parties, songs, games, and other rituals enactive of such culture were the generalization of the enormously comprehensive and fine-grained regime of deep-acted feeling rules Hochschild saw in the airline industry. As Michael Rosen terms it in his ethnographic research into organizations, such rituals were part of the "dramatic performance" and "social dramas" by which companies "dramatically enact" a "*communitas*—a temporary levelling of structured relations." Dramaturgy in corporate culture, that is, managed the divided and distinctly *un*leveled corporate "structured relations" of the times by acting as if those relations were their opposite. The means of such deep acting was precisely the management of the heart. As Harrison M. Trice observes, the cultures of work groups and occupations are by their nature "emotionally charged" and "shot through with strong emotions." Corporate culture, as Van Maanen and Kunda put it, "is structured to channel, mold, enhance, sustain, challenge and otherwise influence the feeling of organizational members—toward the organization itself, others in the organization, customers of the organization, and, crucially, themselves."[66]

The academic literature that has now sprung up in the wake of the "changing our culture" vogue to scrutinize how corporate culture thus structures feeling includes works not only by those already mentioned (Kunda, Rosen, Trice, Van Maanen) but also by Mats Alvesson, Per Olof Berg, Joanne Martin, Christopher Newfield, Lee Sproull and Sara Kiesler, and many others.[67] Especially useful in our context is Joanne Martin's influential *Cultures in Organizations* (1992), which surveys the scholarship to attempt a "metatheoretical" assessment of three main approaches to corporate culture.[68] In what Martin calls "integration" theory, an organization's work practices, symbolic forms (shared stories, rituals, jargon, dress), and "content themes" (e.g., "egalitarianism," "innovation") are presumed to cohere paradigmatically in a unitary corporate culture. For better or for worse, a good firm has *one* structure of feeling. (Corporate cultures are "loose" constructs that turn out to be "remarkably tight" in their "rigidly shared values," Peters and Waterman originally argued in their *In Search of Excellence*.)[69] By contrast, "differentiation" theory studies the same material but presumes that any unitary corporate culture is only apparent. Corporate culture instead always breaks down into dominant and subordinate fractional cultures locked in oppositional contest: management versus the line, engineering versus marketing, male versus female, white versus black, and so on.[70] Yet differentiationism is also the uncanny double of integrationism, Martin points out, because it requires that each subgroup be treated as internally unified. All ambiguity and contradiction, as she puts it, is "channeled outside" the group. Finally, "fragmentation" theory (which Martin formulates with the aid of deconstructive and postmodern theory) asserts that neither the whole nor the fractions of corporate culture are stable constructs. In lived practice, each is fraught with "ambiguities" as well as, most recalcitrantly, "paradoxes" caused by structural contradictions (e.g., a corporate culture that believes innovation means radically changing the system but also maintains that everyone should work for change *within* the system). An adequate conception of corporate culture, Martin argues, requires blending all three perspectives.

I will return to paradox in corporate culture later. For now, we can focus on the integrationist view in tandem with its differentiationist critique. An especially instructive instance of officially integral corporate culture is offered by Van Maanen and Kunda, detailing the repertory of means available to an enterprise to build the appearance of an overwhelmingly unified culture.[71] Or perhaps *enterprise* is ultimately too businesslike a term for the kind of emotionally demanding and all-pervasive culture making that Van Maanen and Kunda find in their recital of companywide retreats, beach trips, communal sports, and so on. We are really on the scene of what they call a "corporate anthropology" of "ritual, myth, and ceremony." When, for example, a participant in a mass company beach excursion (complete with games, songs, and cheers) explains to a curious passerby, "Oh, we're just working on our culture," we sense that we are witnessing the performance of a ritual so mythic that the word *culture* is euphemistic. Back-office and front-office employees are deep-acting the unity of a "cult."[72]

So was formed what I have called the Klein bottle safety valve. If tensions and dissonances were brought on by underlying contradictions in the nature of work, then the optimal solution was no longer to vent them into fantasmatic consumption outside the scene of production. Instead, as witnessed not just in beer- or Chardonnay-guzzling company beach fests but most spectacularly in the "T-group" sensitivity training sessions that flourished for a time in business after the 1960s, the solution was to bring crucial elements of the "hot" fantasy life of leisure *into* the "cold" scene of production itself. There any resulting tensions and dissonances could be variously shouted, wept, cheered, or back-slapped away in retreats from work that in the name of "building our culture" normalized the contradiction of warm/cold efficiency experienced every day *in* work.[73] Such corporate culture may have been a "mere" act, of course, for who has not at times mentally opted out of the particular act one is performing as a pledge of allegiance? But the act was so complete in its hold over all the days and, increasingly, nights of the worker that it is also possible to say that the act was as deep as life. After all, as work increasingly spilled over into leisure hours, and as the larger or more innovative companies began to arrange for a greater part of the personal lives of their employees (culminating in the total-care philosophies of some of today's Silicon Valley firms, with their gyms, social clubs, dance clubs, child care, on-site dental care, and on-site hair salons), it would have been cruel to say to the employee, "Get a life." If the corporate culture of warm/cold service that the worker inhabited was just an act, then *life* could just as well be said to be an act. What alternative life could be pledged outside of work—politics? charity work? hobbies? music? sports? religion? even the family?—that was not just as open to the charge of being all a contradictory act, all an undecidable charade of the displayed and the felt?

We can now glimpse the strategy of meta-management that business needed to "solve" its specific paradox of vision versus supervision within information work. If such a paradox of self-managed versus managed work was caused by incommensurability between the new machines and the old hierarchies, then corporate cultures designed to regulate the larger paradox of warm/cold service work were the perfect way to *meta*-manage this contradiction brought on by hierarchical management itself. By contrast, we can take this snapshot from the early history of IBM corporate culture under Thomas J. Watson, Sr. in the 1930s. Likening Watson's methods to that of John H. Patterson, the founder of the National Cash Register company and Watson's early mentor, *Fortune* Magazine reported in 1932:

> It was Patterson's invariable habit to wear a white stiff collar and vest; the habit is rigorously observed at I.B.M. It was Patterson's custom to mount on easels gigantic scratch-pads the size of newspapers and on them to record, with a fine dramatic flourish, his nuggets of wisdom; nearly every executive office at I.B.M. has one. It was Patterson's custom to expect all em-

> ployees to think, act, dress, and even eat the way he did; all ambitious I.B.M. folk respect many of Mr. Watson's predilections.[74]

Early information age corporate culture in this mold was wholly an extension of top-down management. But by the end of the mainframe era, as Van Maanen and Kunda put it, corporate culture was a "control device" that supplemented traditional means of organizational control with an entirely different style of "culture control." It was a "meta"-control that appeared to float free from hierarchy and well up from everywhere or nowhere. Thus the tech firm they studied, for example, spoke of itself as controlled through "matrix management," "multiple dotted lines," "dense networks," and "structural ambiguity."[75] "In essence," Van Maanen and Kunda conclude, "we regard conscious managerial attempts to build, sustain, and elaborate culture in organizations as a relatively subtle yet powerful form of organizational control. It is subtle because culture is typically regarded as something people can do little about, something that flows from the ordinary problems people face and the characteristic ways they solve them."[76] Corporate culture in the mainframe age managed the unsupervised mentality or vision of information work ("fantasy") by making it part of a greater enterprise of service work that was now all the more productive because it appeared to be self-managed as part of a collective cultural fantasy.

Corporate culture, in consequence, increasingly managed information work through Disney-like effects of pure, shared fantasy: ambience, texture, milieu—in a word, *style*. The consumer's experience of style that had once fantasized the outsider's challenge to technological rationality was now incorporated within work itself as a corporate fantasy of the insider's ultimate *submission* to technological rationality: *Here at* [substitute for Disney the name of any contemporary major corporation or start-up] *we are driven to be competitive not because of top-down orders but because we all share the same beliefs and myths, repeat the same "war stories," dress the same, talk the same, have the same established/trustworthy or young/punk "attitude," and take our orders from the same total style of work.* "Hoods" in high school during the 1950s and 1960s used to say that whatever had style was "boss." Now corporate style was boss in every sense of the term. The boss was no longer just at the top of the hierarchy but distributed everywhere throughout the system and, ultimately, within the deep-acted self as a self-managed "style" of work. No matter the emotional tensions that still arose from underlying paradoxes in the system; it was boss (in the company spirit) to act as if contradictory information systems of vision and supervision were indeed one *loosely* coherent system—and, for that matter, best system—called "style." By normalizing contradictions of feeling through *stylized* feeling, corporate culture in the mainframe era optimized contradictions in production. It was now the Big Unity of corporate culture (as opposed to the Big Split between work and leisure) that "contained" paradox within the scene of combined work/lifestyle that was the totality of production. Paradox remained but could not "gesture" to an alternative ethos.

There is no better dramatization of such a view of corporate culture than the following passage from Robert Howard's "High Technology and the Reenchantment of the Work Place" (1984):

> [E]ven the archetypal beer bust . . . is not some trivial social occasion. . . . The purpose of such rituals, says [one manager], "is to have everyone know where things are going, from the top level people all the way to little Suzy down there stuffing the printed circuit board. She gets information on what the product is for. She has an idea of the whole board."
>
> It is a striking image. Suzy may be "down there" on the assembly line. Nevertheless, she gets some of the same information as the people at the top. She is able to transcend the concrete limits of her particular job and become linked to the purpose of the corporation as a whole. Through this broad-based structure of motivation, she is "locked in mentally" to the future of the company. It is a kind of parallel production process whose purpose is to manufacture not products, but attitudes and expectations.[77]

Little Suzy may just be processing a board with integrated circuits (or processing an order for one, working in inventory control, or entering account data), but she is herself a chip in the larger integrated circuit called corporate culture that "locks" her in "mentally." Soldered into corporate culture, she feels exactly as if she were "transcending" her place in the scheme of things to gain a vision of the whole. Her vision, mentality, or fantasy is plugged into the total vision thing.

Or such, anyway, is the integrationist approach to corporate culture (viewed with the critical irony of the "differentiation" perspective). But we have not yet accounted for all the feelings of work. Even amid the consonance that was the grand elevator music of corporate culture, after all, we hear the occasional "Ford whisper" of dissonance. We notice, for example, the flight attendants in Hochschild's study whose failed "transmutation" led to individual lapses from the company spirit ("going into robot"), rebellion against management (e.g., "shoe-ins" to protest dress policy), and even aggression against passengers ("Well, that did it. . . . This lady asked for one more Bloody Mary. I fixed the drink, put it on a tray, and when I got to her seat, my toe somehow found a piece of carpet and I tripped—and that Bloody Mary hit that white pants suit!"). Similarly, we notice in Van Maanen and Kunda's study the workers who opted out of corporate culture through "rituals of resistance," "subversive practices," "pockets of resistance," and "individual distancing." In the middle of a lecture on company morale at the "Tech" firm, for example, an engineer burst out angrily, "What is all this talk about Tech? I don't see any Tech. What is this we? I haven't met a Tech."[78]

If we now return to the paradigmatic work locales sketched by Zuboff, where we began—the "gleaming glass bubble" of a factory computer control room and the white-collar office with its partitioned cubicles—we can unfreeze them in time to unlock their real drama. The computer room at the bleach plant, it turns out, communicated with the rest of the building through a set of smoothly automated, sliding glass doors arranged like an airlock. One was supposed to proceed through the first set of doors and wait for them to close before the next set opened. But Zuboff recounts:

> This is not what most men do when they move from the control room out into the bleach plant. They step through the inner door, but they do not wait for that door to seal behind them before opening the second door. Instead, they force their fingertips through the rubber seal down the middle of the outer door and, with a mighty heft of their shoulders, pry open the seam and wrench the door apart. . . . Three years after the construction of the sleek, glittering glass bubble, the outer door no longer closes tightly. A gap of several inches, running down the center between the two panels of glass, looks like a battle wound.

Zuboff comments, "I hear not only a simple impatience and frustration but also something deeper . . . a boyish energy that wants to break free; a subtle rebellion against the preprogrammed design." And similarly, the office she depicts is the scene of another subtle rebellion:

> One afternoon, after several weeks of participant observation and discussions with clerks and supervisors, I was returning to the office from a lunch with a group of employees when two of them beckoned me over to their desks, indicating that they had something to show me. They seated themselves at their workstations on either side of a tall gray partition. Then they pointed out a small rupture in the orderly, high-tech appearance of their work space: the metal seam in the partition that separated their desks had been pried open.

> With the look of mischievous co-conspirators, they confided that they had inflicted this surgery upon the wall between them.

The clerks, Zuboff generalizes, were rebelling against the Leffingwellian redesign of the office.[79]

What we have failed to account for so far in our story of the mainframe age, in short, is that people are not just robots (even when "going robot"), and that in addition to corporate culture there was always also the potential for cultures of subversion, dissent, and "rebellion." In such incidents as those narrated by Zuboff, the original Marxian drama or antagonism of "man from man" seems to wait just under the surface of the corporate deep act, looking for a chance to pry apart the doors and partitions of the Klein bottle workplace to dramatize the undergirding split between vision and supervision. There *were* "gestures" from within the paradox of information work toward some unspoken, alternative ethos ("two of them beckoned me over to their desks," Zuboff says, "indicating that they had something to show me"). The differentiation theory of corporate cultures would seek to channel such gestures into sharply delineated, self-aware fractional corporate cultures contesting for dominance (e.g., workers versus management). But we can challenge a too easy contestatory model (and prepare for consideration of the "paradoxical" option) by observing that gestures of individual resistance often did not cohere in any meaningful counter-ethos within the organization—especially given that union culture, the traditional organized alternative to corporate culture, did not have a strong presence in white-collar occupations. The thousand and one acts of daily resistance (what Michel de Certeau calls "tactics") were thus fleeting and individual.[80] The doors and partitions closed again.

But even then, perhaps, not quite. There were two significant exceptions to the rule of "contained" subversion—two segments of corporate culture that by the end of the mainframe age had begun to differentiate themselves so boldly and persuasively from normal business culture that, ironically, they soon *became* the dominant unified business culture (a unity of differentiation, we might say, anticipating the Silicon Valley "monoculture of diversity" I mentioned earlier). The collaboration of these two exceptional cultures determined the next major paradigm of work—networking.

But before naming these dual exceptions, let us catch up with the story of cool. In discussing corporate culture in the mainframe age, I earlier asked rhetorically what alternative life could be found outside of work that was not itself open to the charge of being all an act, all a charade. White-collar workers did not have an answer, but their children thought they did. Outside

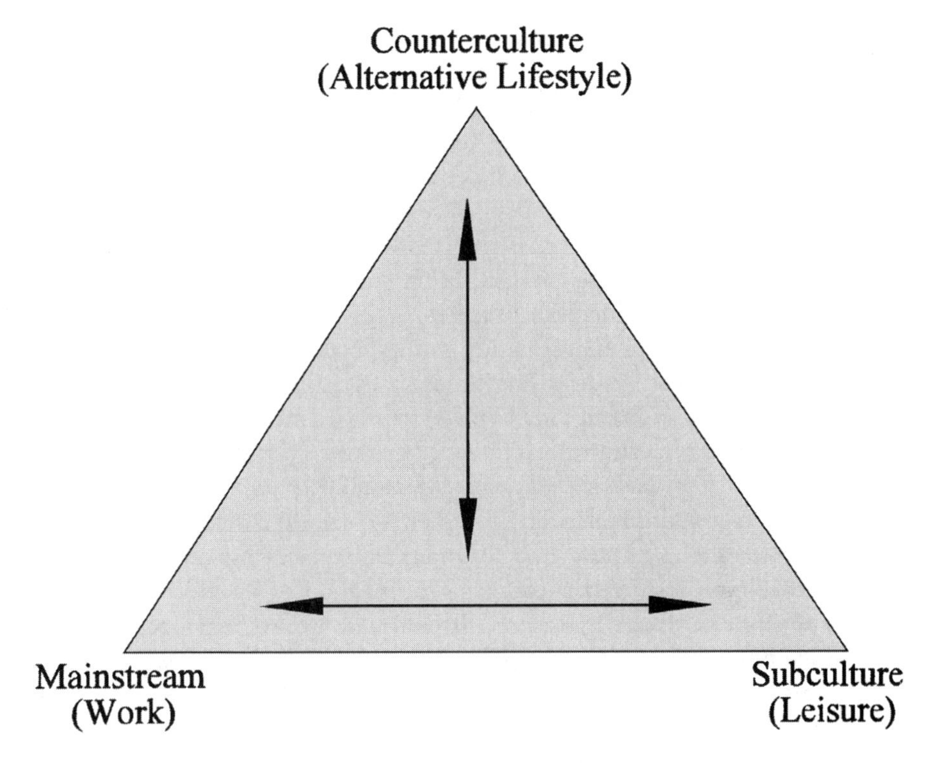

Figure 3.1 Relation of counterculture to mainstream culture and subculture

the workplace in this era there arose a whole rebellion of alternative life. I refer, of course, to youth counterculture, whose gesture toward an alternative lifestyle occurred in the name of a new cool—"hip." What was cool in the age of hip when "technocracy's children," as Roszak called them in 1969, "elected to drop out of the privileged middle class"? when, as Todd Gitlin later recalled, "it was necessary for young people to act unmanageable in order to sidestep the management"?[81]

To frame an answer for our purposes, we can start by assessing the position of countercultural cool relative to the cool I located in chapter 2 in the interplay of mainstream and subculture, work and leisure. If we imagine the earlier relation of "reciprocal appropriation" to be our baseline, then cool from the Beats in the 1950s through the youth movements of the 1960s and 1970s might first of all be imagined as a triangular formation built upon that baseline. Counterculture, that is, inserted itself into the cultural field as a third point of reference above, below, or to one side of the axis between the mainstream and subculture. (See figure 3.1.) On the baseline itself, white-collar culture and subculture continued their complexly symbiotic relation. Those who were professionally cold at work consumed the perceived heat of subculture in their play. Reciprocally—whether the instance was black music from jazz to rap (see Ted Vincent, *Keep Cool*), black youth "masculinity" (Richard Majors and Janet Mancini Billson, *Cool Pose*), or, in an English setting, "motor bike boy" society (Paul E. Willis, *Profane Culture*)—subculture was hot only because it borrowed the look and feel of technological rationality to style itself as cool.[82] Now counterculture added its own style (a whole alternative lifestyle) to the mix, and in doing so extended the dynamics of reciprocal appropriation in the direction of the equally complex phenomenon that Thomas Frank, in his astute and original study of the countercultural moment, *The Conquest of Cool,* terms *cooptation.*[83]

The critical question is where exactly counterculture positioned itself in relation to the baseline of mainstream/subculture and work/leisure. ("Above, below, or to one side," I vaguely said above.) This issue is different from that of the position that counterculture took toward either pole of the baseline considered separately. The relation of disaffiliation between counterculture youth and the adult workaday world, for example, has received more than its fair share of commentary—whether from the perspective of the media, sympathetic participant observers (e.g., Roszak and Gitlin), or political and religious conservatives.[84] After the "discovery of poverty" in the 1960s, the contrasting affiliation between counterculture youth and subculture—as expressed through empathy and/or activism for "the people"—was equally complex if less thoroughly explored.[85] "Although virtually all of them were Caucasian," Jane and Michael Stern recount in their *Sixties People,* "hippies relished their romantic self-image as nouveau red men, living in harmony with the universe, fighting against the white man's perverted society of pollution, war, and greed."[86] The difficulty in linking such counterculture to actual subculture, however, is instanced in such works as Willis's *Profane Culture* (1978), which simply segregates its analysis of "hippies" and "motor bike boys" into two parts.[87] Similarly, although John Clarke, Stuart Hall, Tony Jefferson, and Brian Roberts's "Subcultures, Cultures and Class" (in the Birmingham Contemporary Cultural Studies Centre's important *Resistance through Rituals* [1976]) attempts to relate counterculture to subculture, the result is the least theoretically cogent part of the essay ("there are some problems in deciding whether we can speak of *middle*-class subcultures").[88]

But the important issue in our context is the relation of counterculture to the entire *baseline* relating mainstream to subculture and work to leisure —that is, the relation of counterculture to a *relation,* and therefore to the whole previous dialectic of the insider's versus outsider's accommodation to technological rationality. Since such a formulation may appear overly abstract, it is best to illustrate it by transforming my abstract triangular geometry into the imaginary cartography that the counterculture itself used to triangulate its position relative to all the rest of culture (a cartography that added alternative mental landscapes to the mythopoetic workscapes we previously encountered). Where did counterculture place itself on the map relative to the entire "scene" of work/leisure and mainstream/subculture? On the one hand, counterculture was cool because it was far out. One's head was situated mythically on the road, in the Far East, or (through drugs or meditation) in some interior Zen garden, where it was as if the barest suggestion of waves raked into pebbles was enough to launch a trip to far seas. These fabled locales of counterculture are now clichés to the point of banality, though the particular history of countercultural experiments in Eastern philosophy and dress, drug use, nontraditional living arrangements, nontraditional cuisine, and so on reveals the shrewd, local intelligence necessary to any attempt to rebel through sheer lifestyle. Such lifestyle was counterculture's gesture toward an alternative ethos. And if ever there was a paradox that could gesture in this way, what Roszak called the countercultural "Zen" of "paradox and randomness" seemed the Way.[89] Lifestyle on the road, in the East, or in a bare loft somewhere was about as far from the baseline of work/leisure—on both the points of work and leisure—as could be imagined.

Yet on the other hand, even the far out disliked being so far off the map as to be beyond reach, figuratively or literally, of a plug. We notice, therefore, that alternative lifestyle required *technique* and often also *technology*, even when some of the technology involved was antimodern (organic rather than machinic, for example). To be "on the road" required a car, even if only the "people's" great anticar of the age, the Volkswagen Bug. Indeed, transport to each of the fabled mythopoetic landscapes of counterculture depended on some perverse variation of technological rationality. It is telling, for example, that immediately after speaking of counterculture's "radical rejection of science and technological values," Roszak unflinchingly describes the alternative traditions of Hinduism and Buddhism as blessed by "a vocabulary of marvelous discrimination for speaking of the non-intellective consciousness—as well as a number of techniques for tapping its contents."[90] Meditation, yoga, batik, t'ai chi, and other Eastern methods, in other words, were many things, but one of the things they surely were as received by the West was technique—a way of working some vast, cosmic phantom keyboard. And in other far-out zones of counterculture where the machinery was not so phantom, the technical addiction was even more pronounced. Witness, for example, the technology and technique of drug use as documented by Willis, where "ritual" at the opening becomes "experiment" at the end, as if the uncanny double of counterculture were an industrial pharmaceutical laboratory:

> A whole reverential ritual had developed around different kinds of drug use, and these produced their own material practices and material effects. . . . The serious head smoker inhaled very much more deeply, and kept the smoke in the lungs very much longer, than did the casual user. He was also very much more likely to control the amount of the drug . . . at a high consistent level than was the casual smoker. There was also much more care taken to make the circumstances of smoking more conducive to bringing out the drug's full effect on consciousness and perception. In the case of acid, trips would be prepared and planned for, and supportive contexts organized, in a way quite different from the more arbitrary conditions of innocent experimentation.[91]

As Roszak comments in his later book, *The Cult of Information,* "the premier dope of the era was LSD, which was itself a technology."[92] And above all, as Roszak also recognizes with cutting critique, the music that counterculture relied on to mediate its entire intake of subculture and outtake of self-expression became an orgy of technology:

> The youthful audience that crowded the clubs may have looked grungy in the extreme; it may have come brandishing emblems of disaffiliation from industrial society. But that audience wanted its music explosively amplified and expertly modulated by the best means available. . . . Therefore, the music needed machines. And as time went on over the next decade, the music would need more and more machines as the aesthetic tastes of the period summoned a whole new recording technology into existence that began to replace the performers and their instruments with sound engineering of the most complex kind, and finally with various kinds of digital enhancement. . . . The technicians were steadily taking control of the music. They would make sure it came out sounding *professionally* unwashed, *expertly* uncouth.[93]

"The Fuzz-box; the wah-wah pedal; multi-track tape-recorders; playing of tapes backwards; electronic synthesizers . . . complex light shows involving numerous projectors, motion-picture films, slides, strobes and colour wheels": these are some of the technologies and techniques that Willis adds to such a depiction of countercultural music. In concert, Willis instances, Pink Floyd were "more like engineers than musicians, sitting at great consoles turning knobs and flicking switches rather than actually playing instruments."[94]

Perhaps there is no better way to emblematize the technological rationality latent within counterculture and make it bear upon the age of the mainframe than to go back to the pivotal moment in the late 1950s and early 1960s when Allen Ginsberg—the Beat who most famously made the transition into a 1960s countercultural icon—howled at America/Moloch. Here are the well-known opening lines of Ginsberg's "Howl":

> I saw the best minds of my generation destroyed by
>   madness, starving hysterical naked,
> dragging themselves through the negro streets at dawn
>   looking for an angry fix,
> angelheaded hipsters burning for the ancient heavenly
>   connection to the starry dynamo in the machinery
>   of night.

And from later in the poem:

> yacketayakking screaming vomiting whispering facts
>   and memories and anecdotes and eyeball kicks
>   and shocks of hospitals and jails and wars,
> whole intellects disgorged in total recall for seven days
>   and nights with brilliant eyes.

Though "Howl" is much else, clearly it is in one regard a work about information. In particular, it is a work in which Blakean and Whitmanesque "vision" ("I saw," "brilliant eyes") plugs exactly into the contest of vision and supervision that was the information paradox of the mainframe era (the era, that is, of the real "dynamo in the machinery of night"). Similarly, Jack Kerouac typing "on long, continuous rolls of paper feeding nonstop into his typewriter" and William Burroughs scissoring "apart his manuscripts," as Gitlin describes it in his account of the Beats, were one with Ginsberg in producing a "good deal of beat writing" that "was flat, dry cataloguing."[95] In the age of the mainframe, prophecy, epic, novel, ballad (as in song lyrics), satire, and all the other hip literary forms were the genres, but their mode—or at least alter ego—was the yacketayakking of random information that Thomas Pynchon in *The Crying of Lot 49*, his novel of the countercultural moment, would fabulate as the Tristero and rhapsodize as the "secular miracle of communication."[96] The visions (not to mention drug hallucinations) of counterculture were an uneasy blend of distance from and familiarity with technical rationality in the information age. Ginsberg's Moloch ("whose mind is pure machinery") and America ("Your machinery is too much for me")—like Pynchon's America of Yoyodyne and IBM 7094's—was a mainframe.[97]

So where did counterculture place itself on the map in relation to the baseline of work/leisure and mainstream/subculture? Here there is a dizzying sense of disorientation, as of a whole landscape desubstantializing into mirage and then suddenly shrinking into that tiny monastic cell of contemporary meditation where "ergonomics" is all that remains of yoga and breath control—the cubicle. From this point of view, the "triangle" I began by hypothesizing is not a triangle at all, whose third point is displaced above, below, or to one side of the baseline. Rather, counterculture's relation to the baseline of work/leisure and mainstream/subculture was a fractal, interior geometry that incorporated *within* itself the entirety of that baseline with all its dialectic of assimilation to, yet parody of, technological rationality. This is why it is finally futile to inquire into the relation of counterculture to each pole of the baseline independently ("the establishment," "the people"), for counterculture was a *relation to the relation* of the poles to each other. It was an appropriation of what I have termed the *relation of reciprocal appropriation*. Put simply, counterculture was neither the "insider" (mainstream) nor "outsider" (subculture) of technological rationality. Instead, a whole generation of "technocracy's children" who considered the wattage of a guitar amplifier to be their birthright used technological rationality to carve out a position for themselves as what might be called either "insiders outside" (insiders who *elected* to drop out of work culture to imitate subculture's parody of technological rationality) or "outsiders inside" (outsiders whose mode of protest was imbued so deeply with the techniques or technologies of work culture that parody all but disappeared). Counterculture interiorized in its identity the complex relation between mainstream and subculture. It was an appropriation of reciprocal appropriation, a mimicry of mimicry, a parody of parody. If both mainstream society and subculture quoted each other to identify themselves as cool ("I'm just an accountant, but I'm cool like the 'rebels' I see on film"; and, reciprocally, "I've got no job, but I've got a cool suit and shoes like the 'Man'"), then counterculture quoted the act of quoting ("I'm, like, 'cool'"). Mock country, God, and parents, counterculture said, but don't mess with the bong or the amp.[98]

Thus we come to the two "exceptions" I previously cited to the rule that petty or individual rebellion in the workplace of the mainframe era (rebellion at the level of lifestyle or workstyle) always led to containment. These exceptions were the sites where counterculture—precisely because it was from the first an uncanny incorporation of both technological rationality and its discontents—took root *within* corporate culture to prepare the ground for the next, "revolutionary" industrial paradigm.

One of the exceptions resided in the culture of management itself, where the first signs of a coming revolution in management theory (announced as early as 1960 in Douglas McGregor's notion of "Theory Y") appeared in the so-called "Creative Revolution" of the advertising industry. I borrow here from Frank's *Conquest of Cool: Business Culture, Counterculture, and the Rise of Hip Consumerism,* which is at once surprising and persuasive in its thesis that, contrary to popular myth, hip is not the story of an original cool culture that eventually lost its way through being coopted by the consumer industry. Rather, Frank argues, the notion of cooptation needs to be made more complex to explain how, even before counterculture became the darling of the media in the mid-1960s, it had already infiltrated the workplace culture of the advertising agencies that wrote the script for consumer culture. Paying particular attention to the break with 1950s advertising philosophy made by the Doyle Dane Bernbach agency in the 1960s (responsible for the famous anti-advertising and anti-automobile Volkswagen Bug campaign), Frank documents in compelling detail the way in which countercultural nonconformity and lifestyle rapidly became "the advertising style of the decade, from the office antics of the now-unleashed creative workers, to the graphic style they favored, to the new consumer whose image they were crafting." Cooptation this certainly was, in that "hip" sold consumers on what Frank calls a "perpetual motion machine" of designed-to-become-obsolete goods advertised in the style of rebellion. But it was cooptation from *within* the counterculture ethos itself—or at least the Madison Avenue annex of it. The hipness of the new styles and ads flowered from the fact that advertising producers saw in counterculture a mirror of their own need at the time to experiment with "decentralized, nonhierarchical anti-organizations" that would liberate "creativity." This era brought about the death of the "Madison Avenue" adman dressed all in gray and the birth of the new stereotype of the hip advertising or art director with his hair in a ponytail.[99] Counterculture was the ethos of the "insider outside" or "outsider inside" by which Theory Y, antihierarchical management took hold within the scene of production itself. The eventual outcome would be the changeover of "integrationist" corporate culture to the postindustrial integrationism that now incorporates "differentiation" within itself—that is, to an integral view of corporate culture as *all* differentiation, all a quest for decentralization, innovation, radical restructuring, local team cultures, and so on.

The corresponding outcome on the consumer or lifestyle side of things, as David Brooks argues in *Bobos in Paradise* (2000), was the emergence of the contemporary "Bobo" or "bourgeois bohemian":

> Suddenly massive corporations like Microsoft and the Gap were on the scene, citing Gandhi and Jack Kerouac in their advertisements. . . . Hip lawyers were wearing those teeny tiny steel-framed glasses because now it was apparently more prestigious to look like Franz Kafka than Paul Newman. . . . The bohemian and the bourgeois were all mixed up. It was now impossible to tell an espresso-sipping artist from a capuccino-gulping banker. . . . Most people, at least among the college-educated set, seemed to have rebel attitudes and social-climbing attitudes all scrambled together.[100]

The other exception demonstrating the influence of counterculture is early hacker culture as it began to build a mainstream industry around itself. We recall that "Silicon Valley is just down the (information) superhighway from Haight Ashbury" and that Steve Jobs of Apple "spent a year pursuing transcendental meditation in India," "wore open-toed sandals, had long, lank hair, and sported a Ho Chi Minh beard."[101] As has been widely recognized, so-called "computer liberation" emerged from the "libertarian philosophy" of "radical and communitarian thought . . . [in] the 1960s" and the "general malaise in the under-thirty crowd in the post-Beatles, post–Vietnam War period in the early 1970s."[102] Even as early as 1972, indeed, it was apparent to Stewart Brand, writing in *Rolling Stone*, that "the early hackers of the sixties were a subset of late beatnik/early hippie culture." For the descendants of the dharma bums, in other words, the ultimate mental rush came from a new drug—silicon—whose transcendental, semiconducting power arose only through "doping" its normal, inert state.[103]

This critical episode in the history of the information age now has many histories, including, most significantly (in addition to the works cited above), Everett M. Rogers and Judith K. Larsen's *Silicon Valley Fever*, Steven Levy's *Hackers*, Paul Freiberger and Michael Swaine's *Fire in the Valley*, and Howard Rheingold's *Virtual Community*.[104] But there is no better chronicler for our purposes than Roszak, who follows up on his *Making of a Counter Culture* with a chapter in his later *Cult of Information* entitled "The Computer and the Counterculture." The interface of counterculture with early hackerdom is especially clear in the story Roszak tells of the Resource One computer in Berkeley (a donated XDS-940 timesharing mainframe), intended in the early 1970s to become "a community computer utility" or "urban data base" facilitating grass-roots social and political activism. The addition a few years later to this mainframe-based project of a spin-off, the Community Memory project, founded on "small computer terminals" remotely connected to the XDS-940, was a prophecy of the future. Next to be developed by hacker culture (most famously by the Homebrew Computer Club and the future founders of Apple in the Bay Area) was the personal computer. With the dawning of the personal computer, as Roszak recounts in his shrewd postmortem upon the "utopian" hopes of computer libertarians, "there was an interval in the early 1980s . . . when, in California at least, the guerilla hackers seemed on the brink of making over the Information Age on their own terms."[105]

But "it did not last longer than a few intoxicating years," Roszak adds in the same breath. As Mark Dery points out in his study of computing subcultures, the heady mix of technical and spiritual headtripping invented by computer lib still survives in the "cyberdelic wing of fringe computer culture" associated, for example, with the *Mondo 2000* circle. But the overall fate of the personal computer was to be networked into mainstream corporate computing precisely in time to coincide with the neocorporatist principles that succeeded Theory Y. As Rochlin observes, "The irony of the Homebrew club is that the success of this anarchic collection of independent thinkers and libertarians created a bottom-up demand for de facto standardization whose results were not much different from those traditionally imposed from the top down by large, corporate manufacturers."[106] The informational and managerial "revolutions" thus merged, and the domestication of cyberlibertarianism began in the name of networking. Again, counterculture became the "outsider inside" the scene of production. Again, the eventual outcome was a changeover of "integrationist" corporate culture to an integrationism that incorporated "differentiation" within a new unity: an integral view of corporate culture as *all* differentiation, all information decentralization, all networking. There was thus no contestation that could not be included within the newly emergent, dominant corporate culture, no outside that did not come inside. Indeed, as we will see, the essence of a network is that it effectively colonizes the very idea of an outside. The cultural outside is reconfigured for internal corporate use as "decentralization," "flexibility," "outsourcing," and ultimately "global information economy."

How can one still be cool, we must next ask, when both the subcultural "outside" of older cool (now called "niche markets") and the countercultural "far out" of 1960s cool have been fenced in by networks that integrate differentiation within the corporation?

To answer this question, we must take informating—and the countercultural Age of Aquarius that responded to its technological rationality—into the age of networking proper, where all the communes become local area networks and all cosmic consciousness merely a matter of linking LANs in "wide area networks."

Chapter 4

Networking

As shown by the hysteria at the turn of the new century over "Y2K" in corporate and other enterprise-scale information systems, mainframe hardware and software—some as old as the 1970s or earlier when programmers used two digits for the year—continues to be vital. But since about 1982 networking has clearly taken the lead in setting the agenda of information technology. By "networking" I mean the combination of new information architectures (personal computers, client/server networks, the Internet) and organizational structures that has created what Manuel Castells calls "network enterprise."[1] If "informating" previously generated a thick wrapping of second-order information around automation, then networking now rewrapped the entire ensemble in an even thicker, *third-order* interface perfectly adapted to the boundary-crossing, decentralized, and outward-looking orientation of the new global economy. That interface was the interpersonal, cross-departmental, cross-industry, regional, and ultimately worldwide "outside" of the firm as encountered in networks where one's work came into exchange with *other* people's and firms' work.

In terms of information architectures, first of all, the networking paradigm arose through a twofold rhythm of *convergence* in underlying technologies and *divergence* in understanding what might be called the "philosophy" of those technologies. The technological convergence occurred when computing and communications fused together in three overlapping stages.[2] First was the preliminary decade of the 1970s, which saw the invention or application of such elemental new devices as the microprocessor, digital telecom switch, and optical fiber. On the foundation of these elements, a new generation of compact, modular, and flexible computing and communication mechanisms emerged to concatenate the infrastructure for distributed information work.[3] Increasingly, the model was no longer a massive concentration of processing and switching power at a central location but micro-concentrations all along the line of communications. In computing, the crucial new micro-machine was the personal computer, which was invented roughly between 1974 and 1981.[4] And in communications (specifically, communications networking), the machinery and protocols necessary to link such computers to each other and to mainframes also quickly evolved. In the arena of LANs (local area networks connecting machines within a single site), the leading event was the invention of Ethernet in 1973–76. For WANs (wide area networks connecting machines on multiple, remote sites), there were at least three key developments: the upgrading of telecom networks with digital switches and broadband transmission capabilities; the creation of the ARPAnet or ancestral Internet in 1970 (together with the TCP/IP packet-switching protocol in 1974–77); and the appearance of the modem.

The second stage of technological convergence spanned roughly from 1981 to 1991 when business adopted the personal computer as its own. The watershed year was 1981, when IBM brought to market its business-focused PC personal computer. Soon a fixture of office life, the PC was complemented by such standard-setting, second-generation, personal computer business software as dBase II and Lotus 1-2-3 (the original "killer apps," or paradigmatic software applications). The result was that soon after *Time* magazine named the personal computer its "Man of the Year" for 1982, *Business Week* clarified in August 1983 that the specific identity of that machine was the *office* personal computer. Businesses were in "personal computer shock," the magazine said.[5] By 1984 sales of personal computers already commanded 35 percent of an IT market that just a decade earlier had been 90 percent mainframes.[6] From the mid-1980s on, the next rush was to wire personal computers to the established architecture of mainframes and minicomputers, to each other (in so-called peer-to-peer networks), and, most important, to a new class of "server" computers (souped-up microcomputers anchoring the new "client/server" model of information exchange). In particular, the advantage of client/server LANs—which offloaded significant processing and control functions to the individual desktop while retaining for the server the power of enterprisewide file, database, and messaging systems—proved to be compelling.[7] By 1994, 87 percent of larger firms and 32 percent of small ones had linked their computers together in a LAN.[8] Wide area networking (which widened the communications reach of personal computers by allowing them to connect to servers remotely) was not yet ready for broad business adoption in the 1980s, but was rapidly evolving toward that point. After the breakup of AT&T in 1984, the telecom network became both faster and better suited to data transmission because of competition among service providers and AT&T's own entry into the information service market.[9] In the meantime, the Internet grew to over one hundred thousand host or server machines by 1989, added faster cross-country communication "backbones," and evolved subsidiary and parallel networks/protocols/interfaces (handling newsgroup, ftp, telnet, bulletin board, and e-mail services). By the end of the 1980s modems had also become noticeably faster (jumping from 300 to 9,600 baud).

Finally, the third and climactic stage of technological convergence began in the early 1990s, when office computing—followed by an increasingly strong home computing market —fully merged with the new communication networking technologies. Many developments are germane. Personal computers became ever more powerful, mobile, and capable of communicating (machines began arriving from the factory with modems or Ethernet cards installed). Software kept pace by incorporating communication into its core functions (operating systems, for example, included modules for networking and dial-up connections). LANs meanwhile became pervasive, and WANs came into their golden age.[10] In the case of the Internet alone, commercial use began in 1991 (with privatization of the backbone following in 1995); the World Wide Web appeared in 1992; the Mosaic and Netscape Web browsers arrived successively in 1993 and 1994; and the number of Internet host machines approached 10 million by the beginning of 1996.[11]

The convergence that is the leitmotif of all these developments is perhaps best signalized by one event in particular: the sudden collapse of boundaries between LANs and WANs (necessitating the invention of "firewalls" and other means of reasserting that boundary for security reasons). One factor was that TCP/IP protocol, which had arisen in the Internet arena, now began to colonize LANs in the form of "intranets" (private mini-Internets, often connected to the public Internet through firewall machines or proxy servers designed to screen incoming and outgoing traffic). Another factor was the invention of "tunneling" and "virtual private network" (VPN) protocols that allowed geographically separated users to access company intranets by means of encrypted, private corridors of communication through shared communication space. Such innovations, together with the use of dedicated leased lines for the same purpose, extended intranets into so-called "extranets" connecting a firm to its branches, suppliers, clients, and partners in a total functional grouping.

This technological convergence is now evolving in the 2000s into a further stage that might be called "ubiquitous" or "total environment" computing, that is, broadband networking complemented by "peer to peer" and wireless or small-device networking.[12] The overall result of these successive stages of convergence is that information systems now appear to communicate with each other in such a "worldwide" web of pervasive networking that what is "inside" is also inevitably "outside" and vice versa. "As the rapid gathering, manipulating, and sharing of information become a preeminent process and as company boundaries grow increasingly fluid and permeable," William Davidow and Michael Malone observe in their *Virtual Corporation,* "established notions of what is inside or outside a corporation become problematic, even irrelevant." Or as Rochlin sums it up in his *Trapped in the Net,* "the power of desktop computers, harnessed to the new techniques for global communication and networking, are making possible the creation of new types of large technical systems that are inherently transboundary."[13]

But the convergence of computing and communication networking technologies also harbored a divergence in the way the new networking was received, understood, and deployed. There were two contesting philosophies of networking whose contradiction replayed in updated form the vision versus supervision paradox we saw in the mainframe age. Each understanding had its own ideology, phenomenology, and practice.

One philosophy of networking was decentralization, which understood the addition of communication to computing to be emancipatory—like opening a door to a locked room. To use the ideological idiom common to this view (descended in part from early hacker counterculture), the decentralization thesis held that networks are innately antihierarchical, empowering to the individual user, and therefore democratic. The most spectacular expressions of this view have occurred in the realm of public policy and politics, where each new controversy over online freedom of speech, intellectual property rights, encryption, or privacy has stirred up (in the United States) a wide front of organizations and individuals opposed almost reflexively to any form of control or regulation. But the same cyberlibertarianism, whose politics I discuss later, attended corporate IT as well, where networking was espoused by consultants, the business press, and others as the means to emancipate workers and departments from the stranglehold of traditional management information service departments. "To create a customer-centric, empowered, flatter and more responsive organization we needed a more empowered, distributed and responsive computing architecture," one IT executive and consultant quoted by Don Tapscott says in the rote language of network empowerment.[14] Tapscott himself contrasts the "'master/slave' computing" of the mainframe era with a "true democracy" of networking so expansive that it is nothing less than the new world order itself:

> The crowning achievement of networking human intelligence could be the creation of a true democracy. Technology itself is shifting from mainframe, host, centralized computers to *network computing,* where each computer has autonomy and functions as a peer of the others. Similarly, rather than an all-powerful centralized government, arrogating decisions to itself, governments can be based on the networked intelligence of people. . . . Government as centralized mainframe can be replaced by government as network. And perhaps by combining their intelligence, people can create new levels of consciousness at the local, regional, national, and even international levels.[15]

Decentralization, in short, was the belief that servers really do "serve" their client populations. Control reverts to the individual worker, strict accountability is relaxed, and—to switch from an ideological to a phenomenological idiom—the original excitement of what Zuboff and others called informational "vision" is restored. But this was vision with a difference. The vision first revealed in the mainframe age upon opening a disk and seeing all related files, as I suggested previously, was synoptic because it gave an overview of the total, internal connectivity of an information system. It allowed the user to glimpse the systematicity of system. For this reason the phenomenology of such vision was characteristically vertical ("a helicopter view," one of Zuboff's interviewees said). But vision in a networked environment did not offer the same sense of synoptic overview because the essential experience of networking was a different kind of connectivity: horizontal escape into an "outside" consisting of other, often incommensurable machines and systems linked on an infinite level plane. Instead of a vertical overview, therefore, networked vision provided a sense of what might be called "lateral transcendence" toward an always receding horizon. Networked vision, in other words, was an experience of the connectivity of computerization that never summed up in holistic systematicity. It was *not* like opening a file list on a disk, but like using an Internet search engine and finding thousands of unsystematic results delivered ten or twenty at a time with a button at the bottom of the page marked "More" or "Next."

Crucially, the sense of systematicity was thwarted not just because central management (as in the mainframe era) denied access to the total vision, offering up instead only crumbs of dissociated "facts" and "data," but because the very meaning of "system" changed in the networking age. Castells's conceptual definition of networking is relevant:

> The components of the network are both autonomous and dependent *vis-à-vis* the network, and may be a part of other networks, and therefore of other systems of means aimed at other goals. The performance of a given network will then depend on two fundamental attributes of the network: its *connectedness*, that is its structural ability to facilitate noise-free communication between its components; its *consistency*, that is the extent to which there is sharing of interests between the network's goals and the goals of its components.[16]

With the advent of the networked age, "system" was disaggregated into "connectedness" and "consistency," which no longer necessarily lined up as one browsed or surfed laterally across the networks. Sometimes interconnected components were consistent in their interests and goals, sometimes not. Where system from the time of Taylor through the mainframe age had meant an externally imposed, hierarchically organized, and self-consistent structure of work assignments and procedures aligning technology with technique, "network" defined an entirely new kind of self-organizing and complexity-generating system or, more accurately, unclosed aggregate of systems. Thus, while the crucial opposition at the beginning of the twentieth century had been between preindustrial craft and industrial system, by the end of the century the substitution of "network" for "craft" (both unformalizable knowledge systems, both "associations" that extend outside the firm) had inaugurated a new binary opposition: postindustrial *network* versus industrial system.[17]

The quintessential practice of networking, therefore, now became "browsing" and "chatting." Browsing and chatting are the culmination of a whole range of craftily "connected" but not necessarily "consistent" Internet practices, usages, customs, and netiquettes adapted to populations of users who habitually subscribe to too many incommensurable channels of information to attend to systematically (resulting, for example, in the characteristically fragmented rhythm of discussion in unmoderated newsgroups, chat rooms, IRC channels, and even e-mail correspondence). Such users therefore employ ad hoc or communal methods of maintaining a tradition of understanding (e.g., periodically recirculated FAQs, instructions to the "newbie," threaded discussion forums, shared bookmark lists). And even then—because postindustrial craft retains its lore only as long as a personal "bookmark" or "buddy list" is current—users frequently lose the handle even on those methods (among many other symptoms of entropy are broken links in a bookmark list or homepage, threaded discussions that wander down chains of digression, recurrent "where is the FAQ?" postings in newsgroups, chat or MOO sessions that fracture into multiple, short exchanges syncopated with other exchanges).

Browsing, chatting, and affiliated modes of Net usage, in other words, are not just the casual, quick act of half-attention they are sometimes criticized for being by those who value close reading and critical thinking over easy consumption. From another point of view—that of production—they are the revenge of cubicle consciousness. What goes around comes around, and those whose "vision" had been sequestered by central management into cubicles and the equally partitioned forms, records, fields, pages, and files of information architecture now ride the wave of networking toward a new vision that is not the refutation but the epiphany of dissociation: "More" forms, pages, files, and threaded messages receding toward the horizon in nonsystematic splendor.

Meanwhile, an entirely different philosophy of networking developed alongside the decentralization thesis. This philosophy, which I will call "distributed centralization," understood the addition of communication to computing to be an opportunity for even more pervasive forms of control—like opening a locked room only to let the warden look in. The ideology of distributed centralization could also be defined in political terms. (After all, as Thomas B. Lifson reminds us, the concept of networking historically had political and social connotations: "Americans are very uncomfortable with the idea that networks can be powerful. . . . Networks are predominantly identified with conspiracies, racism, elitism, and market-rigging.")[18] It might be instructive, for example, to make an analogy to U.S. political conservatism in the 1990s, whose campaign against Big Government, seen one way, sought not so much to reduce the level of control over individuals as to redistribute such control into networks of lower level, local centers of control (states, families). But again, I will defer my discussion of the political discourse of networking to a later chapter because, as practiced in business IT, distributed centralization simply folded itself into the ideology that became the aspiration even of much government politics in the era—good management. Distributed centralization was the belief that decentralization could itself be a form of centralized management, an Argus of a hundred eyes.

The most astute and articulate observer of IT in this regard is Gene Rochlin, whose *Trapped in the Net* (1997) argues that the libertarian thesis of network decentralization was always deceptive. "The historical record of the introduction of new techniques, and new technical systems, into factories, offices, and other workplaces," Rochlin says, "is full of parallels that suggest that the democratizing claim is frequently made, and the democratizing effect does indeed frequently manifest itself during the early phases of introduction. But the democratizing phase is just that, a transient phase in the evolution of the introduction of new technology that eventually gives way to a more stable configuration in which workers and managers find their discretion reduced, their autonomy more constrained rather than less, their knowledge more fragmented, and their work load increased." "Almost independent of size, scope, and purpose," he thus notes, "one organization after another moved to centralize information systems purchases and establish organizational or corporate standards, and to replace ad hoc and disorganized management of idiosyncratic LANs with network management under the control of management, usually through a dedicated MIS staff not that different in their power and approach from that of the traditional centralized computer center."[19]

Gradually, then, networking eroded decentralization to the point of wholly ironizing the putative master/slave relation between client and server. As Scott Adams sometimes lampoons in his *Dilbert* comic strip, it is really the *server* (the shrunken mainframe of the downsizing era) that lords it over meek populations of client users reduced to begging their sysadmins for permission to do anything outside the prescribed norm—change a program, access restricted directories, dial into the office over a modem. Phenomenologically, therefore, networking from this viewpoint was so far from being a new mode of vision as to be another form of supervision. But, again, there was a difference. Just as decentralization was "lateral transcendence," so the countervailing centralization of the time was *distributed* centralization. As one information services expert cited by Davidow and Malone says, it was "virtual centralization."[20] Distributed centralization meant the implantation of dynamic packet-filtering firewalls, ever more complex permission levels, increasingly rigorous (and encrypted) login processes, burgeoning access logs that track users by IP number or domain name, proliferating spyware and monitoring programs, and innumerable other presence-points of control and accountability scattered throughout the lateral network.[21]

Wherever there was an emancipated soul working on a decentralized network, there was also a little free-standing particle of supervision (usually several) reviewing the work. Rochlin puts it this way:

> As a result [of networking], it is now possible to control and coordinate process and production without imposing the static and mechanized form of organization of workplace and administration that so characterized the synchronistic approach.
>
> What the computer transformation of business and industry has done is to maintain the appearance of continuing the trend toward decentralization, to further reduce the visible hierarchy and formal structures of authoritarian control while effectively and structurally reversing it. Instead of the traditional means of formalization, fixed and orderly rules, procedures, and regulations, the modern firm uses its authority over information and network communications to put into place an embedded spider web of control that is as rigorous and demanding as the more traditional and visible hierarchy. Because of its power and flexibility, the new control mechanism can afford to encourage "empowerment" of the individual, to allow more individual discretion and freedom of action at the work site, and still retain the power to enforce the adjustments that ensure the efficiency of the system as a whole.[22]

Business commentators and information systems experts back up Rochlin's analysis of distributed centralization. A 1996 article in *Computerworld*, for example, reports the views of one information services consultant:

> From 1985 to 1995, PCs and client server moved too far away from centralized computing. Companies were rebounding from their past dependence on the IS priesthood. Today, companies are turning away from totally decentralized client/server or network computing. . . .Traditional data centers are now being asked to manage departmental/distributed systems in a centralized manner. In part, that's because companies have yet to find systems and network software that do an adequate job of managing decentralized systems in a decentralized manner.[23]

While the machineries had not yet been perfected, clearly the tendency was away from old-fashioned central control toward "managing decentralized systems in a decentralized manner." Why, for example, should a system administrator have to intervene when an employee chooses too simple a password, enters the wrong date in a form, or browses a pornography site, when a simple, standing program on the server can reject bad passwords, a Javascript-enabled Web form can reject bad entries, or a firewall can block particular Web addresses—and can do so in an automated way that neatly sidesteps the authority issues inherent in a situation where the technical "priesthood" usually have much more de facto control over policy than they have been formally given?

Recent developments in corporate IT indicate that distributed centralization will only grow stronger. One early symptom was the mid-1990s move (since largely passed by) toward stripped-down, storageless "network computers" or "thin clients," which would in effect create choke points of control on the desktop. Like dumb terminals of old, such thin computers would have surrendered almost all control to the central server. Similarly, "managed personal computers" began to arrive in the late 1990s complete with the new Desktop Management Interface (DMI) and other specifications allowing IT managers to check desktops remotely for "chassis intrusion" (someone opening the physical computer to install or remove components), to inventory client hardware and software, and to install or configure client software on a mass basis. And meanwhile—just like the hundred eyes of Argus—new Internet-use tracking programs began to monitor employee browsing habits; new legal precedents (in combination with unprecedented archival storage) allowed employers to inspect worker e-mail; and, in general, new devices, procedures, and protocols created what Rochlin calls an "embedded spider web of control," fine as gossamer, unforgiving as steel.[24]

If the emblematic practices of decentralization were browsing and chatting, that of distributed centralization was the multiple login that soon became de rigueur in networked environments: the process of logging first onto one's personal computer, then one's LAN, then one's e-mail account or messaging environment, and so on (not to mention the innumerable Web sites that now require a password or user-specific information).[25] Or, again, consider that with each passing year the act of browsing the Web has incrementally changed in character from the experience of theoretically limitless mobility (where the only friction came from problems in transmission speeds and the entropy of the network itself) toward that of negotiating a thicket of guard pages, including alerts that pop up when making the transition to or from a "secure server," notices of privacy policy, legal contracts that one must "agree" to before proceeding, and so on. On many Web sites in the early 2000s even the advertisements function in a channeling or guard capacity, refusing to lie in passive wait for the user to notice them but actively popping up in a separate window that must be cleared from the screen before the main site can be used. Ironically, networking has not just facilitated democratic access to information; it has also become one of the most fertile breeding grounds the world has ever known for creative, subtle, and distributed ways to *deny, filter,* or *channel* access. Only in the age of networking, after all, does one risk not just forgetting the swarm of passwords and user names that sustain one's networked identity but also being "timed out," having "access denied," being stopped by "Error Code 403: Forbidden" (or a similar message), and encountering various warnings ("relaying refused," "continuous use of our unlimited access Internet service violates your service agreement") from one's Internet service provider. Rochlin comments: "The paradox was nicely put in an article in *Infoworld,* 'In order to construct the flexible, interconnected systems business demands, a fair amount of centralized control must be exercised.'"[26]

Meanwhile, as the contest between decentralization and distributed centralization played out in information architecture in the 1990s, *organizational* architecture, or restructuring, piggy-backed onto IT to complete the overall paradigm of the network enterprise. I refer to the organizational changes surveyed in chapter 1: downsizing, flattening, teamworking, and all the other initiatives that seemed to embrace the new IT so intimately that commentators began simply to fuse restructuring and IT under such names as the "virtual" or "networked corporation." But the notion that organizational restructuring could simply parallel networking requires examination. Here we must pause to follow up the possibility we earlier considered (characteristically ignored in the business bestsellers) that postindustrialism and informationalism may not necessarily coincide.

After all, on what was the faith in the "virtual corporation" founded? The details of the alliance between restructuring and IT had always been fuzzy. In their *Virtual Corporation,* for example, Davidow and Malone begin by assigning to IT the role of cause: "As a result of [technological change], . . . the corporation as we have known it for eighty years will have largely disappeared." But later they qualify, "There is a danger in believing that technological supremacy is enough to revitalize and keep our corporations competitive. . . . This time, technical innovation alone will not save us." Boyett and Conn's *Workplace 2000* settles for the thesis that IT is a partial cause or facilitator: "The ability of large American companies to reconfigure themselves . . . can, at least in part, be attributed to the development of new technology that makes whole layers of managers and their staffs unnecessary." And Hammer and Champy's *Reengineering the Corporation* straddles the line: information technology is "an *essential enabler* . . . since it *permits* companies to reengineer business processes."[27] Recall from chapter 1 all those allegories of the digital in books about restructuring and diversity management ("virtual corporations," "all of us are programmed," and so on). Clearly, one reason for such figures is that they expressed the relation between restructuring and IT with maximum conviction but minimal precision.

Such fuzziness reflected an underlying empirical uncertainty. During most of the 1980s and 1990s, investigators could find no hard evidence relating IT to restructuring in the area where it most mattered—the bottom line. I refer to the scandalous "productivity paradox" of this period when, as documented by Thomas Landauer, Gary Loveman, Richard Franke, Stephen Roach, Paul Strassmann, and others, it became widely known that IT consumed vast amounts of corporate investment but led to no or negative measurable productivity gain.[28] According to a 1992 study by Roach, for example, investment in IT in the service industries rose steeply from 1979 to 1991, while productivity in those same industries was flat or declined narrowly. Similarly, a longer spanning study by Franke in 1987 showed that capital invested per labor hour in the financial industry (one of the most intensely technologized sectors) climbed fivefold from 1948 to 1983 (with the steepest rise in the 1980s), while output per labor hour held flat and output per capital dollar actually dropped by 80 percent.[29] Such striking contradictions left any restructuring rationale that relied on IT as "essential enabler" holding what seemed an empty bag. By rights, the extra shareholder value that accrued upon restructuring should have been justifiable on the grounds of long-term prospects for profitability founded in the last instance on provable productivity increases (which together with increased world markets eclipsed price margins as the driver of profit in the anti-inflationary climate of the times). But if the productivity promised by IT and the flatter, downsized, and flexible ways of working it enabled was either nonexistent or not yet evident (leading many commentators to say that the productivity gains latent in IT could only be unlocked through more or better restructuring in the future), then what lay at the objective foundation of restructuring?[30] How could downsized firms that substituted IT (the dominant, new capital investment) for employees assert with conviction that they could generate more output per labor-hour, let alone grow and innovate? How, in other words, could what became known in the late 1990s as the New Economy—which postulated that the overall combination of restructuring, IT-assisted productivity, and global expansion/competition created a pattern able to sustain wealth generation in steady-state pricing conditions—be explained as anything other than a speculative bubble?

Only with new, broader notions of IT-related productivity (see, e.g., Strassmann's excellent *Information Payoff*), further restructuring, and the striking rise in productivity indexes after 1995 or 1996 did IT at the very end of the century finally seem to fit empirically within the New Economy worldview—at least for a few heady years until the bubble indeed burst in the so-called "New Economy recession" of 2001.[31] In the intervening period between the productivity paradox and the late 1990s boom, there opened a vast lacuna in the legitimation of restructuring, which I suggest—recalling all those "virtuality" metaphors in business literature—may best be analyzed in terms of figuration. Speaking in 1996 of the productivity paradox in IT, William Birdsall notes the importance of "myth" by citing this passage from a report of the Organization for Economic Co-operation and Development: "This is not to say that the mythical quality of high technology in general and of IT in particular has now been dispelled. That will not happen as long as high technology is regarded as an automatic and instant panacea for all manner of economic ills. The essence of a myth is that it can have value in the absence of analytical assessment, and high technology has had great value as political myth."[32] Or we might turn to Martha S. Feldman and James G. March's shrewd study, "Information in Organizations as Signal and Symbol" (1981), which observes that rational choice theory alone cannot account for the prevalent tendency in business organizations to gather and communicate excessive information that has "little decision relevance," is too late for the decision at hand, or is never considered at all, even as an organization requests ever more streams of information. Such dependency on information, Feldman and March argue, can best be studied through an "information behavior" approach that views information processing as in great part a "symbolic" or "ritualistic" *performance* of rational decision making:

> The gathering of information provides a ritualistic assurance that appropriate attitudes about decision making exist. Within such a scenario of performance, information is not simply a basis for action. It is a representation of competence and a reaffirmation of social virtue. Command of information and information sources enhances perceived competence and inspires confidence. The belief that more information characterizes better decisions engenders a belief that having information, in itself, is good and that a person or organization with more information is better than a person or organization with less. Thus the gathering and use of information in an organization is part of the performance of a decision maker. . . . Using information, asking for information, and justifying decisions in terms of information have all come to be significant ways in which we symbolize that the process is legitimate, that we are good decision makers, and that our organizations are well managed.[33]

Such a thesis has recently been reinforced by Andrew Flanagin's empirical study, "Social Pressures on Organizational Website Adoption" (2000), which surveys the "social pressures" influencing companies to start Web sites. Flanagin concludes that social standing and perception (peer pressure from other firms, "perceived leadership in the field and organizational visibility," as well as "self-perception") were critical factors in the adoption of Web IT. Similarly, Flanagin and Miriam Metzger's "Internet Use in the Contemporary Media Environment" (2001) considers the role of "functional images," "shared perceptions," and "symbolic values" of media as well as other factors in determining how the Internet is used. Whether or not information is useful, in other words, it is first of all *mythic or symbolic* of being useful.

Applying this thesis to the technology of information, but varying the vocabulary of rhetorical analysis so as to start with "allegory" rather than "myth" or "symbol," we can say that it was not necessarily the case in the 1980s and 1990s that networking IT determined, enabled, or facilitated restructuring. Those considerations were secondary or still prospective. Instead, IT fulfilled the need for a legitimating *allegory* of restructuring—for digitalism as what I earlier called the great allegory of postindustrial capitalism. Networking IT was what allowed restructuring to represent itself to be just "as" necessary, precise, and developmentally progressive as technology. Networking was the allegory of restructuring as "objective." The dilemma of the "productivity paradox" can thus be rephrased in rhetorical terms as follows. Given the ontologically hollow or empty status of allegory (a "mere" figure), how did the conviction that restructuring depended on the new IT grow to be so axiomatic that allegory indeed became inflated into what rhetorical and literary tradition calls *symbol,* the foundationally full opposite of allegory? How did IT, despite its internal paradox, resolve into a seamless, fused "image" or "icon" able to "embody" restructuring "objectively" (to borrow the New Criticism's terms for discussing poetic imagery)?[34] How, in short, could IT as allegory mimic the transcendental stability or universalism of symbol so as to anchor the otherwise ultra-destabilizing process of restructuring? After all, no company would have wanted to identify itself with IT if the technology in question appeared to shareholders as it really was in the 1980s and 1990s (and as Y2K almost showed it to be): an inveterately unstable, ad hoc, uncongealed tissue of "patches," "security fixes," "service packs," and "upgrades" designed to work around the latest outbreak in what was developing into a perennial feud between decentralization and recentralization.[35] As it were: *Invest in our company; we promise that all the downsizing and restructuring we are doing will create just as much contestation, contradiction, shakiness, instability, risk, and purely speculative value as the network we are so busily wiring and rewiring. In the new millennium, our company will be the first to crash.*

The reason for my use of rhetorical analysis will become clear if we now pick up where we left off in chapter 3 in discussing the "solution" to the earlier, mainframe paradox of vision versus supervision: corporate culture, the place from which all contemporary business allegories and symbols start. As demonstrated with special clarity in such business scholarship as Alvesson's chapter "Culture as Metaphor and Metaphors for Culture" in *Cultural Perspectives on Organizations* or Alvesson and Berg's *Corporate Culture and Organizational Symbolism,* corporate culture is the master figure within which symbolic stories, rituals, jargon, dress, and so on are subordinate figures. In the mainframe age corporate culture optimized the paradox of vision and supervision (and, more broadly, of warm/cold or service-oriented/efficient work) by creating a deep act of unitarian "friendship." Now, at the onset of the network age, corporate culture optimized its updated paradox of decentralization versus distributed centralization by installing a new mode of friendship—one in which the figuration necessary for any deep act could now be enforced not just through cultural but through *literal* programming.

Consider, to begin with, how both poles in the hybrid notion of "efficient service" evolved after the peak of the mainframe age so as to provide the underlying motive for networking and restructuring in the first place. Service moved further in the direction of "flexibility," "quality," "disintermediation," openness of communication, "just in time" design and production, and other implementations of the postindustrial principle that Toffler in 1980 labeled "prosumerism." In Toffler's view, prosumerism occurs when the boundary between producer and consumer is spanned so thoroughly that consumers participate de facto in the design, engineering, manufacture, or distribution of products through instant feedback loops at the point of market research or final sale. A consumer need only express a preference for a certain color, for example, and instantly the firm redesigns its product to meet changing or diversified needs (multiple configurations of the same widget). Prosumerism is decentralized or distributed production; it is self-service production.[36]

In retrospect, Toffler understated the case. Flexibility, instant response, open communications, total quality control, and so on ultimately reconfigured the processes of production so thoroughly that every stage of production could now transact prosumeristically with every other stage whether or not an end-user customer was involved. Every firm or even department within a firm stood downstream from suppliers that had to treat it as their customer (with a direct say in the quality and manufacture of the supplies) and upstream from other manufacturers, distributors, and so on that had to be treated as *its* prosumer customers. As Davidow and Malone put it in two passages of their book:

> On the upstream side of the firm, supplier networks will have to be integrated with those of customers often to the point where the customer will share its equipment, designs, trade secrets, and confidences with those suppliers. Obviously, suppliers will become very dependent upon their downstream customers; but by the same token the customers will be equally trapped by their suppliers. In the end, unlike its contemporary predecessors, the virtual corporation will appear less a discrete enterprise and more an ever-varying cluster of common activities in the midst of a vast fabric of relationships.
>
> Becoming irrelevant are the once obvious differences among suppliers, manufacturers, distributors, retailers, customers, even competitors. At any one time, an enterprise or an individual may play multiple roles. For example, the role of manufacturer might be moved up the chain and placed with the traditional supplier, as with semiconductor companies assuming much of the work once done by computer makers.[37]

The underlying motive for decentralization in both organizational structure and IT was that production itself became decentralized in the service age. It became distributed laterally through networks of codependency—even to the extreme of so-called "coopetition" agreements among competing firms to coproduce or cobrand a product.

Meanwhile, the same retooling of production allowed efficiency—the other pole in the "efficient service" hybrid—to evolve in postindustrial directions as well. "Total quality control," "just in time," "concurrent engineering," and "design to build" are just some of the phrases that describe how the prosumerist principle of intimate relations between all units of production also became a principle of ever tighter control for efficiency. Manufacturers and suppliers had to couple their planning processes and information systems, for instance, to guarantee the increasingly high level of quality control needed for uninterrupted just-in-time assembly. As Davidow and Malone argue, "When a virtual corporation enters into a relationship with a supplier, it takes on a great deal of responsibility to ensure the supplier's success. . . . Customers who wish to receive just-in-time support from their suppliers must be intimately involved in the production of the service. That is because just-in-time support requires suppliers to produce products that are nearly perfect."[38] Processes and products along the whole chain of production became bound with unprecedented rigor to the uniformity of the whole smoothly operating machine.

Indeed, uniformity is the key to the new efficiency, though such uniformity no longer functioned on the model of assembly line industrialism. Instead of enforcing uniformity over all aspects of production through hierarchical, centralized control (as in an old-fashioned, "vertical" organization), the new method was to distribute enforcement locally *throughout* the wide-flung production network via ever proliferating "standards" and "protocols." Some of these standards were proprietary to firms, but an increasing proportion were industrywide or even wider. This explains why, for example, centralization has now become a dirty word in every aspect of the postindustrial economy except in the case of the variously official or de facto standard-setting bodies (industry-specific, national, regional, or international) that propagate fulsome, detailed standards of interoperability (certifying that products are "compliant" with measures of quality, consistency, type, and so on).[39] Such spectacularly visible standards battles of the 1990s as those between different software platforms (e.g., Windows versus Sun versions of Java) and different hardware specifications (in HDTV or DVD, for example) are representative. Even more symptomatic was the rise of the successive HTML and later XML, SOAP, and other specifications that in combination with TCP/IP protocol provided a universal ether of communications between suppliers, producers, and customers (e.g., through "Web services" or automated transactions conducted through the Internet via the new lingua franca of XML)—a sort of virtual spirit medium through which any (choose one: supplier, producer, customer) could automatically place an order or make an inquiry via the Web with any other (supplier, producer, customer). Just as the underlying motive for decentralization in organizations and IT was the decentralization of production itself, so the antithetical motive for "distributed centralization" in those same organizations and IT was the new distributed "standard" of control in such production.

How, then, did corporate culture at the dawning of the network age manage, or meta-manage, the new postindustrial forms of service and efficiency so as to keep the two locked together in a single, harmonious culture of "efficient service"? How could it do so when service and efficiency were no longer fenced within hierarchical organizations but distributed laterally across the permeable boundaries of departments, firms, nations, and whole regions? If the intra-organizational solution was "corporate culture," as we saw above, how could such management by ambience, texture, milieu, and style extend *across* firms to govern distributed production?

No more than a modification—a tweak—of the existing friendship system of corporate culture was needed. Here is where figuration becomes crucial—specifically, figuration that feels so "real" and "objective" that IT could now seem to stand for restructuring in a relation that was not just allegorical but symbolic. Although the tweak necessarily affected every aspect of the new production, the place to begin looking for it is in the great technical "figure" of the time that created a seamless *image* of IT able to fit the new service and efficiency into the representation of a single environment of production (no matter how internally riven between warm and cold work) capable of spanning from firm to firm with the same alacrity as it leaped across networks from computer to computer. I refer to the business deployment in the 1980s and 1990s of the "graphical user interface" (GUI), whose "metaphors" imaged a "friendly," harmonious culture of production seemingly part of the very infrastructure that now networked corporations together. *Friendly corporate culture became "user friendly."* This was the great object-metaphor or symbol of efficient service in the age of distributed production. A terrible—but friendly—new global automatism was born.

Boot up a personal computer in today's networked office environment, therefore, and what do we face? *Not,* certainly, the stark, monochrome, text-only screen of the early personal computer.[40] Of course, even the barest such screen was an interface and thus ineluctably metaphorical. Steven Johnson's *Interface Culture* (1997) is insightful. As Johnson conceives it, interfaces are constitutive rather than merely supplementary representations of what in the final analysis is humanly unintelligible: the flux of high and low voltages within a computer for which even binary zeros and ones are forced representations, or, again, the packetized and relayed flux of information between computers that creates the "infinity imagined" of the network. A computer, in other words, is "a symbolic system from the ground up."[41] Yet even if interfaces are as inescapable as the face one wakes up with in the morning, there is a great difference between such a face in its found state (computer interfaces created through ad hoc compromises between engineering, marketing, and other concerns) and the face one makes up to go to the office. The face of the early business personal computer was definitely not yet made up.

In particular, that face did not present information in what might be called coordinated fashion. The opening screen of the early business personal computer (as opposed to Apple computers) characteristically showed either a bare command line (the infamous "C:\>" prompt) or at best a rudimentary text-based menu (e.g., "Login to LAN," "WordStar," "Dbase," "Change Password," "DOS Command," "Logoff"). In either case, there were few clues about the logical or syntactical coordination of semantic operators, making it easy to forget, for example, that the pathname for a file must begin with "C:\" and not "C:/" or simply "C:". And once one moved from the operating system to application programs, the situation rapidly became even more uncoordinated. Working with an application interface meant pushing through a whole thicket of arcane commands (e.g., CTRL-F5, ALT-4), while also switching between fractured "modes" of work (in word processing, for example, one screen and set of commands for composing text, one for viewing formatting codes, one for creating endnotes, one for print preview). Alternating between applications was an even more arduous exercise in switching modes—the most extreme form of which was hopping across a network to the entirely incommensurable interfaces of other systems. A telnet session with a remote library or research database frequently ended with the user fumbling through random variations of *quit* to find the right logoff command: "exit," "quit," "q," "Q," "logoff," "bye," "CTRL-]," and so forth.

But turn on a networked personal computer today, and the face of information looks quite different. Instead of a character-based monochrome screen, we see the descendent of the bitmap approach first devised in primitive form in the 1960s by researchers at the Stanford Research Institute's Human Factors Research Center and the University of Utah's Computer Science Laboratory. The approach was further developed in the 1970s by the Xerox PARC group, before being adopted in 1984 by Apple for its groundbreaking Macintosh and finally brought into the corporate mainstream in the late 1980s and 1990s in the successive revisions of Microsoft's Windows operating system.[42] We see a graphically bitmapped main "window" whose menu bars and office-themed visual icons (file folders, trash cans, calendars, phones) construct a metaphorical "desktop," the great landscape of the cubicle. Above all, the function of such a desktop is to *coordinate* (and also subordinate) operations and modes. Clicking on desktop icons, for example, initiates sequences of actions or opens up individual windows that, as Johnson points out, are what we now have instead of "modes."[43] Or, rather, such windows obviate the awareness of modes by making mode switching as much as possible a matter of "direct manipulation." Mode switching within an application window thus means simply moving the cursor from text to footnote on the same "page" (where the graphical representation of a page supplies many of the clues needed to navigate between functions). Mode switching across applications is an equally direct matter of clicking on other open windows.

Nor does the modeless coordination of computing stop there, for with the advent of tightly integrated application "suites," individual word-processing, spreadsheet, database, or e-mail programs display what amounts to whole interior desktops complete with cross-application menu bars, templates, embedded program "objects," and "wizards" designed to stitch all the suite into a single, virtual work surface. And since the desktop is now networked, there is in principle no outer horizon to that single work surface. With the arrival of the dominant Windows operating systems and Windows Office suites, for example, the interface of the business personal computer became so tightly integrated with the Web browser that remote resources could be displayed or referenced directly on the desktop as if they were local resources. The Web was the natural outgrowth of the windowed desktop metaphor. Activated with point-and-click simplicity, hypertext and hypermedia became the ultimate means of modeless coordination linking different windows and different machines around the world.

Such is the user-friendly interface of knowledge work in the networked age. But such also is the symbol of a much vaster phenomenon that may be called, to borrow Johnson's terms, the "new cultural form" or "metaform" of the interface as it reaches up from its machinic holdfast to higher levels of human experience. We would be too limited in our understanding of the user-friendly interface of networked computing if we kept our eyes on our desktop, no matter how expansive it has become in coordinating a mighty simultaneity of tasks ("multitasking"). In reality, as both Johnson and the cyberpunk novelist Neal Stephenson (in his nonfictional *In the Beginning . . . Was the Command Line*) have argued, the user-friendly interface is symptomatic of a whole way of relating to culture. In particular, we may add, it is a means for imaging the new corporate culture (the face put on by distributed production to manage its boundary crossings) as itself the *generalized interface* of contemporary culture. To extend Stephenson's comparison between the GUI computer interface and Disneyland, we may say that it is now corporate culture (epitomized in contemporary media by, among others, the global Disney entity itself) that attempts to create a single, continuous interface of social experience stretching from the desktop to Tomorrowland.[44] It is now the distributed environment of corporate culture that "themes" culture like some gigantic desktop of the sort featured in Microsoft Windows desktop themes of the late 1990s: "Science," "Inside Your Computer," "Fashion," "Nature," "World Traveler," and so on.

How exactly does "user friendly" IT symbolize this general cultural phenomenon called corporate culture? In normal working circumstances, we know, *user friendly* simply means "easy to use." (The homepage of the Microsoft Usability Group hangs the following epithet next to its logo, "*usable*: easy to learn, use, understand, or work with.") Of course, there is some complexity hidden within this simplicity. It is inconvenient, for example, that "ease of use" actually contains two criteria, ease *and* use (the kernel code, we may say, of service *and* efficiency). Certainly, it is by no means unanimous that just one proportion between these criteria, and therefore one type of interface, is the most friendly. (The comp.unix.user-friendly and comp.human-factors Usenet groups, for instance, periodically hear from fierce UNIX advocates who claim that the legendarily difficult, command-driven, myriad-featured UNIX interface is friendly because heightened usefulness counts for more than superficial ease.)[45] And even if only ease or only use were relevant, who is to say that either—to parrot the computer industry cliché—is "intuitive"? What would be intuitively either easy or useful about an icon labeled "Recycle Bin" for users in a developing or even industrial nation, as opposed to the postindustrial United States? But in normal working circumstances, clearly, these are quibbles. As was clear from the beginning when Doug Engelbart stunned the audience at his famous demo at the 1968 Fall Joint Computer Conference in San Francisco, the instrumental advantages of the GUI-and-mouse model of direct manipulation is overwhelmingly persuasive.[46]

Yet "user friendly" refused to stay confined to what I have just called "normal working circumstances," and instead rapidly expanded outward in society in the manner of much computer jargon. It is precisely for this reason that we need to look beyond its instrumental rationale to the broader cultural momentum of that rationale. That momentum is evident from the history of the concept "user friendly," which is largely the history of a struggle for legitimation that succeeded so fully that the concept burst from the realm of IT, first into corporate culture at large and then, through the imprinting of such culture on society, into broad sectors of public life. It is instructive that, although the term *user friendly* had entered the language by 1977 *(OED)*, it was not until roughly 1987–92 that the concept firmed up. By this I mean that such major IT firms as Apple, Compaq, Digital Equipment, Hewlett-Packard, Microsoft, Silicon Graphics, and others set up "usability" or "human factors" groups and laboratories staffed by specialists in cognitive and social psychology, human resources, usability engineering, and rhetoric. (A collection of firsthand accounts by founders of such usability programs can be found in Michael E. Wiklund's *Usability in Practice: How Companies*

Develop User-Friendly Products [1994].) But I also mean that, while the new labs were busy establishing usability research as a discipline (building an industrywide set of testing and monitoring procedures, trying different analytical approaches, creating descriptive grammars of interface patterns and functions), they strove to legitimate the basic premise of usability within the IT industry—especially in the eyes of the engineering and management sectors, which were sometimes skeptical. As remembered by early workers in the discipline, for instance, proselytizing was crucial. A standard anecdote in the Wiklund collection of writings by usability pioneers, for example, concerns the make-or-break usability demo during visits by executives or major clients.[47] As the Silicon Graphics usability group puts it, usability researchers worked hard at outreach and eventually achieved companywide credibility: "The Product Usability staff did a great deal to introduce Silicon Graphics to functional usability engineering processes. They got people interested and involved, without making them defensive. Their usability efforts were recognized as valuable and credible. . . . This laid the groundwork for a more formal human factors effort."[48]

Once the IT industry became a believer in "user friendly," then the marketing blitz began to pitch the gospel to corporate customers of IT in other industries. As Wiklund observes, "In the 1990s, usability sells. At least this is the conclusion that emerges from an examination of product advertisements and the manner in which system developers market their wares to industrial users. It seems that everyone has become usability conscious. I credit the software industry for the increased awareness."[49] The accompanying editorial blitz in the IT trade media also helped. "Product sales, particularly through the retail channel, are effected by product reviews in the trade press," the usability experts at the Borland company note. "Most PC-oriented computer magazines consider ease of use a key factor in their product review criteria." The overall result was that "user friendly" quickly became a consensus ideal for business computing. As emblematized by IBM's ad campaign in the 1980s featuring the Charlie Chaplin "Little Tramp" character, business now wanted to put a "human face" on IT.[50] Nor did "user friendly" remain a matter just of business IT. It soon became an epithet of business in general—part of the standard promotional parlance for any product, service, or "corporate image," regardless of relevance to technology. Legions of firms, small and large, thus began to use their Web pages to declare themselves or their products "user friendly."

The final coup was the spread of *user friendly* beyond the corporate sector to all the other social sectors that business ultimately influenced—most strikingly to areas of consumer, family, religious, cultural, and individual life where the individual was refashioned wholesale in the new corporate image as "user" and the elementary principle of sociality likewise reduced to "friendship." By the late 1990s, for example, there were literally hundreds of book titles promising a "user friendly" God, Greek, architecture, family counseling, and so on.[51] Indeed, *user friendly* garnered so much general social currency that it became canonized in government policy of the 1990s on the future of public-sector information technology. I refer to the place of honor given the term in both the U.S. National Information Infrastructure initiative (NII) and the European Union's Fifth Framework Programme for Research, Technological Development and Demonstration Activities (FP5). In working toward its vision of the NII, for example, the Committee on Applications and Technology of the government Information Infrastructure Taskforce (IITF) published a paper in 1994 that includes a section on "User-Friendly Hardware and Software" in which friendliness is cast as one of the key issues of the future (the report as a whole is titled "What It Takes to Make It Happen: Key Issues for Applications of the National Information Infrastructure"). And a 1997 working document of the European Commission, excerpted from its draft proposal for FP5, goes so far as to convene all key issues having to do with IT under the title, "Creating a User-Friendly Information Society."

Perhaps there is no better way to punctuate this review of the cultural propagation of "user friendly" than to note that the concept finally dictated the terms of debate even for what is sometimes called the neo-Luddite critique of IT. An instance is Clifford Stoll's wry, incisive *Silicon Snake Oil* (1995). When Stoll wishes to doubt the utility of networks, for example, he opens with the following rhetorical flourish: "'A tool for what?' I ask my friends." "Friends" here is as deliberate a phrase as "friends, Romans, countrymen" in *Julius Caesar*. Stoll applies "friends" by turns to actual human acquaintances and, ironically, not-so-amicable machines so frequently (sometimes, it seems, on every page) that it is clear his book is fundamentally a meditation on the nature of friendship in the age of "user friendly." Or, rather, the root genre of the book is better described as dialogue rather than meditation. Stoll's critique is at heart a Socratic dialogue in which he convenes human friends of truth (something like "philosophers" in the root sense) to defend ideal amity against mere "talk of friendly, open systems."[52]

"User friendly," in sum, is not just an instrumental value but finally an ideal of the philosophic good life that is as broad in scope as that earlier industrialization of the good life—leisure. Ease of use, perhaps, is no less than the postindustrial repositioning of leisure *within* work. The granularity of our leisure, after all, is now set not just to the scale of our evenings, weekends, or annual vacations, or even to that of our lunch breaks and fifteen-minute work breaks. Rather, just as the total tonnage of the world's bacteria exceeds that of all other creatures by orders of magnitude, so the leisure that now bulks largest is microscopic. The good life is micro-leisure—for example, the ergonomic chair that injects small doses of comfort throughout the day; the rounded corners and muted colors of a cubicle evoking something like a transient rest home for the chronically overworked; the advanced photocopy machine that collates and staples automatically so that we may have a moment of brief, vacuous respite staring into space; and, above all, the computer interface whose ever more technically "sweet," "neat," and (we will return to the term soon) "cool" ways of saving a step here and preventing a mistake there blur the line between ease and use. Life at work is now thick with existential screensavers everywhere —interfaces that cushion the rough corners of work within a fiction of ease even as they simultaneously display the ideal of fearsome efficiency (e.g., all those classic computer screensavers that bounce objects off the screen edges to demonstrate precise Newtonian physics).

All this is to say that "user friendly" requires an explanation that goes well beyond ease of use to the cultural conditions of possibility underlying the *need* to design ease of use into technological rationality. The equivalent motto of the earlier industrial age, after all, might have been "use of ease" (getting the most productive work out of naturally slothful time-wasters). What was the cultural rationale that reversed this equation to make micro-ease a constitutive part of use? What broader explanation reveals how mixing *this* form of ease with use now seemed so right that it was doxologically intuitive? A variation on the tired cliché of the VCR whose clock constantly flashes "12:00" will help make the point. The usual moral is that the VCR needs to be more user friendly so that anyone can set the clock, program a taping session, and "time shift" TV shows for later viewing. But the broader question is why the desire for *this* kind of ease of use arose in the first place. Why was ease of use synonymous with time-shifting leisure? Since when must the *I Love Lucy* sitcoms of our time—unlike the original *Lucy,* which not only came on at a particular time but thematized that time in plots woven around Ricky's homecoming from work—be experienced as just-in-time capsules (or better, cubicles) of modular, flexible leisure? Or, to bring the illustration closer to home—which in our case means taking the scene out of the home and putting it in the office—consider the act of multitasking. Clearly, being able to keep multiple programs open simultaneously and to shift modelessly from activity to activity was the major effect of GUI ease of use. But what cultural rationale identified ease of use with such multitasking in the first place (as opposed, e.g., to the agrarian pattern of picking up different tasks in different seasons or the industrial pattern of single-tasking without interruption)? Why did a particular practice of ease happen to coincide with a particular practice of use at the exact position in social experience where multitasking emerged as intuitive?

To grasp the broader rationale of "user friendly" and thus the reason it was so well suited to symbolizing corporate culture as the generalized interface of culture, consider the section on "User-Friendly Hardware and Software" in the 1994 IITF document for the National Information Infrastructure I mentioned above, which can serve as something like our quick-start manual:

> User-friendly hardware and software always have been important for mass applications of information technology. For NII applications, such as those in health care or education, that are meant for use by broad segments of society user-friendliness will be an important factor in user acceptance. But the impact of user-friendly systems goes beyond simple convenience and marketing to serious questions of accuracy and reliability. User-hostile systems encourage mistakes in using applications, and errors in the information handled by the system.

We can make these observations. First, the "user" in "user friendly" is not just the mass-market or general-public consumer of IT services—that is, the end user. While the section begins by emphasizing "mass applications" and "broad segments of society," it decisively shifts its register midway to focus on matters "beyond simple convenience and marketing." It turns to "serious questions of accuracy and reliability" in data use and management. At this point, the user is revealed to be not so much a consumer or member of the public as essentially a *worker* (or, equivalently, a citizen operating in the mode of worker even when off the clock—e.g., using a computer to become a pseudo-banker managing a personal account or a pseudo–travel agent arranging a travel itinerary). Even in the public sector addressed by the NII, in other words, the end user is not an end in him- or herself but the means to the larger end of the *productive* use of information. While "user friendly" has moved outward from business to general society, the template remains corporate.

Second, "user friendly" is only apparently a matter of ease of use and accessibility (in the phrasing of the IITF document, "simple convenience" and "broad" acceptance). By the time the passage comes to "serious questions of accuracy and reliability," the register again shifts. Ease of use at this point looks uncannily like *controlled* use and accessibility like *accountable* use. The point of "user friendly," then, is to discourage "mistakes in using applications, and errors in the information handled by the system." Whether or not the interface is user tolerant is less important than whether information work can be managed under standards of fault tolerance that ensure zero defects across distributed production environments. From this point of view the purpose of making an interface more user tolerant (able to accommodate multiple ways of working, to correct mistakes, to help users learn the software) is to decrease faults so that distributed knowledge production becomes a single, smoothly interoperable process that can be controlled for quality. Not accidentally, "user friendly" in the IITF document is framed between sections on IT standardization and interoperability. Just before "user friendly" comes up, there are two sections entitled "Information and Data Standards" and "Conversion of Information" (the latter on the need to make print media compatible with new information media), and just after there is a section entitled "Interoperability Standards" (on ensuring "that information can be transferred between different networks, or different hardware and software systems, with accuracy, reliability and security"). "User friendly," it seems, is about adapting for distributed production a pattern that the earlier industrial era would have called systemic or technical "harmony." User-friendly IT is easy to use in the same way that a fine machine part slides easily back and forth on its track: the tolerances are so fine that ease and control appear the same.

But this analogy is not quite right. A third observation is that precisely because "user friendly" is the *postindustrial* culmination of the harmony of friendship we have been calibrating since the days of Taylor, the kind of control and accountability it imposes (as in the case of contemporary "standardization" in general) is finally *not* a matter of constraining something like an industrial age piston to a tight path of movement. Rather, control now takes the form of a specially lax or loose constraint suited to *networked* paths. Here we must turn from the IITF document to the broader lineage of usability research and design work from which "user friendly" evolved. "Constraint," we know, has been one of the major concerns of usability design. In the arena of hardware—the physical and ergonomic interface of information technology—constraint takes the form of physically positioning the user's body relative to the machine, the face to the screen, the hands to the control surfaces. The following account of Kodak's Create-A-Print 35 mm enlargement center (a free-standing machine akin to an ATM for walk-up customers) is representative of the precision of such user positioning:

> Anticipating that the unit might be placed against a wall, the center of the screen was positioned far enough from the side edge so that a customer would have room to stand squarely in front of the screen. The critical distance of 15 inches was determined anthropometrically from half of a 26-inch shoulder breadth plus 2 inches for elbow flex when using controls. The height of the screen was influenced by anthropometric eye-height considerations and the need for space to put the light path underneath it.[53]

Similar decisions about user position inform the designs of a wide range of information technology—for example, laptop computers used in the constrained space of an airline seat.[54] Of course, such design considerations do not at first glance seem markedly different from Taylor measuring the lift arc of a shovel. We would need to look more closely at the machinery in question as well as its context of use to notice what has changed in the networked age. We would need to look, for example, at the very fact that the Kodak machine is meant to be situated in any number of physical areas or that a laptop computer is mobile. Or, again, we would need to consider other kinds of user-friendly hardware—most notably, the hardware of networking itself, with its modular components chained together in flexible patterns. Constraint in the arena of hardware now means physically positioning the user in *multiple, flexible, geographically dispersed* positions.

But it is in the arena of user-friendly software that the difference in postindustrial "constraint" appears most clearly. One of the most original theorists of software constraint is Brenda Laurel, whose *Computers as Theatre* (1991) was based on her experience not just in software design but also in drama, a point I will return to below. Laurel's book contains a section specifically entitled "Constraints." As Laurel sees it, creative and well-integrated methods of constraining the user are the soul of user-friendly software. Such constraints, she clarifies, can take many forms:

> Explicit constraints, as in the case of menus or command languages, are undisguised and directly available. . . . Implicit constraints, on the other hand, may be inferred from the behavior of the system. We can identify implicit constraints when a system fails to allow us to make certain kinds of choices. . . . Some constraints have both implicit and explicit qualities. In Microsoft Excel, for example, menu-based operations are not selectable and the document cannot be closed until the current item has been properly entered on the spreadsheet. If we attempt such an "illegal" maneuver, nothing

> happens at all. We may infer from this "nonbehavior" that we must do something else or do something differently.[55]

A well-constructed interface, in other words, is one that overtly or subliminally forces the user to move experimentally (rather than "intuitively") down a narrowing corridor of hit-or-miss, learn-as-you-go possibilities. This narrowing corridor Laurel visualizes as a "flying wedge." When first encountering an interface, she observes, all is "potentially" possible. But as the user learns the different ways the interface will, and will not, work (as in the Excel example above), the wedge of potentiality gradually narrows to the zone of the "probable" before finishing at the apex of the "necessary."[56] A similar wedge of constraint, we note, defines the Microsoft Usability Group's own approach to helping Excel users. After noticing that users could not always match such menu commands as "consolidate" with the actions they desired because they conceived those actions in other terms (e.g., "combining" or "totaling" ranges), the Microsoft group invented a procedure for eliciting from test users a whole colloquial terminology to be included in the program's augmented *User's Guide*.[57] The synonyms would function like the wide end of a funnel to catch the expectations of users and guide them down to the smaller set of functional terms inscribed in menu bars. Again, the principle is that the user should be guided down a channel from a wide set of initial possibilities (in this case, semantic terms) to narrower options. Generally, indeed, the contemporary GUI interface is replete with multiple or flexible (sometimes customizable) ways to arrive at the same destination. One can invoke a command by hitting an icon, clicking in a drop-down menu, initiating a "wizard," or using a keyboard combination. Whatever the user tries within reason in a well-designed interface is liable to funnel down from the potential to the probable and finally the necessary.

Here we come to the essential difference between friendly, Taylorist control and the more lax mode of user-friendly, postindustrial "constraint." Positioned before a user-friendly interface, the user is *not* like a machine part moving back and forth within a narrow pipe of possibilities. Rather, he or she inhabits a symbolic work environment in which there is room for both decentralized multiplicity and recentralized standardization—both a seemingly wide range of applications, features, and ways of working *and* a pervasive set of constraining forces (distributed at multiple points of decision throughout the system) that consolidate multiplicity into *one* style of working, called multitasking. Now we can grasp the underlying principle of the cultural "metaform" symbolized on the "desktop" of contemporary GUIs. The modeless coordination enacted on the desktop, we realize, makes the "computer as theater" the great symbolic stage of postindustrial, networked constraint. Multitasking users are free to inhabit as many different windows or "scenes" as they wish, representing decentralized locations and protocols, diverse projects, or varying aspects of a single project. While working in a word-processing program on a document describing a product design, for example, one can also have open a spreadsheet to keep track of research and marketing expenses, a groupware messaging program to keep in touch with team members, a Web browser to consult a supplier's extranet, and so on. But all the while, it is really just one main window—the desktop—that is operative, and the working conventions of that window, which are determined through underlying operating system and networking choices made at the corporate and server level (rather than by the individual user), constrain all the flexibility, multiplicity, and dispersion of the postindustrial workday to a single mode of work called "user friendly," expressive of the tight intimacy of prosumerist service and efficiency. In classical drama, the Fates hung over the stage ready to cut the thread of life to untangle the single, necessary plot, while a chorus all the while sang its dark commentary. Now, in the postclassical drama that is every day at work, the sysadmins watch over a desktop crisscrossed with many more threads of multitasked action—a whole Web—designed to ensnare the user in a necessity as ergonomic and inescapable as a gigantic office chair, and the chorus has become a friendly User's Guide or virtual Office Assistant ("the Assistant automatically provides Help topics and tips on tasks you perform as you work—before you even ask a question)."[58]

In practice, therefore, "user friendly" in corporate network environments tends to mean not "ease of use" but "ease of administration"—that is, ease of installing and maintaining common denominator client configurations on a systemwide basis. However friendly an employee might find a competing operating system, word-processing program, or mobile computing device, for example, all such alternative IT is consigned to oblivion by the classic information services department (or sysadmin) response: "We don't support that." While "user friendly" may be a postindustrial *service* concept because it enables the contemporary worker to serve the needs of a corporate economy requiring the interaction of far-flung, distributed partners and processes, it is equally a postindustrial *efficiency* concept because such service is made possible through carefully controlled (coordinated and constrained) standards of modeless interoperability that give "service" the rigorous connotation of what IS departments call "support." "User friendly" is thus the perfect symbol for the way the meta-management of corporate culture can now span across the boundaries of firms and social sectors to become the general form of contemporary culture. Culture as perceived from the postindustrial, corporatist perspective is itself nothing but a distributed yet tightly knit web of perpetual productivity complemented by the consumerist simulacrum of such productivity, the kind of contemporary leisure that "works out" on high-tech gym equipment. (Though they seem worlds apart, the soft-contoured, ergonomic office chairs of today's cubicles are secretly connected—like Jekyll to Hyde—to the angular, metallic, torture machines that office workers rack themselves upon in the fitness gym after hours.) Just *try* giving up an ergonomic chair or a GUI! These artifacts of user-friendliness are ultra-malleable, but they are also absolutely sticky. They glue one to the stronger-than-steel strands—for example, Cat 5 Ethernet or optical-fiber cables—that "support" the world of knowledge work.

We can close this overview of the combined practical and symbolic function of the user-friendly graphical user interface by returning to Laurel's *Computers as Theatre,* which ranks alongside Johnson's *Interface Culture* as one of our most thoughtful meditations on the notion of the interface. The unique aspect of Laurel's book, as I have already alluded to in my theatrical metaphor, is its bold attempt to use Aristotle's *Poetics* to explain coordinated, modeless interfaces as *dramatic* experiences. For Laurel, all the world is, if not a stage, then a desktop on which windows are the proscenium and icons the props. Computing makes intuitive sense to the user, she argues, only if it is perceived—or better, performed—as the dramatic version of modelessness: unity of action. The interface "should concern itself with representing *whole actions with multiple agents.*"[59] It should coordinate all events so that possibilities converge on probabilities and finally on necessity (the exact vocabulary, we now realize, of Aristotelian plot analysis). Nothing inexplicable or anomalous can occur to cause the user to ask, as if witnessing a bad play, "What is happening?"

Crucially, Laurel adds, the most subtle yet powerful form of constraint responsible for dramatic unity is *characterization:* "The most direct material way to influence action is through the shaping of character and thought."[60] The case is clearest in the extreme context of computer games (with which Laurel has extensive familiarity as a designer, researcher, and later co-founder of the Purple Moon Media company that produced "friendship adventure" interactive games/stories for girls). Software that characterizes the user by giving her or him an explicit game role constrains ensuing events to settings, actions, or tools that appear "intuitive" because they are unified around the character (for example, "new girl" at school in the Purple Moon friendship adventure *Rockett's New School*). Such characterization, Laurel notes, is akin to deep or method acting in the theater. But the case is no less evident in the theater of work, where characterization occurs whenever a software interface makes assumptions about the worker's persona that "intuitively" narrow down her or his range of actions. As Laurel puts it, "The character and thought of *human* agents may be designed to a certain extent. . . . Systems must capture and respond to key traits of human agents. They may also transform human traits or even endow a person with new traits in order to enhance his or her ability to act."[61] Everywhere we look in today's GUI interfaces we see examples of such designs upon the human. Opening the menu of "templates" in a word-processing program, for instance, we see options that include "Agenda, Resume, Brochure, Memo, Report"; clicking on a desktop icon, we open a "folder," lift a "briefcase," and so on. These are all subtle ways of scripting us for a particular role. Where, by contrast, is the icon or template for "Whistle-Blower," "Sexual Harassment Complaint," or "Discrimination Grievance," let alone "Unemployment Benefits" or "Welfare Application"? Or consider Laurel's own example: expert system software "agents." Such agents watch over a user's actions, build a background model of what the user is attempting, and then pop up with a "tip" presented by an animated character. Apple, for instance, experimented with an agent named "Eager," and Microsoft in the late 1990s added to its Office suite its infamous dancing paperclip performing the role of "Office Assistant."[62] Such agents engage with the user not just as dramatic character to character but ultimately (because they are instrumental in constraining agency to the imaginary stage set of the desktop) as *director* to character. That these agents are the personification of "friendliness" is no accident. On an actual desktop, paperclips do not smile or dance, but on the virtual desktop they are a whole vaudeville act. That a dancing paperclip can show only a "wireframe" of a smile (as the graphics software industry might call it) is immaterial. The point is that it directs flesh-and-blood workers *how* to smile in a world as grim as a spreadsheet.

We need make only two slight modifications to the terms of Laurel's theory to draw down the curtain on friendly technological rationality as we have seen it evolve from Taylor's time on. One modification is to translate dramatic action into business terms as the action of *work*.[63] To say that "user friendly" creates a unified action is to say that it coordinates all the (multi)tasks of contemporary employment within a single mode of work whose decentralizing/recentralizing and service/efficiency aspects are finally no more contradictory than tavern and court scenes in a Shakespeare history play. The contradiction is still there, but it is normalized through the imposition of an overarching sense of harmony as inescapable as Fate or Divine Providence. But instead of tragic or divine fate—or what is the same in Shakespeare's *Henriad*, history—the fate in question is now "friendly." It is the new Atropos who smiles even while cutting the thread on an employment contract. Our second modification to Laurel's thesis is thus to say that the kind of work that can enforce such harmony through characterological means—scripting everyone in the role of friendly "efficient service"—is no longer simply work. It is a symbolic representation of work as the greatest of "friendship adventures," *culture*. Where industrialism first wrote Marx's drama of alienation as Ford-faced, undramatic work, and where the mainframe era next adapted the play as a deep act of paradoxically cold/smiling corporate culture, now the networked age directs—or rather, programs—the final drama. The deepest act will henceforth be a scripting that binds workers not just to the friendship system of corporate culture but, through their automatic participation in a universal environment of "user friendliness," to corporate culture as the stage of *general* culture, as the new model of general sociality, interaction, and communication. We don't need to be kind, generous, tolerant, accepting, sympathetic, or, in a word, social, anymore. We just need to be user friendly, which is the same as being corporate.

"Culture" will now be figured as a networked world of corporate subjects become pure, distributable objects—teamworkers who have no legitimate identity and emotion except when wired like animated paperclips into the network. Culture will consist of "human agents" who are finally indistinguishable from friendly software agents. The dancing paperclip is the perfect image of such a teamworker/networker: an object that exists not in its own person but purely as a creature of wires designed to connect multiple actions, players, and scenes of information work into a facsimile of "knowledge" (figured in a lightbulb above the clip).

We are now ready to conclude our study of the networking age in the same way we capped our study of the automating and informating ice ages—by taking the temperature of the "cool" of the times. What is cool in an era when all our techniques are bonded to all our technologies through a paradoxically de- and recentralized network of standards, protocols, routines, metaphors, and, finally, *culture* that makes knowledge work simulate an eternal, inescapable friendship? Consider the following speculation in Johnson's book on possible future "subcultures" of the user-friendly interface: "The interface subcultures of the future will offend the traditionalists by being too *difficult*. 'User-hostile' may sound like an odd goal for interface design, but the truth is the field could use a little tough love. No medium has managed to reach the status of genuine artistry without offending some of its audience some of the time. Even under the user-friendly dictates of interface design, you can't make art without a good measure of alienation."

Might there someday be such a subculture and art of the interface—one that is cool to the friendship system even while wired into the culture of that system?[64]

The answer, I believe, is that such a subculture is already here. But it is not a subculture, and not even a counterculture. It is the intraculture of knowledge work. Whether it can also be art is an open question we will need to consider after first studying the intraculture of "cool."

PART III

The Laws of Cool

At lilac evening I walked with every muscle aching among the lights of 27th and Welton in the Denver colored section, wishing I were a Negro, feeling that the best the white world had offered was not enough ecstasy for me, not enough life, joy, kicks, darkness, music, not enough night.

Jack Kerouac, *On the Road* (1957), quoted in Hebdige, *Subculture: The Meaning of Style*

Someday, we'll all agree on what's cool on the Net. In the meantime, the Netscape cool team will continue to bring you a list of select sites that catch our eye, make us laugh, help us work, quench our thirst . . . you get the idea.

Netscape Communications, Inc., "What's Cool?" page (March 1996)

Preface

"What's Cool?"

"What's cool?" the Netscape corporation asked on one of the best known pages of its Web site in 1996–97.[1] The answer offered on the page (quoted as one of my epigraphs above) quickly passes over the "we" who do not yet know what cool is ("someday, we'll all agree on what's cool on the Net") to install a "cool team" charged with generating an empirical definition of cool—a list of sites. These are the sites, the team says (itself now appropriating the first-person plural), "that catch our eye, make us laugh, help us work, quench our thirst . . . you get the idea" (ellipsis in the original).

For just a moment, it is as if we were latter-day Chaucerian pilgrims questing ("searching," they call it on the Internet) not so much for a holy site as for any quotidian site along the way whose sensual, even multimedia gusto can help us reintegrate our lives of work and leisure ("eye," "laugh," "work," "thirst"). Or, again, it is as if the cool team were leading us on an exodus out of the pharaoh's land of routines, procedures, standards, and protocols toward a land of milk and honey far beyond the reach even of company rafting trips, rock climbs, desert hikes, and other such retreats of corporate culture. It is as if . . . you get the idea.

What is cool in the age of networked information? What is cool when even our youngest children know to say "cool" in the presence of high technology?[2] when *cool* and its antithesis, *suck* (as in, "either you're cool or you suck"), have become the two most widely diffused slang terms for approval and disapproval among American high-school and college students (with recognition rates above 90 percent)?[3] when students and workers entering the corporations are as likely to dress their stereos, TVs, and other high-tech gear as themselves in cool, haute-couture black? and, above all, when everywhere we look (at TV shows or films with their own Web sites, for example), a new king cool, *information cool,* has been anointed to carry all the rest into the new millennium?[4] The day is not yet, perhaps, when we will be cool for the circuitry tattooed under our skins or gold processor sockets inlaid in our heads, though the swarming of sleek cell phones and handheld computers may presage such info-fashion. But already our computer "desktop," that workaday interface of information culture, sports a new look far more seriously cool than any of the screensavers of old. Our desktop, as it were, has grown tailfins.[5] At a click, its window opens onto the World Wide Web, whose hyperbolic, even desperate, cool may be taken to be the insignia of information cool.

If Web cool is the tailfin on the machines that take us down the information superhighway, what do we see when we stare into all that chrome? What is information cool?

The answer, I propose, may be looked for within the ellipsis on that classic 1996–97 Netscape cool page. At once gesturing toward the "idea" of cool and withholding that idea, such an ellipsis is a uniquely paradoxical inexpressibility topos—one whose fullness of implied content (site after site after site) is constrained by a silence that is more than simply neutral or practical. The silence is also a proscription, an interdiction. It takes very little pressure upon cool, after all—*asking* someone to define it, for example—to bring out the force of that proscription. Those who insist on asking, the Internet has not so subtle ways of declaring, are definitely uncool.[6]

The proscriptive ellipsis may be taken to be the elementary rhetoric, the mental pixel, of information cool. Inflected in different ways and on various scales, the rhetoric of unproducible knowledge—of knowledge that can never be known and shown simultaneously—is universal to cool online. Of course, there may not always be a literal ellipsis, and the proscription may be less prohibitory than ascetic. But the thought—or rather, unthought—is there. Consider the Project Cool Web site, for example, which in 1997 posed its version of Netscape's "What's Cool?" question. "What is this thing called cool?" Project Cool asked on a page titled "About the Coolest on the Web" explaining its anthology of cool sites.[7] What is "this thing everyone wants/ and no one can quite define"? The answer came in a poem of (in)definition:

> There is no one definition of cool.
> There is no one definition of Beauty
> Art
> Obscenity.
> It's a sort of
> "I know it when I see it"
> type of thing. You can argue
> 'til the cows come home
> that this was or wasn't cool,
> but it's all pretty subjective.[8]

Declaring that there is nevertheless "method to our madness," the poem then goes on to list several formal criteria for cool (writing, design, content, graphics, and so on). Yet it is these definitive criteria that the poem at last throws into doubt:

> We set up some categories.
> They aren't perfect,
> but they seem to work.
>
> We started filling them with
> sites that exist today,
> sites that we like,
> sites we think are cool for that category.
>
> We stepped back and,
> with reason and deliberation,
> deconstructed our gut instinct.

In the last stanza above, *deconstruct* appears to mean "take apart" or "analyze" (one popular misuse of the word). Project Cool is defining cool by taking its "gut instinct" responses to cool Web sites and breaking them into analytical categories. Yet almost immediately any "method" of analysis is destabilized through the introduction of a kind of Lucretian turbulence. According to the gnomic last stanzas of the poem,

> Cool isn't static.
> Today's cuts might not make it
> next week
> next month
> next year.
> The definition evolves . . .
> Coolest on the web is always in flux,
> reflecting current state of the art.

This final evolution of definition into "flux," we may say, is Project Cool's true act of deconstruction. Categorical definition prompts a powerful reticence about, or preemptive deferral of, definition that indicates not so much ignorance as over-full knowledge. There is so much "more" about cool than can be said without falsifying its *je ne sais quoi*. "The definition evolves . . . ," the poem says, ellipsis and all. Step back from the mystery, therefore, even as we approach it. The koan of cool can be put as follows: *we know what is cool, but part of what we know is that we cannot know what we know. Cool forbids it.*[9]

What's really cool, after all? At the moment of truth on the coolest Web sites—when such sites are most seriously, deeply cool—no information is forthcoming. Cool is the aporia of information. In whatever form and on whatever scale (excessive graphics, egregious animation, precious slang, surplus hypertext, and so on), cool is *information designed to resist information*—not so much noise in the information theory sense as information fed back into its own signal to create a standing interference pattern, a paradox pattern. Structured as information designed to resist information, cool is the paradoxical "gesture" by which an ethos of the unknown struggles to arise in the midst of knowledge work.[10]

Since I am less concerned with the deconstructive rhetoric of cool that Project Cool puts in play (albeit in a debased way) than with the cultural equivalent of such rhetoric—a particularly contradictory social posture or stance—it may be helpful to make an analogy to a human science that is practiced in the paradoxes of culture: ethnography or cultural anthropology. The proscriptive ellipsis, I suggest, tells as much upon the elementary structures of cool as the incest prohibition—another universal nix—upon what Claude Lévi-Strauss called the elementary structures of kinship.[11] "Thou shalt not commit incest" and "thou shalt be cool" are strangely correlative laws, even though in some senses cool is precisely the incest of information (information fed back into its own signal).[12] As a preface to an investigation of cool, therefore, we can imagine that we are anthropologists encountering information culture as a new mode of "savage thought." What are the paradoxical laws of cool that give the tribes of the cubicle their secret culture amid corporate culture, their fantasy of knowledge work, their Atlantean cool underneath the sea of information?

Just four themes of cool in the information age—each phrased as an assertion about the life of information, followed paradoxically by its contradiction—will provide an adequate dossier. Cool is, and is not, an ethos, style, feeling, and politics of information.

Chapter 5

The Ethos of Information

Cool Is an Ethos of Information

We may remark, first of all, upon the sheer scale of cool on the Web. Of course, there can be no pretense of a systematic census, even in the merely prospective or incomplete mode of empiricism for which Derrida criticized Lévi-Strauss.[1] Not only do network phenomena transform the notion of systematicity and completeness, as we saw in chapter 4, but the primary instruments that allow us to attempt a census, Internet search engines, are themselves part of the network and thus themselves constitutively unsystematic and incomplete. (Their indexes cannot cover the whole Web, especially the increasing proportion of information served up dynamically from back-end databases; their searching algorithms vary in undocumented ways; their results are structured or presented in forms that cannot always be compared; they cannot find text rendered in graphics; they cannot filter out literal uses of the word *cool*.) The following census, therefore, is offered merely as a propaedeutic; its uncertain indication of quantity has value only insofar as it leads us to discern the central quality concealed in such quantity.

I conducted this census in 1998 when the Web (and Internet) first seemed to have reached a threshold of maturity—in density of servers, users, usages, and domains (including the now all-important .com domain)—sufficient to give a sense of its full potential. On 6–7 July 1998 the major Internet search engines produced the counts for *cool* shown in the table below.[2]

Survey of "Cool" and Related Terms on the Web Using Major Search Engines, 6–7 July 1998

Search Engine. (total pages in database [millions])	AltaVista (140)	Infoseek (30)	Excite (55)	Hotbot (110)	Northern Light (67)
"Cool" anywhere	5,681,310	2,582,284	676,122	1,614,631 (1,314,428 in North America)	1,424,618 (excluding proprietary pages)
In page title	71,444	19,168	N/A	57,773	N/A
In page text (excluding links and images)	2,031,469	N/A	N/A	N/A	N/A
In link(s) on page	456,529	132,630	N/A	N/A	N/A
In URL	41,604	7,029	N/A	N/A	N/A
"Cool links"	68,761	3,209	647,140	65,312	82,541
"Cool sites"	34,980	1,369	647,140	36,006	83,936
"Cool stuff"	34,996	1,121	647,140	64,342	59,405
"Cool pages"	1,205	417	647,140	15,221	26,597
"Cool cool"	17,347	34	46,690	2,952 (exact phrase)	3,952
"Kewl"	97,070	567	11,643	12,167	24,235

These are evidently "lumpy" results (for example, two of the highest rated engines at the time, AltaVista and Infoseek, used algorithms that returned totals for *cool* amounting to about 4 percent and 9 percent of their entire database, respectively, while others returned a much smaller relative number.) A more scientific study would also need to extend the census to successive years, especially since both search engine technology and the academic study of search engines have since evolved (as represented, respectively, in the Google search engine that became prominent in the early 2000s and the methodology that Geoffrey Rockwell and his research group have developed to use search engines to tally cultural phenomena).[3] Nevertheless, this "snapshot" of cool on the Web in 1998 allows us to make some useful observations.

First, there is clearly much more cool on the Web in the form of self-declared cool pages, references to other cool pages, and general appreciation of things cool than might be expected for a word whose usage is nonfunctional. AltaVista's count for *cool* (5.68 million), for instance, tops its count at the time for *useful* (3.95 million) and far exceeds matches on such other specific stylistic/aesthetic descriptors as *beautiful* or *nice*. The frequency of *cool*, indeed, falls in the range of such relevant genus-level concepts as *style* (5.79 million). Only *hot*, among other comparison terms that readily come to mind, exceeds *cool* in matches (12 million in AltaVista) because of its overwhelming presence on sex sites (as in "hot hot bodies" or "hot hot hot XXX hot hot hot").[4] To put things in proper scale, we might consider that such mainstay words of the medium as *computer* and *business* produced counts in AltaVista of 29.8 million and 43.4 million, respectively. This means that *cool* occurs on a Web page once for every five to eight times that such words as *computer* and *business* occur. (And this is just to count the Web itself and not such other provinces of the Internet as Usenet, e-mail, instant messaging, and so on.)[5] It is as if cool were the chorus of information: wherever information and its technology enact the contemporary drama of knowledge, there a chorus sings, "cool."

Second—and here I begin to shade into qualitative analysis—the raw totals understate the case considerably because cool collects disproportionately in those parts of Web pages that have premium value: in the encoded page title, the links on the page, and frequently even the URL or domain name (e.g., "www.coollinks.com").[6] Concentrated in these locations, cool is a rhetoric that is purposely installed as close as possible to the functional code of the Web—the parts of the HTML or Hypertext Markup Language that are not simply content but that allow browsers to navigate and operate upon content. When one connects with a server named "www.coollinks.com," that is, one is connecting with a rhetoric that is not just a para-text or "para-site" of information. It is part of the DNA of networked information itself. Cool is not a chorus standing behind information and its technology, it is a voice like conscience issuing from *within* IT.

Indeed, the idea of "conscience" takes us in the right direction. Third, cool exploits its position close to the bone of networked information to express information culture's awareness of itself. Unlike such comparable terms as *hot, beautiful,* and so on, cool does not describe primarily or even equally the external realm that information purveys. Rather, its primary reference on the Web is the Web itself. Cool is the code (in almost a literal sense, as we have seen) for awareness of the information interface—as if at the moment of cool we stared not *through* "windows" toward the content of information but *at* the gorgeousness of stained-glass windows themselves.[7] As Steven Johnson has said, the desktop today is what we have instead of Gothic cathedrals. It is our interface with infinity (or what I earlier called lateral transcendence), and as such it is not ultimately about the transmission of information.[8] Rather, clarity of transmission is subordinate to the complex opacity of what Coleridge, during the revival of Gothic in the Romantic and Victorian era, called the "translucence" of symbol.[9] Information is refracted in such a way that its always too mundane focus (yet another personal, corporate, or media home page, for example) seems to be lit from within by an unearthly halo of cool whose source lies on some other plane entirely.

Lest we grow too sanctimonious about the crystal cathedral of the desktop, however, we may note that the inner light symbolized by cool comes in for its fair share of irony—though this merely raises to an extra power the awareness of the interface that is my main point. Besides cool awareness of the interface, there is also cool awareness of that awareness. Consider, for example, such inflections of cool appropriated from youth culture as *kewl* (the orthography for "coo-ooool" pronounced with a drawn diphthong to denote one's meta-, arch, or otherwise differentiated sense of cool).[10] So prevalent was *kewl* on the Web by the mid-1990s that icons and buttons emblazoned with the term routinely showed up in downloadable icon collections, ready to be included on any page as a hipper way to be cool. Similarly arch or sardonic were such Web sites of the mid-1990s as Jensen Harris's *Mediocre Site of the Day* and Mirsky's famous *Worst of the Web* (both now extinct), which parodied the generic "cool site of the day." Cool, in short, is not just a form of consciousness (the old-fashioned term for information interface); it is a form of self-consciousness. Cool names the moment of tricky reversal when we see that interfaces are always two-sided: at the moment of cool it is as if the user throws his or her point of view ventriloquially outward into the realm of information and from there peers inward back through the interface at his or her own awareness of that information. Cool is not so much a quality of information as a view or stance toward information, expressed from within information itself. It is a "way of looking" at the world of information that exceeds the utilitarian sense of either presenting or receiving information to include the sense that we (whether the user or the Web page author) are ourselves being presented, displayed, or received in the way we feel about information. Looking at a gaudy, animated cool page, it is really *our* peacock feathers that begin to lift.

Thus the essential quality we are looking for amid the enormous quantity of cool on the Web comes clear. Coextensive and integral with information to the point of being its other face (the alter-face, as it were, of its interface), cool is an ethos or "character" of information—a way or manner of living in the world of information. Too fundamental and inchoate itself to be called an identity, it is nevertheless the formative material of imagined identities promising knowledge workers some hope of alternative lives of knowledge. In the words of a later version of Netscape's "What's Cool?" page, cool gives the knowledge worker the hope of "personality." (Netscape added an "Editorial Policy" subpage to its "What's Cool" page in 1998 that declared, "When it comes to cool sites, personality goes a long way.") In particular, cool gives the knowledge worker a paradoxical personality akin to the pronoun *I* in a statement of the sort, "To know who I am, you've got to see my family, my house, my car, my music collection, my Web bookmark list, my home page, my. . . . "[11] Such an "I" at once steps forward proudly and withdraws utterly into ellipses or etceteras of external objects of knowledge. Personal home pages are thus by themselves just "more" information. But saying the word *cool* adds the sense that someone—some character or personality—is truly at home within all that faceless work of acquiring, exchanging, merging, delivering, and otherwise interfacing information.

But we are not yet in position to discern the exact nature of this character that thinks it is at home in the work of information. Before we can know who is at home in information work, we need to who is simultaneously *not* at home in it.

Cool Is an Ethos against Information

If cool is the ego of information, then clearly it is also an alter ego. In this regard, the most influential cool anthologies on the Web—sites that over the years have become something like churches of cool celebrating a saint's calendar of cool sites of the day, week, year, and so on—have it exactly wrong when they later legitimate their original faith by asserting the information sanity of cool. Witness the 1998 "Editorial Policy" added to Netscape's "What's Cool" page (now confident enough to drop the question mark). The policy statement, as we have seen, first identifies cool with personality. But then it emphasizes the "relevance" and "utility" of content as well as the "clarity" and "relevance" of design: "A cool site can't be so aloof that it fails to address its subject matter. The overt mission of a site is to communicate its message quickly, effectively, and playfully." So, too, the 1998 version of the *Cool Site of the Day* (the canonical cool list originated by Glenn Davis) added "How to Be Cool" instructions starting with the dictum, "Create something useful."[12]

Such revisionism, it may be suggested, protests too much. Although most users continue to view the Web as a utilitarian rather than entertainment medium, this is not to say that cool has anything to do per se with utility of information.[13] Closer to the mark, perhaps, were such sites in the 1990s as *The Useless Pages, Ambit Totally Useless: Sixty-Four Exquisite Sites with No Purpose, The Worthless Page,* and *Deb and Jen's Land o' Useless Facts.* Though hyperbolical, these satirical pages emphasized that the allure of information cool could just as well be said to be useless as useful information. Or, more accurately, to borrow a concept from the Critical Art Ensemble, the allure lies in "the technology of uselessness": "To expand on the suggestion of Georges Bataille, could the end of technological progress be neither apocalypse nor utopia, but simply uselessness? Pure technology in this case would not be an active agent that benefits or hurts mankind: it could not be, as it has no function. Pure technology, as opposed to pure utility, is never turned on; it just sits, existing in and of itself."[14] In the end, the schema of useful versus useless is inadequate, for it is the uselessness of useful information upon which cool rings the changes. Relevance, clarity, and utility are not the virtues of cool *qua* cool except insofar as these traits of efficient information are parodied, misapplied, radicalized, and otherwise used against the grain (as the Russian Formalists might have put it) to "roughen" and "defamiliarize" the notion of efficient information.

Leaving for later the issue of graphic design, let us concentrate first on the content of cool to observe some of the grossest and most obvious uses of information to abuse information. "Cool anthologies," as I have called them, are an especially convenient guide in this respect. The following list is a sampling of anthology sites circa 2000 that includes, besides general-purpose anthologies of the sort I have already mentioned (Netscape's "What's Cool?"; *Cool Site of the Day;* and Project Cool's "Coolest on the Web" or "Sightings"), a cross-section of more casual or specialized anthologies from different parts of the world, different fields, and even different age groups.[15]

- *All-In-One Cool Sites Page*
- *Cool Central*
- *Cool Crescendo Site of the Day*
- *Cool Links of the Week*
- *Cool Medical Site of the Week*
- *Omni Magazine*
  - "CoolScience"
- *Cool Site of the Day*
  - "Cool Site of the Year Awards"
  - "Cool-O-Meter"
  - "Still Cool: Archive"
- *Cool Site of the U.K.*
- *Cool Sites on the Web*
- *Coolest Links on the Net*
- *Project Cool*
  - "The Coolest on the Web"
  - "Sightings"

- *Extra Cool Sites*
- *Netscape Communications, Inc.*
  "What's Cool?"
- *NetWatch: Cool Sites!*
- *Rachel's Cool Site for the Time Being* (by Rachel, a sixth-grader)
- *TnT—Tristan and Tiffany's Cool Stuff for Kids*
- *Too Cool Awards*
- *TUM Cool Site of the Week* (Japan)
- *Various Cool Links to the World*
- *What's Hot and Cool on the Web*
- *Yahoo!*
  "Cool Links"

Such anthologies may be arbitrary in the specific cool sites they feature (though there is a surprising amount of consensus), but collectively they attest to consistency in the *kinds* of sites deemed cool. Rather than attempt a systematic survey of such kinds or genres (a difficult task because only some sites keep archives of past selections), I will take the liberty of creating a mock-up of a typical cool anthology site. The links in the mock-up are organized for elucidation into a set of informal, overlapping genres of cool—each named by my designation but stocked with examples from the indicated, actual cool anthologies (supplemented on occasion by similar examples I myself collected while creating *Voice of the Shuttle*):[16]

Links on a Typical Cool Sites Page, c. 2000*

Ordinary Cool

- *Tabatha's Days at Work* ("Welcome to the automagical Tabatha MPEG movie maker. She has a camera pointed at her which snaps a photo every five minutes") (Martin Richard Friedmann) (C 2 December 1994)
- *The Simulator* (takes you through the day from getting up to driving to work to making fast-food hamburgers, burger after burger, and so on; at each point, there is interactive "choice") (Garnet Hertz) (C 30 August 1997, P 18 August 1997)
- *Online Cameras* ("Wanna see what other ignorant humans are doing right now?") (AvEdis)
- *Carolyn's Diary* (diary in HTML) (Carolyn L. Burke)

* C=*Cool Site of the Day*; N=Netscape's *What's Cool?*; P=Project Cool's "Sightings." Brief descriptive annotations are adapted from my VoS "Laws of Cool" page.

- *The Semi-Existence of Bryon* (from original description: "On this page I keep my diary. Sometimes it is interesting, sometimes not but it is always real. It is updated daily")
- *What Is Miles Watching on TV!* (Miles Michelson) (N 7 March 1996, C 12 April 1995)
- *What's Inside Jeremy's Wallet?* (Jeremy Wilson) (C 21 October 1996)
- *Motel Americana* ("Welcome to Motel Americana. Here, you can re-live the glory days of the classic roadside motel") (Jenny and Andy Wood) (C 11 November 1995, P 28 January 1996)

Ultra-Tech

- *The Amazing Fishcam!* (Netscape Communications, Inc.) (C 10 September 1994, N 7 March 1996)
- *Mike and Anthony's Wired Room* (wired dorm room at Dakota State University tells you the state of its doors, lights, temperature, number of socks in drawer, and so on; visitors can activate a fan to blow on Mike and Anthony)
- *Paul's (Extra) Refrigerator* (technical detail on the current temperature and other status of Paul's other refrigerator; includes extensive meta-technical detail on how the above information is monitored) (Paul Haas)
- *The Tele-Garden* (experiment in telepresence from the creator of the Robot Tele-Excavation Project; users manipulate a robot arm to plant seeds in a garden, water them, and watch them grow) (Ken Goldberg and Joseph Santarromana) (C 25 July 1995)

Ultra-Graphical

- *MkzdK* (Stephen Miller) (C 14 October 1994, 8 March 1997)
- *Crash Site* (Big Gun Project) (C 19 April 1995, P 10 June 1996)

The Experience of the Arbitrary

- *URouLette* (one of the original random site generators) (Jill M. Sheehan) (C 3 September 1994)
- *The Information Supercollider* (random link generator) (Electrical Engineering and Computer Science, Harvard University) (C 25 May 1995)
- *MagicURL Mystery Trip* (random link generator) (Ryan Scott)
- *Michi-Web Random Link to a Random Page* (meta-random site that takes the user at random to sites that generate random links)

Cool Personal Home Pages

- *Paul's a Computer Geek* (Paul Schrank) (P 9 December 1996)

- *The Rest of Me* (Karawynn Long) (C 6 August 1997)
- *Flaunt* (Shauna Wright) (C 8 July 1998)

We can briefly describe these categories as follows.

"Ordinary Cool" designates pages whose obstinately mundane, everyday, personal, or otherwise unremarkable information is rendered cool through the very fact of being put on the Web (with or without any special technical or design prowess, though prowess is a plus).[17] On Martin Friedmann's *Tabatha's Days at Work,* for example, we click on MPEG video files offering fast-motion glimpses of Tabatha going about her ordinary office routine in the early 1990s. Garnet Hertz's *Simulator* similarly provides an interactive visual recapitulation of nothing more interesting than getting up in the morning, choosing one's clothes, driving to work, and making tedious burger after burger (each of which must be assembled part by part). *Carolyn's Diary* and *The Semi-Existence of Bryon* (two of the earliest, best known, and most compelling online diaries before the advent of today's "blogs" or Weblogs) chronicle the lives of persons whose most distinguishing trait from the point of view of the Web is that they put their lives on the Web.

"Ultra-Tech," which overlaps with ordinary cool and other categories, is a convenient way to characterize pages whose use of information technology is so advanced, either absolutely or relative to the matter at hand, that it goes beyond "high tech" to become an experience of high tech *qua* high tech—of advanced information technology liberated from its function of serving anything as useful as information. In this regard, the "Extra" in the title of *Paul's (Extra) Refrigerator* is emblematic. Not only is the topic of this page an existentially "extra" refrigerator (Paul's main refrigerator is never mentioned), but the combination of remote sensor and information devices trained on Paul's spare refrigerator to allow us to remotely monitor its temperature, light, and contents is fundamentally excessive of the utility of IT. Paul's extra refrigerator is cool because it literalizes the way in which a cool page converts information into something that, though nothing but information, is finally as impervious to the transmission of information as the white door of a refrigerator—the very blankness of which becomes in common use the tablet for a secondary layer of nonfunctional information (photos, children's drawings, magnet poems). It is as if, for just a Monty Python–like moment, the intricate, brilliant computing machine on which we spend our working days entering information were converted into a 500-pound meat locker (or digital processing into the mere issue of whether the refrigerator light is on or off). Other well-known cool pages in this genre—the *Amazing Fishcam!* for example—have the same effect of instantaneously dumbing down our computer into an aquarium, iguana habitat, and so forth. Accessing the information on such pages, we learn the remarkable fact that the fish have indeed moved since the last time the Webcam looked, or that, no, the iguana has still not moved.

"Ultra-Graphical" refers to sites that go well beyond the routine use of multimedia to make graphics, video, animation, and sound bear critically on the notion of information itself. Rather than just use multimedia to deliver, illustrate, highlight, or navigate information (the reasons typically invoked to legitimate so-called Web "eye candy"), ultra-graphical pages discard the entire apologetics of imagery on the Web to reveal that one fundamental instinct of such imagery is to impede information. Our first impression upon opening Stephen Miller's *MkzdK* (at the modem speeds prevalent during the site's genesis in the 1990s), therefore, is how long it takes to download all the gorgeous and intricate graphics (not to mention the audio tracks and extra software "plug-ins" necessary to play the multimedia). Our second impression is that while the resulting graphics are nothing but functional (the primary means of moving through the site), they nevertheless suborn all the "legitimate" functions of Web graphics. They neither directly communicate information in the manner of a house photo on a real estate page nor complement text in the manner of ordinary icons, buttons, logos, or pictures used as links. The graphics on this page mark out zones of pure possibility, uttering nothing and compelling us to click on them in blind trust. The few initial text labels that do exist—e.g., "Cosmos" and "Visions"—are not only nearly as cryptic as the name *MkzdK* but lurk *within* the complexity of graphic images. Only after an inordinate amount of time spent looking around, exploring different navigational avenues, and waiting for the media files to download for each such avenue can we finally reach any text capable of informing us what *MkzdK* is "about." Yet at this point whatever description we come up with to characterize such information—perhaps "New Age" or "cosmic consciousness"—is curiously inadequate. The real information—the site's repudiation of our normal, technological-rational understanding of information—has already arrived en route to the information. A shrewd, alternative intelligence lies concealed in the graphics in plain view.

"The Experience of the Arbitrary" names a different way of thwarting the expectation of information. I refer to the experience of indulging in the unpredictability, quirkiness, and sheer aleatory surmise that is everywhere in cool but nowhere more explicit than on "random link" pages that exist for the sole purpose of converting informational relevance and use into their direct opposites: irrelevance and play. Random link pages feature a link that, when pressed, starts a randomizer program (usually running on the server) that launches the user to any of a large number of URL addresses. Click once on the link in Jill M. Sheehan's *URouLette,* for example, and go to a bank's home page, click again and go to someone's personal home page, and so on. The ultimate zen of such arbitrariness, perhaps, is the *Michi-Web Random Link to a Random Page* site, which adds a meta-layer of randomness (it randomly sends us to a random link page). The destinations that random link pages discover are sometimes useful, but only in the way of the proverbial monkeys typing Shakespeare. Random link generators, we may say, take an information technology originally designed to be as certain as 1 + 1 = 2, or at least as probable (in a keyword search) as "Shakespeare + Romeo = *Romeo and Juliet,*" and make it say that 1 + 1 equals anything at all (even, conceivably, *Romeo and Juliet*)—and all this through *overutilizing* rather than—like the monkeys—underutilizing IT.

Finally, we can mention the inevitable "Cool Personal Home Pages." I refer to pages that exceed the cool of simply *being* someone's personal home page (the original cool when the Web was in its infancy). Since the late 1990s, really cool personal home pages have instead showcased pictures, lives, friends, interests, thoughts, pets, and so on, with exorbitant technical sophistication, multimedia splendor, arbitrariness, or some blend of the other categories listed above. These are pages that do not merely inform us about someone, but become whole digital environments through which we wander as if through an autobiographical version of those ground-breaking exploration computer games of the 1990s, *Myst* and *Riven*. Entering Paul Schrank's *Paul's a Computer Geek* site, for example, we are pulled ever deeper into areas of unpredictable interest, including music and design philosophy, expounded with such technical sweetness combined with design panache that they ironize the "Geek" that is Paul's nom de plume. If this is a geek, then there is an *r* missing, as in Apollonian "Greek." So, too, Karawynn Long's *The Rest of Me* lures us ever deeper—"wade," "swim," or "drown" are its three levels of navigation—into a shrewdly designed site that transforms the often flat quality of a personal home page (like turning a page in a photo album) into a multidimensional experience that becomes literally limitless if we register to read Long's online journal. In both instances, advanced resources of the medium are used for purposes that undercut, amplify, or complicate the notion of personal information until it becomes something other than information. Long's site even makes a game of deliberately resisting information. There are two modest nude photos of herself on the site, Long promises, but these are obscured under a Javascript-generated matrix of cards that disappear only as pairs of cards are matched

("To see the first photo you'll have to match twenty-one pairs of animal tracks, some of which are deceptively similar"). The result is a classic memory game that uses all the varieties of computer memory—the cache on the processor, RAM, perhaps also the swap file on the hard drive—for the sole purpose of reducing us once more to faulty human memory.

Wherever we look in the anthologies of cool, in sum, we see an ethos that is *of* information—nothing but information—but also *against* information. On the one hand, information cool is the latter-day version of what I called mock- or camouflage-technology (the outsider's appropriation of workaday technological rationality), except that in this context there is no outsider's position. Employed within information work, cool is so enthralled by information technology that it identifies with it utterly. In the case of *Paul's (Extra) Refrigerator,* it is not just that there is no primary refrigerator; there is finally no Paul either. There is only the extra refrigerator on the scene equating the persona of Paul with an object of purely supernumerary knowledge constructed at the dense convergence point of remote sensor and Internet technologies. Cool is not just mock- or camo-technology, we may say, but deco-technology. It is an experience in which people disappear into the environment—the decor—of useless technology. Yet on the other hand, as the notion of decor suggests, all the while that cool masquerades as information, it also prettily subverts information. Taken to an extreme, deco-technology mocks the technologies and techniques it masks itself in. To switch our analogy from kitchen appliances (refrigerators) to cars: the comparison to tailfins I made earlier is indeed apt, though with one crucial difference. Cool resides precisely in a tailfin effect—let us call it an "information effect"—much like those decorative simulations of functional streamlining that once sprouted in sheet-metal and chrome splendor on mainstream American cars to the horror of International Style design philosophy.[18] Cool is a tailfin of the silicon rather than chrome era that aligns the user with data so hyperbolically that, like big tailfins catching (rather than slicing through) the wind, it actually obstructs the datastream. But unlike the subcultural mock-technology, say, of hot-rod culture from the mid-twentieth century on, information cool at all times remains respectable. "My Web page is cooler than yours," a knowledge worker might say, or again, "I have a 3-gigahertz computer; what's yours?" Yet all the bravado of this particular drag race just comes down to competing users sitting in cubicles before identical beige or charcoal boxes.[19] "Vrooom," the tinny computer speakers enunciate. The possible exception is the corps of Generation X technical wizards or Webheads that *Business Week* in 1999 crowned "Generation $." Working within the corporation but allowed to dress in punk or some other subcultural style (spike hair, tee-shirts, nose-rings, and so on), Generation $ is the Frankenstein's monster that bares to view the raw suture between subcultural and knowledge-worker cool.[20]

Yet even the cultural suture represented by Generation $ (photographed in coiffed and nose-ringed perfection in the *Business Week* article) is at last smoothed over and interiorized within knowledge work. The name of that smoothing—itself paradoxical—is what style becomes in a corporate setting: design.

Chapter 6

Information Is Style

Cool Is a Style of Information

Cool style must figure in any discussion of cool content. If information cool is a form of "deco-technology," as I have suggested, then another way to say that its content is useless is to identify "uselessness" with the conversion of all the workaday techniques, procedures, routines, protocols, and standards of information into an experience of pure style, which in the producer culture of knowledge work is translated as "design" (a style one either produces or can imagine "prosumeristically" producing, not a style one just consumes). After all, even "no design" in the workstyle or lifestyle of knowledge work would still contain enough minimal design—that of the underlying GUI interface—to allow cool pages to be flaunted in ways that exceed the strict bounds of rational content. (Witness the *Ode to Lynx* page that wittily celebrates the bare-bones functionality of the old, text-only Lynx browser by using the GUI to display a graphical screenshot of Lynx.) And "no design" is the rare exception rather than the norm for cool pages, some of which are designed to within an inch of their lives.

In short, formalism—as I have foreshadowed by invoking the New Critics on paradox and the Russian Formalists on defamiliarization—is a necessary approach to information cool. Formalism, above all other twentieth-century artistic and critical movements, suborned the technological rationality of modernity by remolding its functionalist assumptions so profoundly as to imprint them with the distinctive style of "modernism." That imprint came from inscribing the idea of form so deeply into function that it could no longer be discarded like packaging from the product.[1] Form instead became the new belief system called design. It will be helpful, therefore, to put on formalist spectacles for a time to see how contemporary, postindustrial cool on the Web descends with variation from the modernist, industrial era fusion of form and content that the New Critics (as if foreseeing the GUI interface) called "iconic." What is the formal design of information cool?

A first answer will arrive if we simply line up information cool over modernist formalism as if tracing a pattern on a transparency. Formalism of the early to mid-twentieth century revolved around the conviction that while good form required unity, such totality could no longer be crafted on the models of classical symmetry or Romantic organicism—that is, by subordinating or internalizing variations in form. Rather, good form required the exposure of the variations and counter-pressures, even the stark contradictions, that composed it, and in a manner that was not just decorative or historicist (Victorian, some modernists called it) but aggressively functional, like the strutwork of a steel bridge. In the realm of literary criticism, for example, the New Critics argued that good poetry sustains its "unity," "balance," and "harmony" not despite but through "tension," "ambiguity," "irony," and "paradox." And the Russian Formalists similarly argued that good literature unifies ("motivates") all its motifs but also requires elements that "roughen," "make more difficult," or strike atonally against too "automatic" a uniformity.[2]

In our present context the most relevant formalisms are neither literary nor intellectual but graphical and commercial. I refer to the modernist graphic design movements that followed up on the avant-garde experiments of futurism and dadaism (studied by Johanna Drucker in *The Visible Word*) by professionalizing the design trade and welding it to state or corporate industry: first Russian constructivism and the Dutch De Stijl movement in the 1910s and 1920s, then Bauhaus and the affiliated New Typography of the 1920s and 1930s, and finally—after the great exodus of Middle European designers during World War II (especially to the United States)—the generalized Swiss or International Style that dominated business advertising and "corporate identity" campaigns in the 1950s through at least the 1970s or 1980s.

Prompted by political, artistic, and technological changes, this modernist genealogy broke with nineteenth-century graphic styles that, in its view, had

Figure 6.1 Printer's proof for a poster, 1888

From Philip B. Meggs, *A History of Graphic Design*, 2nd ed. (New York: Van Nostrand Reinhold, 1992), p. 138, fig. 10-16. Text copyright © 1992 by Philip B. Meggs. This material is used by permission of John Wiley & Sons, Inc.

composed variation and unity as if on two separate tracks. Variation in the nineteenth century had been accented through busy ornaments, rules, serif typefaces, type sizes, illustrations, and so on, marking out unrelated "zones of activity" on the page.[3] Unity, meanwhile, had been enforced by arranging all elements symmetrically around the central vertical axis, which thus functioned as an externally imposed schema independent of any particular activity on the page (see figure 6.1). To modernist eyes, it appeared that there was no overall synthesis of design in Victorian graphic style—no single perspective (and, not incidentally, no self-aware profession of designers) charged with overseeing the total form of the composition. The fundamental reason was that the one principle necessary to integrate variation with unity was missing. That principle, articulated most famously in Louis Sullivan's dictum that "form follows function," was that both variation and unity must be motivated by a common governor of design internal to each page, poster, leaflet, or other formal frame: content. Only the functional relation of form to content could rationalize the bolding, upper-casing, or placement of a word *here* as opposed to *there* while keeping sight of the need to coordinate effects around a single, desired impact beyond the imagination of Victorian axial symmetry (see figure 6.2). In the final analysis, earlier typographers were not truly designers because they did not design *meaning*.

By contrast, modernist graphic design focused above all on totality of design (*gestalt,* in German). Or, rather, the idea was to look at every page, poster, or leaflet as a whole in which variation and unity were so tightly, if tensely, bound that their very nature—and so the totality of design they constituted—altered. "Variation" became the great modernist design principles of asymmetry, contrast, and tension. Everything depended on edgily unbalanced, striking elements that danced dialectically with their opposites to suggest a merely implicit, dynamic totality. Asymmetry itself came to express the new, modernist sense of symmetry (figs. 6.3, 6.4, 6.5).[4] Reciprocally, unity was no longer central axial balance but what might be called antithetical balance: the perception of individually off-center elements whose very clash of "large/small, light/dark, horizontal/vertical, square/round, smooth/rough, closed/open, coloured/plain" (to use the terms of Jan Tschichold's *New Typography*) marked the negative image of the center—the invisible center *not* there. There could be no symmetry except as instantiated in asymmetry.[5] In practice, therefore, modernist graphic design did not only favor asymmetrical arrangements, slashing diagonals, contrasting primary colors, assertive white space used to set off elements, and wildly disproportionate type. It also subscribed to the so-called "grid system" of unified geometrical arrangement in which invisible (though sometimes

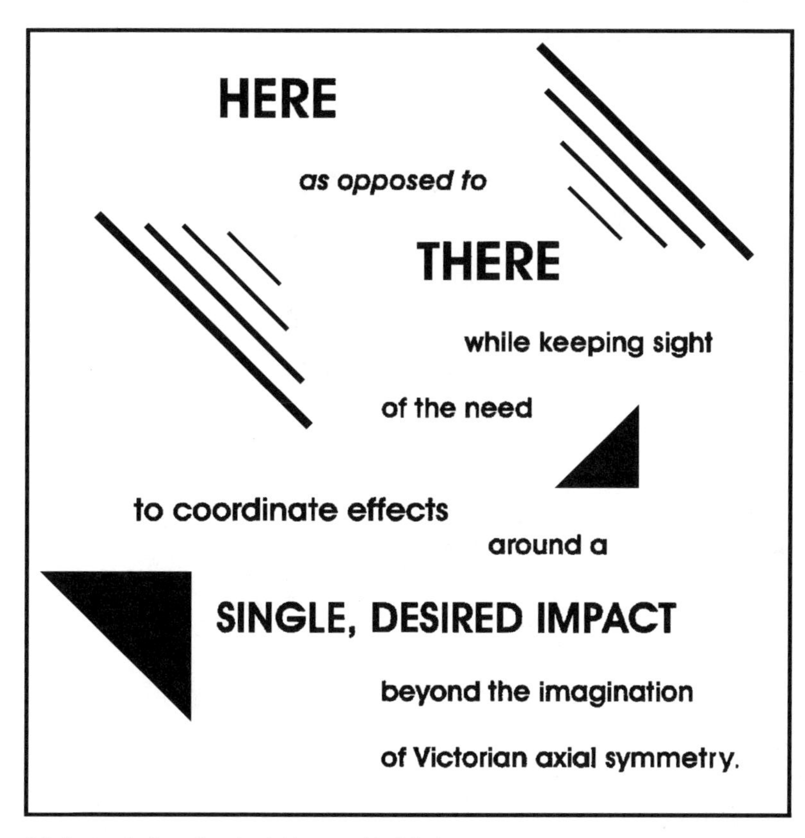

Figure 6.2 Demonstration of modernist typographical design

deliberately marked) columns and rows monitored the overall balance of antithetical elements (see figure 6.6).[6] Like a perfectly poised mobile of disproportionate parts, variation and unity composed a totality of design expressive of the uniquely modern, off-center "unity of life."[7]

Underlying this new design sense was the crucial modernist principle that both variation and unity had to serve a single function—the communication of meaning. Modernist designers, who differed from literary formalists in accepting the commercial need to pitch a "message," sometimes saw themselves less as artists of meaning than as technicians of *information*. The great fetish of design in the lineage of Bauhaus and the New Typography was thus "clarity," which Lázló Moholy-Nagy, Tschichold, and others defined explicitly in informational terms as clean, efficient visual communication for an age drowning in media.[8] Clarity was like the early radio age's quest for purity of signal. Indeed, we might say that by reducing all graphic elements (even individual sans serif letters) to geometrical shapes that commu-

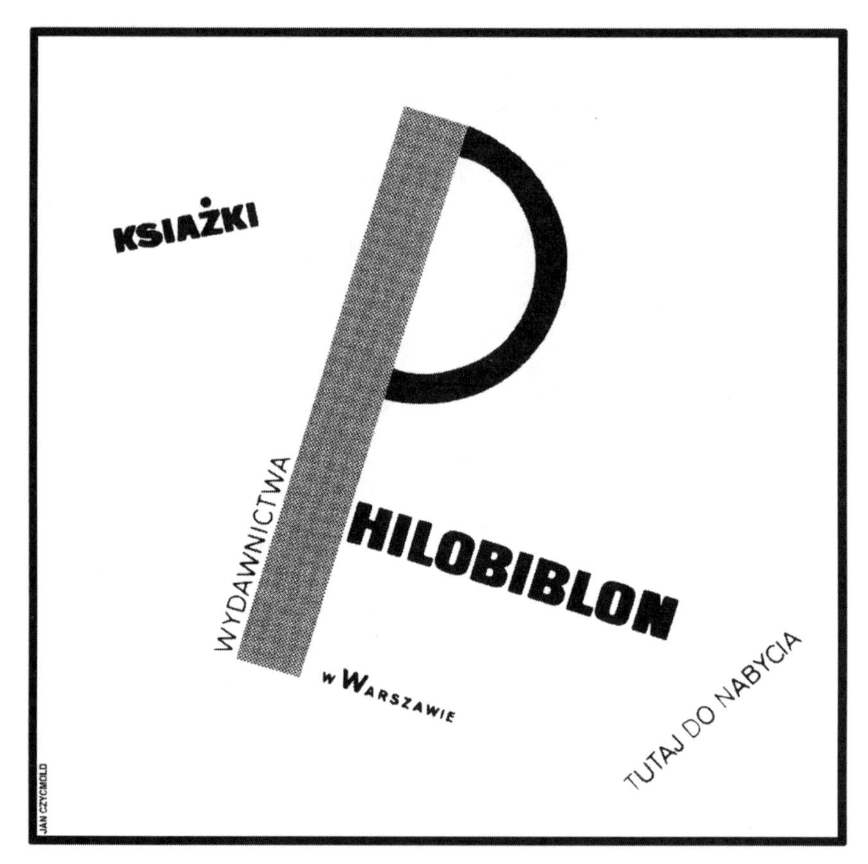

Figure 6.3 Jan Tschichold, display poster for a publisher, 1924
From Philip B. Meggs, *A History of Graphic Design*, 2nd ed. (New York: Van Nostrand Reinhold, 1992), p. 298, fig. 19-24. Text copyright © 1992 by Philip B. Meggs. This material is used by permission of John Wiley & Sons, Inc.

nicated compositionally with other shapes at remote locations on the "grid," modernist design in effect created a visual analogue of telegraphy, radio, and other remote information technologies.[9]

The descent of Bauhaus and New Typography into the International Style that subsequently dominated advertising, the new magazines (e.g., Otto Storch's makeover of *McCall's* and Henry Wolf's redesign of *Esquire* and *Harper's Bazaar* in the 1950s and 1960s), and the corporate identity business has been surveyed in such works as Philip Meggs's *History of Graphic Design*.[10] We need do no more here to indicate the general impact of this descent than note that from the 1960s on virtually any mass-circulation book, magazine, advertisement, or package design with mixed text and graphics showed precisely the sense for asymmetrical arrangement that the New Typography had schematized

Figure 6.4 Tschichold's cover for *elementare typographie* insert, 1925
From Philip B. Meggs, *A History of Graphic Design*, 2nd ed. (New York: Van Nostrand Reinhold, 1992), p. 298, fig. 19-25. Text copyright © 1992 by Philip B. Meggs. This material is used by permission of John Wiley & Sons, Inc.

as early as 1928 (see figures 6.7 and 6.8). Such was now simply "good design"; it "looked right."[11] Similarly, almost any work *about* design printed from the 1960s through at least the 1980s recites exactly the modernist mantra of asymmetry, overall unity, and informational clarity. In his "Principles of Design" (1980), for instance, Jon Lopez notes the requirement for "tensions"

VORZUGS-ANGEBOT

Im VERLAG DES BILDUNGSVERBANDES der Deutschen Buchdrucker, Berlin SW 61, Dreibundstr. 5, erscheint demnächst:

JAN TSCHICHOLD

Lehrer an der Meisterschule für Deutschlands Buchdrucker in München

DIE NEUE TYPOGRAPHIE

Handbuch für die gesamte Fachwelt und die drucksachenverbrauchenden Kreise

Das Problem der neuen gestaltenden Typographie hat eine lebhafte Diskussion bei allen Beteiligten hervorgerufen. Wir glauben dem Bedürfnis, die aufgeworfenen Fragen ausführlich behandelt zu sehen, zu entsprechen, wenn wir jetzt ein Handbuch der **NEUEN TYPOGRAPHIE** herausbringen.

Es käm dem Verfasser, einem ihrer bekanntesten Vertreter, in diesem Buche zunächst darauf an, den engen Zusammenhang der neuen Typographie mit dem **Gesamtkomplex heutigen Lebens** aufzuzeigen und zu beweisen, daß die neue Typographie ein ebenso notwendiger Ausdruck einer neuen Gesinnung ist wie die neue Baukunst und alles Neue, das mit unserer Zeit anbricht. Diese geschichtliche Notwendigkeit der neuen Typographie belegt weiterhin eine kritische Darstellung der **alten Typographie**. Die Entwicklung der **neuen Malerei,** die für alles Neue unserer Zeit geistig bahnbrechend gewesen ist, wird in einem reich illustrierten Aufsatz des Buches leicht faßlich dargestellt. Ein kurzer Abschnitt „**Zur Geschichte der neuen Typographie**" leitet zu dem wichtigsten Teile des Buches, den **Grundbegriffen der neuen Typographie** über. Diese werden klar herausgeschält, richtige und falsche Beispiele einander gegenübergestellt. Zwei weitere Artikel behandeln „**Photographie und Typographie**" und „**Neue Typographie und Normung**"

Der Hauptwert des Buches für den Praktiker besteht in dem zweiten Teil „**Typographische Hauptformen**" (siehe das nebenstehende Inhaltsverzeichnis). Es fehlte bisher an einem Werke, das wie dieses Buch die schon bei einfachen Satzaufgaben auftauchenden gestalterischen Fragen in gebührender Ausführlichkeit behandelte. Jeder Teilabschnitt enthält neben **allgemeinen typographischen Regeln** vor allem die Abbildungen aller in Betracht kommenden **Normblätter** des Deutschen Normenausschusses, alle andern (z. B. postalischen) **Vorschriften** und zahlreiche Beispiele, Gegenbeispiele und Schemen.

Für jeden Buchdrucker, insbesondere jeden Akzidenzsetzer, wird „Die neue Typographie" ein **unentbehrliches Handbuch** sein. Von nicht geringerer Bedeutung ist es für Reklamefachleute, Gebrauchsgraphiker, Kaufleute, Photographen, Architekten, Ingenieure und Schriftsteller, also für alle, die mit dem Buchdruck in Berührung kommen.

INHALT DES BUCHES

typ. tschichold

Das Buch enthält über 125 Abbildungen, von denen etwa ein Viertel zweifarbig gedruckt ist, und umfaßt gegen 200 Seiten auf gutem Kunstdruckpapier. Es erscheint im Format DIN A5 (148× 210 mm) und ist biegsam in Ganzleinen gebunden.

Preis bei Vorbestellung bis 1. Juni 1928: **5.00** RM
durch den Buchhandel nur zum Preise von **6.50** RM

Bestellschein umstehend

Figure 6.5 Tschichold's publicity leaflet for *Die neue Typographie*, 1928 (original printed in black on yellow)

From Jan Tschichold, *The New Typography: A Handbook for Modern Designers*, trans. Ruari McLean (Berkeley: University of California Press, 1995), p. xxi. Image copyright by Brinkmann & Bose Publisher.

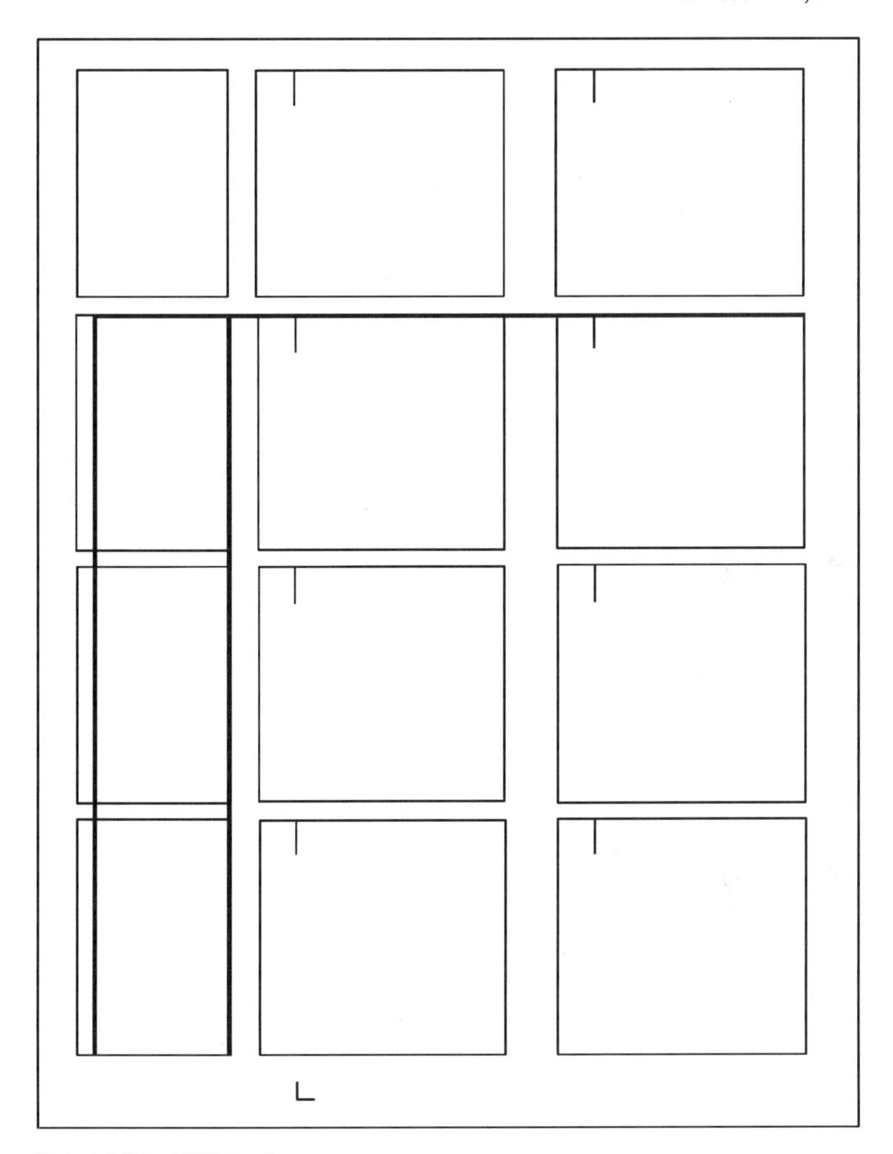

Figure 6.6 "The Grid System"
From Paul Rand, *A Designer's Art* (New Haven: Yale University Press, 1985), p. 195. Reprinted by permission of the Estate of Paul Rand.

formed "by confrontation of parts with the page and with each other," then emphasizes that "the key to successful design is harmony: All elements must be brought into agreement and must be considered in relation to each other and to the whole. Harmony is derived from coherence in design, from clarity of thought transferred by the designer to his work through a singleness of aim."[12]

Figure 6.7 Tschichold, **"How Blocks Used to Be Arranged in Magazines:** Schematic, thoughtless centring of blocks. 'Decorative,' impractical, uneconomic (= **ugly**)"
From Jan Tschichold, *The New Typography: A Handbook for Modern Designers*, trans. Ruari McLean (Berkeley: University of California Press, 1995), p. 210. Copyright © 1995 Regents of the University of California, © 1987 Brinkmann & Bose Publisher.

Donis A. Dondis's *A Primer of Visual Literacy* (1973) puts the case with systematic precision. A visually literate design requires both asymmetrical contrast and unified harmony as follows:

Contrast	*Harmony*
Instability	Balance
Asymmetry	Symmetry
Irregularity	Regularity
Complexity	Simplicity
Fragmentation	Unity
Intricacy	Economy
Exaggeration	Understatement
Spontaneity	Predictability
Activeness	Stasis
Boldness	Subtlety
Accent	Neutrality
Transparency	Opacity

Figure 6.8 Tschichold, **"The same blocks, correctly arranged in the same type-area.** Constructive, meaningful, and economical (= **beautiful**)"
From Jan Tschichold, *The New Typography: A Handbook for Modern Designers*, trans. Ruari McLean (Berkeley: University of California Press, 1995), p. 211. Copyright © 1995 Regents of the University of California, © 1987 Brinkmann & Bose Publisher.

Variation	Consistency
Distortion	Accuracy
Depth	Flatness
Juxtaposition	Singularity
Randomness	Sequentiality
Sharpness	Diffusion
Episodicity	Repetition[13]

The effect of the resulting gestalt, the very definition of "visual literacy," is clarity: a "clearer comprehension of visual messages" in which "with the speed of light, visual intelligence delivers multiple bits of information."[14]

We should dwell, however, on at least one event in the later twentieth-century reception of modernist design that bears directly on our topic. This event—exemplified in Dondis's phrase, "multiple bits of information"—was the assimilation of clarity not just to information in general but specifically to information technology. The leading players in this process were the technology and media corporations of the 1950s and 1960s whose packaging, advertising, and corporate identity campaigns acquired the flair of International Style under the tutelage of such influential American designers as Paul Rand and Eliot Noyes (and in Europe, Josef Müller-Brockman). At the time, such companies marked the leading edge of corporate design. Though many firms in manufacturing, transportation, and other industries were already following the earlier example of the Container Corporation of America in sponsoring European design, it was above all firms in communications, electronics, and office technology—industries associated with the transmission or reproduction of information—that gave the new principles of design their most visible showcase. These firms included CBS, ABC, Westinghouse, Olivetti (in Europe), and, most influentially, IBM, which had a uniquely strong impact on the design styles of other corporations. Rand and Noyes made International Style pervasive at IBM. Rand, for example, designed the now iconic IBM logo, as well as a host of packaging, poster, trade show exhibit, annual report, and other company materials. Similarly, Müller-Brockman, founder and co-editor of *New Graphic Design*, the definitive journal of the International Style, became design consultant for IBM Europe.[15]

Clarity of design in the modernist style thus molded itself to the new information technology—to the point (especially after the impact of Apple-based computer graphics software) that the discourse of design simply internalized its affiliation with IT.[16] Steven Heller and Seymour Chwast's *Graphic Style: From Victorian to Post-Modern* (1988), for instance, begins by defining design as a target-aware "transmission code." "The graphic designer," they say, "is basically organizing and communicating messages—to establish the nature of a product or idea, to set the appropriate stage on which to present its virtues, and to announce and publicize such information in the most effective way. Within this process, style is a transmission code, a means of signaling that a certain message is intended for a specific audience."[17] In general, as Matthew G. Kirschenbaum has explored in his innovative dissertation, *Lines for a Virtual T/y/o/pography: Electronic Essays on Artifice and Information*, one of the most powerful aesthetic phenomena of recent decades has been the consolidation of graphic design and information display into a common "spectrum of tropes, icons, and graphic conventions that collectively convey the notion of 'information' to the eye of the beholder."[18]

The consolidation became even more pronounced in the flourishing subbranch of graphic design known as "information design" or "information architecture," which specializes in explanatory, diagrammatic, statistical, cartographic, and other information-dense forms. Graphic design is "rational imagery," Jacques Bertin said in his *Semiology of Graphics* (1967), and added in a preface to the English translation (1983) that such design is the supplement specifically of computerization.[19] Similarly, the first sentence of

Edward Tufte's *Visual Display of Quantitative Information* (1983) declares that "excellence in statistical graphics consists of complex ideas communicated with clarity, precision, and efficiency"—an argument that unfolds its proto-cybernetic personality only when we then realize that the book is concerned throughout not with the communication of "complex ideas" but with the communication of *data* (as in Tufte's second sentence, which begins, "Graphical displays should show the data").[20] "Data," the concept and the word, is the fetish of Tufte's book. There is to be functional visual code and only code, Tufte insists; all else is "non-data-ink" and "chartjunk" to be rigorously expunged.[21] So, too, Peter Bradford and Richard Saul Wurman's *Information Architects* (1997; a showcase for information exhibits that includes Web sites) celebrates the emergence of "information design" as an independent art form. By the time of the Web, in sum, graphics and digital information became part of the same integral design. Both were aspects of the single, great canvas now subsuming all the pages, posters, leaflets, packages, and so on that the modernist designers had created. That canvas is the generalized information "interface."

Now at last we are in position to look at cool design on the Web, much of which copies *mutatis mutandis* the "look and feel" of modernist graphic design. In his perceptive "Avant-garde as Software," Lev Manovich has argued generally that contemporary "new media" is deeply instinct with the principles of the original avant-garde: "constructivist design, New Typography, avant-garde cinematography and film editing, photo-montage, etc." of the 1920s era are now "materialized" and "naturalized" in computing interfaces.[22] I will extend Manovich's argument to the case of the World Wide Web in particular by starting with one exceptionally direct, if limited, piece of evidence of the persistence of modernist design.

Opening Paul Rand's essay "Design and the Play Instinct" (which follows "The Good Old 'Neue Typografie'" in his retrospective *A Designer's Art*), we encounter a series of visual paradigms demonstrating how good design blends formal discipline with the "play principle." Meditating on the grid paradigm, for example, Rand uses the page layout of his essay itself to play off asymmetry against the grid's "orderly and harmonious distribution of miscellaneous graphic material."[23] Similar meditations on visual paradigms from both Western and Eastern art follow, and the essay concludes by discussing the *chasen* whisk of Japanese tea ceremonies, which Rand transforms into the Eastern imago of International Style (fig. 6.9):

> Some years ago in Kyoto I was fortunate enough to witness a young Japanese craftsman make the chasen you see here. The chasen is a whisk used

Figure 6.9 *Chasen*
From Paul Rand, *A Designer's Art* (New Haven: Yale University Press, 1985), p. 201. Reprinted by permission of the Estate of Paul Rand.

> in the tea ceremony and is cut from a single piece of bamboo with a simple tool resembling a penknife. Both the material and manufacturing process (which took about one-half hour) are the quintessence of discipline, simplicity, and restraint. The invention of such an article could not possibly have been achieved by anyone lacking the ability to improvise and the patience to play with a specific material: to see the myriad possibilities and to discover the ideal form.[24]

The *chasen* is as pure in its variation within unity, we may say, as sans serif type.

Now opening Darrell Sano's *Designing Large-Scale Web Sites: A Visual Design Methodology* (1996), we encounter a practicum on Web design steeped in explicitly modernist principles—to the point, for example, where quotations from well-known earlier designers furnish the chapter epigraphs. (Sano's bibliography also includes ample references to the design tradition. Sano himself was an interface designer at the Netscape company, Silicon Graphics, and Sun Microsystems.) The first epigraph in the book is a passage from Rand's *Designer's Art* emphasizing the totality of design: "Graphic design is essentially about visual relationships—providing meaning to a mass of unrelated needs, ideas, words, and pictures. It is the designer's job to select and fit this material together—and make it interesting."[25] Perhaps not surprisingly, therefore, we discover in Sano's book several reproductions of an apparently identical *chasen* (presented in the same proportions and same orientation as Rand's). The *chasen* appears on the Web site for a "Japonesque Designs" online store from which Sano draws many of his examples (see figure 6.10).[26] As illustrated on the pages of this site, Sano writes, "the use of asymmetrical page layouts corresponds nicely with the aesthetic of Asian art." Thus is the affiliation of Web design with the asymmetrical modernist tradition at once clinched and—through an orientalism precisely like Rand's—displaced.[27]

What this detail of the *chasen* highlights is the massive influence of modernist design on the now flourishing corps of Web design practitioners and consultants (allied with the interface designers and programmers responsible for the underlying GUI/browser environment). The majority of Web authors do not, unlike Sano, borrow in any deliberate way from Bauhaus, the New Typography, or International Design. The influence runs deeper. Once the decision is made to "design" at all—that is, to add to a page a *sense* of design epitomized by cool—then the entire modernist heritage (mediated through print media, print-influenced desktop publishing conventions, and even print-influenced titling or framing effects in video) flows onto the Web.

Thus we may consider alongside Sano's book such other works about Web design as Crystal Waters's *Web Concept & Design* (1996). Waters includes a section titled "Flexing the Grid: An Example" that in essence recapitulates Rand's argument about play in the grid and includes a sample page from Waters's *Typo News* Web site showing asymmetric composition balanced within a five-column grid structure ("the main body of the text doesn't follow a straight column down the page, but it does use the column guides to keep the balance and proportion of the text consistent").[28] For designers like Waters, no direct citation of modernist principles is necessary; "International" Style now simply *is* "World Wide" Web style. Again, consider Vincent Flanders's well-known Web site (and later book, coauthored with Michael Willis), *Web Pages That Suck*. Subtitled "Learn Good Design by

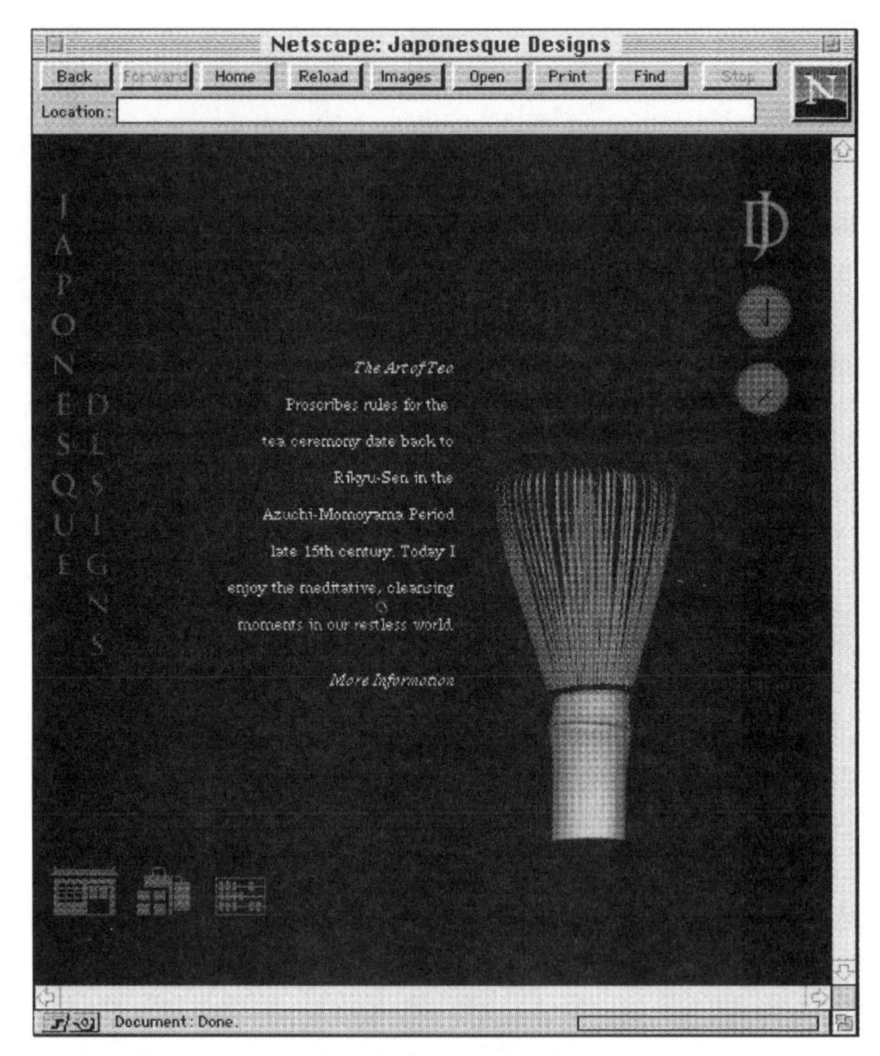

Figure 6.10 "Asymmetrical page layout using tables with no border"
From Darrell Sano, *Designing Large-Scale Web Sites: A Visual Design Methodology* (New York: John Wiley & Sons, 1996), p. 191. Copyright © 1996 by John Wiley & Sons, Inc. This material is used by permission of John Wiley & Sons, Inc.

Looking at Bad Design," the site and book answer the question "What makes a site cool?" by satirical inversion, exhibiting what amounts to a rogue's gallery of uncool pages and techniques. While Flanders and Willis animadvert on some problems specific to the online medium (wasting bandwidth, for example), their general point is that pages "suck" if they become too busy with background images, Java and Javascript, mixed fonts, drop shadows, and so forth. Pages suck, in short, if they ignore both the sans serif or white space minimalism that the New Typography had identified with clarity (the original conservation of bandwidth) and the careful coordination of asymmetry that the New Typography had called gestalt. Exemplary is Flanders and Willis's trenchant critique of designers who are seduced by their graphics programs into promiscuous imagery. Besides the issue of bandwidth, they say, "there's the whole concept of aesthetics. It's hard to fight the urge to make really complex images because, quite frankly, complex images look cool. Once again, it takes talent to make them fit in with the look of a web page."[29] At core, the aesthetics invoked here is modernist. Cool is only cool if "fitted" into a functionally clear, total design.

Actual "cool pages," as opposed to works about cool, provide the QED to my argument. We might look again at Paul Schrank's *Paul's a Computer Geek*, for example. Whether we view the home page of this site or one of its second-tier pages (e.g., the "Design" page exhibiting selected past and present graphic designers), we observe a particularly striking use of such New Typography features as asymmetrical layout and diagonal forms (all accented by an icon labeled "Gestalt" at the bottom of the home page) (figs. 6.11 and 6.12). So, too, perusing Shauna Wright's *Flaunt* site (one of the personal pages that have made the cool anthologies), we find pages whose "left-" and "right-centric" vertical panels resemble the layouts of Tschichold.[30] Of course, the asymmetrically placed vertical panels on such pages—particularly when used to list contents or hold a navigation bar—are just a cooler version of the standard Web "magazine" format mimicking the dominant format of print magazines since the 1960s (which in their turn mimicked such layouts as Tschichold's celebrated leaflet publicizing his *New Typography*). Compare, for example, the *C/Net News.com* page on any day with Tschichold's leaflet (fig. 6.5). Or, to accommodate a variant of contemporary magazine formatting: consider the paradigm of the double-page spread with graphics and text contoured together into a single visual ensemble. We have only to look at *The Rock-n-Roll Gallery Collection of Richard E. Aaron* (a *Cool Site of the Day* selection in August 1998) to see how the Web duplicates Otto Storch's redesign of *McCall's* using New Typographical "typo-photo" composition.[31]

Ultimately, the variety of cool graphic designs on the Web is so ample that any further conventions I instance here could only exemplify a small part of the overall resemblance to modernist design. It may be best to generalize, then, by descending to the level of the underlying code. As originally conceived, HTML or Hypertext Markup Language (like the SGML or Standard Generalized Markup Language of which it is a much reduced subset)

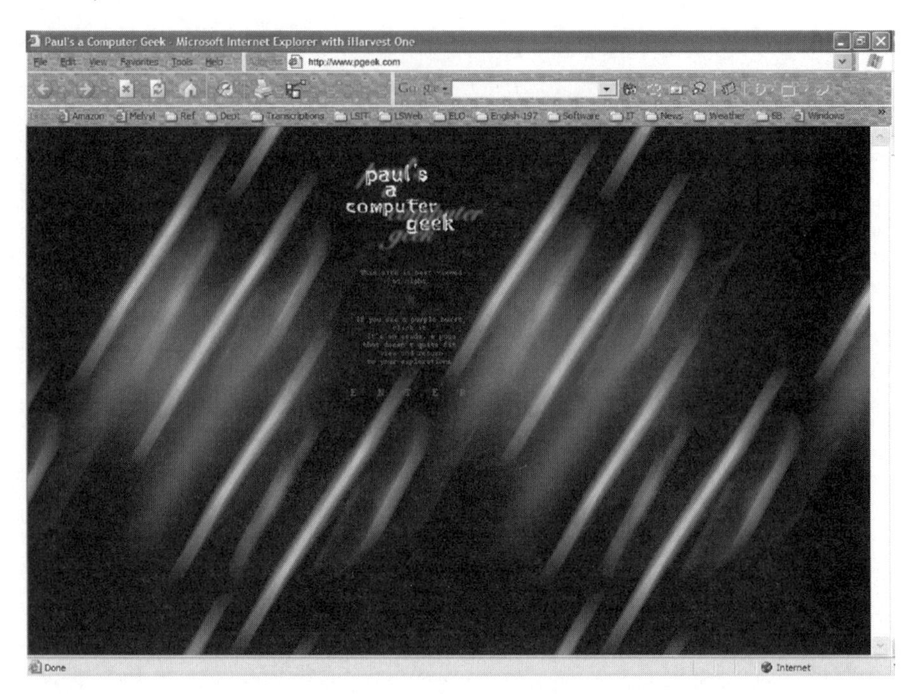

Figure 6.11 Screenshot from Paul Schrank's *Paul's a Computer Geek* Web site, 2000
Retrieved in various versions, 1997–99; last retrieved 3 December 2000. <http://pgeek.com> Used with permission of Paul Schrank.

was primarily a *logical* rather than layout-oriented tagging language. HTML, for example, used tags (enclosed in angle brackets) to declare simply,

<H1>**This is a level 1 header**</H1>

or, again, <EM>**This text is emphasized**</EM>. It was up to the particular browser program to choose the actual display font, type size, color, and screen position for any tagged content. (A few redundant tags such as <B>**boldface**</B> instead of <EM>**emphasis**</EM> addressed display properties more directly, but were supposed to be used infrequently in order to allow the browser the greatest freedom in rendering content suited to the user's hardware and software environment.)

But as HTML evolved into its 2.0, 3.0, 3.2, and 4.01 specifications under the governance of the W3C (World Wide Web Consortium), two complementary developments increasingly gave designers more direct control over the display of their work. First, the major browsers (by this time, Netscape's Navigator and Microsoft's Internet Explorer) shifted at least some of their emphasis from improvising maverick, proprietary tags (e.g., Netscape's infamous <BLINK>) to respecting the official HTML specifications. This al-

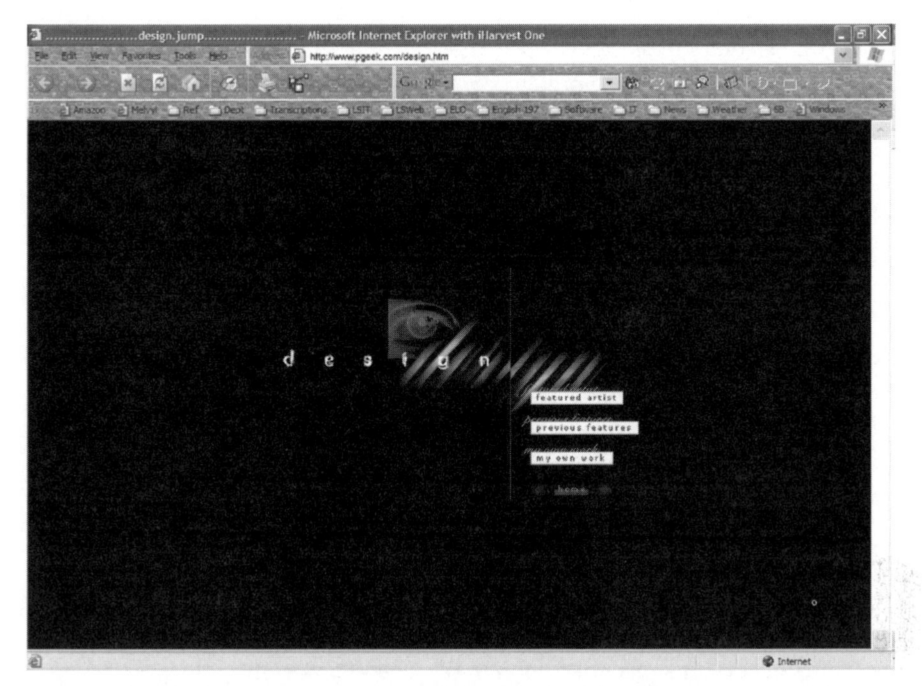

Figure 6.12 Screenshot from *Design* page on Paul Schrank's *Paul's a Computer Geek* Web site, 2000
Retrieved in various versions, 1997–99; last retrieved 3 December 2000. <http://pgeek.com/design.htm> Used with permission of Paul Schrank.

lowed designers to sketch on a more stable, standard canvas on which they could predict—or at least control variations within—the actual display of a page (though Netscape and Internet Explorer continued to be incompatible when it came to more advanced HTML features). Second, the reason the browsers were able to converge upon official HTML while still accommodating the need for design that had originally driven their invention of rogue tags was that the official code standard itself—like the late Roman empire incorporating the barbarians—brought more design features into the fold. Designers now gained official ways to exert finer control over basic typography—the choice of fonts, sizes, type colors, and type alignments—and thus to implement bolder, more sophisticated, and often asymmetrical and sans serif typographical arrangements. They gained the ability to "wrap" text to the left and right around images, which provided some limited ways to flow text around photos in a "typo-photo" ensemble approximating the asymmetrical schema we saw in Tschichold (figs. 6.7, 6.8). Later evolutions of HTML, indeed, offered even more flexibility in positioning images relative to each other and to text—for example, through the use of superimposable, floating "layers." Perhaps most important, designers finally gained control over "grid" layout. In particular, the previously dominant HTML linear layout device of the "list" (the bulleted or numbered series of items that was the most advanced formatting feature of early Web pages) expanded into two dimensions. With the addition of tags for "tables" in HTML 3.0 (and later, "frames" in HTML 4.01), list design could be exploded into full-blown grid design such that each page or section of a page became a matrix of columns and rows. A vertical navigation menu, for example, could be laid out asymmetrically on the page by including it in a column or frame to the left or, for more intricate effects, within a "nested" table-within-table.

By the time CSS or Cascading Style Sheets arrived in 1997–98 to complement HTML 4.01 (by allowing designers to create a separate style sheet providing much more extensive, granular, and sitewide control over the appearance and positioning of typographical elements), the coding conventions for modernist "total" design were all in place. HTML now allowed designers to work in something like a desktop publishing environment that could make the Web look more like familiar magazines, newspapers, TV/video, and other preexisting media (a fine example of what Jay David Bolter and Richard Grusin call "remediation").[32] If we factor out animation and video (which have their own complex relations of remediation to film, still photography, and print media), then what is left when we survey the coding possibilities added to each successive HTML specification is a collection of tags and attributes (and style sheets) whose primary intent was to import New Typographical layout principles wholesale onto the Web.[33] Through the channel provided by HTML, "good design" could simply flow into contemporary interface design like water flowing down the path of least resistance. The entire tradition of modernism and International Style poured into online cool, "Bauhaus" into "home page."

But to say that cool on the Web thus merely absorbed modernist design once a conduit was laid in the code is to understate the phenomenon drastically. Above all, the Web represents the contemporary knowledge worker's active *thirst* for design—his or her eager, restless, even desperate need for a design sense akin to that honed by modernist forebears. In this regard, the fact that cool knowledge workers are swathed in "designer" labels everywhere in their consumer life—in their clothes, bottled water, gym equipment, coffee grinder, and so on—is the least of it. The other half of the equation is the overwhelming need for design that such workers express in their *producer* life. Until recently, this need was fulfilled by corporate International Style only in highly nonegalitarian ways. Lavish corporate lobbies and well-appointed executive suites were a design Olympus from which issued all the logos, rituals, symbols, and other insignia of corporate culture governing mere mortals, while at ground level, just the subtlest, rounded contouring of cubicle furniture, Euro-styled desk lamp, or injection-molded computer casing gave any hint of design. But with the arrival of a new generation of GUI interfaces and, most spectacularly, of the Web, for the first time the prosaic space of the monitor at the dead center of the life of knowledge work became selectable, configurable, *stylable*. With form now an option at the functional heart of knowledge work, a vast, pent-up desire for design and style in the workplace was released. Where previously knowledge workers could satisfy their need for design only in consumption, now they exercised the power of design in production, too, as if completing the circuit of earlier twentieth-century lifestyle (production driving consumption) in a more total experience of "lifedesign" (production and consumption integrated in the activity of design). Today, everyone wants to be a designer: not just graphic artists, architects, or fashion designers, but—in different yet related senses—engineers, programmers, database designers, Web page authors, and even clerical workers using the latest formatting features of GUI word processers to create new memo styles.[34]

Why the need for such a total life-gestalt of design? If the fundamental agenda of the ethos of information is proto-identity or "character," as I have argued, then that of the *aesthetics* of information is collective character—culture. Design is now the primary discipline—far more so than education—through which knowledge workers receive their culture. Whether the culture received is high or popular is not the issue here (design is receptive to both). Rather, the issue is whether the culture received is to be organized according to one or the other of the two following schemas. One is for culture to be disseminated in what is now the dominant postindustrial fashion from the culture of production to that of consumption. In this way any holism that arises—let us simply call it the "good life"—comes only through the infiltration of production values into all facets of consumption (extending workplace culture into leisure at home). The other schema of the "good life"—which is what Web design helps knowledge workers imagine, if not fully enact—is a reaction against postindustrial life. It is the coequal dissemination of the life of consumption into production—like keeping a sports page open on one's desktop next to a spreadsheet. Design is precisely the discipline that promises to fuse the production and consumption of culture so as to integrate undecidably the lives of work and nonwork. When put on the desktop, as opposed to being kept on a pedestal in the executive suite, design is the last hope—or fantasy—of the knowledge worker for a cultural holism hearkening back to preindustrial times.

In sum, it is the craving for *form* in information—the need for a Web page to be more like a glossy consumer magazine than cargo off a truck—

that gives the information worker the hope of being truly in-"formed" with character. Only such a craving makes even the most routine documents of work seem to open up with a mere click of the browser into whole vistas of culture beyond the ken of corporate culture. There could be no better role model for such seeming independence from corporate culture (positioned *within* the cubicles of such culture) than the hip designer—with whom anyone who today says, "Cool!" while gazing at a Web page or other high-tech artifact now subconsciously identifies. For who if not the designer in our time exemplifies the *professional who has culture?*

Cool Is an Antistyle of Information

No sooner do we see the unmistakable imprint of modernist design in cool, however, than we immediately note its antithesis. Cool is also fundamentally antidesign.

It will be useful to start by looking back to antidesign in print graphics. In this context, *antidesign* names a particular movement of the late 1960s, primarily Italian, that played up "distortions of scale and form, the shocking use of colour, [and] visual puns or the undermining of an object's functional value" to protest the cooptation of "good design" by commercial interests.[35] More generally, the term may serve to label a whole confederation of graphic design movements from the 1970s on—for example, the California New Wave, "deconstructivism," the Memphis school, New Wave Typography, and Postmodernism.[36] Though varied in their styles, media, tone, and stance toward commercial interests, such movements had in common the need to contravene International Style through such means as deliberately fragmenting a composition; crowding the frame with superimposed or overlapping elements that eschewed "clarity"; blurring, twisting, stretching, distorting, or repeating text elements to near illegibility; deploying diagonals and other atectonic forms "deconstructively" to highlight unresolvable rather than dialectically poised contradictions; and so on. Illustrative of such tendencies is the complex work of April Greiman, who trained at the Basel School of Design in Switzerland but later became prominent in the California New Wave (figs. 6.13, 6.14). Another example is that of Wolfgang Weingart, who taught Greiman in Basel and was a founding figure in the break with Swiss style and in New Wave typography (fig. 6.15).

Many of these antidesigners were assisted in their experimentation by new computer graphics programs that emerged during what Anne Morgan Spalter calls the "second period" of computer graphics from the 1970s to 1980s (the genealogy that extends from MacPaint on the original Macintosh

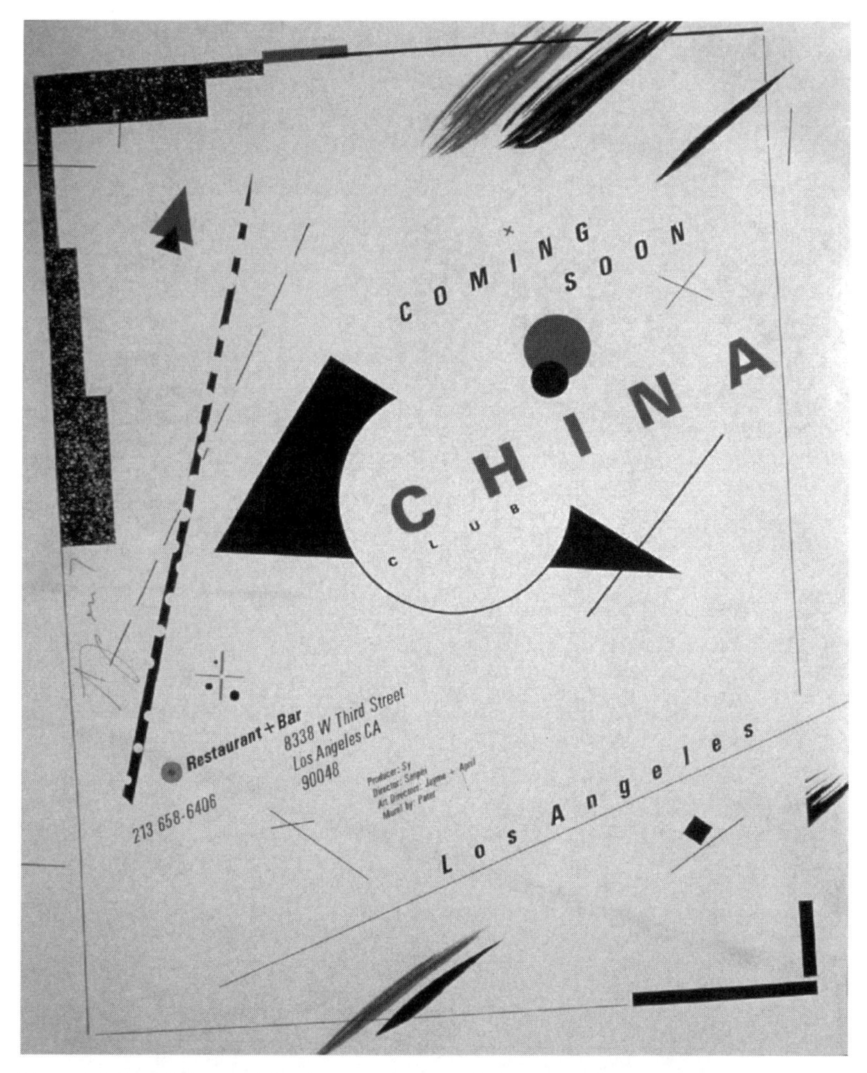

Figure 6.13 April Greiman, China Club advertisement, 1980
April Greiman Made in Space.

Figure 6.14 April Greiman and Jayme Odgers, California Institute of the Arts (CalArts) poster, 1979
April Greiman Made in Space.

Figure 6.15 Wolfgang Weingart, exhibition poster, 1977
From Wolfgang Weingart, *Wege zur Typographie [My Way toTypography]* (Baden, Switzerland: Lars Müller, 2000), p. 468. Reprinted by permission of Wolfgang Weingart and Lars Müller Publishers.

computer to the now industry-standard Adobe Photoshop and its whole sub-industry of third-party "filters" and "plug-ins").[37] These programs and the computer-focused technologies of vision they drew attention to established what amounted to a new visual vocabulary.[38] In Greiman's 1986 poster for the Los Angeles Institute of Contemporary Art, for example, bitmapped or raster imagery emerges from the status of tool to become the dominant design element (fig. 6.16). Obvious "jaggies" in the typeface emphasize an iconography of pixelation that repeats in pixelated textures throughout the work as well as in the basic compositional concept of jumbled, superimposed layers (an updating of the avant-garde modernist collage principle in which each layer, like a pixel, is an independent image-element to be switched on or off autonomously). The prominent role of the computer in such work allows us to recognize that the negative impulse in antidesign to repudiate International Style—to deface the pattern of asymmetry, unity, and clarity—was also a positive impulse to open design to new digital media, forms, and tools that challenged, rather than simply extended, "good design." Just as experimental typography had started the twentieth century by registering the collision between print and such new media as photography and film, so it finished the century by registering the collision between print and digitization. The typo-photo principle of composition that by midcentury had restabilized type relative to photography now faced the challenge of pixelation, which changed the rules of the game by reconstituting the nature of both type and photography.[39] Dissolving the two into a common digital dust of vision, it created the conditions for an integration so thorough—or at least so altered—that it ironically undermined the modernist ideal of integration based on clearly differentiated *unlike* elements.[40]

Phrased eschatologically, the crisis triggered in print design by digitization marks what Matthew G. Kirschenbaum calls "the other end of print," by which he means not just the digitization of print ("the 'end of print' is . . . a phrase routinely invoked in the context of electronic media") but specifically "radical demonstrations of the communicative limits of the textual condition" conducted in graphic design. Speaking of the paradigm-setting antidesign of David Carson in the music magazine *Ray Gun*, Kirschenbaum observes,

> This *post-alphabetic* aesthetic is one that appears precisely at the point of print media's imperative to formalize a representation of its own putative demise. That is, it is an aesthetic that is intensely self-reflexive in its attempt to depict, and at some level iconify, the material conditions of print's communicative exhaustion. The body of graphic design work associated with Carson, *Ray Gun*, *Emigre*, Cranbrook and CalArts therefore bears close scrutiny by students of the new media, for it dramatizes that aspect of the relationship between print and electronic textualities driven by the need of the former to assimilate and contain the ruptures of the latter.[41]

Figure 6.16 April Greiman, Los Angeles Institute of Contemporary Art (LAICA) poster, 1986
April Greiman Made in Space.

Figure 6.17 David Carson, "Too Much Joy"
From *Ray Gun*, 1992. Reproduced in Lewis Blackwell and David Carson, *The End of Print: The Graphic Design of David Carson* (San Francisco: Chronicle Books, 1995). Design © 1995 by David Carson.

Or, put another way, whether print *can* "assimilate and contain the ruptures" imposed by digital media is the open question posed by compositions of antidesign as ferocious as those collected in Carson's retrospective portfolios of his work (the first of which was Blackwell and Carson's *End of Print* [1995]; see figure 6.17). After all, if we look at *Wired* Magazine—which by the late 1990s had become the advertisement of the technical, thematic, and commercial impact of computerization on graphics—we see a Carson-like "illegible" ideal of typography that indeed seems to take the International Style to its antipodes.[42] At least as laid out on the magazine's most self-conscious showcase pages during its earlier, more experimental years, very little of such typography is balanced or unified in the modernist style. Rather, asymmetry knows no bounds. Colored text floats over or under garishly clashing backgrounds, type moves every which way and even tunnels in narrow, horizontal corridors from one double page spread to the next, autonomous graphic elements jostle each other in cut-and-paste patterns wholly unaware of any grid, lines and box frames seem to exist less to steer reading than to allude to some ineffable, abstract principle of circuitry, and so on. Correlatively, very little in such typography is "clear." The only thing that comes clear is the fundamental *un*clarity of the "vision" of information in an age when that vision is caught in the contradiction between synoptic overview and fragmentary views. The underlying message of antidesign, in other words, is that good design cannot be equated with informational clarity because information itself is profoundly unclear.[43]

If we now turn from print media to the Web, we discover a mode of antidesign that is often less acutely self-conscious but more existential. This is because antidesign is part of the medium's condition of possibility (and not just, to use Kirschenbaum's term, of its demise). The Web is a creature of networked, distributed information, and that fact alone shakes the notion of gestalt to its foundation. The networked nature of its medium imposes on the Web the following two antifoundational axioms of design: first, content is not form; second, form (unmoored from the fiction of governing content) is never stable, simultaneous, or total.

The first axiom follows from the logical structure of HTML as it evolved to suit the physical parameters of networked, client-server communication. Those parameters decree that designers who mount pages on a server cannot know except in broad terms what display format lies at the other end of the wire on the client machine. Monitors differ in physical screen size, display resolution, color depth and temperature, and sometimes also aspect ratio, and even if all monitors and their supporting graphics cards and software drivers were the same, different operating systems, browsers, and individual browser configurations on the remote client machine would affect the dimensions of the display window, the type font and size, the presence or absence of images, and other features. As we have seen, HTML is a markup language whose original, intended function—though less puritan in this regard than its parent SGML—was to describe rather than form content. Even with all the new tags for absolute positioning, sizing, coloring, and so on that we have reviewed, HTML authors must in the last instance acknowledge that the responsibility for forming or "rendering" content belongs to the particular browser and machine at the other end of the connection (as is clear to any Web designer who has ever spent hours trying unsuccessfully to trick a particular browser into displaying something against its grain).[44] Effectively, the Web manages content and form as if on two separate tracks; all the tricks of later HTML are merely incomplete workarounds for this fundamental bifurcation in the design firmament.

The ramifications of this bifurcation are myriad—so much so that the split between content and form must be considered the very engine of formal innovation on the Web (just as the principle of the unity of form and function drove modernist innovations). The fecundity of these ramifications may be suggested by imagining the following, hypothetical "manual" of standard formal problems in HTML and their most common improvised solutions (plus, to take the logic one step further, the problems *caused* by such hacks). Just three initial sections of such a manual (out of many more that could be added) will be representative:

> *Design Problem*: To the horror of designers accustomed to print media, Web pages are not really "pages" at all because they open outward indeterminately on both the right and bottom depending on the size of the user's browser window. Even more strangely, the lack of determinate boundaries is not just extensive (extending to the right and bottom) but what might be called intensive as well. Interior distances between text or graphic elements on a Web page stretch and contract with a kind of plasticity unknown to physical design surfaces. Such dynamism in both external and internal space makes it impossible to define the proportions of a Web composition in the manner of a print-based "grid" with precisely calibrated columns and rows.
>
> *Hack*: Lock all content into a fixed-width "table" structure set at a width of approximately 760 pixels (to accommodate smaller monitors running at a horizontal screen resolution of 800 pixels).[45] Alternatively, put the entire design or the most definitive portion of the design (e.g., the logo) in a single, large image that remains the same size no matter how far the browser window extends to the right or bottom.
>
> *Secondary Design Problem*: Fixed-width, 760-pixel tables and graphics appear on wider monitors to be either unintentionally asymmetrical or too severely symmetrical. That is, when tables or graphics are locked to the left, the screen stretches out on large monitors asymmetrically to the right; when such material is aligned down the screen center, left and right margins sprawl outward on either side like the wings of a butterfly pinned for dissection.

Design Problem: The lack of any bottom to a Web page makes it difficult to compose on the vertical axis, since the user may have to scroll down a variable number of lines (depending on the display) to see the whole view.

Hack: Either ignore the vertical axis when designing a page or attempt to size content to fit approximately on a single screen (especially in the case of such crucial design elements as introductory "splash pages," top-level home pages, or vertical navigation bars).

Secondary Design Problem: The latter, one-size-fits-all solution not only breaks up content into screen-size chunks that exaggerate the "jump and go" rhythm of hypertext (often in contravention of logical rhythm) but also leaves almost no room for interesting design in the vertical space allowed for on smaller monitors (e.g., 600 pixels minus 80 pixels or more for a browser's menu bars). Every page, in essence, would be something like a "banner"—those narrow advertisements whose constricted vertical dimension testifies to the preciousness of vertical space on Web pages.

Design Problem: The internal arrangements of tightly composed table, frame, or other gridlike design structures on a Web page become scrambled when the user's browser is configured with a type size that is too large. The text in a narrow, vertical navigation menu, for example, might disappear at the right, spill over onto an unintended second line, or cross over into another cell in the visual grid.

Hack: Hard-coding fonts and type sizes into a Web page helps, but users can still configure their browser to override such specifications in favor of a default font and type size. Therefore, the only guaranteed solutions are the heavy-handed ones of defining fixed font sizes through a cascading style sheet affiliated with the Web page (a tactic more resistant to variations in browser configuration) or putting the text in question into graphic images that the browser must render intact at the specified size or not at all.

Secondary Design Problems: Pages that govern font size through cascading style sheets make it difficult for users with visual impairments or with large monitors set to high screen resolutions to vary the text size to improve basic legibility. A page that puts ordinary text in graphics incurs an overhead of increased download time, difficulty of revision, unsearchability (searching on the Net is primarily text-based), and lost audience (including not just the impatient or technologically obsolete but also users with physical handicaps requiring the assistance of text-to-voice interfaces). In general, the more the designer gimmicks a page, uses advanced HTML, or deploys excessive graphics, the smaller the number of users who can predictably ac-

> cess the page with disparate hardware, connection routes, operating systems, browsers, and browser versions.

This abbreviated manual of HTML design problems indicates how pervasively, deeply, and corrosively the lack of fixed spatial dimensions defaces the concept of design on the Web. Indeed, we may take it as a rule that the finer the Web design (e.g., designs exhibiting intricate tables, framesets, or layers), the more the designer has had to fight the conditions of the medium—until the designer is no longer drawing on a screen, but on something "just like" a fixed newspaper or magazine page, except, of course, much thicker, heavier, and more fragile.

The second antifoundational axiom of Web design—that form on the Web is never stable, simultaneous, or total—can now be seen to be simply another way of saying that when form is separated from content it becomes dynamic. While my observations have so far concerned the dynamism of spatial form, the addition here of the notion of "simultaneity" highlights the fact that dynamism on the Web is also profoundly temporal. The reason is that the online environment is indeterminate not just due to varying hardware and software but also because of the unpredictable geographical distances involved. The extensive and intensive stretch of space on a Web page, indeed, may be conceived as a representation of the underlying stretches in the social geography of the Internet, where "near" and "far" have no easy physical analogues but nevertheless do matter. Depending on how far away a client machine is and at what time of day it is trying to connect to a server, the number of intervening servers, gateways, and routers ("hops"), as well as the amount of competing message traffic, can vary widely. The result is a temporal disturbance as powerful as any of the spatial deformations surveyed above. This disturbance contributes to the standard complaint that Web pages take "too long" to download—a perception that is only partly about duration (and often independent of the speed of one's connection to the Internet). The other component in this perception is the *unpredictability* of the speed, pace, and rhythm of the Web. Online temporality, in other words, amounts to antidesign. My meaning can best be demonstrated by a Web site I constructed in 1995 to make visible the design potential, but also the fundamental irrationality, of Web temporality: the *Lyotard Auto-Differend Page.* Constructed on the basis of one of the few HTML means available at the time to manipulate Web content dynamically and temporally ("client pull"), the site consists of a number of text "tracks" composed of quotations from Lyotard's works. Viewing the first quotation in a track automatically pulls in the next page of quotations after a prescribed interval. The interval is timed roughly to match the amount of content or to indicate a theme (e.g., the warning on the "Terror" track says, "you have 20 seconds to read this page . . . 15 . . . 10 . . . ," and so on). Yet as I noted in the "Philosophy of This Page" essay on the site, the ability to design such temporality on the Web is extremely limited—so much so that ultimately it is as if the designer cedes the act of design to the geographical and social collectivity that is the Internet itself. The only design, in other words, is the pure contingency or historicity of social time:

> Unlike server-push, client-pull works by breaking and opening the connection between client computer and remote server for each successive page. This means that there is a randomness or aleatory quality built into the timing of client-pull universes that bespeaks the intervention of the Internet itself. All Internet transmissions, of course, are aleatory in their timing and route; but client-pull makes this fact more than normally visible. Client-pull makes it possible to reflect on the fact that each of our communications is paced by simultaneous demands made on the network by other communications—by the time-sensitive collectivity that constitutes historicity. Beyond the individually totalizing conventions of *my* communication or *your* communication, in other words, lies a surprise that emerges from the inventiveness of the interaction of *our* communications.[46]

Even in the best of circumstances, therefore (e.g., if one has a broadband connection), the temporality of presentation on Web pages is uncoordinated, illogical, and unpredictable. Text and images download at different paces; the layout of a page on the screen is variously immediate or deferred depending on the nature of included table structures and whether images are coded with explicit size descriptions; and even revisiting the same page produces a different temporal experience based on whether the page was previously cached on the user's hard drive or service provider's specialized caching equipment. This is not even to mention the extension of the Web into animated, video, audio, and database-driven presentations. Such advanced dynamic features have led to the use of a proliferating set of supplementary plug-in programs (e.g., Flash), scripting languages (e.g., Javascript, Active Server Pages), "middleware" programs written in such languages to mediate between databases and the Web, or other event-controlling languages and protocols that create a busy, composite environment of dynamic Web page effects. Yet what is striking about all such features on the Web is that the finer the degree of temporal design they permit, the grosser the mistimings they showcase when undesigned events on the Internet or on the local machine intervene—for example, the Java applet that takes seven seconds to start because the code interpreter must be initiated, the Javascript-coded movement or transformation of images on a page that works only after all the images are downloaded, the "real time," "streaming" video or audio that skips unpredictably, and so on. At no point up to and including the present broadband age has the temporal experience of opening a Web page matched the crispness and simultaneity of opening a page in a book or magazine—leading to the contradiction that while information may be delivered faster over the Internet, design as such is delivered faster in a physical book.[47] Even when, or if, the fabled era of unlimited bandwidth is reached, it may be predicted that the kind of simultaneity staged by online media will still differ experientially from that of print media. Just as radio or TV differed from print in temporal feel (immediate, simultaneous, but until recently not usable at a time of the user's choice), and just as print itself differed in similar ways from pre-electronic oral media (print, for example, is not simultaneous in creation and reception), so "real time" online media may not converge on a single, universal intuition of time but instead on multifarious new experiences of time (sensitized, for example, to the distinction between synchronous and asynchronous communication).[48]

What is a cool Web page as a matter of design, then? The answer is now clear, though also paradoxical in a way that shatters the glass of any transparent notion of clarity. On the one hand, cool pages are those that recognize the spatiotemporal disturbances of the medium but then accommodate those disturbances through clever visual metaphors or coding techniques that create the façade of a whole harmonium. Thus, for example, a cool page understands how difficult it is to control external and internal white space and so elects to turn "liquid" rather than lock content into fixed-width designs. A liquid page, in the vocabulary of Project Cool, is one that appears equally attractive on narrow and wide screens, contracting or expanding with such easy grace that we are fooled by a digital version of *trompe l'oeil* into thinking that the source code and the visual rendering (content and form) naturally converge in a unitary page.[49] Just as the brush in the hand of a master calligrapher seems to flow at once unpredictably and with inimitable precision, choosing from infinite possibilities the one perfect, instantaneous apprehension of form, so in the hand of a master designer HTML is a tool to be brushed as much against as with the grain to create unaccountable "elegance" (one of Project Cool's favorite descriptive epithets). Indeed, it is no accident that in describing cool design on the Web we should here reach for a metaphor from an older medium. Cool design on the Web is commonly attended by a wit that delights in covering up the gap between ambitious design schemes and limited design control by framing the whole composition within an arbitrary visual metaphor drawn from older media—for example, a Web page presented as a notebook complete with spiral binding, a TV with channel controls, a jukebox with selections that light as the cursor clicks on the buttons, and so on. Such metaphors naturalize the limitations of the new medium by disguising them within those of older media (a jukebox console, for example, explains why there should be a fixed-width menu of selections).[50]

On the other hand, the pure whimsy of such visual metaphors intended to "explain" accidental formal solutions on the Web may far exceed any impression they leave of control and legitimacy. The ultimate impression the Web leaves, in other words, may well be that there is no good design that is not also inescapably just a mixed metaphor. The *really* cool pages are thus those that understand the disturbances inherent in the medium so well that they do not attempt to accommodate them within a fiction of elegant harmony but instead make disturbance their medium. Really cool pages *play up* antidesign in an irresolvable contest of visual metaphors exposing the constitutive contest of metaphors that *is* the interface.[51] Rather than being seamless examples of *trompe l'oeil,* they are seam-full collages designed to expose the innate craftiness—the subterfuge and imposture—of Web craft. Consider again, therefore, all those early Webcam pages whose cool consisted in populating computer screens with fish or lizards—a feat of technical skill whose pleasing quality derived only apparently from achieving the illusion of an aquarium or terrarium. From another point of view, the pleasure of such pages stemmed precisely from the witty, unresolvable *frisson* they created in the clash between computers and livable habitats, machines and nature. The aquarium or terrarium, perhaps, was the looking glass of the cubicle. The caged knowledge worker sitting immobile in front of a glass screen was the real iguana trapped in an unnatural, vitreous world in which people were taken out of their cultures and dropped into a corporate culture where everything is mixed metaphor (e.g., "outplacement," "golden parachute"). Similarly, consider all the early cool Web pages whose backgrounds eschewed the white space look of modernist design (e.g., all white or all black backgrounds) in favor of "textures" representing something a computer absolutely cannot be: wood, stone, water, fire, gold, fur, scales, and so forth. The ideal nomenclature for computing equipment, perhaps, should not be the hybrid Latin fantasies that have been all the commercial vogue in trademarks (e.g., Intel's Pentium and Celeron, Dell's Inspiron, Compaq's Presario) but instead the original fantasies of classical hybridity: Gryphon, Chimera, Centaur, Minotaur, and so forth. As thematized in one of the paradigmatic works of hypertext fiction in the 1990s, Shelley Jackson's *Patchwork Girl*, the new digital medium bares, rather than hides, its existential seamfulness. Or, to take one more example, consider again the Web site titled *What Is Miles Watching on TV!* What came clear when one tuned into this site was that its images of whatever Miles happened to be watching on TV were really nothing at all like watching TV. Nor was it even like watching Miles in the act of watching TV, since Miles was nowhere to be seen. If this page was cool, as the cool anthologies declared, it must have been because it was one of the Web's many equivalents of Magritte's *Ceci n'est pas une pipe.*

Such failed *trompe l'oeil* reveals that it is about the very GUI interface that we must at last say, "This is not a pipe." I take a page here from Neal Stephenson, who after publishing his Pynchonesque novel *Cryptonomicon* exploring the parallel histories of computing and code breaking since World War II, put online (and later published in print) a short monograph titled "In the Beginning . . . Was the Command Line" (included on the Web site promoting the novel). Commenting about the original commercial GUI, Stephenson writes:

> The overarching concept of the MacOS was the "desktop metaphor," and it subsumed any number of lesser (and frequently conflicting, or at least mixed) metaphors. Under a GUI, a file (frequently called "document") is metaphrased as a window on the screen (which is called a "desktop"). The window is almost always too small to contain the document and so you "move around," or, more pretentiously, "navigate" in the document by "clicking and dragging" the "thumb" on the "scroll bar." When you "type" (using a keyboard) or "draw" (using a "mouse") into the "window" or use pull-down "menus" and "dialog boxes" to manipulate its contents, the results of your labors get stored (at least in theory) in a "file," and later you can pull the same information back up into another "window." When you don't want it anymore, you "drag" it into the "trash."
>
> There is massively promiscuous metaphor-mixing going on here, and I could deconstruct it till the cows come home, but I won't. . . . So GUIs use metaphors to make computing easier, but they are bad metaphors. Learning to use them is essentially a word game, a process of learning new definitions of words such as "window" and "document" and "save" that are different from, and in many cases almost diametrically opposed to, the old.[52]

"Click" on the "link" on this "page," an ordinary Web page might thus say, unconscious that pages do not click in either the aural or tactile senses and that links in any case would produce a sound more like "clink." But cool pages know this. Like Web pages that suck (though differentiated from them by some unknown quotient), really cool pages play upon a scrambled code of metaphors that is no sooner decrypted from one medium or form into a remediated analogy than it is immediately (or rather, as Bolter and Grusin put it, hypermediately) reencrypted into further remediated analogies. The design of the GUI interface, in short, is that uncanny doppelgänger of modernist clarity, a cipher. Quick, answer me this riddle: what kind of "desktop" has a "window" that contains "pages" allowing us to "scroll" through "portals" into ever more windows, pages, scrolls, and portals? The Enigma machine that Alan Turing and others battled during World War II, perhaps, is still with us. Only now the cryptological enigma is that *anti*-universal Turing machine called the World Wide Web. Designer: "here is my intricately crafted page." User: "it came out scrambled on my screen and I keep getting Javascript error messages."

Thus in the altercation between designer and user the deep design—which is to say, antidesign—of the Web at last shows itself. Both the spatial and temporal conditions of the Web scramble design, and the result is to destabilize the underlying *social meaning* of design. The more the Web designer attempts to freeze the composition on the screen as if it were a display affixed to a spatial whole that can be delivered temporally to the user all at once, the more that designer resists an even deeper design imperative in the medium—the need to make design as fluid as possible so that it can pour across the wires into the unpredictable receptacles, rhythms, and ultimately lives of others. The deep design of the Web is the *distribution* of the authority of design from the content-designer to the user-designer (collaborating with legions of hidden programmer-designers) who configures the local machine and browser. This is why it is not possible, when discussing the Web, to be wholly unambiguous about which perspective—the Web author's or user's—we are assuming. Not only is an ever larger number of users becoming literal Web authors and designers (as ordinary word processing, desktop publishing, database, and other office products evolve Web publishing capabilities), but the very concept of using the Web implies a degree of designing the Web. Whether or not the overall quantity of design remains constant (an issue that it would be interesting to measure empirically), the proportion that official designers control—by contrast with the glory days of the great modernist designers—wanes.

Chapter 7

The Feeling of Information

Cool Is a Feeling for Information

As may now be apparent, the ethos of information cool I discussed in chapter 5 is intimately related to the style of such cool I studied in chapter 6. To gaze upon the character of the informed is also to encounter the character of the designer. The move I now make similarly marks not so much a clean transition as a continuation along a single curve of inquiry. Given the paradoxical ethos and style of cool, we may ask, what is the *feeling* of that paradox? How does online cool complete the genealogy of industrial and postindustrial feeling that we have followed? How does the "Fordized face" of the automating era, already having morphed into "service with a smile" in the informating era, now morph yet again in the age of networking? The answer, as we will see, extends the issue of style. Cool style at the beginning of the twenty-first century completes the severe revision the twentieth century had already initiated in old traditions of affiliation between artistic styles and modalities of feeling.

The heart of the problem lies in determining whether the cool "feeling of paradox" is in fact a structure of feeling at all rather than, equally intuitive, a *lack* of feeling. On the one hand, information cool is robust with feeling. Cool, as we remember Netscape's "What's Cool?" page saying in its best circus barker's voice, will "catch our eye, make us laugh, help us work, quench our thirst." Cool on the Web is a heady brio, gusto, rush, thrill, *feeling* of information. Rather than being the dull, dim anomie that David Shenk calls "data smog" ("increased cardiovascular stress," "weakened vision," "confusion," "frustration," and so on in the face of information), it is a keen or glossy heightening of sensation well expressed in all the "glows," "drop-shadows," and other visual effects that Web page designers now routinely add like a glory around ordinary text.[1] The Netscape page is just one anecdotal evidence of the emotional halo in which cool burnishes information. A fuller study of the emotional tone of the Web (setting aside the more flagrantly emotive discourse of newsgroups, e-mail, and chat) would need to take a much broader sample of cool pages, scrutinize the subset explicitly called cool, assess the remaining quotient of cool on pages not specifically identified as such, and then use discourse analysis and empirical psycho-social research to calibrate the evidence of feeling against some historical or social baseline (measuring, for example, the pattern of affective terms on the Web against a sample of women's and men's magazines). Since such a study is beyond the scope of this book, I will employ a tactic intended just to take the practical emotional temperature of cool—like thrusting one's hand into the water to see if it is cold, hot, or lukewarm.

The best single body of documentation for studying cool feeling on the Web is the Project Cool site, whose extensive historical archive of cool "Sightings" provides ample material for discourse analysis. Since 1996, Project Cool has gathered one "cool" site each day, supplying a brief and colorful annotation for each after November 1997. Like most general audience or commercial cool anthologies, of course, Project Cool's archive is excessively PG-rated in its language of cool. Ideally, we would want to correct this bias by attending to the edgier (but less well-documented) youth subculture fringes of cool where feeling has either a rawer or noir quality. But if we allow for this astigmatism of sensibility (perhaps mentally alternating each of Project Cool's epithets of cool with "rad," "extreme," or some more current adjective of subcultural edginess), then the pattern that emerges from the archives is revealing. Taking a continuous, one-year sample from the Project Cool archives for September 1998 through August 1999 (collected on the site in one-month segments under the title "Previous Sightings"), we observe the following structure of feeling.

"Cool," Project Cool first of all believes, should definitely be said with feeling. Here, for example, are excerpts from the archive for December 1998 (numbers refer to date of month):

5 *Maximov Online*
Today's Sighting is a portal to another region displayed
in a beautiful interface.

6 *Hamsters*
Who says a site can't be fun. Today's Sighting certainly is.

14 *Zen*
While we're not quite sure what to call Today's Sighting,
we are sure that it was quite enthralling.

15 *Hacker Tax Credit*
A simple and elegant Liquid design is used to convey a
message at Today's Sighting.

20 *U.S. Kids Compute*
Today's Sighting, while visually appealing is also just
plain fun to look at and we won't even mention informative.

21 *The World of Bacardi*
Today's Sighting pushes the envelope of what can be
done. Oh what a fine envelope it is, too.

22 *AndyArt*
Featuring tons of inspiration for interfaces, Today's
Sighting also gives back to the Web.

23 *spaceman*
A clean, modern look at Today's Sighting showcases content
you don't necessarily need the language for.

24 *Norad Tracks Santa*
It's too bad that not every website goes to the trouble of
Today's Sighting to be both fun, informative, and seasonal.

25 *Castle Arcana Christmas Calendar*
Today's Sighting is almost a coloring book, almost. It's
also a lot of fun.

30 *White House News Photographers' Association*
The evocative, moody presentation wins at Today's Sighting.

In its adjectival excess, such writing injects from the start a breathless "oooh" or "ahhhh" into browsing intended to elevate the ground tone of affect, much like turning up the brightness of the display. Indeed, "oooh" is not far off the mark. "Oh what a fine envelope it is, too," Project Cool choruses about the Bacardi site on 21 December, using the classic poetic device of the vocative to summon up inspiration. The specific spirit thus invoked—not classic but Disney—is "fun." Fun is the primary mood of cool according to Project Cool. Of the approximately 360 sites designated cool in the period from September 1998 to August 1999, no fewer than 30 are explicitly denoted as "fun" (or a close variant, such as "playful," "made us laugh," and so on).[2] Besides the fun in the above excerpts, for example, we can adduce such other rote samples as: "Some of the best fun we have is playing with presentation. Today's Sighting obviously had lots of fun" (21

January 1999); "Today's Sighting is nothing more than fun. That's all, nothing more" (13 June 1999); and "Cartoonish, humorous, retro, fun" (30 July 1999). Fun then fans out in the archives into more adult colors as well, as in the praise for the "enthralling" Zen site or the "evocative, moody" White House News Photographers' Association site (14 and 30 December). We are indulged elsewhere in our surveyed period with a connoisseurship of sites "dreamy," "pleasant," "colorful," "positively swimming," "enjoyable," "flavorful," "whimsical," "charming," "captivating," "delightful," "curiously strong," and "wild."[3]

Cool information, in other words, is not just colorless bits, keystrokes, files, and passwords. Nor is it just the neutral colors of the cubicle (tan, suede, gray, or sage) except insofar as such earth shades can be assimilated in fantasy to the archetypal desert landscape of knowledge work I previously depicted: a wildscape on which the "strong" and "wild" roam and where even the barest pleasures (a smoke, beans simmered on a campfire, a quiet sunset, or the knowledge worker's equivalent—a Starbucks coffee) become, by contrast with the total desiccation of the cubicle, an explosion of the senses. Or if sublime landscape is the wrong precedent, then we might look to eighteenth-century "picturesque" landscape to appreciate all that is "charming," "delightful," and "whimsical" about cool. After all, we should not be surprised that the epithets used to excess in the wake of the Enlightenment to describe detached, visually framed scenes of enjoyment should reconvene in the age of knowledge work to describe an equivalent visual frame—the computer screen.

Cool Is an Apatheia of Information

Yet, with equal certainty, cool in Project Cool's archives and elsewhere is anything *but* emotionally expressive. Here I take my cue once more from the "Editorial Policy" appended to Netscape's revised "What's Cool" page in 1998, which states that the "personality" of cool should be hedged in by severe barriers to expressiveness: "When it comes to cool sites, personality goes a long way. Even a site devoted to toast can offer fascinating information that's packed with humor and punch. On the other hand, exciting information should steer clear of hype and cliche (unless of course, it's appropriate to the site's objective). We look for sites that use language that is engaging not obnoxious, informative not boring." Like finding the toy packed into a Cracker Jack box during the mainframe 1960s and saying "cool!" opening a Web page in the network age is cool if the page is "packed with humor and punch." But immediately the "other hand" of discipline exerts control. In plain contravention of "packed with humor and punch," Netscape recommends in its best hall monitor's voice that "exciting information should steer clear of hype and cliche." "Be bold, be bold . . . Be not too bold," we are thus told as if we were Edmund Spenser's Britomart watching a pageant masque (the Renaissance version of what the Web calls "eye candy").[4]

Or we can look to Restoration and neoclassical times for a closer precedent. Like eighteenth-century poetic "wit" with its famously paradoxical portraits of character (or like the picturesque somewhat later, as well), cool is an emotional state so torn between incitements and proscriptions to passion that it is oxymoronic, even manic-depressive, in feeling.[5] Here, for example, is Alexander Pope on the characters of women: "Reserve with Frankness, Art with Truth ally'd, / Courage with Softness, Modesty with Pride" ("Epistle II: To a Lady"). Here is John Denham on the river Thames: "Though deep, yet clear, though gentle, yet not dull, / Strong without rage, without oreflowing full" ("Cooper's Hill"). And here is Netscape on cool: "engaging not obnoxious, informative not boring." It is not accidental, perhaps, that those tricksters of the Web—the authors of *Web Pages That Suck*—are satirical in the grand old neoclassical tradition. What is *Web Pages That Suck,* after all, but the postindustrial *Dunciad?* Cool is caught in a paradox of feeling and *not* feeling so severe that the only clear expression for it is the meta-feeling of neoclassical satire, which, as in the best of Pope, is a feeling of edgy, above-it-all mockery that somehow also accommodates keen affects of sympathy, mourning, and love (as in the portraits of individual women in "Epistle II: To a Lady"). To recur to the precedent of modern times, cool is the "irony" of paradox savored by those critics of modernity so well studied in neoclassical wit: the New Critics.

Thus, if we return to the Project Cool archives, we see that the full structure of feeling sketched there is one that inhibits as much as it releases feeling. Two of the entries during the year I surveyed are emblematic. One is for the cool site of 17 February 1999, about which Project Cool says, "Capture a feeling? Today's Sighting definitely does that. It also gives great coverage to a passion." The site thus honored is *The Jalopy Journal,* which is devoted to hot rod cars (represented in sensuous, high-quality images). The other is for the cool site of 25 May 1999, named *.ttf,* which offers downloadable text fonts. Project Cool writes: "Color and a feel of precision make this download site stand out as Today's Sighting." Read in tandem, these two sites testify to the signature of cool as distinguished from such previous forms of paradoxical emotional detachment/attachment as irony. There may be tremendous "passion," but passion feels strangely like detachment because it is "captured" within a specific kind of attachment—the enormous need of twentieth-century cool not so much to consume as to be consumed by a particular class of affect-objects. That class consists of technologies and increasingly techniques. That Project Cool pays homage to a hot rod site is nostalgic for all the twentieth-century technological objects that subcultures had used to front their cool assimilation/repudiation of mainstream culture. In contemporary information cool, however, libidinal investment in technology converts with unprecedented ease into the pure eroticism of *technique.* While technology and technique may both be necessary objects of cool passion, in other words, now there is an excess of passion for technique. The really cool sites for Project Cool are those like the *.ttf* site that allow us to enjoy the "feel of precision." Cool is feeling that is muted by the technical. It is a technical feeling or feeling for the technical.

This is to say that the heart of Project Cool throbs for *design,* which can now be understood in a fresh context. Design is how we can be dominated by instrumental rationality and love it, too. When one reads through the Project Cool archives, the overall impression that emerges is not "fun" but the containment of fun and other feelings in a "design sense" that compresses emotionality within narrow, even minimalist parameters—for example, the all-black or all-white backgrounds of many faux-modernist cool sites. The predominant name that Project Cool gives such design sense (the equivalent of what engineers and programmers call technically "sweet") is "elegant." If *fun* appears in 30 of the site descriptions in the year I surveyed, *elegant* appears in 21. What is important to Project Cool is not just fun but fun expressed elegantly in design technique: for example, "liquid" HTML tables adaptable to browser window size, sophisticated Javascript rollovers and other dynamic HTML, well-chosen colors and backgrounds, deft Flash animation. Elegance is technique become the fetish of enjoyment. The description of the *PBS Online* site on 10 January 1999, for example, subordinates all affect within the pleasure of elegant design: "a useful, elegant and quite eye-pleasing design." Similarly, the description of the *Belles de Jour* site on 10 February 1999 praises a "pleasingly designed directory." Or consider the following hedonism of design (phrased with a touch of wit) in regard to the *HM Prison Service* site: "I could stay and examine the elegant design for quite some time" (2 June 1999), or again the hedonism of pure technique regarding the *Wayback* site: "we took great pleasure in its animations and effects" (27 November 1998). Combined with such other *à la mode* terms of design worship on the site as *chic, graceful, stunning, exquisite,* and so on, the "elegant" affective discourse of Project Cool testifies powerfully that emotions in the information age are not by preference communicated through smiles or frowns (which are barred by the physical circumstances of networked computing), or even through the ironic "emoticons" that for a time flourished in e-mail. Emotions are instead vested in the design of the interface itself. We now style or posture our feelings through the technical design of the browser in a way that supplants the predecessor social techniques of emotional management: rituals, habits of dress, hair styles, slang, and so on. Once the cool dressed and walked just *so*. Now the cool dress their pages in complex tables, colors, and fonts while letting DHTML (dynamic HTML, responsible for certain motion and interactive effects) do their walking for them.[6]

We can now understand why, when measured against historical artistic styles and their modalities of feeling, cool seems so diminished an experience. We remember that once the great public genres and style of affective experience—for example, tragedy, comedy, the beautiful, the sublime, and so on—corresponded with modes of feeling that, however much zeroed out at crucial moments in *apatheia* (serene passionlessness, as in the aftermath of tragic purgation), were painted in broad emotional gestures. Big laughs, big anger, big tears, big terror. By contrast, what is remarkable about the public Web (as opposed to the flame wars that break out in less fully public quarters of the Internet) is the severe restriction of acceptable modalities of feeling. There are a few pages on the Web that are tragic or sublime, including such superbly designed sites as the San Francisco Exploratorium's *Remembering Nagasaki* and such crudely designed, yet no less moving sites as Tim Law's *Stillbirth and Neonatal Death Support* (pages I have regularly directed students to as counter-examples to cool). But the tragedy or sublimity of such pages—even if, as in the case of the Exploratorium site, they are sometimes called cool by the anthologies—is fundamentally incommensurable with cool. Cool feeling may be everywhere on the Web (cool, after all, ranks among the most totalitarian aesthetics ever created). Yet there is so little feeling in cool feeling. Before all the horrors and despairs offered up on even ordinary journalism Web sites, cool is wordless, or at best responds, "That's uncool." (In the vernacular of student discourse: why not kill someone? Well, it's not cool.) And for extremes of pleasure rather than horror, the Web can only direct us in a sly whisper to the back alleys of the information superhighway where "hot" is just as much a stylized mask of feeling—that is, to XXX sites. In the information world, we may say, this is all that is fun: :> This is all that is sad: :< And the only mask of lust is: **X**. All terror, anger, lust, joy, and so on thus bleed out of cool to manifest with compensatory, even artificial, fervor in personal e-mail, alt. newsgroups, chat, hate sites, porn, and other parts of the Internet that sequester themselves from postindustrial knowledge work by being intractably "unproductive."

Cool may not be exactly a "Fordized face" of no emotion, then, but as the alter-face of the interface of information, it is just as constrained. Like the "mirrorshades" of cyberpunk science fiction, information cool is a kind of high-tech Fordization of the face we might call "designer emotion."[7] At base, its rictus of compulsory fun is *automatic* feeling, a holdover of the age of automation within postindustrialism. Who is cool in the information age? Street performers gathered at the great tourist and transit centers of contemporary urban culture know. Look at that street performer painted all in silver who enacts the "feel of precision" in the form of a robot whose limbs move in synch to ventriloquized machine sounds: "zzzzzzzzz," "hmmmm," "brrrrrrrrrr." *There* is the spitting image of all the information workers who gather around the performer on illusory holiday from their own routines dedicated to nothing other than the feeling of precision. And so the entire genealogy of cool we have followed in the twentieth century at last downloads onto our desktop: the "cool pose" of subculture as Richard Majors and Janet Mancini Billson title it in their book on the detached rage of young black men in America; the cool counterculture that borrowed the pose of subculture to protest angrily in the streets while elsewhere being "laid back"; and now cool Webheads (rather than Deadheads) who inhale the most mind- and emotion-numbing, yet also consummately professional, of all consciousness-changing drugs—information. These are the generations of cool.

There is, however, at least one other name for the cool pose of information that we have yet to consider. This name—equally familiar to subculture, counterculture, hackers, and now cubicle warriors—is *attitude* (as in "bad attitude"). Attitude is the incomplete politics of cool.

Chapter 8

Cyber-Politics and Bad Attitude

Cool Is a Politics of Information

"This is a cool holiday," a young Bay Area software entrepreneur observed on the Fourth of July, 1997, to Josh McHugh of *Forbes* magazine. "It's the day we celebrate overthrowing the government." In his teens, McHugh notes, the entrepreneur had once manually typed the entirety of Thoreau's "Civil Disobedience" into an Apple computer and posted it to an electronic bulletin board. But now his preferred means of civil disobedience is "strong cryptography." Strong cryptography—or how the high-tech population secede from the nation—is his business product; but it is also his belief. Strong cryptography, to cite the title of the resulting *Forbes* article, is an instance of "Politics for the Really Cool."

Most users of online media are likely to have encountered—and in practice at least minimally endorsed—"politics for the really cool." One does not need to be overtly political, after all, to feel a vicarious thrill of revolution while downloading copyrighted music at work (e.g., using Napster before 2001), viewing previously restricted satellite images on the Web, or simply typing in a false phone number to protect one's privacy when placing an online order. Even purely fictional representations of digital information technology, such as the 1990s films *The Net* and *The Matrix,* implicate their audiences in this way. When we root for a heroine trying to protect personal data from government agencies or a hero overthrowing information enslavement, we participate in politics for the really cool. But to study the logic of such politics, it will be useful to turn our attention from casual experience and entertainment to works of more deliberate advocacy. I refer to "cyberlibertarianism," the dominant—some say exclusive—politics of information on the Net.[1] Not the same as politics *in* information (the extension of political discussion, lobbying, and campaigning into new media), opposed to the definitely uncool politics of *regulating* information (information and communications policy), and hard to square with the standard alignment of left versus right (organizations like the Progress and Freedom Foundation, discussed below, can seem both right and left in ordinary political terms), cyberlibertarianism is the belief that the technological and social covenants of networked information are a new form—or reform—of politics. As Jon Katz wrote in the oracular opening of his *Wired* magazine article "Birth of a Digital Nation" (1997), "the world's information is being liberated, and so, as a consequence, are we." Or again, from the same opening prophecy: "I saw the primordial stirrings of a new kind of nation—the Digital Nation—and the formation of a new postpolitical philosophy."[2]

A representative list of leading activist organizations, magazines, and manifestos of cyberlibertarianism, together with some critical commentary to give us perspective, would include the following:[3]

- *Organizations.* Some of the best known organizations are the Electronic Frontier Foundation (EFF), Center for Democracy and Technology (CDT), Progress and Freedom Foundation (PFF), and Computer Professionals for Social Responsibility (CPSR). The first three were all started in the 1990s; CPSR began in 1981 as an organization concerned about "the dangers posed by the massive increase in the use of computing technology in military applications" and later broadened its scope to such more typical cyberlibertarian issues as those addressed by its "Cyber-Rights" and "Privacy and Civil Liberties" working groups.[4] Characteristically, these organizations publish news and commentary, provide extensive online resources, watch over or advocate government regulatory actions, and support legal actions related to information policy.
- *Magazines.* The most famous is *Wired,* whose sometimes literally neon (not just purple) prose on all things cool and digital has colored the general social awareness of cyberculture—as witnessed in the magazine's commercial success and the high visibility of its writers and editors in other venues.[5] Also important are such Internet technology news sites as *C/Net News.com* and *ZDNet News,* whose coverage of information tech-

nology is objective rather than hip but nevertheless intently follows the main cyberlibertarian issues of privacy, censorship, and so on.

- *Manifestos.* Although a great deal of cyberlibertarian discourse occurs in such informal channels as Usenet newsgroups, listservs, e-mail letter campaigns, and (in the earliest days) BBSs, the best known manifestos are those published by the activist organizations, *Wired,* and other high-visibility advocates. The organizations, for example, feature "mission" or "about" statements on their Web sites that amount to cyberlibertarian declarations of independence. In addition, they regularly issue full-scale position papers—perhaps most famously John Perry Barlow's "Declaration of the Independence of Cyberspace" (on the EFF site) and PFF's "Cyberspace and the American Dream: A Magna Carta for the Knowledge Age" (co-authored in 1994 by the blue-ribbon panel of Esther Dyson, George Gilder, George Keyworth, and Alvin Toffler). Similarly, *Wired* has published such well-known article-manifestos as Barlow's "Economy of Ideas" and Katz's "Birth of a Digital Nation" as well as its sequel "Netizen: The Digital Citizen" (all duplicated on the *Wired News* or *HotWired* Web site). Also relevant are independently published works by the leading members of the so-called "digerati," including not only the leading contributors or editors of the above-mentioned organizations and magazines but also the founders of such legendary BBSs as The Well.[6]
- *Critical Commentaries.* While there is not yet a canon of critics to match the nascent canon of advocates, we can usefully set alongside cyberlibertarian discourse at least a sampling of critical perspectives (including those written from a non-American viewpoint). Some incisive examples are Richard Barbrook and Andy Cameron's "Californian Ideology"; William F. Birdsall's "Internet and the Ideology of Information Technology"; Paulina Borsook's "Cyberselfish"; Stephen Doheny-Farina's chapter "Seeking Public Space on the Internet" in his *Wired Neighborhood* (whose cover imitates the look of *Wired* magazine); Roberto Verzola's "Towards a Political Economy of Information"; and Langdon Winner's series of piercing essays on the politics of information and technology (e.g., "Cyberlibertarian Myths and the Prospects for Community," "Mythinformation," "Peter Pan in Cyberspace: *Wired* Magazine's Political Vision," and "Techné and Politeia").[7]

Using this short list as our data set (and supplementing it with other resources as needed), we can sketch the basic cyberlibertarian argument—what amounts to a Bill of Information Rights as follows.[8]

Freedom Is American (Specifically, Left Coast American)

I begin with the unspoken premise of cyberlibertarianism. As Barbrook and Cameron observe from a British standpoint, cyberlibertarianism (which they call "Californian Ideology") "was developed by a group of people living within one specific country with a particular mix of socio-economic and technological choices." Cyberlibertarianism does have international reach, as seen, for example, in Privacy International (an organization based in London with an office in Washington, D.C.), the Global Internet Liberty Campaign (which issued a report titled "Regardless of Frontiers: Protecting the Human Right to Freedom of Expression on the Global Internet"), and the EFF, which on occasion has invoked Article 19 of the United Nations Universal Declaration of Human Rights on "freedom of opinion and expression."[9] Nonetheless, its political baseline is clearly and abundantly American (or more broadly, Western) on all key points of history, assumptions, paradigms, and vocabulary. It is not accidental that cyberlibertarians frequently recite an honor roll of "Jeffersonian" democratic ideals, the Stamp Act, the Magna Carta (reaching across the Atlantic for an English precedent), and so on.[10] Cyberlibertarianism is founded on the unquestioned proposition that "freedom" as spoken in American is a universal human right taking precedence over competing political systems that vest independence in the first instance in collective entities (clan, religion, state). The almost total avoidance of a discourse of "duties" (to one's state, to one's religion) in favor of a discourse of "rights" is symptomatic.

Freedom Is Freedom from Government

To emerge from its unspoken premise into positive statement, cyberlibertarianism must go by way of negation—two negations in particular that subtract latter-day notions of "America" to reveal the allegedly original, core values of the nation. The first negation may be discerned in the cyberlibertarian distaste for collectivist institutions. We do not get very far thinking about the Americanism of cyberlibertarianism before running into a profound disconnect with the very notion of "America" as a collective *nation*—the precise disconnect that allows cyberlibertarians to elide their geopolitical site and masquerade as citizens of the world. In the United States, freedom was originally constructed on the framework of two separate collectivist powers: the dissenting church and the rebel state. But *neither* term of this founding national compact (both cast by the "digital nation" as obsolete) currently sanctions the dissent of cyberlibertarianism. Much the reverse, since both—as epitomized in the Communications Decency Act (CDA) of 1996 that cyberlibertarians demonized as the worst kind of political *cum* moral-religious repression—oppose the pure condition of stateless, agnostic information.[11] Especially after the CDA (and more so with each succeeding regulatory battle against politicians intent on censoring online speech, limiting encryption, or taxing e-commerce), it has become *de rigueur* among cyberlibertarians to cast the institutional powers of the nation as the enemy of freedom. As Barlow puts it in the opening of his "Declaration of the Independence of Cyberspace": "Governments of the Industrial World, you weary giants of flesh and steel, I come from Cyberspace, the new home of Mind. On behalf of the future, I ask you of the past to leave us alone. You are not welcome among us. You have no sovereignty where we gather."

It is only at first glance surprising, therefore, that cyberlibertarianism seems uncannily aligned with the small government mantra of the Republican 104th U.S. Congress. This is especially the case with the Progress and Freedom Foundation, which Winner notes was started in part by "Newt Gingrich and his associates." Getting government out of the way is a precondition of the PFF's primary emphasis on establishing a "Third Wave" environment for business (thus, for example, the "major project" announced on its "Mission Statement" page in 2000 "aimed at identifying public policies to limit government interference in the market for digital broadband networks").[12] But even cyberlibertarian organizations that evolved from countercultural roots tend to approach government in a manner strikingly similar to contemporary conservatism. At its most extreme, indeed, the hostility of cyberlibertarianism to government is exceeded only by the most virulent, survivalist forms of libertarianism. However, because the Internet seems to offer a "frontier" or "territory" allowing it to push endlessly beyond the pale of regulation, cyberlibertarians do not need to bunker down in survivalist fashion in some thinly populated mountain state. They have all the New World of the "Mind," and especially the mentality of the Silicon West, in which to roam. Barlow thus says in his "Economy of Ideas" that "one of the aspects of the electronic frontier which I have always found most appealing—and the reason Mitch Kapor and I used that phrase in naming our foundation [Electronic Frontier Foundation]—is the degree to which it resembles the 19th-century American West in its natural preference for social devices that emerge from its conditions rather than those that are imposed from the outside." The authors of the "Magna Carta" add, "We are entering new territory, where there are as yet no rules—just as there were no rules on the American continent in 1620, or in the Northwest Territory in 1787."[13]

Freedom Is Freedom from Big Business

As recently as mid-1996, Paulina Borsook could accurately complain: "Technolibertarians rightfully worry about Big Bad Government, yet think commerce unfettered can create all things bright and beautiful—and so they disregard the real invader of privacy: Corporate America seeking ever-better ways to exploit the Net, to sell databases of consumer purchases and preferences, to track potential customers however it can." But even as Borsook wrote, controversy over the great gold rush of firms onto the Web was beginning to accompany controversy over the CDA. By the late 1990s netizens habitually stood not just against big government but also—articulating their second, resounding negation—against big business in its personality as consumer industry (leaving aside other dimensions of business we will need to consider later). Specifically, cyberlibertarians elevated to the status of a cardinal issue the monitoring of personal data by online firms. This was especially true after 2000, when the controversy over the Doubleclick firm's handling of personal data advertised the embarrassing inability of online commerce to regulate itself and raised the specter of an antidote that cyberlibertarians thought was just as poisonous: government legislation.[14] The search was on, then, for purely voluntary or technological protocols for limiting the commercial use of personal data (e.g., the P3P initiative in 2000) as an alternative to government regulation. Not for nothing were such leading, early cyberlibertarians as Barlow (Grateful Dead lyricist and a founder of the EFF) weaned on the grassroots, gift-economy ethos of counterculture.[15] To thwart consumer business through the equivalent of grassroots rather than regulatory action was to fight the good fight. It was to defend "consumer privacy" as one of the very "pillars of Internet freedom."[16]

Information Wants to Be Free

We can now emerge from cyberliterarianism's negations into the light of its great positive propositions. Consider for a moment the dangers of a purely negational position. While the negations of cyberlibertarianism subtract modern notions of "America" to reveal supposedly original values, they do so at the risk of creating a vacuum of authority. On what authority does cyberlibertarianism ground its unquestioned faith in American freedom, coupled with its equally unquestioned conviction that government and business (despite checks and balances and antitrust) are too totalitarian to represent that freedom? Barlow asserts in the second paragraph of his "Declaration of the Independence of Cyberspace," "We have no elected government, nor are we likely to have one, so I address you with no greater authority than that with which liberty itself always speaks." The lack of detail in this statement about what kind of authority liberty actually "always" spoke is symptomatic. In the original U.S. Declaration of Independence, the People were the authority of liberty. Or, rather, it was God who authorized the beliefs of the People ("We hold these Truths to be self-evident, that all Men are created equal, that they are endowed by their Creator with certain unalienable Rights . . ."). The founding structure of authority behind American freedom might be represented: God → People → Government. God authorized the people, who in turn authorized representative government (and later shareholder-owned big business alongside big government). But according to cyberlibertarianism, the Internet is a constitutionally decentralized cosmos with no God or transcendental sysadmin, and the government (like certain businesses) is Satan. Both the ultimate source and delegated form of authority are thus delegitimated. The founding structure of authority that once organized the idea of the People is stripped away, leaving the People in search of a new authoritative configuration for their freedom.

We can now see where the great, positive assertions of cyberlibertarianism come from. They appear to stem from the "We the people" of the founding fathers, but in fact they arise *ex nihilo* in the process of restructuring the People's authority as follows: Information → People → Network. In place of God, let the transcendental sanction of freedom be a free-floating, abstract, yet also strangely personified Information (as in the expression, "information wants to be free"). Not just technological determinism, in other words (which Barbrook and Cameron call the "nearly universal belief" of cyberlibertarianism), but *information determinism* rules.[17] The abstract principle of information is so determinative that it would be a trivial operation to replace each of the transcendental terms in the founding statements of American democracy with "information" or "information technology." For example: "We hold this information to be self-evident, that all men are created with equal access to information, that they are endowed by information technology [the decentralized nature of the network] with certain unalienable Rights. . . ." Symmetrically, in place of Government or Big Business, let there be the radical federalism and gift economy of the Network. How should such issues of governance as flame wars, spamming, and other forms of inappropriate use be decided? The cyberlibertarian party line is "let the Network decide." Let the network discuss the issues, reach a consensus, and enforce its decision not through central power but instead through the massed impact of thousands of individual messages urging sysadmins to pull the plug on a transgressive user or change a pernicious usage policy. As Barlow puts it in the "Declaration of the Independence of

Cyberspace": "Where there are real conflicts, where there are wrongs, we will identify them and address them by our means."

Between Information as the new transcendental authority and the Network as the new political representation of that authority, behold a new, wired People. Like some chat-room participant assuming a pseudonymous identity, this People only *looks* like "We the people." Shunning the authorized collectivism of the founding "we" ("one nation under God"), this is a self-authorized People owing its authority to nothing other than being "well informed" and delegating only so much of that authority to the Network as can instantly be withdrawn once more into the hidden ("private," "anonymous," "lurker") omnipotence of the self. Specifically, cyberlibertarianism posits three interlinked selves for its People that, while overlapping with similar retro-patriot identities in mainstream U.S. politics of the late twentieth century, also justify specific information policies: the individual, the consumer, and the entrepreneur.

Freedom Is Individual

The first identity of the People that we normally encounter in cyberlibertarian discourse is the "individual." Exemplary is the "Mission Statement" of the Progress and Freedom Foundation, which speaks in the name of the "human" and "Mankind" but clearly presumes that the essence of universal mankind is "individual":

> The Foundation believes that the digital revolution portends fundamental cultural, economic, political and social changes. While these changes bring challenges, they also create the potential for a new era of human progress. The foundation embraces the idea of progress—i.e., the belief that Mankind has advanced in the past, is presently advancing, and will continue to advance through the foreseeable future. And it believes that the sort of progress brought about by the digital revolution is inherently favorable to enhanced human individuality and freedom.

Similarly, we may cite the following passage from a section of the "Magna Carta" that bears the distinct impress of one of its co-authors, Alvin Toffler: "The complexity of Third Wave society is too great for any centrally planned bureaucracy to manage. Demassification, customization, individuality, freedom—these are the keys to success for Third Wave civilization." Perhaps not surprisingly, therefore, the "Magna Carta" later conspicuously quotes Ayn Rand as its "libertarian icon":

> It is the proper task of government to protect individual rights and, as part of it, formulate the laws by which these rights are to be implemented and adjudicated. It is the government's responsibility to define the application of individual rights to a given sphere of activity—to define (i.e. to identify), not create, invent, donate, or expropriate. The question of defining the application of property rights has arisen frequently, in the wake of oil rights, vertical space rights, etc. In most cases, the American government has been guided by the proper principle: It sought to protect all the individual rights involved, not to abrogate them.[18]

As Winner observes, a "key theme" of cyberlibertarianism is "radical individualism. Writings of cyberlibertarians revel in prospects for ecstatic self-fulfillment in cyberspace and emphasize the need for individuals to disburden themselves of encumbrances that might hinder the pursuit of rational self-interest."[19]

The action items that attend the individualism of the People are online "free speech" (or anticensorship) and "privacy," which together stand at the head of virtually all cyberlibertarian agendas. Witness, for example, the mission statement of the Center for Democracy and Technology: "With expertise in law, technology, and policy, CDT seeks practical solutions to enhance free expression and privacy in global communications technologies." Free speech online means the right of every individual to be an uninterruptible transmission source. As the EFF says, "The new networked society has created a platform, which will allow every person to speak their mind and query the world to create their own point of view."[20] Privacy (together with strong encryption) is the symmetrical right of every individual to withhold transmission when desired.

Freedom Is Consumerist

A corollary of cyberlibertarianism's vigilance against monitoring by consumer businesses is that the People are consumers. This hardly seems to need additional statement. But as we will see, positioning the People as consumers has profound ramifications. That cyberlibertarianism actively rather than passively poses the People as such is undoubted. The CDT, for example, says succinctly, "Internet privacy should be addressed from a consumer perspective."[21] Or again, here is the overview statement on the EFF Web page on "Defining Digital Identity":

> Are you what you read? What you wear? What you listen to? Do you define yourself by the company you keep? Your friends? Your family? Or are you simply your fingerprints and DNA?

> In a networked society, information about us travels at high speeds around the world. Most likely, it's being stored in the databases of companies you've never heard of.
>
> And this very private information—your digital identity—is not just being captured on the Internet. It's being collected in the physical world also.
>
> Buying patterns are being data-mined from your supermarket club card. Biometric body scans capture everything from your inseam measurement to your bust size when you're buying a new pair of jeans. "Smart" workspace programs monitor your every keystroke for your employer's security. And soon, your local ATM may be using retinal scans for user authentication.

Setting aside the single anomalous sentence on "smart" workspace monitoring (a topic I return to later), it is clear that this EFF document primarily characterizes wired people as shoppers for books and jeans, users of supermarket club and ATM cards, and so on. In a similar vein, one of the key tenets of the PFF "Magna Carta" is "demassification," a concept that replaces the notion of mass consumer culture with what might be called individualist consumer culture: "Turning the economics of mass-production inside out, new information technologies are driving the financial costs of diversity—both product and personal—down toward zero, 'demassifying' our institutions and our culture. Accelerating demassification creates the potential for vastly increased human freedom." Freedom, in other words, is the ability to buy a pair of jeans cut to suit oneself, down to the exact roll and tuck of flesh. Yet, apparently, it is also the right not to be tracked in corporate databases down to that same roll and tuck.

The action items that indulge the identity of the People as consumers include the whole range of cyberlibertarian initiatives related to online commerce and information technology business. Besides consumer privacy, for example, there are the issues of antitrust (associated with the inveterate skepticism of large parts of the cyberlibertarian community toward the Microsoft company), taxation of online commerce, "common carriage" (the obligation of telecommunications and cable companies to carry competing online services on a nondiscriminatory basis), and so on. The common thread is the well-being—or, in the antique sense, "wealth"—of the netizen as consumer.

Freedom Is Entrepreneurial

As we saw in the case of the strong encryption entrepreneur in the *Forbes* article, that good information politics can be good business is not a contradiction for cyberlibertarianism because its view of business cleanly divides the viewpoint of the producer from that of the consumer. "U.S.A." becomes just another dot-com, and the pioneering spirit of the early patriots is alive and well, despite such Valley Forges (or, perhaps, Silicon Valley Forges) as the bear market in technology stocks that began in 2000. In the words of Barbrook and Cameron, cyberlibertarianism "promiscuously combines the free-wheeling spirit of the hippies and the entrepreneurial zeal of the yuppies." Or as they ventriloquize, cyberlibertarianism is "the opportunity to become a successful hi-tech entrepreneur. . . . Big government should stay off the backs of resourceful entrepreneurs who are the only people cool and courageous enough to take risks." Particularly aggressive in promoting information entrepreneurship is the PFF, whose "Magna Carta" is emblematic in its sequence of topics. After setting the stage with a preamble and an exposition on the nature of cyberspace, the "Magna Carta" proceeds to "The Nature and Ownership of Property" and "The Nature of the Marketplace" *before* coming to "The Nature of Freedom." To its authors, the history of information technology is the process by which "the hacker became a technician, an inventor and, in case after case, a creator of new wealth in the form of the baby businesses that have given America the lead in cyberspatial exploration and settlement." These authors include George Gilder, whose bestseller *Wealth and Poverty* "helped popularize and politicize the ideas of the Chicago school [of supply-side economics] during the early days of the Reagan administration," and Alvin Toffler, whose thought on "third wave" economy and information society strongly influenced Newt Gingrich in the mid-1990s.[22]

Perhaps the most severe critic of the way the greenback colors the cyberlibertarian red-white-and-blue is Paulina Borsook, a contributing writer at *Wired* who also published in *Mother Jones*. "I grew up in Pasadena, California," she recounts, among the liberal technologists at Caltech and the Jet Propulsion Laboratory.

> So it came as a shock, when, 20 years later, I stumbled into the culture of Silicon Valley. . . . Although the technologists I encountered there were the liberals on social issues I would have expected (pro-choice, as far as abortion; pro-diversity, as far as domestic partner benefits; inclined to sanction the occasional use of recreational drugs), they were violently lacking in compassion, ravingly anti-government, and tremendously opposed to regulation. . . . These high-tech libertarians believe the private sector can do everything. . . . They decry regulation—except without it, there would be no mechanism to ensure profit from intellectual property, without which entrepreneurs would not get their payoffs.[23]

All hail the nation of the almighty start-up. All hail the initial public stock offering that is now the entrepreneur's only means of imagining a "public."

The action items of cyberlibertarian entrepreneurship touch upon the following areas of dispute listed in the "Magna Carta": "ownership of the electromagnetic spectrum"; "ownership over the infrastructure of wires, coaxial cables and fiber-optic lines"; and new "accounting and tax regulations" favoring rapidly depreciating information capital rather than "heavy industry." In all these cases, the puzzle for cyberlibertarianism is to knit the countervailing forces of established business, start-up businesses, and government regulation together in such a way as to promote a hybrid ideal: the power of ownership *and* the power of free "competition" coupled with "innovation" (that is, a chance for new or small entrepreneurs to "creatively destroy" established ownership in the telecom, cable, or software industries). Another crucial issue, mentioned by Borsook, is intellectual property law. On this front, which we can take as representative of hybridized ideals, cyberlibertarianism clearly wants to have its cake and eat it, too. As the "Magna Carta" sees it, strong intellectual property rights law is essential to information freedom *because* of, rather than despite, the individualist nature of the new kinds of information property: "Clear and enforceable property rights are essential for markets to work. . . . The dominant form of new knowledge in the Third Wave is perishable, transient, *customized* knowledge: The right information, combined with the right software and presentation, at precisely the right time. Unlike the mass knowledge of the Second Wave—'public good' knowledge that was useful to everyone because most people's information needs were standardized—Third Wave customized knowledge is by nature a private good." If the new forms of intellectual property are all "private goods" even when they are in the realm of "public good," the "Magna Carta" continues, then perhaps the propertization of such knowledge can be regulated in ways adapted from the first to the decentralized nature of the new knowledge. Perhaps the purposes of making private goods available for public good can be served by packetizing intellectual property so that smaller, real-time bits of it can circulate in a more fluid system of market exchange. The "Magna Carta" thus quotes one of the best known, alternative visions of intellectual property, Barlow's "Economy of Ideas":

> One existing model for the future conveyance of intellectual property is real-time performance, a medium currently used only in theater, music, lectures, stand-up comedy and pedagogy. I believe the concept of performance will expand to include most of the information economy, from multicasted soap operas to stock analysis. In these instances, commercial exchange will

> be more like ticket sales to a continuous show than the purchase of discrete bundles of that which is being shown. The other model, of course, is service. The entire professional class—doctors, lawyers, consultants, architects, etc.—are already being paid directly for their intellectual property. Who needs copyright when you're on a retainer?

Intellectual property in the information age, in other words, should be predicated on a notion of property that cannot be tied to any public body as such because it is moment by moment served up by the individual. Such a reconceptualization of intellectual property—and even such recent revisions of "public good" theory as the "open source" movement—undermines established copyright law, but it also at last reestablishes the principle of owned knowledge at different levels—for example, knowledge capital sponsored by subscription, micro-payments, or (in the case of open source) "value added" commercializations of common code.

Freedom Is Communal

One last cyberlibertarian proposition may be mentioned as the summation of these statements about the identity of the People. Dedicated to the negation of big government and big business, yet positive that individuals, consumers, and entrepreneurs can invent new ways to constitute a public, cyberlibertarianism places all its faith in the idea—both retro-federalist and retro-countercultural—of "community." Community is the perpetually loose, elastic public that is the civil society of networked individuals, consumers, and entrepreneurs—like a great, buzzing hive of bees, except without the queen.[24]

The ideal of "the wired neighborhood" and of "classical communitarian anarchism"—as Doheny-Farina and Winner put it, respectively—has evolved intact from the era of BBSs (paradigmatically, The Well), newsgroups, and e-mail listservs to a technological environment that in the early 2000s included chat, instant messaging, Web discussion forums, and such peer-to-peer file-sharing programs as Napster and Gnutella.[25] We can draw a direct line of descent, for example, from the passionate communal life of The Well (as recounted in Howard Rheingold's *Virtual Community*) to the CPSR Cyber-Rights Working Group, whose "Cyber-Rights Home Page" promotes the following "basic rights": "the right to assemble in online communities; the right to speak freely; the right to privacy online; the right to access regardless of income, location, or disability."[26] Here, the first right is community. Or, again, consider the following two, more recent cyberlibertarian huzzas for community:

> Open, decentralized, abundant, inexpensive, user-controlled and interactive, [the Internet] is the first medium that allows anyone, anywhere to find or create communities of interest, to publish to audiences around the world, to engage in global commerce, and to participate in government and civil society across borders of time and distance.[27]

> No one knows what the Third Wave communities of the future will look like, or where "demassification" will ultimately lead. It is clear, however, that cyberspace will play an important role knitting together in the diverse communities of tomorrow, facilitating the creation of "electronic neighborhoods" bound together not by geography but by shared interests.[28]

The action item that follows from the community principle is, above all, "access." This was especially true in the late 1990s when the mainstream press and the Clinton administration began to trumpet the "digital divide" between the wired and unwired (even as contradictory evidence mounted to suggest that the divide was rapidly closing in the United States along class, ethnic, and gender lines).[29] Besides advocating privacy and freedom from censorship, cyberlibertarianism promotes broad and egalitarian access to the network as the precondition of free assembly. All users must have access to the new online democracy, and (the specifically communal aspect of the issue) such democracy founded on the frictionless ability of netizens to communicate and organize must be like a town-hall meeting freed from the limitation of any actual locality or town. As the Center for Democracy and Technology says: "CDT is working to foster widely-available, affordable access to the Internet. We believe that broad access to and use of the Internet enables greater citizen participation in democracy, promotes a diversity of views, and enhances civil society. We work for public policy solutions that maximize, in a just and equitable fashion, the unique openness and accessibility of the Internet and preserve its vision as it evolves with ever more powerful broadband technologies." Similarly, the EFF works "on policies that encourage the government to stimulate the development of experimental, precompetitive, network technologies and to fund the development of applications that are of use to 'low-end' users, who are traditionally underserved by advanced digital media."[30]

We can end by thinking of the question Winner asks in the title of his well-known article of 1980, "Do Artifacts Have Politics?" In the case of contemporary networked information technology, the answer to the question would certainly seem to be yes. To the shores of the new, virtual America come the wired, *not* poor, huddled masses (or, rather, "individuals") yearning to breathe free. Cyberlibertarianism holds high the torch of liberty—even if, as in the case of the campaign of "blacked out" Web pages in 1996–97 protesting the CDA, that torch negates as often as it posits.

Cool Is a Nonpolitics or Antipolitics of Information; It Is "Bad Attitude"

From another viewpoint, however, cyberlibertarianism is a flawed politics or, more extreme, no politics at all.

Let me take up the first half of this proposition first. That cyberlibertarianism has been described by its critics as flawed is not by itself surprising, since it is now conventional to say as much about any politics (especially after the period from Vietnam to Watergate). A standard rhetorical trope in political criticism, therefore, is to convert the external measure of political legitimacy (whether a particular politics is right or wrong in its goals) into an analysis of internal flaws (phrased morally, its hypocrisy or scandal), and vice versa. An ethical flaw can always be found in politics because politics may be said to be a structure of flaws or internal rifts. Politics is not so much about taking or influencing action as transposing difficult issues sitting intransigently at the intersection of incommensurable communities and needs into a structured discourse of action ("so moved"). Although such discourse has real performative force, it is untrue insofar as it poses as "a" singular action rather than a contestation of actions, counteractions, and failures to take action.

By referring to cyberlibertarian politics as a "structured discourse," I am suggesting a particular analytical approach. Since we will soon come to the anthropology of cool I earlier promised, we may take our paradigm for cyberlibertarian "myth" (as Winner calls it) from structural anthropology. Consider, for example, Lévi-Strauss's "Structural Study of Myth," which argues that myths are "a kind of logical tool" designed simultaneously to acknowledge society's existential contradictions (e.g., life/death, nature/culture) and to prevent impasse by displacing such fissures into analogous contradictions that mediate and domesticate the original trauma. Thus while the starkest contradictions of lived experience are never "solved" ("father, why do people die?"), social thought incessantly revolves around and within those contradictions, weaving ever more encompassing, intricate meshes of contradictions that have the effect of spanning them, dressing them, ornamenting them, and so making them *habitable* truth. It is as if the great tiles in the mosaic of social existence were perennially shattered. But reproduce the breakage in a micro-texture of "crackle" (a decorative ceramic finish equivalent to craquelure in painting)—reproduce the force of necessity, in other words, as an interior dynamics of arbitrariness and play—and the shards no longer seem quite so sharp-edged. The sheer tesserae of existence are assimilated in a society of broken things embracing each individual broken creature in a holistic environment of incompleteness—a complex, nuanced, emergent, *civilized* myth (we know it is fragile; do not press too hard)—that makes it possible to learn to live. We might look for a visual emblem of such cruel yet gentle crackle—at once a dissection and a holism—to the diagram of the "totemic operator" in Lévi-Strauss's *Savage Mind* (fig. 8.1). Along the vertical axis of the totemic operator runs the cardinal social contradiction between the "individual" and the "species." But *between* the poles grows a fractal crystal of mediating oppositions that make the original collision between the solitary and collective body as livable as one's own literal body (represented in a totem system that links animal species to individual animals and body parts).

Of course, as *The Savage Mind* also makes clear (especially in its chapters on time and history), the inherent pathos of "structure" is that it is never complete. In temporal terms, structure never remains complete. Something unexpected—a famine or genocide that obliterates key positions in the system, the discovery of new worlds and peoples—always ruptures the fearful symmetry of the crystal to let in the howling unknown. We will come to such incompleteness in cyberlibertarian myth. But for the moment, it is instructive to keep the idea of crystalline structure intact. Cyberlibertarian politics is a myth constructed around not just one but *three* cardinal contradictions in social identity varying upon Lévi-Strauss's "individual" versus "species." These axes—the underlying structure of cyberlibertarianism—form another sort of totemic operator, as seen in figure 8.2, where arrows indicate contradictions between paired identity formations. Individualism tenses against communitarianism; entrepreneurism lines up against consumerism (which even in "demassified" form derives from "mass"); and Americanism stands out from globalism. Common to all these contradictions is the logical engine I have inserted in the middle of the diagram. This relational engine is itself a contradiction: market vs. gift economy. Market exchange relates America to the world, the entrepreneur to the consumer, and (as is increasingly evident in Web "communities" sponsored by advertisements) the individual to the community. Unless, of course, the *opposite* of the market—that is, gift exchange—mediates the polarities. Even when the identities in the diagram are bound together in capitalistic relations,

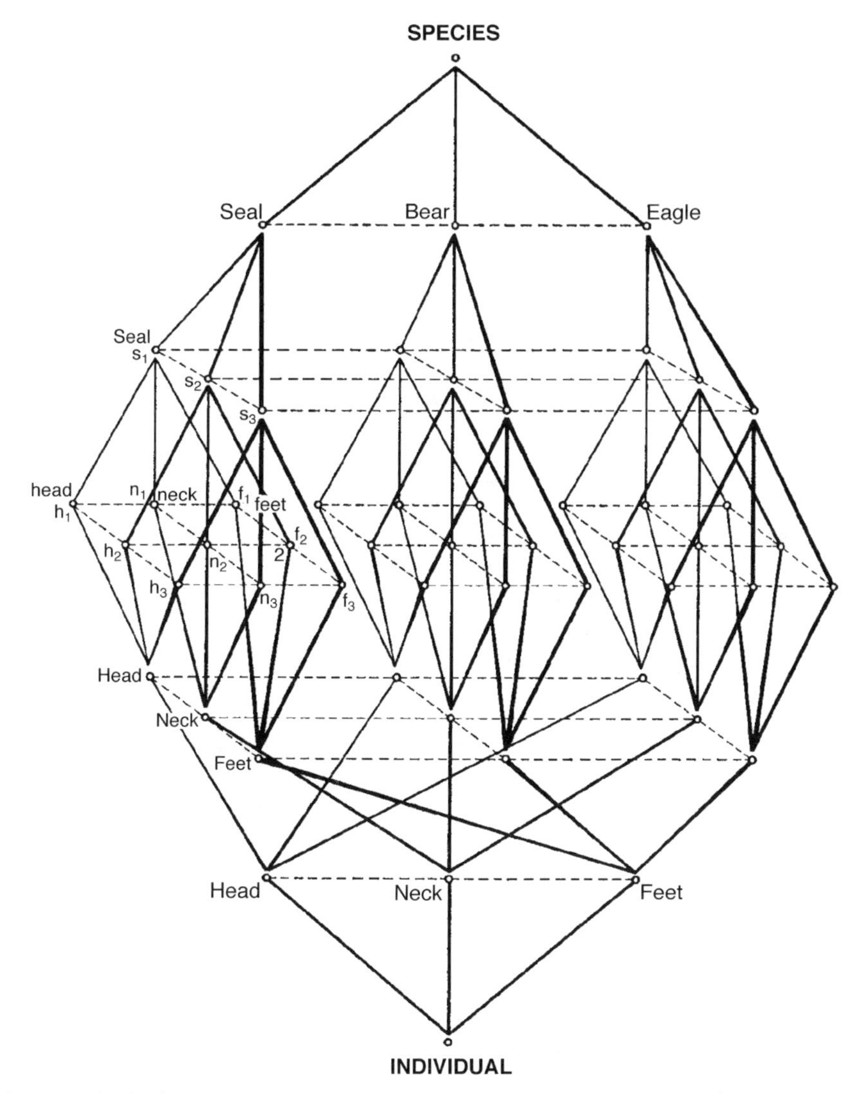

Figure 8.1 Claude Lévi-Strauss's totemic operator

From Claude Lévi-Strauss, *The Savage Mind* (Chicago: University of Chicago Press, 1966), p. 152.

capitalism has not so far been able to explain those relations without leaving a significant remainder reminiscent of the countercultural and communitarian origins of cyberlibertarianism. As demonstrated by software firms that give away their products for free or dot-com firms in 1997–2000 that perennially "burned" more venture and market capital than they recouped in sales, information capitalism is still a loose regime hospitable to the potlatch of

Figure 8.2 The structure of cyberlibertarianism

the gift (as well as outright black economies of pirated software, music, and images).

To flesh out or, as I put it, "dress" this bare circuit diagram of the digital nation is the task of the cyberlibertarian discourse we earlier surveyed. This discourse, as we saw, generates from the underlying contradictions a series of agenda items proposing laws, conventions, standards, and protocols designed to make the contradictions habitable—that is, to legitimate particular positions upon them. If we were to complete our diagram in a manner analogous to Lévi-Strauss's totemic operator, we would thus now need to draw an intricate, interior web of oppositions all restating or displacing the core tensions between the individual and the collective (and between market exchange and gift economy) in terms of palatable political stands. (The opposition between America and the world, for example, looks much more acceptable when posed in the form of protecting free speech and intellectual property against Asian autocracy and piracy, respectively.) The whole ensemble—as complex in its wiring, perhaps, as the digital version of a totemic operator, an integrated circuit—is the "logical tool" through which cyberlibertarianism processes the world.

Rather than attempting to construct a needlessly complex diagram, however, we will find it more useful to return to the critics of cyberlibertarianism. These writers variously characterize cyberlibertarianism as too technodeterminist, selfish, U.S.-centered, California-centered, conservative, or liberal. But all these criticisms are at base diagnoses of the "flawed" contradictions of cyberlibertarian identity along with the fine web of particular agendas, issues, and positions that make that structure of identity crackle, that make it *political*. One of the main rhetorics in such criticism, indeed, is what might be called the "blazon" or ornate listing of contradictions that reproduces in words something like Lévi-Strauss's totemic operator with all its meshwork. Here, for example, are Barbrook and Cameron on the compound contradictions of cyberlibertarianism:

> This new faith has emerged from a bizarre fusion of the cultural bohemianism of San Francisco with the hi-tech industries of Silicon Valley. . . . The Californian Ideology promiscuously combines the free-wheeling spirit of the hippies and the entrepreneurial zeal of the yuppies. This amalgamation of opposites has been achieved through a profound faith in the emancipatory potential of the new information technologies. In the digital utopia, everybody will be both hip and rich. Not surprisingly, this optimistic vision of the future has been enthusiastically embraced by computer nerds, slacker students, innovative capitalists, social activists, trendy academics, futurist bureaucrats and opportunistic politicians across the USA.

Even cyberlibertarian champions offer a similar analysis, though with the connotation of bipartisan breadth rather than contradiction. In "Birth of a Digital Nation," Katz writes:

> From liberals, this [postpolitical] ideology adopts humanism. It is suspicious of law enforcement. It abhors censorship. It recoils from extreme governmental positions like the death penalty. From conservatives, the ideology takes notions of promoting economic opportunity, creating smaller governments, and insisting on personal responsibility.
>
> The digital young share liberals' suspicions of authority and concentration of power but have little of their visceral contempt for corporations or big business. They share the liberal analysis that social problems like poverty, rather than violence on TV, are at the root of crime. But, unlike liberals, they want the poor to take more responsibility for solving their own problems.[31]

If we examine the mesh of contradictions in cyberlibertarianism, we notice in particular some difficult issues (each masking one or more underlying cardinal contradictions).

Piracy: Good or Bad?

As we have seen, many non-American critics see a contradiction between the global persona of cyberlibertarianism (more generally, of the Internet itself) and the reality that such culture has often been dominated by U.S. technology, language, and customs. At least some of these accusations cannot be dismissed as standard anti-Americanism because they touch so nearly on divergences within American values themselves. One of the most significant examples pertains to intellectual property. On this front, the perennial tension in U.S. cyberlibertarianism between acclaiming private property and crediting a utopian gift economy is as nothing compared to the amplification of that tension on the international scale. Other nations with a different understanding of the relation between the individual and the collective regard the U.S. stance on intellectual property as deeply duplicitous. This is especially the case in the disagreement between the United States and other (e.g., Eastern) nations about "software piracy." Roberto Verzola, writing from the Philippines, is acute on this issue. "Piracy: good or bad?" he asks, and argues (in a passage worth quoting at length):

> It is to the interest of developing countries, both the agricultural and the newly-industrializing economies, to dip freely into the world's storehouse of knowledge and adopt technologies where they might be useful for the country's development. When it was still a developing country, in the 18th and 19th centuries, the U.S. was one of the worst pirates of British books and publications. When it was trying to catch up with the U.S. and Europe, Japan also freely copied Western technologies. Taiwan did the same. So did Korea.
>
> Yet, the U.S. and Europe would lead us to believe that piracy is morally wrong. They do not want us to pirate their books, their software and their designs.
>
> They say we pirate their intellectual property rights. Yet, they continue pirating our intellectuals. Advanced countries think nothing of pirating our best scientists, engineers, technicians, and other professionals. They patent or copyright the works of these intellectuals and then sell them back to us at high prices. They also pirate our genetic resources. Their scientists roam the world pirating biodiversity resources like microorganisms, plants, animals and even human DNA. They then claim monopoly ownership over the genetic information they extract, patent them, and sell them back to us at high prices.
>
> When the U.S. sent spy satellites in space, countries complained that the U.S. was taking away strategic information and violating their sovereign control over their own territories. The U.S. insisted that it was free [to] get this information whenever it wanted, even to sell them back to those countries, if they were willing to pay for them. U.S. commercial satellites then started beaming video programming into other countries. When . . . [some cultures] considered the video content objectionable, the U.S. invoked the concept of "free flow of information" to insist that it had the right to beam these programs. Yet, when local people developed a taste for U.S. programs, captured these satellite broadcasts, and distributed them locally, the U.S. started complaining [that] people were receiving and copying their broadcasts without paying for them. According to their twisted logic, this was a violation of their intellectual property rights. . . .

> In short, information acquisition has been defined so that when it is bad for the interests of the U.S. and other advanced countries but good for us, it is called "piracy" and "freeloading," but when information acquisition is good for their interests and bad for us, it falls under labels like "free flow of information" and "common heritage of mankind."

Privacy, Free Speech, or Freedom of Information?

Cyberlibertarians treat these pillars of cyberlibertarianism as if they were quarried from the same block. Yet there are some deep philosophical rifts. Consider, for example, the cyberlibertarian crusade for encryption. The possibility that strong encryption may protect privacy only at the cost of freedom of information—"our society's highest traditions of the free and open flow of information and communication"—never comes up.[32] Encrypted or (a related issue) anonymous communication may sometimes be necessary to foster the conditions of personal safety necessary for free speech (a third term in the problem). The right to withhold information or to withhold it from certain parties, in this sense, is the other side of the coin of the right to say what one wants. But does that mean encryption is *always* identical with free speech and thus the right of all parties (including, for example, gangsters and government officials)?[33] And even if it were, what is the relation of free speech to freedom of information, the latter of which would seem to mandate a kind of perpetual and involuntary act of speech? If "information wants to be free," we could ask why all (or, at least, much) information shouldn't be free to all. Why should one's personal e-mail be private by default (as opposed to by selection) if the "personal" e-mail of government officers suspected of misconduct is public?[34] There are no natural or intuitive answers here, only negotiated, legislated, and judicially decided standards that triangulate the relation between privacy, free speech, and freedom of information at a particular, dynamic point. For example, the precise equilibrium between these three concepts at the present time might perhaps be represented as follows: Privacy ← Free Speech → Freedom of Information. If the principle of free speech currently mediates between the opposed logics of absolute privacy and absolute freedom of information, then recognizing this may help us understand why acts of free speech now rely exclusively (more than in the era of American revolutionary print journalism, when free speech was originally defined) on *media*. The media are the theater in which the private is broadcast publicly (freedom of information) and, reciprocally, where public information is sent like a targeted missile directly into one's living room (the locus of privacy).

At stake, ultimately, is how cyberlibertarianism understands the fundamental relation between the public and the private that is constitutive of modern polity. Cyberlibertarianism shows little tolerance for the idea that the social contradiction of public versus private demands compromise of a sort that cannot adequately be negotiated in media alone (in the absence of the other civil, legal, and political institutions that cyberlibertarianism scorns even more than it scorns traditional media), while compromise in this regard—or, less euphemistically, being compromised—is elsewhere the condition of joining any society governed by the great bureaucratic institutions of our times: governments and corporations. The issue comes closest to awareness in cyberlibertarianism's perennially torn stance toward hacking—that is, in its recognition that the potential conflict between the right to open up and to close off information requires improvising such boundary concepts as the "ethical hacking" of businesses and government agencies. (And this is not even to mention the vexed area of intellectual property, where a similar effort to concoct a boundary zone between free and closed information results in concepts like "shareware," "copyleft," "open source," and so on.)[35] By contrast, the issue is furthest from awareness in the blind eye that cyberlibertarians turn toward different understandings of free speech and privacy in other countries. Whatever one's view of online censorship or regulatory intrusion in the international arena (e.g., the privacy laws in Europe that exceed any U.S. standard of online privacy, or the apparently surprising move of the British government in 2000 to enact the Regulation of Investigatory Powers [RIP] Bill allowing the Home Office and the MI5 security agency to demand encryption keys, track online activity, and intercept e-mail), there is no adequate ground for discussion if it is simply assumed that free speech, freedom of information, and privacy are pieces of a single, homogeneous freedom: take it or leave it.

Privacy or Individualism?

Related to the above contradiction is the unstable balance between cyberlibertarian privacy and individualism. There would at first glance seem to be no discrepancy between the two. As we have seen, cyberlibertarians believe the individual to be the primary unit of political, social, and economic life, and privacy therefore to be about the protection of the individual. However, the rift between individualism and privacy comes to view when we consider that the individuality of users is exactly what online businesses say they wish to cultivate by *transgressing* privacy through "cookies," databases, and other means of tracking the behavior of their customers. "Welcome back to our Web page, Alan Liu! Based on your recent purchases, may we suggest the following new products? You may also be interested in our affiliated company, which features music and videos that other customers from your profession, age group, locale, tax bracket, and company have enjoyed." Of course, cyberlibertarianism rejects such individual tracking by consumer businesses and—even more vehemently—by the government. Cyberlibertarianism is in-your-face individualism, except when it would rather lose itself in the anonymous mass.

Left or Right?

There are other contradictions, major and minor, that one can criticize in cyberlibertarianism. But here I will sum up the flaws of the movement in what critics have viewed as the great sump of such flaws: indiscriminate, unrealistic, unfixed, and unethical mixing of left and right politics. Barbrook and Cameron, Borsook, Winner, and others, for example, are never more eloquent than when excoriating the wired for crossing the wires of liberalism and conservatism.[36] Discussing the political ambivalence of the "virtual class," or ultimate exemplar of the bipolitical New Class we earlier heard called the "contradictory class," Barbrook and Cameron thus produce an extended example of what I termed a "blazon" of contradiction:

> Living within a contract culture, the hi-tech artisans lead a schizophrenic existence. On the one hand, they cannot challenge the primacy of the marketplace over their lives. On the other hand, they resent attempts by those in authority to encroach on their individual autonomy. By mixing New Left and New Right, the Californian Ideology provides a mystical resolution of the contradictory attitudes held by members of the "virtual class." Crucially, anti-statism provides the means to reconcile radical and reactionary ideas about technological progress. While the New Left resents the government for funding the military-industrial complex, the New Right attacks the state for interfering with the spontaneous dissemination of new technologies by market competition. Despite the central role played by public intervention in developing hypermedia, the Californian ideologues preach an anti-statist gospel of hi-tech libertarianism: a bizarre mish-mash of hippie anarchism and economic liberalism beefed up with lots of technological determinism. Rather than comprehend really existing capitalism, gurus from both New Left and New Right much prefer to advocate rival versions of a digital "Jeffersonian democracy."

The essential charge behind such criticisms, perhaps, is not that cyberlibertarianism is fickle with regard to any established political camp. It is that in playing at a "Jeffersonian democracy" no more substantial than the paper or, rather, screen it is written on, cyberlibertarians are insufficiently engaged with the broader and real constituencies of such camps. The reason that cyberlibertarianism can so easily spin mainstream positions around 180 degrees, after all, is that its politics swing upon the frictionless pivot-point of an immaculate "individual" attached by no more than a single wire to such community paradigms with entrenched political histories as the left's "inner-city neighborhood" or the right's "family values." It is hard to fix the state of liberty idealized by cyberlibertarianism when the political battle lines are displaced into a virtual state where the relation between the individual and the collective can be terminated at the press of a key.

Cyberlibertarianism is thus indeed a politics as flawed or internally rifted as any other contemporary politics. According to its critics, it is even more so. But our analysis is not complete unless we now also distinguish cyberlibertarianism from other politics, including not just the mainstream left and right but—the true rival concealed behind the posturing of cyberlibertarianism against mainstream politics—*alternative* politics. Let me turn, therefore, to the second part of my initial proposition: that cyberlibertarianism is not just flawed in its politics but, by distinction with the most relevant of the alternative politics, is no politics at all.

It is at this point that we need to break open the crystal of the movement's political structure, as I put it, to admit change. The great historical change we have been following in this book is that which has transformed the whole system of relations between the individual and community, entrepreneur and consumer, and America and the world. That change is postindustrialism, which skews such relations, as I have argued, around yet another cardinal axis not even glimpsed in the cyberlibertarian worldview—the relation of the individual to the workplace. The postindustrial relation of the individual to the restructured multinational, team-based, and "flat" corporate workplace now increasingly determines the very notion of the individual and its collective formations. In a contemporary society where the sun—which is to say, the computer screen—never sets on the empire of work from time zone to time zone (or, within personal life, workday to worknight), the notion of a "Jeffersonian democracy" based originally on Jefferson's vision of a decentralized, agrarian republic seems singularly hapless.

It is thus crucial to compare cyberlibertarianism to two alternative politics that also take to information technology with a vengeance but that do so with more pointed awareness of the overall climate of postindustrialism: the NGO (nongovernmental organization) movements and labor activism (especially labor groups affiliated with the NGOs). While these politics may or may not be just as flawed in their vision as cyberlibertarianism, their structure of thought is oriented in such a way as to discern whole arcs of experience invisible to the "digital nation." They see postindustrialism as a problem, while for cyberlibertarians postindustrialism may have problems (relating to excessive commercialism), but in itself is just an environment through which to move and grow. It is just an "incubator" of the digitally enfranchised individual consumer and entrepreneur (alluding to the so-called "incubator" companies that nurtured dot-com startups in the late 1990s).

I address first the sphere of the NGOs—specifically, the wired NGOs that do not merely "use" information technology to network their geographically diverse members and initiatives but integrate the *idea* of information networking into their rationale of global alliance, access, and equality.[37] As remembered in Rheingold's chapter in *Virtual Community* on "Electronic Frontiers and Online Activists," NGOs once seemed simply to walk the same trail as cyberlibertarianism. *Grassroots* means "rhizome," after all, and it was not surprising that alternative politics in the 1980s and 1990s (the same era that saw the emergence of the EFF and other cyberlibertarian groups) so rapidly seeded itself rhizomatically on a decentralized Internet that it perceived in ways undifferentiated from cyberlibertarianism at that time.[38] As François Fortier observes in *Civil Society Computer Networks: The Perilous Road of Cyber-politics* (1996; an online dissertation based on field studies in South America and Vietnam), the characteristics of NGO networks, including "low capital and usage costs, synchronous or asynchronous interactivity, difficulty [of] control, and fungibility in time and space" have "a particular potential for contributing to the expansion of grassroots network structures, exchanging information and allowing logistical coordination of otherwise scattered social organizations."[39]

But the trail forked somewhere along the way, and cyberlibertarianism and the NGOs diverged. We can get a good sense of the path the NGOs took if we consider two of the most established meta-organizations devoted to supporting NGO movements with information technology. One is the Institute for Global Communications (IGC), which began in the United States in 1987 and now includes under its umbrella PeaceNet, EcoNet, WomensNet, Anti-RacismNet (as well as LaborNet from 1992 to 1999). The second meta-organization, co-founded by the IGC and others in 1990, is the Association for Progressive Communications (APC), which in 2000 included nineteen member NGO networks based all over the world.[40] In addition, we can round out our view of the wired NGOs by sampling the online networks that arose in 1999–2000 specifically to organize protest at the meetings of the World Trade Organization (WTO), International Monetary Fund (IMF), or World Bank (in Seattle and Washington, D.C.) and at the U.S. Democratic and Republican presidential conventions (in Los Angeles and Philadelphia). These include D2KLA, the Direct Action Network, and the Philadelphia Direct Action Group, together with city-based "indy media" centers focused on these occasions (e.g., the Seattle, Washington, D.C., and Los Angeles Independent Media Centers).[41]

Here are representative passages from the online mission statements or slogans of these NGO cyber-networks:

> IGC shares the vision to actively promote change toward a healthy society, one which is founded on principles of *social justice,* broadly shared economic opportunity, a robust democratic process, and sustainable environmental practices. We believe healthy societies rely fundamentally on respect for individual rights, the vitality of communities, and a celebration of diversity.
>
> The Mission of IGC is to advance the work of progressive organizations and individuals for peace, *justice,* economic opportunity, human rights, democracy and environmental sustainability through strategic use of online technologies. . . .
>
> Our initial network, PeaceNet was founded in a garage in Palo Alto, California in 1986. Since then, IGC has expanded its capabilities to bring Internet tools and online services to organizations and activists working on peace, economic and *social justice,* human rights, environmental protection, labor issues and conflict resolution.[42]

> The Association for Progressive Communications is a global network of non-governmental organisations whose mission is to empower and support organisations, social movements and individuals in and through the use of information and communication technologies to build strategic communities and initiatives for the purpose of making meaningful contributions to equitable human development, *social justice,* participatory political processes and environmental sustainability.[43]

> A Call for Celebration and Action for Global, Social, Economic, Racial and Environmental Justice! No More Business As Usual![44]

> NOW IS THE TIME TO TAKE ACTION FOR DOMESTIC AND GLOBAL JUSTICE![45]

> CALL TO ACTION FOR DOMESTIC AND GLOBAL JUSTICE[46]

What is clear in the cyber-politics of the NGOs is that the dominant, even shibbolethic imperative is "social justice" (also referred to as "domestic and global justice"). Equally clear is that "social justice" has almost nothing to do with "individual rights" as advocated by cyberlibertarians. Of course, there are some areas of overlap where the NGOs focus on cyberlibertarian themes. Witness, for example, Privacy International, "a human rights group formed in 1990 as a watchdog on surveillance by governments and corporations." Or, again, there was the *Internet Rights* listserv set up by GreenNet and LabourNet to protest the U.K. Regulation of Investigatory Powers (RIP) bill, which soon broadened its focus (as reported by the APC) to the discussion of "the privacy and free speech rights of users everywhere."[47] But the major issues for the NGOs are uneven development (between Western and Eastern nations, as well as between North and South America), women's rights, environmentalism, and—a topic we will focus upon below—labor rights. Information politics is important, but only as part of the core belief that social justice requires a *general* commitment to freedom of information and free speech (freedom to *be* informed and freedom *to* inform the world). Not a politics unto itself but part of a broader social justice, in other words, the Internet is "freedom of information" and "free speech" in the originary countercultural sense of the Berkeley Free Speech movement and other movements of the 1960s: it is "power to the people," not "power to the individual."

If social justice is the common cause, then what the NGOs link arms against on the street (and network against on the information superhighway) is social injustice—specifically, a common injustice that is less the "government" inscribed in their nom de guerre than a new, postindustrial dominion. That dominion, against which the NGOs organized their famous actions against the WTO, IMF, and World Bank in 1999–2000 and after, is global capitalism. Representative of NGO invective against global capitalism are the following two mottoes on D2KLA's Web site: "No More Business As Usual!" and "Human Needs Not Corporate Greed!"[48] Equally formulaic is the following boilerplate language from the Direct Action Network's call to action against the U.S. political party conventions in 2000 (echoed verbatim on such affiliated NGO pages as the Philadelphia Direct Action Group's "Call to Action for Domestic and Global Justice"): "It is no secret that both the Republicans and the Democrats are in the pockets of multinational corporations. Rich corporations increasingly control our political system, the media, culture and the economy. A handful of powerful people in a handful of powerful nations continue to dominate both economically and militarily leaving the vast majority of the world impoverished and under the gun."[49] Global capitalism is the perfect, antithetical rival of the NGOs, after all, because it is just as subversive of government sovereignty and just as dependent on decentralized, cross-border networks. Global capitalism is thus at once the uncanny mirror of contemporary alternative politics (with its merged radical networks and pseudo-corporate "affiliations") and a phantasmal enemy against which decentralized alternative politics can—in retro-political, industrial-era style—"unite!" Witness, therefore, such instances of monstrous doubling as the APC's project of creating business-savvy NGOs. On its "Managing Your NGO" page, it offers an extensive set of business "tools" (spreadsheets, worksheets, checklists, analysis forms, case studies) to assist NGOs in preparing business plans, managing finances, signing contracts, and developing marketing strategies. NGOs, too, need to "balance sustainable business practice with their missions."[50]

But the NGOs do not target all aspects of their archenemy global capitalism indiscriminately. Their special animus is directed toward what I have termed "producerism" rather than consumerism. The homogenization of world consumer culture by multinational enterprises such as McDonalds, Nike, or The Gap is far less important than the way these and other firms were alleged to affect the lives of global workers *qua* workers—whether French farmers, non-Western factory hands, exploited women and other employees, or "permatemp" and restructured workers. As a consequence, the NGOs have been strongly affiliated with various labor activist, workers' grievance, and other anticorporatist movements. We can witness, for example, the past role of LaborNet within the IGC, or the participation of traditional labor unions in NGO protests against the WTO. Or again, we may note the constant news on NGO home pages about labor grievances ("Travelodge workers won reinstatement," "South Korean site by steel workers claiming unfair dismissal under threat from multinational").[51] So, too, there are the links from the Web sites of the IGC or its member groups to one of the leading anticorporatist Web organizations, San Francisco–based *Corporate Watch* (renamed *CorpWatch* in 2001), whose Web site was established in 1997 by the Transnational Resource and Action Center.

We can thus usefully turn at this point from the NGOs in general to their labor arm, as well as to the unions that have consented to join arms. Most germane are the labor groups that not only have embraced the Internet through high-quality Web sites (characteristically featuring crisp newspaper-style formats, multimedia, interactivity, and sophisticated database backends) but also have made the Internet the icon of labor's newest theme: the information industry as industry. In this regard, nothing is more paradigmatic than labor's campaign against the acme of New Economy industry—the industry of silicon. In the land of silicon—especially in Silicon Valley and

Seattle—labor has not only organized the so-called "invisible workforce" of blue-collar and lower level service workers within high-tech firms (e.g., assemblers and janitors), but has embarked on the same course at the level of professional-technical workers. Such labor activism opens to view a cyber-politics unlike any seen in cyberlibertarianism. We can compass this brave new world of cyber-politics in four main portfolios.

Information Technology and the Environment (Especially the Worker's Environment)

Despite clichés about the "paperless office" and "green" IT, labor activists charge, high-tech industry is yet another instance of global corporatism riding roughshod over the environment. IT only looks green, that is, when one's gaze extends no farther than a manicured Silicon Valley high-tech "campus" or research "park" (the picturesque terms of preference for the topography of IT). By contrast, if one surveys the factories and water tables around the world where high tech is actually manufactured (not to mention the landfills where ever burgeoning masses of computers with toxic components are deposited), then silicon is definitely not "green." Witness, for example, the Corporate Watch "feature" presentation (a suite of Web essays and interviews accompanied by an overview editorial) titled "The High Cost of High Tech." Corporate Watch interviews the director of the SouthWest Organizing Project, which battled the Intel Corporation's expansion in New Mexico on grounds of "environmental degradation and worker exposure." It cites the claim of the Silicon Valley Toxics Coalition (SVTC) that the "more than 700 compounds . . . used to make just one computer" create "extreme health hazards" for electronics assemblers. It also publishes a compelling contribution by Ted Smith, executive director of the SVTC on "The Dark Side of High Tech Development." Smith presents a litany of concrete charges against high-tech industry for environmental destruction, endangerment of workers, corporate dominance over local communities, and so on. The Web site of the SVTC itself contributes just as strongly to the argument, demonstrating a mix of local and global environmental concerns. As indicated in its name, the SVTC is local to the Silicon Valley environs where it first became known for exposing the contamination of the area's water tables. But the interest of the SVTC in "dirty" IT now extends around the world. The SVTC home page of 11 July 2000, for example, leads with a news brief on the attempt of the United States to weaken the European Commission's draft directive, "Waste from Electrical and Electronic Equipment" (known as WEEE). Under the heading "High-Tech Campaigns and Projects," the page then lists "Clean Computer Campaign," "Global Expansion of High-Tech Industry," and "Global Semiconductor Health Hazards Exposed." Most ambitiously, the SVTC has initiated a worldwide network called "Campaign for Responsible Technology" to track "the global expansion of high-tech electronics manufacturing."[52]

Information Technology and Ergonomics

What about white-collar workers in cubicles far removed from the blue-collar terrain of "dirty" high-tech assembly plants and developing nation landscapes of dumped computer parts (or, at least, far removed once Silicon Valley began exporting most of its manufacturing work and so-called "recycled" products to other parts of the world)?[53] Here, the relevant concept is "ergonomics," which is the environmental critique of the cubicle. Katherine Hayles and Albert Borgmann have written powerfully of the inescapable "embodiment" or "ancestral environment" of virtual experience.[54] Where such embodiment comes home to roost in the workplace, we may say, is the air-conditioned, pristine, even clinical purity of the high-tech cubicle. The bodies in those cubicles may not sweat or bruise, but as proven by the recent history of workers' litigation and federal regulation, they "stress." "Stress injury" is the embodiment (and transference, too) of labor grievance in the postindustrial age. The endless mobility of information and capital, in other words, is paid for by paralyzing workers in bodily positions or repetitive motions that refute the thesis of mobility in a symptomatology of pain-stiffened wrists, necks, and backs. Speaking up for such knotted bodies are groups such as the Coalition on New Office Technology (CNOT). Significantly, CNOT's home page not only registers medical symptoms (e.g., repetitive strain injury and carpal tunnel syndrome) but also political remedies. CNOT prescribes a "Computer User's Bill of Rights" that would recognize such high-tech workers' rights as these: "work without pain," "a reasonable workload," "a workers compensation system that works," "adequate breaks from computer use," and "an ergonomic workstation." Also representative is Paul Marxhausen's Web page on "Computer Related Repetitive Strain Injury" (recommended by CNOT), which provides information on the medical, self-help, and practical aspects of high-tech stress injuries; its annotated bibliography includes such works on the politics or economics of ergonomics as Penney Kome's *Wounded Workers: The Politics of Muscoloskeletal Injuries* and Amy Clipp's *Job-Damaged People: How to Survive and Change the Workers' Compensation System.*

Information Technology and Workplace Privacy

By contemporary legal standards, we know, the cyberlibertarian credo of "Jeffersonian democracy" is wholly irrelevant to the U.S. workplace. If workers' bodies can be inspected by employers through mandatory drug testing, then certainly their information habits are wide open. Orwell's "1984," in other words, is really "workplace 2000." This is so well established in law (as of the time of this writing) that it has become a frequent topic of the mainstream press. *Business Week* observed in an article published in 2000, for example, that firms have recently been motivated to boost their monitoring of employees because of concerns over trade secrets, worker productivity, and legal liability (for "sexually explicit, racist, or other potentially offensive material"). Information technology is both the means and object of such monitoring:

> When it comes to privacy in the workplace, you don't have any. Time and again, courts have sided with employers when it comes to spying on employees. Your boss can monitor the time you spend on the phone and can eavesdrop on your voice mail. Employers can review your computer files and copy and read your e-mail. They can secretly tag along when you're surfing the Internet. They can even put video cameras in washrooms—though not in the stalls.
>
> Employee surveillance has mushroomed recently, and you can blame it on the Internet. Nearly three-quarters of U.S. companies say they actively monitor their workers' communications and on-the-job activities. That's more than double the number four years ago, according to an American Management Assn. poll of its members, mostly midsize-to-large organizations that together employ a quarter of the U.S. workforce. Some 54% track individual employees' Internet connections; 38% admit to storing and reviewing their employees' e-mail. Three years ago, only 15% monitored e-mail.[55]

Similarly, Jeffrey Rosen observes in his *Unwanted Gaze: The Destruction of Privacy in America* (2000) that companies now subject their employees to increasingly rigid monitoring and acceptable use policies—so much so that one e-mail policy cited by Rosen sets out to prevent corporate liability for sexual or racial harassment through a blanket prohibition. E-mail, the policy declares, shall not be used "for gossip, . . . for emotional responses to business correspondence or work situations" or "in any way that may be seen as insulting, disruptive, or offensive by other persons, or harmful to morale."[56] What the "unwanted gaze" enforces, in other words, is a holdover from the industrial era—the "Fordized face."

The stance of labor activists and anticorporatist groups on this issue (allied in this instance with the ACLU) is the ultimate "watchdog" position: we need to watch the watchers. Thus *Disgruntled: The Business Magazine for People Who Work for a Living* published online articles and interviews on the perilous state of workplace privacy. Similarly, the ACLU comments in an "Issue Summary" linked from its "Workplace Rights" page (authored by the ACLU National Task Force on Civil Liberties in the Workplace):

> The Constitution does not apply to the workplace. In the 18th Century, when the Bill of Rights was adopted, only the government was seen as a major threat to individual rights. Today, many if not most Americans are more vulnerable to violations of their rights by employers than early Americans were by the government. . . . This concern is particularly urgent now. In today's fast-paced, highly competitive economy, many corporations and some public agencies are experimenting with new technologies and management practices that pose a serious threat to employees' rights. Electronic monitoring and discrimination based on genetic test information are two examples of such practices.

Another such commentary upon workplace privacy is Jennifer Vogel's "The Walls Have Eyes," which PBS published online on its *Working Stiff* site. "They can do just about anything they want as long as it's not done in a discriminatory manner," Vogel says. "This level of workplace paranoia hasn't existed since Henry Ford sent committees around to workers' homes to see what newspapers they read."[57] So, too, there is the Privacy Rights Clearinghouse's "Fact Sheet #7: Employee Monitoring: Is There Privacy in the Workplace?" (1993; rev. 1997), which notes the "virtually unregulated" ability of businesses to inspect worker activity before addressing specific questions about the monitoring of phones, computers, and e-mail.[58]

High-Tech Industries and Labor Organization

We now come to the home turf of labor activism: union organization and strike actions. In the phrase of Amy Dean, the oft-cited executive director of the Silicon Valley–area South Bay AFL-CIO Central Labor Council, organizing in Silicon Valley is how labor can show that it can be a "new labor movement."[59] High-tech industries have the reputation of being "unorganizable" because professional-technical workers distrust unionization, work in smaller concentrations (no huge masses to organize in any one locale), and change jobs often. Not so, labor unions and NGO-related labor groups argue—both because professional-technical workers are not the only relevant workforce (assemblers and janitors matter, too) and because the class of professional-technicals includes within itself a "proletariat" with organizable grievances.

An eloquent spokesman is David Bacon, who worked at National Semiconductor before heading the United Electrical, Radio and Machine Workers of America (UE) Electronics Organizing Committee from 1978 to 1983. In articles that have been widely circulated among the online labor and NGO networks, Bacon demonstrates that the history of labor action in Silicon Valley is coextensive with the rise of the IT industry itself. It began in the early 1970s with the failed actions of UE-affiliated worker-organizing committees in high-tech manufacturing plants, proceeded to the success of the Santa Clara Committee on Occupational Safety and Heath in the late 1970s (which raised awareness about risks to high-tech workers), first peaked in the UE-led actions of the early 1990s against the Versatronex plant ("the first plant in Silicon Valley struck by production workers, and the first plant where a strike won recognition for their union"), and then led to the widely publicized janitorial worker strikes in Silicon Valley in the late 1990s and early 2000s.[60]

As Bacon's history indicates, one of the main concerns of labor organizers in high-tech industries has been the "invisible workforce" of blue-collar workers.[61] But by the late 1990s, after Bacon's history breaks off, labor organizers were also rallying the proletariat, as I called them, among white-collar technical workers themselves. These are the "permatemp" technical workers that such IT firms as Microsoft employed indefinitely on a temporary basis without full perquisites (through the agency of middleman temp firms). It was against Microsoft that WashTech (the Washington Alliance of Technology Workers, affiliated with the Communications Workers of America) won its landmark suit in 2000 declaring the company a "co-employer" rather than just contractor of such temps and thus responsible for benefits and stock options.[62] Against one of the paragons of the high-tech New Economy, labor won an entree into the new millennium version of Big Steel or Big Oil. New labor drew a line in the silicon sand.

In short, when labor activists and others concerned with high-tech labor issues look out over the populations of workers possessed by IT corporations, they ask with incredulity, "What knowledge empowerment?" "What flat organization?" *Corporate Watch,* for example, quotes the following blanket condemnation of corporate high tech from Jerry Mander, a critic of technology and globalization:

> Multinational corporations are decentralizing operations and jobs around the world, but at the same time, they are intensifying their centralized control over these decentralized operations. . . . For all the hype in the media about how the new technologies will enhance democracy, what we are getting is not individual empowerment but a new empowerment for multinational corporations and banks, with respect to workers, consumers and political systems.[63]

From the labor perspective, information technology makes the first-person plural in "We the People" a corporate concept.

Now we can return to cyberlibertarianism, whose almost total lack of interest in the issues of the wired NGOs and labor groups is striking. To an almost unbelievable degree, cyberlibertarianism is oblivious to topics that in any reasonable definition—whether or not one agrees with the NGO or labor position upon them—would seem to be essential to a general cyber-politics. The partial exception that proves this rule is the Computer Professionals for Social Responsibility group (CPSR), which is older than the other cyberlibertarian organizations. CPSR's evolutionary path is telling. As its former chair, Terry Winograd, reflected in 1996, CPSR initially stood against the military use of information technology but later divided its focus between "social justice" and "individual rights." Winograd worries:

> There is, of course, no need for an organization to focus on only one problem, but when there is more than one, it creates stresses—people feel that there is no common direction, and often will disagree with each other on whether a particular issue or action is appropriate (for the organization, or at all). CPSR is in that situation—some people who see Social Justice as the problem want CPSR to sign on to a wide variety of causes that may be antithetical to people who see Individual Rights as the problem. . . .
>
> CPSR can't be all things to all people. If we take a stand for social justice issues in general, then we have to be willing to alienate people who don't agree. If we focus on individual rights then we have less appeal for people who think that the key work to be done is getting collective social control over resources that are now dominated by profit motives. It is possible that some issue will come along that will unify and motivate people across these boundaries . . . , but we can't just sit and wait for that.

> I'm a product of the sixties, and for me Social Justice is the key problem for our society at this point in history, with the others taking a secondary place (though often they may lead in compatible directions). That was typical of many CPSR members in the early days, but I don't know about the mix today.[64]

Whatever the current state of the CPSR, we may say in response to Winograd, the *general* cyberlibertarian "mix today" is clear. "Me, myself, and I," cyberlibertarianism says in its mantra of justice for the individual consumer or entrepreneur. Almost completely lost to view is social justice as understood in the collective sense by NGOs and labor groups. With the exception of the social justice working groups in the CPSR, cyberlibertarianism is loudly silent on the cyber-politics of the environment, ergonomics, labor action, and—especially surprising in light of its stated beliefs on privacy—workplace privacy.[65] Thus, for instance, the agenda of the Electronic Frontier Foundation hardly overlaps with any of the NGO issues—and then only in ways that underscore how little such issues matter. While the EFF's file archive on "Privacy, Security, Crypto, & Surveillance" holds a sub-archive on "Privacy—Workplace Monitoring & Employer/Employee Privacy Conflicts," this sub-archive contained just six files and one additional subdirectory in mid-2000. By comparison with the rest of its parent archive or other such fat EFF archives as "Censorship & Free Expression," this is nothing. So, too, the EFF "Privacy, Security, Crypto, & Surveillance" archive contained in mid-2000 a sub-archive on "Privacy—Surveillance" whose eighteen files and four subdirectories included not a single item about privacy in the workplace. Other major cyberlibertarian organizations are just as uninterested in NGO and labor activist cyber-politics. The Center for Democracy and Technology's briefing on "Democratic Values for the Digital Age," for example, explicitly identifies "consumer privacy"—but not workplace privacy—as one of the "pillars of Internet freedom." So, too, the Progress and Freedom Foundation, which by comparison with the EFF and CDT is noticeably pro-business and pro–New Economy ("Third Wave"), is oblivious to NGO issues.[66] Cyberlibertarianism is the land of the brave and free individual. Workers need not apply.

In his "Birth of a Digital Nation," Katz led the wired people into the promised land as follows: "I saw the primordial stirrings of a new kind of nation—the Digital Nation—and the formation of a new postpolitical philosophy." Katz's follow-up essay for *Wired* eight months later, "Netizen: The Digital Citizen" (December 1997), then debated the finer point of whether the chosen people had really crossed the virtual Jordan into a new world of alternative "postpolitics" (as "Birth of a Digital Nation" had concluded) or instead remained committed to the established political system (as suggested by an intervening survey commissioned by *Wired* and the Merrill Lynch Forum).[67] But Katz's primary point remained that politics for the really cool is indeed some form of politics. Notwithstanding (or perhaps because of) the item in the *Wired*/Merrill Lynch Forum survey that showed the "connected" and "super-connected" to believe that "Bill Gates has about as much influence over the fate of the nation" as the U.S. president, netizens in Katz's view are among "the most informed and participatory citizens we have ever had or are likely to have." Thus he adds, "The common stereotype of the Internet as a haven for isolated geeks who are unaware of important events occurring outside their cavelike bedrooms can now be exploded as an inaccurate myth."

But after comparing cyberlibertarianism to NGO and labor activist cyberpolitics, it would also be possible to frame a contrarian thesis. Let us hypothesize that in the contemporary moment "politics for the really cool" is actually not any kind of meaningful politics at all because emerging from "cavelike bedrooms" is too low a standard. Katz's implication is that to leave the stereotypical bedroom of the "geek" rigged for online fantasy games, chat, porn, hacking, and so on is automatically to step into the proverbial street where—as for the counterculture of old—battle can be joined with big government and consumer business. Yet in contemporary life there exists outside "cavelike bedrooms" an even more encompassing cave that netizens not only do not emerge from but (as in Plato's paradigm) do not see on their political horizon at all. That universal cave is the cubicle. Blind to the cubicle, cool people can spelunk as subversively as they like yet never find the exit to the street. They are "postpolitical" only in a sense that ironizes Katz's term. Playing at politics like actors in American revolutionary (or, what is the same in Silicon Valley history, hippie) costumes, they engage in a purely postmodern simulation of politics: a retro-politics of free speech, privacy, and so on enacted on old stage sets of antigovernment and anticonsumerist protest. They act up everywhere but in the workplace where such issues now make the most difference in tangible quality of life. By comparison with the cyber-politics of the NGOs and labor groups, they are apolitical. Even the Progress and Freedom Foundation's "Magna Carta" speculates that cyberlibertarian individualism is a direct cause of apoliticism: "But a 'mass movement' for cyberspace is still hard to see. Unlike the 'masses' during the industrial age, this rising Third Wave constituency is highly diverse. Like the economic sectors it serves, it is demassified—composed of individuals who prize their differences. This very heterogeneity contributes to its lack of political awareness. It is far harder to unify than the masses of the past."

But we cannot be unduly harsh. I am using deliberately overstated language in saying that cool politics is "no politics" because it avoids acting up in the corporate workplace. Such a stark phrasing serves to make the point, but it begs the question whether by comparison the NGOs are themselves truly political. Are the NGOs and labor groups any less limited in their views, however different those views may be? Are NGO-organized street actions against the WTO any less a matter of pure "play-acting"? And in general, what is "political" as opposed to symbolic in the postindustrial age? These are questions of such scope that they cannot be addressed here. What we can do is to conclude upon a slightly different assessment of "politics for the really cool"—one that does not hold such politics to an excessively artificial, black-or-white standard of the political versus the nonpolitical.

Let me thus at last venture the thesis of "bad attitude." However much cyberlibertarianism may be apolitical or ambivalently political in the broader scheme of things, it is certainly full of passion, energy, and activism on its chosen issues. Apolitical is not the same as apathetic or anaesthetized. A more adequate description of the strange politics/no-politics of cyberlibertarianism is Bad Attitude, which is how cyberlibertarianism *does* show up in the workplace.

I refer to a "political effect"—a style or pose of politics as pure "acting up"—that official cyberlibertarian politics (as publicized on the earnest sites of its major organizations) rarely admits but that the unofficial hinterland of the movement glories in. It would be possible to witness this "political effect" by focusing just on the colorful fringe or marginal expressions of cyberlibertarianism where niche Web sites display an online version of NGO-style street theater. Here, for example, are some juicy razzes from "Lizard's All-Purpose, Multi-Functional, Free Speech: Civil Disobedience, Enemies List, and Survival Guide," which "Lizard" created during the battle over the Communications Decency Act and left standing as a "monument":

> We won in New York, too. Let's see. . . . six judges have looked at the CDA. Six judges have said, in effect, "You've got to be shitting me." And, thanks to those six judges, I can still write that where delicate widdle kiddy-widdies might see it and get all corrupted. So, to those who would tell me that I have no right to speak freely, I must reply, politely and with all due respect, "Neener neener neener."

> I have no clue who "vandalized" the US Dept. Of (in)Justice Web Site. I have no idea what his or her politics are, or on how many points other than freedom of speech we'd agree. Regardless, I have only this to say: *Way To Go, Dude! (Or Dudette)*. . . .
>
> I'm dividing this document into three parts, sort of like Gaul. The first part is basic techniques to make sure your message gets through, and the messages you want to read get through—regardless of what the State tries. If there's any internet at all, these ideas *should* work.
>
> The second part is, basically, "*How To Piss Off Censors, Busybodies, Bluenoses, and Government Spooks, Usually Within The Letter Of The Law.*" Any suggestions I *know* to be actually criminal, I'll note as such. Consider them to be . . . er . . . warnings of what not to do. Yeah. I mean, how could you know a particular act was illegal unless someone told you, right? (Netscape needs a "wink wink, nudge nudge" tag).
>
> But seriously, folks. This isn't about hacking or acting up for the sake of being k00l or rebellious for the hell of it. This is about preserving the first truly free society against the invaders. Do whatever you feel is just or moral to either keep freedom alive, or keep the barbarians from getting any use out of their conquest. Your own values, morals, principles, and standards must be your guide. We're a lot better off working as individuals—an organization with a leadership can be compromised, but free-acting individuals can never be infiltrated, disbanded, or corrupted from within.
>
> The third part is a "who's who" of would-be book-burners, fascists, censors, and other such scum.[68]

This, clearly, is what is called in the vernacular, *attitude*. This is the finger that the cool give to the establishment in a sort of downsized version of the two-finger peace sign of counterculture. Or, again, this is the cool's best postmodern, media culture, *Saturday Night Live* imitation of Jeffersonian discourse: not "We hold these Truths to be self-evident" but "seriously, folks"; not "That whenever any Form of Government becomes destructive of these Ends, it is the Right of the People to alter or to abolish it" but "Neener neener neener," "wink wink, nudge nudge."

As Katz generalizes the pose of political bad attitude in his "Birth of a Digital Nation":

> The digital young, from Silicon Valley entrepreneurs to college students, have a nearly universal contempt for government's ability to work; they think it's wasteful and clueless. On the Net, government is rarely seen as an instrument

> of positive change or social good. Politicians are assumed to be manipulative or ill-informed, unable to affect reform or find solutions, forced to lie to survive. The Digital Nation's disconnection from the conventional political process—and from the traditional media that mirror it—is profound.

The most extreme proponents of such attitude, perhaps, are the contemporary vandals at the gates of virtual Rome: the hackers who make their graffiti within the systems not just of "wasteful and clueless" government but also of its apparent opposite, efficient and smart business.

But I do not wish at the conclusion of this chapter to focus just on the most extreme or individualistic "actings up" of cyberlibertarianism where the edge of the movement bleeds into the confrontational and irreverent street style of the NGOs. Instead, I wish to return to the ordinary, implicit politics I earlier called the "casual experience" of cyberlibertarianism; to the heartland of contemporary America and, more broadly, of the knowledge economy at large where cool people do act up—but oh so secretly, subtly, and undecidably (suspended between passiveness and activism, despair and hope). That heartland is not at the outer but what might be called the inner edge of cyberlibertarianism. It is the cubicle.

How do knowledge workers act up in their cubicle? We remember that Henry Ford knew that the ordinary assembly line worker acted up all the time. This is what his rules and company police were for. Similarly, we know that contemporary knowledge workers act up all the time—sending inappropriate personal e-mail on the company system, browsing Web sites unrelated to work, engaging in deliberate inefficiency or "time theft" ("soldiering" in the parlance of Ford's day), and even sabotaging or hacking company information systems. Michel de Certeau's word for it in his *Practice of Everyday Life,* which in part examines the way common workers pull the wool over their employer's eyes by "diverting" time and resources, is *la perruque* (an instance of what he generally calls "tactics"):

> *La perruque* ["the wig" in French] may be as simple a matter as a secretary's writing a love letter on "company time" or as complex as a cabinetmaker's "borrowing" a lathe to make a piece of furniture for his living room. . . . In the very place where the machine he [the worker] must serve reigns supreme, he cunningly takes pleasure in finding a way to create gratuitous products whose sole purpose is to signify his own capabilities through his *work* and to confirm his solidarity with other workers or his family through *spending* his time in this way.[69]

Our word for such an oblique tactics of resistance is *cool*. Ordinary knowledge workers in their cubicles are cool when they exhibit some degree of the bad attitude we earlier glimpsed in that crevice opened between cubicle walls by the women information workers in Zuboff's *In the Age of the Smart Machine* (see my discussion in chapter 3). The "friendliness" of corporate culture we have studied is the negation, but never the extinction, of such bad attitude.

I choose the expression "bad attitude" for the strange politics/no politics of the cool because of a specific precedent. That precedent is *Processed World* magazine, which was published in the 1980s and early 1990s in San Francisco by an editorial collective of "post–New Left, post-situationist" twenty-somethings who awoke one day after counterculture had peaked to find themselves living in the belly of the beast. They were office "temps."[70] What *Processed World* believed in, above all, was "bad attitude," which was its motto and also the title of a 1990 retrospective anthology of the magazine's articles, readers' letters, and illustrations.

What is bad attitude in the "processed world" of information work? Leafing through the *Bad Attitude* anthology, we might at first think that the magazine's mission was exactly akin to that of the NGOs that evolved their post-counterculture identity in the same period. As the introductory essay in *Bad Attitude* by Chris Carlsson and Adam Cornford recounts: "*Processed World's* creators shared a varied political background and outlook that continues as a non-doctrinaire hybrid of traditions and theories. They have in common an opposition to capital and wage labor, nationalism and governments, and a profound wish for a world whose people cooperate freely to satisfy collective need and enlarge individual possibility, without the compulsions of money or hierarchical authority."[71] And certainly, the various "actings up" of the *Processed World* collective in the streets of the Bay area during the 1980s resonate with the street theater of later NGO actions:

> Every Friday, writers and editors would head out to the streets to hawk [the magazine]. . . . Collective members donned papier-maché costumes. These, like VDT [video data terminal] head-masks labelled "IBM—Intensely Boring Machines" and "Data Slave," or an enormous detergent box whose familiar red-and-yellow sides read "Bound, Gagged, & TIED to useless work, day in, day out, for the rest of your life?" attracted immediate if often puzzled attention from passersby. Sellers pranced around on the sidewalk in these get-ups, yelling "Processed World: The Magazine With A Bad Attitude!" (*Bad Attitude*, p. 9)

But these resemblances aside, there was a crucial distinction between the sensibility of *Processed World* and the NGOs and labor movements we surveyed. Where the NGOs carried on the old tradition of "out of doors" extra-parliamentary protest—staging actions on the public street as well as on the public information superhighway—*Processed World* concentrated its protest *indoors* in private enterprise and the cubicle. Above all, the magazine was known for recommending acts of trivial sabotage by individual knowledge workers. Thus Christopher Winks suggests sabotaging business jargon by substituting "anti-lexicons" of Freudian slips, "sound poems," "abstract patterns of letters and words," or "scandalous xero-graphic collages" for business jargon (*Bad Attitude,* p. 49). A letter to the editor published by the magazine recommends that secretaries simply stop editing managers' communications to expose their innate idiocy: "Now I regularly send memos out systemwide with sentences like 'Thank you for your patients,' and 'Newer construction are listed for rent,' and 'Local environs are well appearing'" (*Bad Attitude,* p. 47). Thus, too, one of the single most famous *Processed World* articles was "Sabotage: The Ultimate Video Game" by "Gidget Digit" (Stephanie Klein), which got its author fired when a copy was found on her desk at work. The article promotes "white-collar opposition" through theft, "time theft," and sabotage.[72] The reprint of the article in the *Bad Attitude* anthology is accompanied by illustrations, letters from readers, and other material suggesting concrete ways to throw the sabot into the gears of high-tech—stapling floppy disks, pouring coffee on keyboards, inserting metal objects in cooling slots, opening computer cases and tampering with chipsets, running a magnet across storage media, writing hard on floppy disk labels with ballpoint pens, and so on (*Bad Attitude,* pp. 59–66).[73]

Just as important, as these examples indicate, *Processed World* recommended not just any kind of cubicle sabotage but *secret* sabotage. How does one destroy data on a disk with a magnet, for example? By wearing a secret magnet within a ring on one's finger (*Bad Attitude,* p. 60). As in the case of de Certeau's *la perruque,* the goal was not to create a public statement that would get one fired. The goal was instead to make a private, subtle, even purely symbolic statement that came in under the radar horizon. If "bad attitude" was any kind of politics at all, it was stealth politics. A defining feature of the magazine, therefore, was its reaction to more public, even terroristic, acts of sabotage against information technology—for example, acts by the antiwar BEAVER 55 group that destroyed equipment and data at a Hewlett-Packard facility in Minnesota in 1970 or by the French CLODO group (Committee to Liquidate or Divert Computers) that destroyed programs, data, and storage media at a Toulouse software firm in 1980. Writing about these actions in her "Sabotage: The Ultimate Video Game" article, Gidget Digit argues:

> The implications of the repressive and socially negative ways in which computers are used need to be explored. However, in their emphasis on massive destruction, groups such as the above direct themselves too much against the technology itself (not to mention those groups' authoritarian internal structure). They do not pursue the positive aim of subverting computers, of exploring the relationship between a given technology and the use to which it is put. In this sense, pranks and theft, often carried out spontaneously and almost always individually, are more radical than the actions of those who group themselves around a specific political ideology. (*Bad Attitude,* pp. 64–65)[74]

What *Processed World* stands for, as opposed to political ideology, Carlsson and Cornford say in their introductory essay to *Bad Attitude,* is "a nondissident majority in the office world, and a different adaptation to office life, not confrontational but perhaps passively resistant."

> By serving as a forum for "ordinary" workers, *Processed World* reinforces the often suppressed truth that subversive wisdom, and knowledge about power/class relations, flow from people's daily lives and not from an ideology or group of experts. As a radical publication filled with art and humour *PW* emphasizes the importance of immediate enjoyment, both for surviving this insane world, and as an alternative to the deadly serious political discourse and emphasis on self-sacrifice typical of opposition politics. (*Bad Attitude,* pp. 11, 7)

The whole phenomenology of the magazine is characterized by interiority and everydayness. Politics is not about a noisy, collective action on the street. It is instead an action so immured within the cubicle, within one's individual workstation, and ultimately within the most interior of all cubicles, one's own head, that it is as if global capitalism could be fought through totally nonglobal (private) forms of anticapitalism (theft and sabotage).

We may say that finally the true register of "bad attitude" politics—the only medium in which a wholly interiorized, private sensibility of rebellion could manifest—was "style." A particular style of (resisting) work became the personal pose, psychic stance, *attitude* of the worker as she or he contravened the dominant corporate culture of friendliness. Carlsson and Cornford write:

> Like most even remotely successful thngs in this image-dominated world, *PW* is best known for its style. Fortunately, in the magazine's case, style is an integral aspect of substance. *Processed World's* style is summed up in its notorious "Bad Attitude," the indignant and undying creative spirit in workers that refuses to conform altogether to the absurd demands of the job. By turns sullenly passive, gleefully satirical, and covertly defiant, the Bad Attitude walks a tightrope between the persistent effort to preserve one's psychic integrity and the necessity of participating at least minimally in the worker role. In its most basic form, then, the Bad Attitude is a kind of go-slow action against what sociologist Arlie Hochschild has called "emotion work," the maintenance of a friendly, cheerful façade demanded by the smile-button culture of the service sector. . . . *Processed World's* Bad Attitude both expresses its producers' immediate emotional outlook and tries to spread this attitude as the necessary precondition for more radical and direct forms of refusal. (*Bad Attitude,* pp. 12–13)

In particular, "style" was the melange of crudely produced articles, letters, and satirical graphics with which *Processed World* filled its pages as if in a send-up of glossy, graphically harmonious corporate annual reports. Inspired by the Situationist method of *détournement* ("this means ripping products of the ruling society—from tools and technologies to words and images—out of their approved contexts and putting them to new subversive use"), *Processed World* practiced a "wild style" in which conventional prose, graphics, and concepts from the business world were collaged together in scandalous ways. The effect was an exaggeration of the normal worker practice of posting satirical cartoons on cubicle walls (e.g., "a xeroxed cartoon with 2-inch lettering reading 'They can't fire me! Slaves have to be sold!'"). The pages of *Processed World* were filled with flamboyant cartoons and collages that knew no bounds of taste. One graphic, for example, shows the infamous photo of the South Vietnamese general firing a pistol into the head of a prisoner under the large title, "The One Minute Manager." Another graphic published in response to *Time* magazine's selection of the personal computer as "Man of the Year" in 1982 shows the "Woman of the Year": an artificial woman with marionette limbs and a computer monitor cinched around her waist like a chastity belt. "Crudely produced," I said about *Processed World.* Ultimately, this is not a judgment just about the technical and financial limitations under which the magazine was created. It is also a recognition that the jarring, disconnected, edgy, even underground-comics style of the magazine—in which the purpose of collage is to bare rather than hide seams—is in itself a refutation of the seamless, harmonious interfaces of "user friendly" corporate culture. As slick and well-produced as *Wired* magazine would later become, its style also ultimately descended (through such graphics progenitors as David Carson) from similar bad attitude (*Bad Attitude,* pp. 15–16, 54, 73, 192).

The lesson that the cool now can take away from *Processed World* is that they may not be able to change the conditions of their life or even express a political need for such change. But they can change their Web pages into something so cool in style that even as they communicate information, they also break or even sabotage it. Look at all those cool Web designs that defy the modernist dictum of clear message by implementing a digital makeover of *Processed World*'s style: mixed media metaphors, near-illegible textual and graphics F/X, animations that impede the delivery of content, and so on. Look, in sum, at the signature of the ethos of the unknown in the work of knowledge. The Founding Fathers signed the Declaration of Independence in flourishing Hancocks below the text; the cool sign their declaration of commingled in/dependence in a **drop shadow**.

PART IV

Humanities and Arts in the Age of Knowledge Work

[T]he men would spread out around a clump of trees where coati had been seen, shoot arrows into the branches, throw pieces of wood or clumps of earth, shout and make such a racket that the animals became completely panic-stricken and could only think of one thing—to flee from the noise. They would scurry down headfirst. But at the bottom would be a man, his left forearm bound in a thick cord made of plant fibers and women's hair. . . . When the coati reaches the ground, the man pushes him against the trunk with his protected left arm, takes him by the tail with his right hand, swings him in the air, and smashes him against the tree with all his might, which breaks his skull or his spine. With this technique, the hunters can also capture the coati alive and use them as guard animals in the camps.

Pierre Clastres, *Chronicle of the Guayaki Indians* (1972)

As you would expect, the Google Search Appliance provides the accuracy, speed and ease of use that have made Google the favorite of web searchers worldwide. When employees have instant access to the best information, they get more done. That makes them happier and makes your company more productive.

"Google Search Solutions," Google, Inc.

Preface

"More"

There is much "more," as I said in my introduction. If we were to press the phantom "more" or "next" button at this point in our argument, we would come to additional topics in the study of knowledge work and information culture. Clearly, for example, my cultural critical approach has meant that this book has concentrated on the "identity," "subjectivity," and even "fantasy" of knowledge work in ways that bypass psychological or psychoanalytic exploration.[1] Clearly, too, I have not dealt in depth with the effect of information culture on any particular aspect of the group identities I observed generally in part 1, which scholars in "new media studies" have begun to explore under the topics of digital gender, race, and ethnicity.[2] These latter portfolios of study have the potential to do more than just confirm established modes of identity criticism in the humanities. They promise to add unpredictable spin to such criticism. For example, how should the cultural critical narrative of "marginalized" minority groups be inflected to take account of "decentralization" as the new, dominant social ideology? Or, to reverse the direction of the question, how should decentralization, which is espoused by business consultants as if it were the cutting-edge discovery of the new millennium, be inflected to take genuine account of peoples with centuries of experience in practicing local or community knowledge outside the "center"? Albert Borgmann—whose *Holding On to Reality: The Nature of Information at the Turn of the Millennium* I have read with widening admiration in the latter stages of my own writing—gives one glimpse of the promise of this line of thought by grounding his expansive explanation of information in part upon the prehistorical, "ancestral environment" of information once habitual to Native Americans in his home state of Montana.[3]

Perhaps most important, my study has focused on the United States and thus does not give due attention to the "world" even while making the "World Wide Web" a case study in postindustrial cool. In broader view, knowledge work, data production, and contemporary business culture are global in their implications. We do not even need to consult such world-spanning scholars of information society as Manuel Castells or Armand Mattelart to glimpse this fact. We need only watch the opening of any ordinary TV news show to see in the now clichéd computer animation of the spinning globe the very image of information as world.[4] From the vantage point of information-mongers, that is, the world is identical with the span (and spin) of information mediated by knowledge workers. Or, put another way, zones in the world opaque to information (what Castells calls " 'black holes' of marginality") are only precariously part of the world at all.[5] No study of knowledge work, therefore—let alone one centered in a nation like the United States that exerts such influence over the content, media, protocols, and even language of knowledge work internationally—can now ignore the constitutive role of such work in globalism. The contemporary globe, perhaps, is not so much a preexisting object as a standing wavefront of simulation generated by knowledge work as an *idea* of globalism—named, for example, "new world order," "global market," or "World Wide Web."[6]

Knowledge work and information culture must thus be evaluated by the measure not just of individual weal or national wealth but of global commonwealth. The important question is: what standard of well-being might emerge from the seemingly closed circuit of "well informed" citizenry in the West that can transform the U.S.-led vision of the world as all "global competition" into a less fundamentally cruel imagination of wealth in common?[7] Or, to put it another way, a book like this one on the alienation of knowledge workers *within* postindustrial society requires a companion text like Castells's that also listens to regions of the world alienated *outside* postindustrial society. Knowledge workers who grow cool when threatened by networked corporate culture share a secret quotient of identity with peoples who grow nationalist or fundamentalist when threatened by the "West" in general. There is thus rich research to be done on the puzzle of why knowledge workers generally think there is nothing more *uncool* than nationalist and religious revivals—why the cool (high-tech, media-saturated, futurist) and the "uncool" (anti- or last-generation tech, anti-media, traditionalist) see right past each other despite their homologous resistance to postindustrialism.

Nevertheless, I hope I have touched upon a sufficient number of topics to render knowledge work and information culture in their very "subject" or, to borrow from David Whyte's *The Heart Aroused: Poetry and the Preservation of the Soul in Corporate America,* "soul" (see epigraph to part 1). After all, the recognition that knowledge workers *have* a soul, which is a way of naming that secret quotient of identity withheld from (or within) the global corporate network, is what is fundamentally at stake in grasping the *culture*—as distinguished from society, economics, or politics—of postindustrialism.

Let me thus at last offer the anthropology of cool I promised in order to start upon my concluding topic. That topic is the cultural education of the cool and, correlatively, the future of the humanities and arts in the information age. The generations of cool destined for the cubicles, let us imagine, only appear to accept a future in which they sit docilely in front of cold computer screens mutely clicking and mousing. In reality, they are wild and sit by hot campfires at night where their soul can sing. What they sing is older than the rock music that the cool used Napster to appropriate in the late 1990s. It is older even than the jazz that an older generation of cool (in the first colloquial usage of that term) vented in the nightclubs of the 1930s.[8] To hear that song—a cry of "more!" as primeval as the forests—I will listen to the anthropology of an "archaic" tribe of the forest and make of it a fable for the tribe of the cubicle. My essential question will be: can the cool be educated in their singing so that the song of culture they raise will be adequate to an age of knowledge work that is neither prehistorical nor (even more mythical) "posthistorical," but in the last instance truly historical? And will the humanities and arts—specifically, humanities education coupled with the artistic ideologies, institutions, and practices that make up the aesthetic education of the cool—be up to the task?

Chapter 9

The Tribe of Cool

The Tribe of the Forest

I previously drew upon the work of Claude Lévi-Strauss. Here I turn to another French anthropologist, Pierre Clastres, whose *Society against the State* (1974) is dedicated to the proposition that there are societies without politics. Most Indian tribes of the Americas, Clastres says, "are headed by leaders and chiefs," but "none of these caciques possesses any 'power.'"[1] "One is confronted, then, by a vast constellation of societies in which the holders of what elsewhere would be called power are actually without power; where the political is determined as a domain beyond coercion and violence, beyond hierarchical subordination; where, in a word, no relationship of command-obedience is in force" (pp. 11–12).[2]

Yet for Clastres, flat societies without hierarchy are not necessarily utopian. If individuals and groups do not have power, then this may mean they are disempowered by an overall system of equality that is in itself all powerful. They are subject, in other words, to the true power of their world: the very laws of sociality that enforce the sharing of power. Such subjection, at times, stirs resentment.

Witness one of the most poignant chapters of Clastres's book, "The Bow and the Basket" (pp. 101–28; supplemented by his earlier *Chronicle of the Guayaki Indians*). The South American Guayaki (or Aché) people, Clastres reports, are nomads whose harsh environment demands an unusually strict sexual division of labor.[3] The men (known by the bow they fashion with their own hands and keep for life) hunt and also gather in a forest that makes both activities difficult. The women, by contrast (each represented by a unique basket woven upon coming of age), transport belongings, make baskets and pottery, cook, and take care of children (pp. 103–4). Missing is any middle ground, such as agriculture, where the labors of the sexes could mix or, at a minimum, share a space and time. Who is more powerful in this society? No one politically among the men and, of particular interest in this chapter, neither sex over the other. That the men do the most immediately "important" work of providing all the food does not make them more powerful than the women. If anything, Clastres argues, the power relation could just as easily be said to be the reverse because of the combined effect of two fundamental social laws in Guayaki society. One is the alimentary taboo that forbids a man from consuming "the meat of his own kill" (p. 114). Forced to give food to others and receive food in turn, men are perforce socialized. The correlative law is polyandry, according to which Guayaki women—who demographically number half as many as the men—can take two husbands (pp. 116–17). Such is the other foundation of the men's sociality: they must share their wives. In both the registers of consumption and marriage, that is, the men are bound to each other by laws of exchange that cannot be gainsaid by any individual even if, as in the case of polyandry, there is considerable private resentment against those laws: "the men almost always harbor . . . feelings of irritation, not to say aggressiveness towards the co-proprietor of their wife" (p. 117).

Hence, Clastres argues, one of the most striking *cultural* facts about the Guayaki: the custom by which the men sing their strange night song.

> One might think the men had fallen asleep. But seated around their fire, keeping a mute and utterly motionless watch, they are not sleeping. Their thoughtful gaze, drawn to the neighboring darkness, shows a dreamy expectancy. For the men are getting ready to sing, and this evening, as sometimes happens at that auspicious hour, they will sing the hunter's song, each man singing separately. . . . Soon a voice is raised up, almost imperceptible at first, coming as it does from within, a discreet murmur that refrains from enunciating anything distinct, for it is engaged in a patient search for just the right tone and the right discourse. But it rises by degrees, the singer is sure of himself now, and suddenly, the song rushes out, loud and free and strong. A second voice is stimulated and joins with the first, then another; words are uttered in quick succession, like answers

> always given in advance of questions. All the men are singing now. (pp. 101–2)

And what is it that the men sing, each on his own but all together? It is easy to tell generally. The songs all celebrate individual pride in hunting, prowess, and potency. But, Clastres notes, it is not so easy to tell specifically. The songs increasingly diverge from the linguistic norm until, at the peak of performance, each man speaks not just in a gender-specific language but in a dialect of one:

> The language of the masculine song . . . is highly distorted. As its improvisation becomes progressively more fluent and rich, as the words flow out effortlessly, the singer subjects them to such a radical transformation that after a while one would think he were hearing another language: for a non-Aché, these songs are strictly incomprehensible. With regard to their thematic composition, it basically consists of an emphatic praise which the singer directs at himself. In point of fact, the content of his discourse is strictly personal and everything in it is said in the first person. . . . *cho rö bretete, cho rö jyvondy, cho rö yma wachu, yma chija* ("I am a great hunter, I am in the habit of killing with my arrows, I am a powerful nature, a nature incensed and aggressive!"). And often, as if to indicate how indisputable his glory is, he punctuates his phrase by extending it with a vigorous *cho, cho, cho* ("me, me, me"). (pp. 112–13)

The function of such "endo-language," Clastres concludes, is to transgress against the normal function of language: communication or *sharing*. Everything else important in life must be shared, the Guayaki men say. Let a time be set aside, therefore, when language—the medium of sharing par excellence—is *not* shared: "The men's song, while it is certainly language, is however no longer the ordinary language of everyday life, the language that enables the exchange of linguistic signs to take place. Indeed it is the opposite. If to speak is to transmit a message intended for a receiver, then the song of the Aché men is located outside language. For who listens to the hunter's song besides the hunter himself, and for whom is the message intended if not the very one who transmits it?" (p. 122).

Though Clastres does not say this, we can collate his work with Lévi-Strauss's *Elementary Structures of Kinship* to make a comparison to that most fundamental transgression against the laws of sociality—incest. The song of the Guayaki men is the incest of language, an endo-"message" or endo-"transmission" that—at least in fantasy, in a gesture there by the fire at night—can be withheld from exchange even within what is otherwise the freest and most malleable of all media of social exchange, language.

The Tribe of the Cubicle

The relevance of the "archaic" tribe to our argument is not that contemporary knowledge workers are lonely men working far into the night in the programming, engineering, and other technical sectors of information work. Spanning the managerial, professional, skilled clerical, and even to some extent blue-collar populations, knowledge workers are now significantly more diverse than in the era of the 1950s "organization man" (despite inequalities that a gender-oriented study, just as one example, would need to explore). The relevance instead is that the tribe of the cubicle, like the tribe of the forest, also inhabits a sociality of mandatory disempowerment. We can summarize the laws of this sociality—the postindustrial sociality of "neocorporatism" and informationalism discussed in previous chapters—as follows:

- *The Law of "Nature."* The Guayaki are the people of a fierce, penurious forest; their social laws reflect the ancient "law of the jungle." Knowledge workers are the people of "global competition"; their social laws reflect the "economic laws" of the new world order. Global competition, in other words, is how postindustrialism reconstitutes an apparent natural law upon which to found harsh social laws.
- *The Law of Mobility.* In particular, global competition legitimates restructuring, downsizing, outsourcing, and the replacement of career workers with permatemps—the social law of a new nomadism.
- *The Law of Modularity.* Mobility across jobs and professions is only the outward manifestation of a more existential nomadism within jobs and professions. In the "team" and "flat" organization, positions, roles, and even pay (tethered to team, company, or stock performance) shift dynamically and interchangeably. Like the cubicles they inhabit, workers are modular units to be assembled and reassembled in complex, ever differing configurations without innate identity.
- *The Law of Random Access.* Corporate "diversity management" programs display another aspect of postindustrial modularity: "we" are a rainbow not so much of ethnicities, races, genders, nationalities, religions, ages, and so on, but of "talents" and "skills" to be quickly assembled and reassembled in competition against similarly ad hoc consortiums of talent around the globe. Whatever the messy entanglements that make us "us"

(our personal histories, lineages, communities, bodies), our traits can be accessed at random—like a CD-ROM over which a laser plays to seek out specific pits and bits.

- *The Law of Exchange.* The bottom line of competition, mobility, modularity, and randomness is capital, which in the New Economy finds its perfect allegory in information. In the dominion of the Guayaki, food and spouses must be exchanged; in that of knowledge workers, information must be networked. Such is the postindustrial fulfillment of Taylor's scientific method, which first showed how intellectual capital embedded in individuals and communities could be extracted in programmable, manageable, and exchangeable forms.
- *The Law of Cool.* Finally, there is the law, or perhaps antilaw, of cool, which I have theorized as the "gesture" of ambivalent, recusant oppositionality (not quite a "statement," "expression," or even "representation" of defiance) within knowledge work. "I am a great hunter," the Guayaki men sing in a lyrical gesture that is both the corporate logo of their hunter-gatherer society and a protest against that society. "I'm cool," twentieth-century subcultures and countercultures once said in gestures of clothes, hair, music, cars, and so on that similarly mimed and protested the techno-industrialism of mainstream culture. "I'm cool; I'm wired," postindustrial workers now add in *their* equivocal gesture (as I have defined information cool) of "information designed to resist information."[4]

These, or something very like these, are the laws of cool. These are the laws of knowledge work that now make it *mandatory* to be cool or at least—trained by cool consumer cultures that are really a lifelong, parallel education system teaching us how to live under the dominion of producer culture (the "corporate culture" of the new workplace)—to *aspire* to be cool. We can put the case in its most general form as follows. To go to work today is to face the demand for a fearsome new rationalization: industrial efficiency and productivity *plus* postindustrial flexibility and decentralization. Such neo-rationalization, however, feels unreasonable, or "paradoxical," because it is heedless of the "archaic" and/or "residual" biological, prehistorical, agrarian, craftsman, and even early twentieth-century industrial rationales of behavior codified in the habits ("habitus") of communities and individuals.[5] (To give a concrete example intimately familiar to many knowledge workers: "lifelong learning" demands perpetual reeducation at "Internet speed," but bodily, family, community, ethnic, gender, and other duties of everyday life pose an incalculably great counter-demand that makes it almost impossible, for instance, to study at night.) Therefore, there are only two equivocal ways that the archaic and unreasonable can protest their submission to the new rationalization. One is to quit and move to another job, which exactly reproduces the conditions of mobility, modularity, and random access that support the "flexibility" and "decentralization" of postindustrialism.[6] The other way is just as conflicted: to express in lifestyle and, increasingly, in what I have called "workstyle" the enormous reserve of petty kink that *Processed World* called "bad attitude" but that now appears with mind-numbing regularity in popular culture, the media, and the Web as "cool."

Cool is the protest of our contemporary "society without politics." It is the *gesture* that has no voice of its own and can only protest equivocally within the very voice of the new rationalization. It is the incest of information that secretly "nixes" the exchange of information. Structured as *information designed to resist information,* cool is the paradoxical gesture by which the ethos of the unknown—of the archaically and stubbornly unknowable—struggles to stand in the midst of knowledge work. In the maze of cubicles and even more amazing information architecture of networks, an estranged, small voice—picking up riffs and styles ad hoc from whatever resource it can find—sings "cho, cho, cho."

The Idea of a "Technical" Culture: Technique and Slack

Picking up riffs and styles ad hoc from whatever resource it can find, I said just now. Actually, this is not strictly true, and the reason why points forward in my argument. What is untrue is the phrase *ad hoc.* Wholly improvisational cool may seem, but the very notion of "improvisation" as we now understand it means free play based antithetically on underlying *technical rigor,* like a jazz riff played on a demanding instrument. Cool, in other words, has something deeply to do with "technique," whose efficient alignment with technology has been the premise of both industrialism and postindustrialism, but whose exact nature we must now probe more deeply. Above all, cool shows that there is no such thing as the "exact" slaving of technique to technology. Rather, technique is always also a way to express the archaic interval, lag, play, or "slack" between a people and their society. Such slack, we may say, is the fundamental gesture within all cool gestures.

To see what I mean, consider one further similarity between the Guayaki and the cool: the way both tribes have dual, split understandings of technique.[7] Here, according to Clastres, is the Guayaki technique of the bow: "Scarcely having reached the age of four or five, the little boy receives from his father a little bow that matches his size; from that moment he will begin to practice the art of shooting the arrow." At manhood, his "first concern . . . is to make himself a bow; henceforth a 'productive' member of the band, he will hunt with a weapon shaped by his hands and nothing but death or old age will separate him from his bow" (p. 106). (Similarly, the "nine- or ten-year-old girl receives from her mother a miniature basket, the making of which she has followed with rapt attention. Doubtless she carries nothing inside, but the gratuitous posture she assumes while walking, her head lowered and her neck straining in anticipation of its effort to come, prepares her for a future that is very near. . . . As the first task required by her new status [as a woman], and the mark of her definitive condition, she then makes her own basket [p. 106].) We do not know from Clastres the precise nature of the technique of making a bow or shooting the arrow, but we surmise that there is relatively little room for play in such technique, since the instruments have few moving parts, the target moves in instinctive ways, and the boundary between survival in the forest and hunger is razor sharp.[8] As the instrument of efficiency and productivity in this society, the technology of the bow does not tolerate play in the aim, slack in the string, or lag behind the prey.[9]

Crucially different is the Guayaki technique of song, which Clastres's description clearly identifies *as* a technique: "Soon a voice is raised up, almost imperceptible at first, coming as it does from within, a discreet murmur that refrains from enunciating anything distinct, *for it is engaged in a patient search for just the right tone and the right discourse"* (p. 102; my emphasis). Both the bow and the song aim to hit exactly the "right" target, then, and both do so in a way that makes technique *de rigueur* not just to function but social identity. "I am a great hunter," the bow and the song proclaim together. But where the bow expresses identity as the ruthless coefficient of the forces of production, the song does something slightly different. The technique of song plays upon, within, around, and above the rhythms of necessity even as it sings of nothing *but* necessity (hunting). It is a cultural "second nature" that in a slack, twilit moment between two needs—the need to hunt and the need to sleep—improvises upon necessity in the idiom of the uncanny double of necessity—technique. Technique reveals its true potential in song as the same-yet-different *representation* of necessity: not just bound to the laws of nature and their social enforcement, but also bound to swerve and jink in reflection upon those laws.

Now we have a basis upon which to understand the hijinks of cool. On the one hand, cool acknowledges a regime of strict technique akin to that of the Guayaki bow. As I argued previously, industrialism was based upon the notion of a socializing method—Taylor's "friendly" scientific method—

that aligns technique perfectly with the modern bow of necessity—technology. Technology and technique, I said in chapter 1, are aligned under the assumption that the most fundamental value is efficiency or productivity. The governance of such alignment is management; and the institution of that governance is the corporation.[10] Postindustrialism is grounded on the notion of an even more "user friendly" socializing method, corporate culture, that can accomplish the same task using the infinitely more flexible bow of information technology. Thus arises, I said in chapter 4, an era when all our techniques are bonded to all our technologies through a network of standards, protocols, routines, metaphors, and (corporate) culture that makes knowledge work simulate an eternal, inescapable friendship or systemic harmony—an exquisite, intimate coupling between technology and technique, machine and (as we now say instead of "human") "user."[11]

On the other hand, the very traits most valued by postindustrialism—flexibility and decentralization—open up within necessity the space for a *virtuosity* of technique akin to Guayaki song. After all, running a "traceroute" program to chart the hops between one's local computer and any destination on the Internet will show that even the shortest jaunt down the information superhighway to a geographically neighboring Web server is routed through several or more continent-spanning gateways, routers, and servers in a surprising pattern.[12] Given such flexibility in the infrastructure of postindustrial efficiency and productivity, it is no wonder that cultural superstructure now has more maneuver than ever. When a knowledge worker uses a search engine to find a target Web page, that is supposed to be like the twang of a bow. But when after a long download the page finally arrives with cool graphics, drop shadows, fancy animation, complex scripting, plug-in requirements, and other marks of excessive style (not to mention all the intervening cool pages a worker might have strayed to during the search), that is like the twang of song. Cool arises as a play of culture—a *style* of work—within what I have called the paradox of information: within the fissure, slippage, or slack, that opens up between the ethos and counter-ethos of postindustrial information work, between knowledge work and the "ethos of the unknown."

In the tribes of both the Guayaki and the cool, then, the song of culture is only possible on the basis of a certain slack that is so far from being extraneous to technique—like noise in a signal—that it is integral to it as its very virtuosity. What is the fundamental nature of such slack? Consider this physical analogy. Tie two weights together with a string. Then on a frictionless surface (let us say "ice" to stay with my allegory of the modern "ice ages") push one weight. If the string is taut, the second weight will follow smoothly. But if the string is slack, there will be an interval when the second weight does not move at all, then a sudden jerk as it starts in motion, and simultaneously a retrograde jerk on the first weight that creates cross-rhythms, perhaps even standing waves, in the connecting string. The string, as it were, sings.

For the Guayaki, the weight placed in motion is the progress of civilization; the trailing, second weight is the individual hunter; and the slack in the string is the "archaic" lag between older biological or social imperatives and the new sociality ("I am a great hunter," the archaic male insists; "*We* are great hunters," polyandric society corrects). For the cool, a similar cultural physics applies. The first mass is modern industrial or postindustrial progress; the second is the worker; and the slack in the string is a whole series of "archaic" (or, as Raymond Williams put it, "residual") lags between older social orders and the new world order.[13] Contemporary slack arises when the bodies, personal identities, and group habits of knowledge workers exert upon "progress" the profound inertia of tacit group understanding combined with canny (even witty) personal resistance, when the archaic infiltrates the contemporary using techniques that disguise it as a style of work (only seemingly "cutting edge") called cool.[14]

The concept of "disguise" is indeed relevant here. How, we might ask, can slack flourish within the constantly supervised workplace of ultraproductivity? Most discussion of slack has focused on Gen X "slackers," particularly from the perspective of lifestyle. The notable exception to lifestyle analysis of Gen X is Bruce Tulgan's *Managing Generation X* (1995), which brings work into view by advising businesses on making the most of "star Xers." Yet even this book relies on clichés about Xer lifestyle in the process of compulsively refuting them. (Gen Xers only appear to lack attention span, long-terms goals, humility, loyalty, and so on, Tulgan argues. Actually, he believes, these traits are the hallmark of New Economy flexibility and decentralization; since Xer's come to such a "postmodern" mindset natively, enlightened firms should learn how to appreciate them.)[15] For our purposes, then, it will be useful to make a fresh start—and, indeed, to expand the discussion beyond Xers to the longer genealogy of peoples resistant to rationalization—by avoiding lifestyle stereotypes entirely and starting with workstyle, where slack is a technical factor in work processes.

We can take our cue from two of the analysts of information work process I previously consulted when discussing networking. In his authoritative *Information Payoff: The Transformation of Work in the Electronic Age* (1985), Paul A. Strassmann observes that there is an enormous amount of invisible slack in routine knowledge work. On the one hand, such slack is clearly inefficient: "The real world of the office is full of delays, miscommunications, errors, and changes. . . . For instance, tracking a simple four-page bulletin informing the field sales force about a minor change in pricing policy can reveal a surprisingly complex train of events." But on the other hand, Strassmann argues, trying to rationalize all office work to the point where even complex tasks are programmed into information technology is "an unreasonable objective" because many tasks are unprogrammable. Indeed, unprogrammability is crucial to "informality," which amounts to no less than a whole parallel work flow that gets things done precisely by circumventing unrealistic or inappropriate standards, procedures, protocols, and programs. In one office Strassmann studied in minute detail, for example, only 12 percent of transactions could be accounted for as part of the formal system of work; 53 percent "were part of a formal system but required a great deal of discretion, training, and experience"; and 35 percent "were not systematized at all and required a great deal of initiative and personal skill."[16]

Gene I. Rochlin, whose *Trapped in the Net* turns on his intimate knowledge of such "expert operators" as air-traffic controllers, argues further that inefficient and informal slack is actually part of the real "knowledge" in knowledge work. Slack, in his view, is a crucial reserve of learning, judgment, and adaptability that must be protected from excessive dependence on information technology: "But safety in the expertly operated systems we have studied depends not only on technical redundancy against possible equipment failures and human redundancy to guard against single-judgment errors, but also on that wonderfully scarce resource of slack—that sometimes small but always important excess margin of unconsumed resources and time through which an operator can buy a little breathing room to think about the decision that needs to be made, and in which the mental map can be adjusted and trimmed." And again: "Of particular concern is the degree to which what is destroyed or discarded in the relentless pursuit of technical and operational efficiency is not waste or slop, but 'slack,' the human and material buffering capacity that allows organizations and social systems to absorb unpredicted, and often unpredictable, shocks."[17]

Now we can see how archaic slack fits into the contemporary workplace of ultra-productivity. Once slack not only percolated throughout the continuum of work but could repose occasionally in broad view by the coffee machine or water cooler (the unregulated spaces and times purposely reserved for it, such as the coffee break, the hallway, the lunch room, and so on). But knowledge work is now very different. The continuum of work is ever more supervised and monitored by information technology (to the extreme, for example, of keystroke-monitoring programs and ceiling cameras). Meanwhile, the spaces and times where slack can legitimately collect itself are being aggressively colonized by the whole regime of *official* informality: team meetings, team-building exercises, company retreats, dress-down days, and so on (no more coffee breaks; let's have refreshments at team sessions instead). Contemporary slack thus has nowhere to go to be itself. Its only recourse is to *disguise* itself within the processes, procedures, and techniques of information technology so as to be able to fly in under the radar of that same technology, there to appear in the guise of the informal qualities the corporations now purport to value: collaboration, creativity, "thinking outside the box," and so on. We might take for our New Economy paradigm the incalculable inefficiency and irrationality involved in creating very cool Web pages. As every cool Web designer knows, even with the assistance of the most sophisticated authoring software (let alone the hand-coding that many Web authors still swear by) advanced Web design is in the final analysis archaic. It is a regression to hand craft. Web "technique" consists in endless trial-and-error tweakings of design, innumerable revisions to accommodate different browsers and screen sizes, serendipitous discoveries that cannot easily be recreated (e.g., the brilliant graphic created in Adobe Photoshop in thirty steps that no one can exactly imitate even with the aid of the master image file and its archaeology of "layers"), and at every level encounters with what can only be called programmed irrationality (inconsistencies, bugs, and undocumented features in the underlying operating systems and client-server software upon which Web sites depend).

Slack, then, is not just a resource useful for optimizing knowledge work (Tulgan's and Rochlin's assumption) but also a hidden reserve that insinuates into each of the stray and informal intervals of work an archaism making possible an *act of judgment*. I mean by this more than the functionalist judgment that makes it possible "to think about the decision that needs to be made" and how "the mental map can be adjusted and trimmed," in Rochlin's words. I mean also critical judgment. Slack is when the archaic body, personal history, group habits, and so on render judgment—through the profound yet inchoate wisdom of cool—upon the very notion of postindustrial "adjusting and trimming" (restructuring, downsizing, flexibility, just-in-time, and so on). In short: "I work here, but I'm cool." Most uncannily, such judgment seems to come not from any transcendental perspective removed from knowledge work but from deep within the belly of the beast: from a technical perspective (getting the pixel width and color palette just right, for example) that is itself all about adjusting and trimming. This is why "cool!" exclaimed about any new toy, dress, high-tech gadget, or Web page always hovers ambiguously between the subjective and objective ("I'm cool"/"it's cool"). "Cool!" is the song of a subject robbed of any voice except that of the technological object.

Here, then, is the song of cool as I have learned to hear it in contemporary culture. In the register of consumerism first: *This tee-shirt with a corporate logo or this TV show with its advertisements: what does it have to do with "me"? It's just a medium, like air or water. "I" am the one who knows how to wear, sample, assemble, mix, or filter that medium with "style." I am a great hunter.* And the matching song of producerism: *These standards, routines, procedures, protocols, and programs that rule my life: what do they have to do with "me"? They are just my environment. "I" am the one who knows how to browse, search, sample, mix, or filter them with cool technique. In the forest* [network] *of information technology, I am a great hunter.*

Not "cho, cho, cho" but "cool, cool, cool." Such is the song of the people of technique. Such is the riff of those who live and breathe nothing but New Economy rationalization yet believe there is room in such rationalization for the archaism of slack. Once workers wanted a "decent wage for a decent day's work." Now "decency" is not the ethos under which knowledge workers make an equivalent demand for dignity. Knowledge workers want to be cool.

Chapter 10

Historicizing Cool

Humanities in the Information Age

But the gesture of "technique" by which knowledge workers express their need to be cool, as I described it, is not enough. At this point, I resume my tandem speculations about the future of humanities education and the future of aesthetics. Though I have cast my argument in this book primarily in the descriptive voice (what was and is), I now adopt the more anxious mode of prophecy. Futurology is facile insofar as it presumes that *will be* has the same indicative status as *was* and *is*. But the mode facilitates reaping from *was* and *is* a recommendation—more monitory than premonitory—of what "could be." Indeed, such a recommendation based on the hybridization of past and present is the unique contribution the humanities can make today to a society whose other major prophecies (in business and information technology) base their recommendations almost solely on the present while caricaturing, stereotyping, or excluding the past.

Let me start with just the first of the futures coupled in my inquiry into humanities education and the arts. What will be the future role of humanities education in the age of knowledge work? And a particularly fraught sub-issue: what will be the role of such education now that the older, universalist mode of humanistic inquiry has been inflected toward difference, flexibility, and contingency by those movements that are the uniquely academic version of postindustrialism: poststructuralism and cultural criticism? Here is a puzzle to consider: What's wrong with the picture I have so far drawn of the analogy between the Guayaki and the cool, the tribe of the forest and the tribe of the cubicle? The virtue of any powerful analogy is that it illuminates only partly through the comparisons it makes possible. There is also a second, more advanced level where the analogy reveals even more by the contrasts it harbors in reserve. Such is the case with the fable of the tribes I have told, which if pressed further for comparisons would likely demean both terms of the comparison. To answer the question I pose about the function of humanities education at the present time, we must now assert a difference as well.

It all comes down to what we mean by *archaic,* the term under which I have up to now grouped the prehistorical Guayaki and (as postmodern theorists say) posthistorical cool together. The ease with which I have equated these two realms is not innocent of the main subject of my critique: postindustrialism. Part of the world-view of postindustrialism—at least outside the academy—is that there are no fundamental differences between the past and the present, only technical differences like the contrast between black-and-white film and today's high-sensory, special-effects cinema. The past and the present are presumed to be comparable on a single plane of technologies, techniques, and the efficiency of their alignment; with the predetermined result that the contemporary almost always outmatches the past (since the goal of generalized efficiency is modern). The exception is when the contemporary raids the past for analogies to help it become even more efficient, a process that characteristically neutralizes the pastness of the past. Thus, for example, a recent book by Alan Axelrod (previously the author of *Patton on Leadership*) is entitled *Elizabeth I CEO.* "You can learn that being a leader is being a leader," the book says, "whether your enterprise is a renaissance kingdom, a small business, a major corporation, a corporate department, or a three-person work group with a job to do." The indifference of (and ultimately to) history here, whatever plentiful historical details the book otherwise offers, is sublime.[1] So, too, some years ago Miyamoto Musashi's seventeenth-century Japanese martial arts treatise, *A Book of Five Rings,* became a business bestseller. Managers were samurai for a day.[2] Just as relevant, computer interface designers habitually ransack the past for such "metaphors" to place on computer desktops as Microsoft's screensavers of the late 1990s starring Leonardo da Vinci or the "1960s."[3] As Baudrillard, Jameson, Jencks, and other theorists have commented, postmodernism—the cultural arm of postindustrialism—is all about applying the past as a simulation, pastiche, or façade to the present.[4]

What is wrong with such a screensaver picture? The answer is that not only do historical differences intervene between the pre- and posthistorical, but a crucial part of that difference is the consciousness of history itself (as once codified primarily in institutions reliant upon literacy). The Guayaki do not have historical consciousness as such, though they have something just as rich and interesting: "myth." But cool knowledge workers today have been taught that they have a history. That is, they have been taught a history that, while it consists of the same primal elements as myth—origins and ends, creations and destructions, the living and their ancestors, selves and others—is understood differently. Above all, history since the Enlightenment has been understood to organize the abiding elements of myth within the same neutral and inhuman firmament of temporality from which modernity at last derived its timeclock of efficiency. The Deistic "clockmaker" god of the Enlightenment, Napoleon famously consulting his watch during battles, and Taylor with his stopwatch all punched the same clock.[5] Mythic origins and ends, creations and destructions, the living and ancestors, and selves and others, in other words, were processed by the order of rationalization into the uniform difference between the primitive *past* and the modern *present*. *Our* myth is thus Progress; *our* sense of primal creations and destructions, selves and others, is ultimately the "us" against "them" of global competition.

We might think of it this way. Mythic difference has always had an absolutist tendency to accelerate opposing terms as far and rapidly away from each other as possible. We are truly "we" only when we oppose a starkly, even absurdly other "them." The principle of sped-up global competition is already there in myth, in other words, though held in check by tricky (and "trickster") mediations that split the difference between "us" and "them" to mark the potent, ethical center of the universe. All that was then necessary was to harness such accelerative difference to the neutral logic of temporality, which, lacking any social or ethical checks, could simply enlist primal differences for the purpose of perpetual speed-up or Progress. Rapid design and prototyping, fast assembly, quick turnaround, just-in-time supply chain: this is how ancestral identity has been coopted to the hyperbolic "'we' must beat 'them'" of global competition. "We" end up a functionalist concept indistinguishable from New Economy production and its primary units of identity: teams and corporations.

But that's not it exactly, or at least not all of it. Progress is not the only history that the cool have been taught. In prehistorical societies, the anthropologists and religious philosophers tell us, there are special times in life—just as there are special places of taboo—that do not obey the rules of normal reality. In these zones of *illud tempus*—whether it is a "liminal" coming of age or just the strange time of singing before the Guayaki go to sleep—identities come into being that are not so much part of the progress of life as transcendences or regressions to a whole different order of life (often haunted by the co-presence, the simultaneous urgency, of ancestors): *I am no longer who I was; I am my own Other.*[6] Something like this happens in history, too, where people are taught historical consciousness through a refrain of dynastic changes, revolutions, wars, strikes, protests, migrations, and other decision points where new identities come into being (or ancestral identities reclaim their rights) in ways that seem to advance the progress of civilization only by interrupting it: *The king is dead; long live the king!* Recently, cultural criticism has added to the humanities the sense that there are also smaller, more decentralized moments of revolution—"subversion" and "the practice of everyday life"—that enact the same drama of historical otherness within progress. What such petite revolutions lack in catastrophe, they more than make up in the capillary fractures they introduce everywhere in known reality. Or, rather, perhaps subversion is not about "reality" at all so much as "representation." In what the New Historicists call "representations" and the cultural anthropologist Clifford Geertz calls "interpretations," the order of myth crowds into history in images, fictions, and symbols that create a time-out in official reality—a theatrical play or cockfight, for example—where tales other than the dominant story of progress can be told.[7] Not just in 1776, 1789, 1848, or 1917, cultural criticism says, but in the work days and holidays of everyday there is an *illud tempus.* Such is one expression of the different perspective that cultural criticism, as a specifically academic version of postindustrialism, introduces within the normative postindustrial world-view.

All of this, then, the cool are taught—especially as they ascend to levels of education where learning "skills" is subordinated (at least in the humanities) to learning the history of societies, literature, art, film, religion, and so on. The great tragedy of contemporary education, therefore, is that "taught" does not mean anything like "learned." I refer to something that has so far remained merely implicit in my whole discussion—the gigantic distance between the cool and the educational system. The distance is so absolute that we would do best to think of it as constitutive. Cool is an ethos that starts as early as daycare and primary school (to judge from raising my own child), matures in high school (as documented in Marcel Danesi's *Cool: The Signs and Meanings of Adolescence* [1994]), and becomes adept in college (as instanced in the paradigmatic student we all know at the back of the lecture hall: slouched in his or her chair, feet up, dressed in a tee-shirt advertising an alternative or heavy-metal band, headphones on, looking up at the ceiling or down at the ground). Cool is nothing if not closely bound to the schooling system. Yet cool is anything *but* identified with schooling as such. Rather, it is a parallel system of learning—or just as accurately, antilearning—that turns away from an educational system it believes represents dominant knowledge culture, toward a popular culture whose corporate and media conglomerates, ironically, *are* dominant knowledge culture. As I put it earlier, the cool seek an "ethos of the unknown"—a knowledge of and for themselves recused from the society of knowledge work. But the first walkabout they take in quest of that ethos is mental truancy from the culture of school, their most proximate institution of knowledge work. The cool seek the ethos of the unknown through a tactics of unknowing making them ever more vulnerable to the conglomerates that truly prey on the unknowing.

Here we may make another comparison to archaic peoples, but this time to those residing within modern society itself. As my choice of the word *walkabout* above suggests, the cool are a kind of aboriginal who for a season silently withdraws from modernity into the unknown outback. Or to borrow from William Faulkner's story, the cool are Ike McCaslin, who on his version of a walkabout must "relinquish" compass, gun, and other technological artifacts of modernity before the Bear—the sublime of the archaic unknown—will make itself known.[8] Only, the cool today take their Walkmans, MP3 players, and handheld and laptop computers with them when they "go walkabout."[9] They relinquish nothing offered by the technologies and media of consumer culture because their instinct is that such culture—full of sensation instead of knowledge, mass appeal rather than teamwork—is the only regressive, hence apparently archaic, zone left in a world given over to knowledge work and its proxy, school.

The contradiction in looking to high-tech consumer culture for refuge from the knowledge work that produces such culture is not lost on the cool themselves. Their response to the contradiction—at once their fiercest, most genuine critique to date and the symptom of a profound defeat—is *irony*, our great, contemporary "Fordization of the face."[10] Of course, when measured against such prior instances of irony as neoclassical "wit" with its unerring sense of contrast between epic history and modern meanness, cool is almost unbelievably narrow in tone, incapable of modulation, cruel without compensating pathos, indiscriminate, inarticulate, and, above all, self-centered or *private*. Another way to say this is that at the moment of cool, knowledge workers (not to mention students training for knowledge work) regress to "adolescence," which is less a dismissive epithet than a structural description of individual as opposed to social archaism. Even when knowledge workers have graduated and gone to work, "cool" is how they instantly retreat to their mental "room" instead of joining the broader, *public* history of peoples resistant to rationalization.[11]

More's the pity, then, that the schooling system cannot persuade the cool that education itself is at least as genuine a champion of the ethos of the unknown and archaic as popular culture. More's the pity, in other words, that humanities education cannot demonstrate that its archaism goes well beyond the obsolescence, inefficiency, and irrelevance with which it is routinely charged by postindustrial business to that special kind of respect for the unknown fostered by historical awareness. Besides imitating the sciences by searching for unknown discoveries (a practice that fits well with the postindustrial religion of innovation), humanities education also truly does *re*-search: it recovers as much as discovers archaic history in a quest of remembrance, reflection, and judgment. It calls for critical timeouts from a world of the "known" that pales in significance to what is unknown *within* the known. More's the pity, in sum, that what business calls inefficiency and inattention to contemporary needs, but what education itself (at least in its best moments) believes is humane learning, cannot demonstrate that historical critique is a credible, public way of going walkabout. After all, as Corynne McSherry has shown in her insightful book on the special nature of academic "intellectual property," *Who Owns Academic Work?* education has historically had a set-aside status as the domain of "public" knowledge opposed to "private" enterprise (even if the boundary between the two has now become ever more unstable and uncanny).[12] The greatest part of "public," we may add, is history, which is what continually lapses from the span of copyright or patent into the commons of knowledge curated by the academy. More's the pity that the academy cannot show that caring about public archaism, and not just private regression, is cool.

How might academic critique based on historical consciousness join hands with cool irony focused on the latest style, band, or tech? Given the divorce between "professional" and "lay" reading that Guillory explores in his "Ethical Practice of Modernity: The Example of Reading," we might think, academic critique is always at risk of attenuation into abstract, bloodless skepticism, while cool irony is always at risk of materializing around the clichéd, media-dependent, yet nevertheless experientially visceral sense of the immediate.[13] Together, they might be something.

Here, then, is a prophecy about humanities education. In the relevant future, I believe, education will increasingly need to teach that *cool is a historical condition.* Cool has a history, and its core experience, however inchoate and unknown to itself, is *of* history. A first step would be to follow the lead of those cultural critics who have investigated the history of media, advertising, entertainment, fashion, and other forms of consumer culture. We might thus imagine the creation of a systematic curriculum in mass culture designed to give some ballast to the many mass-culture examples that instructors already infuse in their practical pedagogy (e.g., ad hoc comparisons to TV, journalism, movies, music, entertainment).[14] But teaching the history of consumer culture can only be a beginning. My contention throughout this book has been that the crucial inquiry now concerns "producer culture"—that is, the values of the postindustrial workplace that have increasingly spilled over into everyday life. Postindustrial "corporate culture" challenges humanities education because it is at once integral with the basic educational goals of "competence" or "skills" and usurpative of the other presumed educational goal of "culture." To teach a broader sense of culture in the age of corporate culture, therefore, must mean above all teaching that the contemporary instinct for technical competence need not be oblivious to the sense of history that is the primary means by which the humanities at once reinforce and critique culture. Technique cannot be surrendered up to the forces of productivity as a matter of purely elementary skills and competencies extrinsic to serious humanistic study. Rather, the line drawn in the nineteenth century between Cardinal Newman's "idea of the university" and the new polytechnic universities (exacerbated in the twentieth century by what C. P. Snow called the split between the "two cultures") must now be blurred to accommodate the possibility that there can be something deeply humane, and historically aware, about technique. "How to Use a Computer" perhaps should be a course in every humanities department taught by its most philosophically broad, theoretically advanced, and/or "cultural critical" faculty member (even if technologically undistinguished or even hopeless).[15]

After all, theorists have been intent since at least the time of the Russian Formalists on showing that the humanities can be methodologically technical (raising the ire of those who accept the need for technical "jargon" in every field of contemporary knowledge except the humanities). Ultimately, the harvest of this effort must be to equip educators to reverse the field by addressing the humanity of technique. The best way to do so is to bring to technique an awareness of archaic and historical techniques. The sense of technique and the sense of history can be integral with each other if both can be shown to play upon the perpetual tension between the archaic and the new.

Here are the kinds of questions to be asked in such an approach:

- How might knowledge workers be educated both in contemporary information technique (the collection, verification, and collation of data; comparative and numerical analysis; synthesis and summarization; attribution of sources; use of media to produce, manipulate, and circulate results) and in archaic and historical knowledge technique (e.g., memorization, storytelling, music, dance, weaving and other handicraft, iconography, rhetoric, close reading), with the ultimate goal of fostering a richer, more diverse, less self-centered sense of modern technical identity?

- What and how did people "know," for instance, when cultures were dominated technically by orality, manuscripts, or print?
- How did aspects of older technical regimes survive, adapt, and even flourish in succeeding knowledge regimes, such that, for example, oral culture today appears not just in the "secondary orality" of audiovisual culture but also in the e-mail, threaded discussions, chat, and other talky media of information culture?[16]
- In what equivalent ways will the culture of literacy survive in the age of browsing?
- How, in other words, is the progress of knowledge constituted from broad, diverse, and always internally rifted negotiations with historical knowledges, such that every "cutting edge" or "bleeding edge" innovation creates in its shadow not just a dark hemisphere of obsolete peoples ("residual," "subcultural," "throwaway") consigned to the social margin, but also a repurposing and recirculation of the knowledges of the people of the margin (the true bleeding edge)?

"The street finds its own uses for things," William Gibson famously said, referring to the wizardry with which today's neo-archaic "street" people across the global metropole repurpose high tech—for example, by appropriating communications gear or pharmaceuticals for criminal or otherwise deviant purposes.[17] But Gibson's cyberpunk motto also has a flip side: high tech finds its own uses for the street. High tech at its best, in other words, knows how to enfold within itself the methods, protocols, interface design, and often even mystique (or "myth") of the archaic.[18] Certainly, for example, the point-and-click interface of the GUI accommodates an older, manual stratum of human technical skill than the command-line interface with its total reliance on literacy. Or, more complexly, consider the trope of the oriental martial arts in cyberpunk imagination (e.g., the Japanese throwing star that is the totem of Gibson's protagonist in *Neuromancer*, the karate-obsessed hackers in *The Matrix*, the kung-fu fight genre in arcade computer games, and perhaps even the business cult, mentioned above, that arose around Musashi's *Book of Five Rings*).[19] Digital orientalism is only partly a symptom of deference to the postindustrialism of Japan, Inc., China post-WTO, and the other Far East economic "tigers." It is also a way to embed within high tech a sense for archaic—in this case, feudal—technique. As Darko Suvin has said about *Neuromancer*:

> Although Gibson's views of Japan are inevitably those of a hurried if interested outsider who has come to know the pop culture around the Tokyo subway stations of Shibuya, Shinjuku, and Harajuku, I would maintain there is a deeper justification, a geopolitical or perhaps geoeconomical and psyhological logic, in his choosing such "nipponizing" vocabulary. This logic is centered on how strangely and yet peculiarly appropriate Japanese feudal-style capitalism is as an analog or, indeed, ideal template for the new feudalism of present-day corporate monopolies: where the history of capitalism, born out of popular merchant-adventurer revolt against the old sessile feudalism, has come full circle.[20]

From the viewpoint of the West, that is, digital orientalism is the way to injection-mold archaism into high-tech civilization with a seamlessness for which the West currently has no other viable paradigm (with the possible exception of the Euro-medievalism of swords and sorcery that so fixates digital gaming culture). Whether Eastern or Western, feudalism is a means of imagining decentralized networking. Whatever the cruelties of its allegory (especially its romanticization of violence), feudalism more capably fuses the postindustrial and archaic horizons of experience than such alternative allegories as the disintegrated federalism of post-Soviet Russia, where high-tech archaism tends to appear to the West merely as falling Mir orbiters and decaying infrastructure.

Phrased in the archaic idiom of the most cool students and gamers of our time, then, my essential questions become these: How might humanities education best draw upon the deep—yet also deeply inchoate—reverence for historical technique simulated in what Willam Warner has called "kung-fu" hacking?[21] How can humanities education use historical knowledge to teach that in any adequate technical culture, technique—whether a karate strike or cyperpunk hack—is never just about efficiency? If the imaginary dojo is the only school in which the generations of cool are now willing to enroll without irony, how can humanities educators be their sensei?

Questions like these suggest a whole research and teaching agenda for the humanities of the future, for which I can here give only a local, sample portfolio. The portfolio consists of four of the Web-based educational projects I have been involved with in the last decade. One is my *Voice of the Shuttle: Web Site for Humanities Research* (initiated in 1994–95 and redesigned as a dynamic database site in 2001), which gathers humanities-related Web links in an organizational scheme intended to offer an academic alternative to mass-market "portals." Besides mapping the established terrain of the humanities by discipline, period, and so on, *VoS* takes special notice of those areas where the contemporary humanities engage such new or affiliated fields as "cultural studies," "cyberculture," "media studies," "technology of writing" (e.g., hypertext, hypermedia), and "science, technology, and culture."[22] The site thus argues implicitly that if the humanities are a kind of weaving—alluding to the myth of Philomela behind the "voice of the shuttle" title—such weaving is not just archaic. It is archaism in negotiation with contemporaneity.

The second project is the *Romantic Chronology* (started in collaboration with Laura Mandell in 1995–96).[23] This site uses database-driven Web technology to offer a view of what literary history might look like in the decentralized, dynamic era of the network—in particular, how a Romantic era famous for its own brand of dynamic decentralization (egalitarianism, revisionism, revolution) appears from the vantage point of our own age of creative destruction. What perspective do we gain on the work of William Wordsworth or Mary Shelley, for example, when we track it interactively alongside innumerable other works, writers, artists, and historical figures of the time in an information-rich matrix that makes the relational database not just a means but a paradigm of knowledge?

The third project is one I began in 1998 while reading in the literature of the new business: *Palinurus: The Academy and the Corporation (Teaching the Humanities in a Restructured World)*.[24] Named after Virgil's fallen helmsman, *Palinurus* offers bibliographies, reports on controversies, and other resources focused on the burgeoning, yet fraught, relation between universities and corporations. While educators are asleep at the wheel, the site suggests, business and businesslike university administrations have steered toward a new understanding of "knowledge" that humanities educators would do well to triangulate alongside their historical knowledge, so that both kinds of understanding can be drawn on the same map of the "global" together. To quote from the rationale of the project: "The site provides an interface through which educators can learn about the new world of business, and business reciprocally about contemporary higher education. Historical resources flank the 'new' for both business and academia in order to provide some perspective on the current rush to 'obsolete' or 'throwaway' the past in favor of 'workplace 2000.'"[25]

Finally, the fourth project is the NEH-funded *Transcriptions: Literary History and the Culture of Information* and its spinoff, the "Literature and Culture of Information" undergraduate specialization in the English major at University of California, Santa Barbara. *Transcriptions,* which I started in 1998 with several colleagues (especially Christopher Newfield, Carol Pasternack, Rita Raley [as of 2001], and William Warner), as well as graduate and undergraduate student assistants, has been the main focus of my practical work during the latter part of writing this book.[26] *Transcriptions* is a curricular development and research project that focuses on the relation between contemporary information culture and literature, including the literatures of such previous "information revolutions" as the ages of orality, writing, print, and early electronic media. The rationale of the project closely parallels my thesis here. I quote from the "About Transcriptions" statement on the Web site:

> The goal of [the Transcriptions Project] is to demonstrate a paradigm—at once theoretical, instructional, and technical—for integrating new information media and technology within the core work of a traditional humanities discipline. Transcriptions seeks to "transcribe" between past and present understandings of what it means to be a literate, educated, and humane person.
>
> Put in the form of a question: what is the relation between being "well-read" and "well-informed"? How, in other words, can contemporary culture sensibly create a bridge between its past norms of cultural literacy and its present sense of the immense power of information culture? . . .
>
> The Transcriptions curriculum hybridizes two broad themes that . . . are stronger in their composite than in isolation. One theme consists of the social and cultural contexts that presently make information so overwhelmingly powerful. For example, Transcriptions includes courses that give humanities majors a look at why large corporations in the U.S. now call themselves "learning organizations," "knowledge industries," or "information industries." . . . Or again, project courses address the status of media and literature in the present era. . . .
>
> The other theme centers on the historical contexts that allow contemporary culture to think about past culture (and specifically, literature) as itself an evolving technology of information. Thus the project includes courses on the immense changes that occurred in the notion of literature when oral and manuscript cultures evolved into early-print culture, when modern ideas of authorship and copyright arose in tandem with modern publishing and archival practices, etc. A course titled Scroll to Screen, for example, follows the evolution of literature from the age of scrolls to that of the World Wide Web page in order to examine the deep philosophical, cultural, and aesthetic assumptions that underlie the notion of reading a "page."
>
> By "transcribing" in this way between information culture and literary history, the project forges a middle ground between the present and the past where literature—and those who love it—can learn how to engage thoughtfully with information culture. The goal of the project's curriculum, in other words, is to argue against the reductive, often-heard notion that the

> age of the book is past and that the future belongs to information technology and digital media. Instead, the project is dedicated to the proposition that both the "well-read" and the "well-informed" can benefit from serious intellectual engagement with each other. Students of literature who participate in the project learn some of the contexts needed to participate professionally in information culture; and what they bring to such culture in return is the reservoir of values that a knowledge of the humanities imparts.

The undergirding thesis of these four projects is that while the humanities must begin to teach alongside literacy and critical analysis the technical skills needed to flourish in today's workplace (*Transcriptions* courses, for example, train students not just in using the Internet for research but in authoring Web resources), such "competence" is most valuable, both to individuals and to society, when wed to a full sense of the "technical" relationship between contemporary society and history. Technique is a medium not just of progress but of history. Orality, writing, print, and so on: these old "information revolutions" recently reviewed in a postdigital light by Albert Borgmann, Michael E. Hobart and Zachary S. Schiffman, and others have something to say to workplace 2000. They suggest that students can grasp the technologies and techniques of current knowledge work to best advantage only when they have the full resources—the craft, assumptions, context, and critical perspective—of past knowledge at their disposal. As in the case of the undergraduates in one of my *Transcriptions* courses who built a Web site about a medieval cathedral seen as an information machine (pilgrimages were the slow Internet of the time, they suggested, and the "sites" of the cathedral and shrines on the way to the cathedral were the interface for a vast network of knowledge about the world), the present and past have an uncanny way of illuminating each other.[27]

Nor is it just a *concept* of the relation between past and present knowledge techniques that is at issue. The negotiation between history and contemporaneity I call for is at stake also in the very process—the practice—of humanities education. Perhaps the single most compelling lesson learned by participants in the projects listed above is that simply discovering how to teach and learn older knowledge techniques in collaborative, networked, and information technology–rich environments is the most rewarding challenge. Consider, for example, that the standard paradigm for humanities research and teaching has been individualistic in ways that curtail the full impact of new technologies. In the humanities, research is primarily the activity of the solitary scholar reading in a library carrel or writing in an office or home study (with occasional, equally solitary research trips to resource collections elsewhere in the world). Attending conferences and presenting papers, the social side of humanities research, is most often a conceptually separate activity—the presentation of research. So, too, humanities teaching ordinarily consists of a solitary instructor facing an individual student or assembly of monad students. In a strictly linear circuit of communication, the teacher asks and the student responds. Or the student asks, and the teacher responds. Or, again, the teacher assigns; the student writes; and the teacher responds. "Discussions" and "seminars" promise commonality in the production of knowledge, but in their best moments even strong seminars cleave to the Socratic model of intense one-to-one dialogue between instructor and student (with the one-to-one relation shifting successively to other students to foster a sense of shared conversation). And to fulfill the function of certification, of course, seminars are constrained in their worst moments to the reality of solitary testing and evaluation (individual assignments and tests, with any team assignments a minor factor in the grade). In such a paradigm of scholarship, new technologies are not irrelevant, but they are marginal. Having a faster computer for word processing or solitary Web browsing in one's office as an instructor, for example, leads to incremental changes in the quality, quantity, and speed of research and teaching, not to paradigm change. The essence of networking is bypassed.

By contrast, the kind of projects I instance above have had to reconfigure humanities research and teaching in ways that borrow something from the sciences and social sciences (with their grant-driven, group-based projects) as well as postindustrial business (collaborative "teams") to improvise different ways of working.[28] A simple example: from 1994 to 1998, *VoS* was a solo, unfunded effort. But thereafter, the effort of maintaining a "portal" for the humanities proved unrealistic for an individual. Funding was raised to permit the participation of graduate and undergraduate research assistants. Increasingly, *VoS* became a managed, team-based labor. Or to cite a more complex example: the *Transcriptions* project involved multiple faculty members working with a changing pool of graduate student research assistants, undergraduate (and later nonstudent) graphic design specialists, and a department staff responsible for budgetary, technical, and other support. Such group-based, networked projects bring elements of postindustrial labor to the humanities in a way that challenges humanists to imitate business management and also to contribute something distinctly humane to such management. The underlying practical question for such projects is the following: How can scholars implement networked collaboration in ways that make sense for the humanities? How can networked collaboration be made to work in an academic environment that is in cardinal ways different from business? (Examples of differences: student research assistants are never just employees; collaborators work both on and off site with an uncontrolled variety of equipment and software; security measures must accommodate a larger sense of open access and intellectual freedom; there is never enough technical support; there is enormous tolerance for failure as a symptom of learning.) In such networked humanities projects, the paradigm is no longer just the teacher and student facing each other in an intense, bipolar circuit of interaction mediated by the "text." Rather, the paradigm changes to one in which the teacher and student stop looking through the text just at each other, turn shoulder to shoulder, and both look at a different kind of project they are building together—one that, as in the case of a Web site, allows them to look *through* it to a public able to look in reverse at *them*. In this situation, new technology does make a difference. The humanities are exposed, for better or worse, to a networked society.

It is difficult to express in short compass what a difference the new paradigm I have sketched makes in the felt experience of humanities teaching—a difference that is by turns both frustrating and liberating. Nothing in my own experience suggests that the humanities should do away with the solitary, scholar-writing-alone or teacher-faces-student model, since at its best it continues to encourage a certain kind of close and deep reflection that is hard to imagine in a different scenario. But much is to be said for complementing the traditional humanities model with the technologies and techniques that represent the new networked model. There exists in the process and technique of education a broad, blank canvas on which education can paint, not so much a timid imitation of postindustrial business (teams, networking, just-in-time) as a particular *view* of the new millennium of work—one enriched by the sense of history.

Here I will link from conclusions drawn from my own projects to those of Jerome Christensen, whose practical experience helping to found the Johns Hopkins Center for Digital Media Research and Development allows him (in the chapter of his book *Romanticism at the End of History* entitled "Using: Romantic Ethics and Digital Media in the Ruins of the University") to ask how the liberal arts in the university can "survive and prosper as a hopeful anachronism" in the age of professionalism and corporatism. His answer, in part, is that in the field of multimedia production, university studios or labs can exploit their uniquely diverse pool of talent, special economies of scale (smaller but more diverse audiences), and tax-exempt status to occupy a high-value "niche" in production that is impractical for major corporations to fill. Crucially, such niche production in Christensen's vision uses advanced techniques in ways that remember old, artisanal work patterns. "The university studio or lab," he says, "can prosper at the artisanal level of multimedia development as long as niche production, niche marketing, and interstitial maneuverability are appreciated tactics in the continued ethical project of finding users." "Digital media production answers to the Romantic dream of a craft that would be sensuously pleasurable and socially useful—one that the impatient intellect would not soon be tempted to quit." Christensen thus imagines a historically aware mode of labor that is at once mindful of the past and vital because it can be renovated by—and help renovate—our contemporary work of "flexible specialization." "Always at hand," Christensen says, "craftsmanship is unbound to any particular form of social organization; a craft can be effectively practiced regardless of the mode of production."[29]

Chapter 11

Destructive Creativity

The Arts in the Information Age

Any inquiry into the future of humanities education of the sort pursued in the last chapter is incomplete, as I have suggested, without a parallel inquiry into the formal and informal aesthetic education of the cool. The special potential of the arts in the age of knowledge work may now well be to complement the humanities lesson that "cool has a history" with the crucial inverse of that lesson: *history can be cool.* The process of history—or, viewed another way, of socially significant mutations, remixings, and destructions, together with their representations (including, but not limited to, narrative)—can itself be cool in a way that propagates outward beyond "teaching" and the "academic" into the main zones of cool. These include the zones of edgy aesthetic experience where avant-garde art, music, film, fiction, poetry, dance, games, fashion, design, Web design, and so on, at once renovate and critique general culture. In short, cool *à la mode*—the cool of the instantaneous present—can no longer be the exclusive obsession of knowledge workers. In the age of "creative destruction," the sense of history will also need to be cool. In a manner much deeper and more disturbing than *la mode rétro,* historical awareness—transformed by the arts into a uniquely kinked, nonmainstream understanding of the mutational processes relating creation to destruction—will need to embody what I have called the ethos of the unknown.[1]

After all, how can humanities education reach beyond the walls of the academy to make a difference in the general culture of "cool" when it cannot even seem to penetrate the cool sang-froid shielding its own students? The answer can be suggested in the form of another prophecy. In the near future, I believe, the humanities will need to ally strategically with the contemporary creative and performing arts, beginning with the arts on campus. Where the gap in disciplinary protocols between humanities and arts departments seems too far to span in a single leap, at least some initial measures might be taken that are within the normal institutional reach of the humanities. For example, such older, text-based humanities fields as literature and history might ally with such newer humanities fields as film studies or media studies—fields, that is, in which historical/critical inquiry blends more easily with pragmatic approaches to the world of avant-garde and commercial art (as when artists or directors become guests in classes, participants in conferences, or employers of interns). Especially relevant are those film or media studies programs (along with just a handful of other humanities programs at this point) that are now incorporating digital "new media studies."[2] Ideally, however, humanities departments would also collaborate directly with art studio, design, visual arts, creative writing, music, and other creative or performing arts programs (including the new breed of "media arts and technology" programs) as these are now exploring new media and collaborating in their turn with engineers and scientists. The creative arts, we recall, were conspicuously absent from the consciousness of the humanities during the latter's recent "interdisciplinary" gold rush toward the social sciences (anthropology, linguistics, sociology), sciences (e.g., complexity theory), and professional disciplines (e.g., critical legal studies). Yet in the institutional organization of the liberal arts, the humanities and arts continue to be conjoined no matter how much the logic of the word *and* in such university nomenclature as "humanities and fine arts" has emptied of meaning. Precisely this conjunction now acquires fresh cogency as the shared exploration of new media opens what amounts to a land bridge between the continents of the humanities and arts.[3]

The case for establishing a working alliance between the contemporary humanities and arts can be made in two reciprocal directions. In one direction, the humanities clearly stand to benefit from adapting some of the technologies, techniques, and general stance toward society of the arts on campus. Technology is the easiest issue to grasp, though perhaps deceptively so. If from the viewpoint of producers cool now appears as "design," while from that of consumers it appears as high-tech or special-effects "entertainment," then clearly one approach the humanities can explore is how to render its historical knowledge with the coolest technological designs and effects being improvised in art studio, architectural, or music digital labs.

Imagine presenting literary, cultural, historical, or philosophical worlds in high-resolution, multimedia, interactive, or virtually real form (e.g., in the breathtaking manner of the architectural and archaeological sites that have been rendered as virtual walkthroughs).[4] An instructor, for example, could take students into an immersive CAVE virtual reality environment to walk the Lakes with an interactive William Wordsworth "bot."[5] Or again, imagine playing the humanities as a digital "game" (akin to the "Ivanhoe Game" of literary interpretation created by Johanna Drucker and Jerome McGann, or the Villa Diodati MOO and Frankenstein MOO environments associated with the Romantic Circles project) and thus tapping into the increasingly powerful culture of gaming among students.[6] Imagine that the humanities could actually be the sensei in what I called the virtual dojo of cool. (What *The Matrix* is really about, after all, is the education of the protagonist, Neo, in the grand old sense of *bildung*—but refreshed to accommodate the just-in-time knowledge of our times. Need to learn how to do kung fu or fly a helicopter? Download the mindware.)

Of course, there are impediments to putting "Hamlet on the Holodeck," to borrow the title of Janet H. Murray's perceptive book. Many of these are practical. Developing instructional technology of the sort I indicate, for example, requires a substantial initial investment of resources and expertise. Moreover, the first results will likely be experimental courses that do not automatically "scale" to large numbers of students and across the curriculum. Whether such technologies are deployed primarily within the physical classroom or in tandem with remote learning, they almost certainly will not initially support the more simplistic, cost-driven models of "distance education."[7] In addition, sustaining and disseminating such pedagogical experimentation will require a high level of overall institutional commitment, organizational innovation, and sometimes academic/industry collaboration of the sort long encouraged by the EDUCAUSE association (created from the merger of Educom and CAUSE). If nothing else, institutions will need to question the assumption that only science and engineering faculty need start-up funds for labs, equipment, and technical support staff.

Beyond the practical impediments, there are also significant obstacles rooted in the formats, approaches, and underlying philosophies of humanities teaching. Of most concern, perhaps, is the doggedness of humanities instructors who are sometimes more than just unadapted to new classroom technologies. In imitation of Socrates rejecting the invention of writing in the *Phaedrus*, they are actively set against such developments. They believe that digital technology is at best a distraction and at worst the agent of a pervasive techno-modernity that both the "traditional" and "revisionary" humanities now agree is the object of critique. But the third-person *they* is fallacious here. The resistance I indicate is so deeply ingrained that it is not just alleged "Luddites" but also "early adopters" of new technology (myself included) who harbor lingering suspicions that a 3D simulation of the Sistine Chapel or an artificial intelligence Hamlet program (whose Turing test must be whether it can successfully confound the digital clarity of "to be or not to be") could only be so much sugar-coating, only a way to make knowledge superficial enough to be swallowed by the consuming masses. Or worse, isn't busy, high-tech distraction the Trojan horse through which business-oriented techno-rationality can invade the last redoubts of humane culture in the academy? This redoubt is the humanities, where teachers and students rehearse pre-industrial ideals of "discipleship" and "apprenticeship" by practicing all the old speech and writing knowledge technologies (e.g., "reading," as intoned in the reverential, hushed voice of the New Criticism or deconstruction).

Ultimately, though, my own experience with undergraduates who report that they have trouble even *seeing* "plain text" amid all the cool multimedia and interactive presentations on the Web indicates that in our current, competitive media environment the attitude of the humanities toward "superficial" presentations of knowledge must change.[8] The now unpalatable "surface" coating of the traditional humanities themselves—their interface—needs to be reworked so as to be turned around (in the way that every interface has two sides) into not just a lookalike of contemporary techno-modernity but also an effective means of thinking critically about modern industrial and postindustrial life. This is especially true when we remember that from Horace to Sir Philip Sidney one of the classic legitimations of the humanities—specifically literature—was precisely that it added a sugar-coating to truth. It "taught" or "informed" by "delighting."[9] The tendency of text-centered scholars to dismiss new media and browsing as merely facile practices of knowledge is itself facile without serious consideration of the unique kinds of thought—and also antithought—native to the delightful new media.[10] The issue is not surface versus depth, in other words, but surface versus surface. Which surface or interface should the humanities adopt to teach and delight most effectively? There is no necessary reason why interfacing through rigorous practices of close "reading" cannot continue in a virtual reality space, though there is also no reason why old standards of rigor invested in particular knowledge interfaces cannot themselves be critiqued by new interfaces in which quick, hypertextual jumps are often more incisive than the most patient sequential or hierarchical probing.[11]

The issue of technique is just as interesting as that of technology, though less easy to observe because it exists in a milieu not just of evolving machinery and software but also of evolving practice. An illustration of what I have in mind is the computing "studio" I developed in my English Department for the Transcriptions Project in 1998. The studio (which features computers in a U-shaped ring around one end of the room and a seminar table with digital projector at the other end) was intentionally designed to allude to the ambience, configuration, and social milieu of my campus's innovative Art Department digital studio—that is, not the usual "computer lab" mimicking data-processing back-offices of the mainframe age but a more intimate, flexible, and collaborative environment akin to an artist's or architect's studio.[12] The kind of work conducted in this environment attests that the arts can teach the humanities not just how to do particular things, but how to do things in general. They can set the example of a whole paradigm of collaborative technique whose genealogy may be quite old (extending to the eras of discipleship and apprenticeship I referred to earlier) but which has now come into adaptive synchrony with "networking." Put simply, the arts seem predisposed to know how to do things in a decentralized and networked collaborative environment. It is not accidental that such artist "collectives" as the University of California Digital Arts Research Network (DARnet) have led the way in deploying information networking as both a medium and theme in their networked, peer-to-peer, "agent," or collaborative database projects.[13]

But the arts may have the most to teach the humanities in regard to what I called "general stance toward society." As I learned through a sometimes awkward apprenticeship in extramural fund raising (related to various digital projects) as well as through involvement in my university's Public Humanities Initiative, the role of the humanities with regard to the public is now one of our most fraught, but also most potentially rewarding, concerns.[14] While there have been various intriguing initiatives across the nation designed to stake out a position for the academic humanities in public life, the arts offer a particularly compelling, complementary model.[15] As witnessed in the number of artists and creative writers who cross fluidly between academic and public domains, the arts have traditionally had more direct and flexible access to general culture than the professional humanities. Particularly striking in our context is how such access is now being refreshed by the information revolution. In extending their work into new media, university-based artists (and their colleagues outside the academy) are producing tools, content, and students of relevance to the high-tech and entertainment industries responsible for shaping information culture.[16] Perhaps of special interest to humanities scholars, such artists are "relevant" in a way that does not necessarily surrender critical perspective. However contrarian or "shocking" the arts may be, they often have far greater appeal to (or earn more outrage from) audiences and donors than the professional humanities. Of course, in a fuller study this rosy view of the arts would need to be qualified, for even casual observation indicates that the life of the arts on campus is not always easy in relation to either the campus or the public. Indeed, there may be a reverse yearning by the arts for the relative stability of the larger humanities programs in the ecology of the university, as well as for the very insulation of the humanities from the whims of funding agencies and the public. Nevertheless, in an era when administrators, legislators, and businessmen increasingly tell the humanities to be more "entrepreneurial," the alternative model of public engagement and fund raising embodied in the contemporary critical "artist" as opposed to "entrepreneur" is worth exploring.

The humanities thus have something to learn from the contemporary arts about how to reach the great, lay audience of the cool. But—turning to the second, reciprocal direction of my prophecy—just as the humanities can learn from the contemporary arts, so these arts (both on and off campus) have something precious to gain in return from the humanities. They can gain the constitutively historical rationale that may now be the only legitimate foundation for avant-garde aesthetics in the new millennium—for an avant-garde of mutations, remixings, and destructions that, far from being exhausted in the aftermath of modernism and postmodernism, has a new claim to life in opening the eyes of the everyday avant-garde of "cool" to the full, humane, and therefore also historically aware potential of its "ethos of the unknown." We may live in the millennium of the "posthuman," in other words, but even the posthuman can be humane if the cool are reminded that *post-* does not transcend either the prehistorically and historically created "body" (which Katherine Hayles foregrounds in her critique of the posthuman) or the body politic of society.[17] Teamed with the humanities, the arts can have a say in that. They can have a say in embodying human history within what would otherwise be a merely free-floating, inhumane posture of cool.

Here there opens to view a topic so expansive and compelling that it really deserves a book-length examination of its own. We will need to be content, therefore, with just scouting the main line of argument. Remember, as I suggested earlier, that the rationale of contemporary knowledge work is "creative destruction," with the emphasis on "creative" and almost no serious reflection on "destruction." The question of a future aesthetics, we may say, is the question of the general legitimation of art in such an age of creative destruction. What is the function of the creative arts in a world of perpetually "innovative" information and knowledge work? Of course, the multifariousness of the forms, media, practices, and views of the contemporary creative arts (including literature) is remarkable. Even in quickly perusing some of the international digital arts events each year, for example, one's breath is taken away by the sheer range and vitality of the arts both inside and outside the academy as they adapt various Bauhaus, pop, minimalist, conceptual, and subsequent "post-" and "neo-" art principles to new media. So, too, one need only scan the voluminous *Directory* of resources on the Electronic Literature Organization's Web site or listen in on the organization's conferences and online events to appreciate the multiplicity of ways in which creative writers are using digital media to try out new genres, writing processes, and publishing methods. No adequate account of such variety can be rendered here.[18] Nor can there be adequate discussion of the other, seemingly paradoxical side of the equation: that despite its splendid variety, so much of contemporary art and literature has a similar look and feel descended distantly from the collages and cut-ups of the modernist avant-garde—for example, assemblage, pastiche, sampling, hypertext, appropriation, mixing, creolization, or, to cite one of the dominant metaphors of recent literary history as well as hypertext fiction, "patchwork."[19] As I have said, it is all mutation, remixing, and destruction. We cannot here redub the whole debate about the transition of modern into postmodern art, that is, and then update that discussion to take account of the current transition toward what Lev Manovich has called the "language of new media" and the aesthetics of the "database."[20]

What is possible in our present compass is just to address the *ideology* of art in the age of knowledge work. By this I mean the governing explanation that each age posits to marshal and explain its aesthetics, and which our own information age is now improvising in its own idiom. In classical times, for instance, one of the governing ideologies of art was mimesis, or the normalization of art and literature as the "mirror" of truth (though a mirror powerfully distorted to favor "universal" rather than realistic truth). In the era of Romanticism the governing aesthetic was determined by a struggle between what M. H. Abrams memorably called the "mirror" and the "lamp"—that is, between outer-directed mimesis and inner-directed subjective expressiveness. Of course, individual artists are diverse in relation to the unifying aesthetic rationales of their time. But such rationales are nevertheless important in the way that recent ideologies of corporate culture are important: they set the agenda and starve or reward individual practitioners. Ideologies may not govern the particular instance, but they define the overall parameters in which particular instances arise, compete, rebel, and legitimate themselves.

Aesthetics in the age of knowledge work is now being defined by several contenders for a dominant ideology that have not yet fully articulated themselves or negotiated their intramural relationships.[21] For example, in a fuller study we would need to explore the new media aesthetics that together make up what might be called a new "picturesque" for our time. I mean by this the aesthetics of mutation and remixing that recreate through new technologies something like the art of quintessential hybridity and chance originally native to the eighteenth- and early nineteenth-century picturesque, which Sidney K. Robinson in his perceptive *Inquiry into the Picturesque* (1991) calls an art of entropic "mixture."[22] One such neo-picturesque aesthetic may be found in complexity or self-organizing art, in which algorithmic, networked procedures churn in aleatory or self-organizing fashion to produce emergent patterns. An example is George Legrady's haunting *Pockets Full of Memories* project, which existed both in a physical installation at the Centre Pompidou in Paris in 2001 and in a Web instantiation. Visitors to the installation scanned in objects they had about their body (utensils, pictures, coins, body parts) and answered a brief set of questions about the nature of the objects. Then a "self-organizing map" or neural-network array dynamically positioned the scanned images on a digital mural in unpredictable, surprising, and emergent patterns of proximity. Similarly neo-picturesque is what can be termed "agent" art. An example is Robert Nideffer's *Proxy* project, which connects software avatars in a peer-to-peer networking system to produce semi-autonomous "agent" behavior. Or, again, we may consider a further variant of the neo-picturesque that might be named "transcoding" art (to adapt Manovich's idiom in *The Language of New Media*). In its most recent and ambitious forms, transcoding is not just the "mixing" or "sampling" that has influenced so much contemporary popular media, but database or data array art—that is, art that explores the creative and cultural possibilities of relational databases, multimedia production programs (e.g., Director, which can hold data objects in timeline arrays), or XML encoding. The distinguishing trait of the new database or databaselike media is that they contain information in structured, logically related "chunks," allowing it not only to be queried or read by machine but to be extracted algorithmically from embodied presentations and re-presented in other formats and remixed combinations.[23] Other digital aesthetics that partly overlap with the "new picturesque" might also be mentioned, including the aesthetics of "gaming," whose playful blend of narrative, interactivity, and simulation is now inspiring artists to adapt the paradigm for aesthetic purposes.[24]

Rather than try to survey the variety of possible new media aesthetics, however, I will take the shortcut of bringing the argument to its most extreme verge by looking beyond the new picturesque of mutation and mixing to the ultimate form of such mutation and mixing: what may be called the new *sublime* of "destruction."[25] Or more fully, we can call this sublime "destructive creation," the critical inverse of the mainstream ideology of creative destruction. The most austere and terrifying implementation of such an aesthetics is what I alluded to in my introduction under the name *viral aesthetics*.

Consider again how cool occurs in those edgy zones where avant-garde aesthetics, as I said, at once renovates and critiques general culture. In actuality, such a description of the contemporary arts and literature is now bankrupt. In the age of corporatized "creativity," the modernist and originally Romantic premise that critique goes hand in hand with "renovation," which is to say "innovation" and "originality," is now dysfunctional as an overarching aesthetic, no matter how functional creativity may be at lower levels of ideology (e.g., as motivation for individual artists and authors, as an argument for funding, or as the rationale for an arts festival). Reading through Hannes Leopoldseder's excerpts from two decades of his contributions to official catalogues for the Ars Electronica festival and Prix Ars Electronica, for example (collected in the foreword to the retrospective volume, *Ars Electronica: Facing the Future*), gives one a sense of how productive artistic "newness" continues to be at the level of the practical circuit of art making, exhibition, publicity, and so on, but also how monopolistic, predictable, and thus paradoxically unable to do new work it has become at the higher level of a public ideology—even, or perhaps especially, in the case of the newest "new media art." The following is a sampling of Leopoldseder's catalogue text from various years of *Ars Electronica*:

> (1979) Thus this event for electronic arts and new experience assumes a character of incalculability, of risk, of daring to try something new. . . . Ars Electronica poses a challenge to artists, technicians, cultural critics, and ultimately to the public encountering new forms of expression in art.

> (1980) Ars Electronica is intended to set signals for the future. Not only as an attempt to link tradition and avant-garde, but also as a cultural experiment seeking to influence the cultural awareness of the public in new ways.

> (1982) Ars Electronica begins 6,308 days before the year 2000, it ends 6,300 days before the year 2000. . . . The days in between belong to the future, to the stimulation of forms of art for the information society.

> (1984) Innovations and changes have one thing in common, a new raw material: not gold, not steel, not petrol, but information and knowledge. Information is the currency of the new age.
>
> (1986) Computer culture calls for a new alphabet, a new language, a new way of thinking. . . . Computer culture allows new experiences in art and culture.
>
> (1987) Computer culture thus characterizes a new step in culture. . . . Computer culture induces radical changes in the history of culture. All the cultural techniques hitherto used—reading, writing, calculating—may be taken over as such by the computer. Computer culture asks for a new evaluation of human capacities, asks for a change in thinking, enables a new start.
>
> (1989) In 1979 Ars Electronica appeared as a time warp into the future. . . . From the beginning, Ars Electronica has been open to signals from the future, open to experiments.
>
> (1991) For Ars Electronica, a festival that has from the beginning always understood itself in a relational network of art, new technologies, and society, a new era has begun.
>
> (1996) We stand at the dawn of this new era. There is much that cannot yet be seen or identified, much is still hidden; no one really knows yet where the digital revolution will lead in a new century.
>
> (1997) "Where is the new place of art?" . . . one new place of art will be cyberspace. The Prix Ars Electronica intends to provide a forum for forging this new place of art.[26]

What is clear in listening to this drumbeat of the "new" is that if Ars Electronica intends to "influence the cultural awareness of the public in new ways," then it must fail precisely to the extent that it is unable to articulate an ideology of artistic "newness" that is discernibly different from the *old* ideology of newness the public knows best as the driving engine of industrial and postindustrial capitalism. Insofar as the avant-garde is indeed exhausted and dead, in other words, then the "make it new" credo that has been its signature since even before the era of Ezra Pound is its palpable corpse, no matter what "new media" attempt to reincarnate it. The truly new art now propagating within that corpse is a viral aesthetics that at once mimes and critiques knowledge work so as to circumvent the corporate tumor that "creativity" has become. Viral aesthetics invents an alternative mode of productivity resident within the other, dark lobe of contemporary creativity—to coin a term, *destructivity*.

I am influenced here by Dario Gamboni's expansively conceived work *The Destruction of Art: Iconoclasm and Vandalism since the French Revolution* (1997), which teaches how intimate destructivity has always been to creativity in the history of art even before we arrive upon the contemporary era. Though at first glance Gamboni's detailed study of art destruction from the French Revolution through the fall of the Berlin Wall might seem too historically miscellaneous to yield a sustainable general thesis, the synthetic conclusions he arrives upon are quite powerful. Gamboni shows that the history of art is incomprehensible without a matching history of "de-arting," the general form of the variously motivated (political, religious, aesthetic, and so on) devaluations of art among which spectacular instances of physical destruction and vandalism are merely the exclamation points.[27] As the history of architectural "preservation" or "restoration" shows, Gamboni argues, the notion of conserving art is sometimes paradoxically equivalent to the wholesale destruction of existing art. (John Ruskin put it this way: "Neither by the public nor by those who have the care of public monuments, is the true meaning of the word *restoration* understood. It means the most total destruction which a building can suffer: a destruction out of which no remnants can be gathered: a destruction accompanied with false descriptions of the thing destroyed" [quoted in Gamboni, *Destruction of Art,* p. 217].) Art devalued by individuals or regimes as nonart; art destroyed for being pagan or primitive; art torn down for its politics (e.g., statues of Lenin after the fall of communism); art slashed or sprayed in museums for a variety of reasons; even art "thrown away" because it is supposedly not recognized as art: all these acts of de-arting are the other side "of the same coin" of creative art. The destruction of art, Gamboni concludes, accompanies art "like a shadow," like a "repression" (*Destruction of Art,* pp. 331, 336).

Of particular relevance to us is the de-arting that may be seen to be the immediate precursor of contemporary artistic destructivity: modernist (and postmodernist) auto-iconoclasm, or the attack of art on itself. In particular, as we learn from Gamboni's chapter "Modern Art and Iconoclasm" (as well as from other accounts of modernist art), it was the avant-garde arts from the early to late twentieth century that prepared the ground for current viral aesthetics by articulating an increasingly generalized ideology of auto-iconoclasm—that is, a stance not just against individual artists and works but against art *in toto*.[28] This hostility faced both externally toward the institutionalized reception circuit of the arts (the galleries, museums, and so on, together with their values) and internally toward the act of art itself. Here it will be useful quickly to review Gamboni's argument in order to come to a point of divergence that will allow us to see the even rougher beast now slouching toward Bethlehem. Gamboni's primarily historical exposition of twentieth-century art to the 1980s suggests the following sequence of escalating auto-iconoclasms.

First, there was the great, angry modernist repudiation of past art, which in its extreme forms imagined that true art could only arise from the immolation of old, especially nineteenth-century, art. As Gamboni notes, for instance, J.-K. Huysmans in the late nineteenth century suggested burning down the great buildings of Paris to express the view that "fire is the essential artist of our time and that the architecture of the [nineteenth] century, so pitiful when it is raw, becomes imposing, almost splendid, when it is baked" (quoted in Gamboni, *Destruction of Art*, p. 236). We can add as chorus the angry young poets and critics of the early twentieth century—T. E. Hulme, for example, or Pound. The originality of modern art, apparently, had to be *ex nihilo*, or at least to draw from safely distant classical or Eastern sources. In the era of the world wars of the twentieth century, art—at least in its loudest manifestos—was a scorched-earth policy.

In a second phase, the destructivity of modernism became *professional* and thus apparently part of the other modernism of industrial capitalism. We earlier witnessed the process by which graphic design in the era of the New Typography fomented a new profession led by angry young designers who sternly repudiated nineteenth-century design. Such professionalization was part of the general process by which the arts in the early twentieth century uncannily molded even their most annihilating "critique" to the exact contours of mainstream industrial competition. As Gamboni reflects about Huysmans's inflammatory iconoclasm, such a statement was "typical of a writer and aesthete actively involved in the new 'dealer-critic system'" in which competition made "innovation a central element of aesthetic value, thus encouraging the collective strategies of groups and '-isms' and inaugurating what would be termed a 'tradition of the new'" (*Destruction of Art*, pp. 256–57). Just as the signature of modern capitalism was "Inc.," we may say, that of modern art was "-ism." Modernism and all its subordinate "-isms" (right up through the dizzying proliferation of postmodernisms associated with particular architectural firms that Charles Jencks parades in his books on postmodern architecture) was at once the mimesis and critique of a modernity whose trademark was its difference from the past.

Third, modernist art universalized its iconoclasm as a critique not just of past art but of the very idea of Art as it had become institutionalized in museums, galleries, and the art establishment—that is, the dealers, managers, and patrons that stood in as a proxy for the larger circuit of capitalism within which artists were at once complicit and critical. As Gamboni says, "With Dada, the rejection of past art was radicalized into an overall condemnation of art as part of the values and civilization that the First World War revealed to be false and destructive, thus requiring above all desecration and 'iconoclasm'" (*Destruction of Art,* p. 259). Marcel Duchamp's Readymades are now the canonical paradigm of such generalized (at once artistic and industrial) auto-iconoclasm—e. g., the infamous industrial snow shovels or urinals posing as art objects in the service of opposing art objects. Emblematic is a note that Duchamp wrote c. 1911–15 and later included in his *Green Box* of 1934: "RECIPROCAL READYMADE / Use a Rembrandt as an ironing-board" (Gamboni, *Destruction of Art,* p. 261). Huysmans may have imagined an apocalyptic auto-da-fé of art; but Duchamp imagined instead just an iron representing everyday, commercialized life.

Finally, modernist art and its post-1960s successors (conceptualism, happenings, installation art, performance art, and so on) installed auto-iconoclasm at a level of pragmatics—of the *act* of art making—that even more thoroughly radicalized universal destructivity. The notion of art dissolved into a performative display of art making that included the process of unmaking, or what Gamboni calls "the destruction of art as art" (*Destruction of Art,* p. 264). As early as the 1930s, Picasso had said: "Previously, pictures advanced to their end by progression. Each day brought something new. A picture was a sum of additions. With me, a picture is a sum of destructions" (quoted in Gamboni, *Destruction of Art,* p. 264). Later artists self-consciously displayed such progressive destruction as their very art "object" (or conceptualist anti-object). Take, for example, Robert Rauschenberg's *Erased de Kooning Drawing, 1953.* As Gamboni describes, Rauschenberg had asked Willem de Kooning, "whom he regarded as 'the most important artist of his time,'" for an important drawing that he could erase. During four weeks of work, Rauschenberg then erased de Kooning's drawing, leaving just "traces of ink and crayon" (Gamboni, *Destruction of Art,* pp. 268–69). This was *his* work, or act, of art. So, too, consider the "happenings" in the 1960s that specialized in the performance of destruction—for example, Arman's (Armand Fernandez's) *Conscious Vandalism* (1975), in which the artist used a club and axe to destroy a "lower middle class interior" (Gamboni, *Destruction of Art,* pp. 266–67). Many similar acts of destruction and artistic auto-destruction in the 1960s and 1970s could be cited.

But it is the mechanization or technologization of such "action art" from the 1960s on that is especially prescient in our context. Exemplary is the work of Jean Tinguely, whose *Homage to New York* (1960) featured a Rube Goldberg–like machine designed to automate the auto-destruction of art (Gamboni, *Destruction of Art,* pp. 273–74). The crowning instance of such technologized iconoclasm was the work of another artist of the 1960s, Gustav Metzger, whom art historian Kristine Stiles, one of his leading explicators, has placed at the head of the school of "Destruction Art."[29] In a series of conceptual statements supplemented by demonstration works and events (most famously, the "Destruction in Art Symposium" he organized in London in 1966), Metzger propounded an art of spectacularly corrosive self-destruction culminating in a vision of high tech–assisted destruction. As Metzger explained in his "Manifesto Auto-Destructive Art" (1960):

> Auto-destructive art re-enacts the obsession with destruction, the pummelling to which individuals and masses are subjected.
>
> Auto-destructive art demonstrates man's power to accelerate disintegrative processes of nature and to order them. . . .
>
> Auto-destructive art is the transformation of technology into public art.[30]

One of Metzger's best known "works," therefore, was an "acid action painting" that involved flinging and spraying acid on three nylon canvases of various primary colors.[31] But another purely conceptual, unrealized work Metzger proposed in a 1965 lecture at the Architectural Association perhaps best captures the way he foreshadowed artistic destructivity in the high-tech age. The "project I would like to consider," he says,

> is in the shape of a 30 ft cube. The shell of the cube is in steel with a non-reflective surface. The interior of the cube is completely packed with complex, rather expensive, electronic equipment. This equipment is programmed to undergo a series of breakdowns and self-devouring activities. This goes on for a number of years—but there is no visible trace of this activity. It is only when the entire interior has been wrecked that the steel shell is pierced from within. Gradually, layer after layer of the steel structure is disintegrated by complex electrical, chemical and mechanical forces. The shell bursts open in different parts revealing the wreckage of the internal structure through the ever changing forms of the cube. Finally, all that remains is a pile of rubble. This sculpture should be at a site around which there is considerable traffic.[32]

As Stiles observes, Metzger's destruction art may be seen in one regard to witness the Holocaust. Metzger's work is "the constant public and social reminder of destruction, its agents, processes, and results. Precisely twenty years after Metzger was sent to England in 1939 at the age of twelve, when his family was arrested by the Gestapo in Nuremberg, he formulated his theory."[33] But in our context, Metzger not only looked back to the then state-of-the-art gas ovens and Panzer tanks of the World War II era but also uncannily *forward* to the state-of-the-art technology of the corporate regime. As exemplified in the "cube" packed with "electronic equipment" that can be "programmed," Metzger's engine of auto-destruction was in essence a computer seen from the perspective of a system administrator. It was the premonition of a technology whose very operation leads to self-corruption or, through contamination over the network, to the especially uncanny form of self-corruption (the utilization of a host system's resources against itself) triggered by viruses.

Cued by thus fast-forwarding Metzger to the age of computer viruses, we are now ready to move beyond the scope of Gamboni's argument to the contemporary scene. To do so requires making one simple, but immensely powerful, alteration. We need only delete the prefix *auto-* from the description of art as "auto-destruction" or "auto-iconoclasm." The most avant-garde arts of the age of knowledge work break out of the confines of the arts to perform "destructivity" in corporate and other dominant social sectors directly. Not just the gallery or arts festival, in other words, but also the office (whether corporate, government, or academic) now becomes the target of iconoclastic art. Or, more accurately, the prefix *auto-* is not simply discarded in the new avant-garde arts of information. Rather, the idea of "auto-destruction" is transmitted into domains of society external to the arts as "viral"—that is, as a destructivity that attacks knowledge work through technologies and techniques internal to such work. The genius of contemporary viral aesthetics is to introject destructivity within informationalism. This, we may say, is very cool.

Here is a sample exhibition of the viral arts and literatures of the information age, beginning with works that still respect the prophylactic "frame" separating art from society and moving on to works that break out of that frame to flood the network with viruses or viruslike attacks.[34]

Exhibit 1. Exhibit 1 is the work of the artist Joseph Nechvatal. In the early 1980s Nechvatal produced physical media works that recombined and recomposed "found" media images. The results, as Barry Blinderman describes, were "intimately scaled graphite drawings comprising saturated, interwoven line tracings of pictures culled from newspapers and magazines." "Irrational juxtapositions of images and scale," Blinderman continues, "were submerged into an all-over abstract network."[35] Such art was not just recombinant, but also conceptually destructive. "I tend to degenerate archetypal media images," Nechvatal said in 1984. "I rip off images from the media . . . then destroy/transform them in the interests of unintelligible beauty."[36] Moreover, Nechvatal's (de)compositions alluded to the general destructivity of contemporary technologies usually fêted for their innovation and creativity. As the artist wrote in 1983, "images of mass annihilation wrought by technology now provide the major context for our art and our lives. With profoundly disturbed psyches, modern people encounter their existential fear in the atom, for when technology relieved much of man's fear of nature it replaced that fear with one of technology itself."[37]

Beginning in the late 1980s, Nechvatal migrated into the digital realm by specializing in "computer-robotic assisted acrylic on canvas paintings."[38] His method is to create "digital maquettes which fuse drawing, digital photography, written language, and externalized computer code," then to use a robotic painting machine to transform the maquettes (or small models, in this case digital) into high-resolution acrylic paintings. As the robotic painting machine passes once over the canvas, pigments are "mixed computationally in real time" and an "airbrush type delivery system" yields a "smooth and lush" finished surface.[39] Yet no matter how smooth and lush the gloss, "finish" in Nechvatal's work is a deceptive concept because the destructivity he sought to express migrated in a unique way into the digital realm. During his tenure during 1991–93 as Louis Pasteur artist-in-residence in Arbois, France, Nechvatal, with the technical collaboration of Jean-Philippe Massonie, initiated the "virus" art projects for which he has become well known.[40] In these projects (Virus Projects 1.0 and 2.0 and his more recent "vOluptuary: an algorithic hermaphornology" [*sic*] series), he unleashes what he calls "computer viruses" on his original image files or digital maquettes, transforming and altering those images in unpredictable ways. Unlike the usual breed of computer viruses, however, which bear only an incomplete or casual resemblance to embodied viruses (as indicated by their incoherent species classification in hacker idiom as "viruses," "worms," "Trojan horses"), Nechvatal's viruses are actually "cellular automata" whose use in "artificial life" research analogizes embodied life.

Cellular automata are a form of digital behavior (usually represented in patterns on a screen) in which complex phenomena emerge from purely local interactions. Instead of a program that from the "top down" determines what appears on a computer screen, cellular automata presuppose that there is no "god" program. Rather, there is only a simple algorithm that from the "bottom up" does nothing more than instruct individual pixels on the screen how to react to phenomena in their immediate neighborhood. For example, an instruction might be the equivalent of: "Look around and see if the adjacent pixels are switched on or off, and in what color. Then average the conditions [or perform some other calculation] and, depending on the results, move yourself eight pixels to the left, turn yourself blue, or duplicate yourself."[41] The simultaneous interaction of many such discrete, local behaviors over time (in multiple iterations of the algorithm named "generations" in imitation of organic propagation) results in surprising, "emergent" patterns at higher levels of organization. Indeed, in the most interesting cases, complex behaviors emerge that appear actually to "live"—that is, to create local formations that maintain themselves, move or "glide," reproduce, and die into stasis.

Applying the principles of cellular automata, Nechvatal's virus projects start with his digital images, set a cellular automata program to work on them, and then capture the result at generation *n* after the algorithm has eaten away at the original images and transformed them (and their acrylic realizations) into something not just visually interesting but often hauntingly beautiful. The "finished" works expose to view the action of interminable mutation. In *vOluptas 2.0 @ 7.5 min.* from the Virus Project 2.0 and "vOluptuary: an algorithic hermaphornology" series, for example, an original image of aggregated red balls—something like molecules in the lattice structure of a crystal—has been acted upon by a cellular automata "virus" to create after 7.5 minutes of iteration a drama of decomposition and recomposition at four different scales of vision from the microscopic to macrocosmic (labeled 1–4 in the key to figure 11.1).[42] The phenomena glimpsed at these perceptual measures may be described as follows:

1. Most obviously, the original has rotted away microscopically until individual pixels of white contaminate the whole like mold cultures starting in a petri dish. Or perhaps even the metaphor of mold forming *over* the image is too sanitary to describe the true depth of the damage. If we look closely, we note that the eye-catching white pixels merely distract us from the deeper pixel-rot: not just the pseudo-organic red balls but the entire visual continuum in which they are embedded is decomposing as if from the inside out into individual pixels of various colors.
2. Such decomposition into atomic dust, however, is merely the opening act in a larger narrative of recomposition. Observing at a slightly larger

Figure 11.1 Joseph Nechvatal, *vOluptas 2.0 @ 7.5 min.*, from Virus Project 2.0 and "vOluptuary: an algorithic hermaphornology" series. (For a high-resolution color image, see <http://polaris.english.ucsb.edu/ayliu/nechvatal>)

Technical collaborator: Stéphane Sikora. Image courtesy of Joseph Nechvatal. Key to measures of vision mentioned in my discussion (approximate):

1=[·]
2=[–]
3=[–––––––––––––––]
4=[––]

scale, therefore, we see that individual pixels that have been freed from the original image do not just wander off in pure entropy. Rather, they form small clusters or patches (of black, red, green, and so on) that resemble the "jaggies" of type fonts displayed on an early computer screen. Or to overlay upon Nechvatal's viral metaphor an astronomical trope, we may say that the interstellar dust formed by blasting apart old stars recombines into myriad new solar systems.

3. Indeed, the grosser our scale of vision becomes, the more we see that recomposition is the story of the picture. At a still more expanded scale of perception, the solar system–like clusters of pixels create larger nebulae or swirls as if beginning to accrete into a galaxy.
4. And at the largest macro-scale of perception (marked out by the vertical bands of distortion seeming to sweep across the whole image), it is as if the entirety of the scene were being rescanned for some great, transcen-

> dental Photoshop in the sky that will subsequently crop, resize, mask, filter, and so on through all the other recombinant effects in our contemporary graphics repertory.[43]

Hauntingly beautiful, I called *vOluptas 2.0 @ 7.5 min.*—or, to apply a phrase from Nechvatal himself, a work of "unintelligible beauty." But really, perhaps, we are dealing not with the pleasure of beauty (despite the work's title, a play on Latin *voluptās*, "pleasure") but instead with the sublime or its near relative, the tragic. It is not stretching too far to say that *vOluptas 2.0 @ 7.5 min.* is indeed what I called it above: a "drama" of decomposition and recomposition. If the original, crystalline image of a well-ordered world is the protagonist of this work (its Lear, we may say), then there is a fateful way in which the antagonists on the scene—the viral agents of decomposition and recomposition—are ultimately *integral* with the protagonist in a way that is classically tragic. It is as if all the action of the work followed the Aristotelian plot: decomposition and recomposition are a "reversal" that do no more than "discover" the hero's deepest, darkest fear, which descends upon him not from outside but in the final analysis from within as a tragic flaw. *vOluptas 2.0 @ 7.5 min.* thus reveals a frightful, if fractal, symmetry in relation to its original image of order. It is "self-similar" through and through: the disassembled pixels, clusters and patches, nebulae, and so on, are simply a strange, or estranged, way of reseeing the red balls at different scales. The balls, that is, were from the first the giant archetypes of pixels, and the mission of Nechvatal's "viral" art is no more than to defamiliarize those archetypes so that we see revealed the recombinant possibilities of destruction and re-creation hidden within the illusion of their gridlike order.

In classical tragedy, we remember, it was Fate that acted upon the inner, tragic flaws of heroes to disassemble their tidy worlds into the primary, even tidier orders nested secretly within them—for example, the strictly binary, utterly cruel order of life vs. death. In Nechvatal's work, however, a little, godless algorithm plays the part of Fate, disassembling order into free pixels whose apparent anarchy is merely a symptom of a purer knowledge of the brutal clarities of order: on/off, white/green, and so on. In such a drama, what level of order, macro or micro, is "creative," what "destructive"? Nechvatal's art refuses to say, which is to say that it ultimately includes the binary of "creative" vs. "destructive" *within* the recombinant logic of on/off, white/green, and so on, in which everything is up for grabs. We are all Lears whose life is staked on the creation of a certain vision of order; we are all therefore also hosts for multitudes of destructive agents that imagine the possibility of other orders. "Creative" or "destructive" is not a decision that can be made from a *deus ex machina* perspective; it is an equivocation that is part of the inner logic of the system.[44]

Exhibit 2. The second exhibit is *Agrippa (A Book of the Dead)*, an art book with text by William Gibson and etchings by artist Dennis Ashbaugh published in a limited edition by Kevin Begos Publishing (1992).[45] We might prepare for reading this book by recalling the "look and feel" of Gibson's better known cyberpunk science fiction novels, beginning with *Neuromancer* (1984) and continuing with variation into such titles as *Virtual Light* (1993). This look and feel—the "interface" of Gibson's fiction, as it were—is like the famously noir, glistening, and multi-ethnic streets of near-future Los Angeles depicted in Ridley Scott's film *Blade Runner*. It establishes not just the setting but somehow the substance of the fiction, as if the plot were ultimately just a vehicle for communicating the "look" that the works are really "about." In this light, the following, most commonly quoted passage from *Neuromancer* (what we can call the "city lights" passage) is misleading: "'Cyberspace. A consensual hallucination experienced daily by billions of legitimate operators, in every nation, by children being taught mathematical concepts . . . A graphic representation of data abstracted from the banks of every computer in the human system. Unthinkable complexity. Lines of light ranged in the nonspace of the mind, clusters and constellations of data. Like city lights, receding. . . .'"[46] In one sense, of course, Gibson's fiction *is* about a Camelot of cyberspace in which his protagonists move like knights, or samurai, of pure light. But it is significant that the definition is uttered in quotation marks in the novel. It is not the voice of Gibson or of one of his characters. Instead, it is the "voiceover" of a "kid's show" that his protagonist momentarily tunes into (*Neuromancer*, pp. 51–52). If we stay tuned beyond the kids' hour, the other side of Gibson's look and feel emerges from the shadows: not cyberspace but the street, and not bright but noir (or more accurately, noir interlaced with the hallucinatory neon of "strip" boulevards). That such noir is merely a surface effect learned by rote from noir fiction and film in the twentieth century is indisputable (in the way that the main characters in *Neuromancer*, it is often noticed, are deliberate knock-offs of film cowboys, samurai, gangsters and their "molls," and so on).[47] But what gives such surface effect substance is the deep feeling Gibson instills in *superficies* for the generalized destructivity of society—for the way dominant social, economic, political, and military institutions use technology to hollow out people and habitats to leave empty shells of identity.

Thus, for example, all the main characters and many of the secondary characters in the novel are hollowed out by traumas of past destruction that leave behind only survivor-identities stubbornly clinging like lichen to life's existential surface. Case, the protagonist, begins the novel literally hollowed out: the nervous system that had allowed him to interface with cyberspace had been deliberately "damaged . . . with a wartime Russian mycotoxin" to punish him (*Neuromancer,* p. 6). Molly, the other main character, had lost her one true love to a Yakuza assassin. Look now into the mirrored lenses grafted over her eyes and you will see anything and everything reflected there except herself. She may be the epitome of intensity and "centeredness" during combat, but her superb body awareness is also a zen of withdrawal from the here and now. In the very kick, slice, and bleed of the instant, she is far gone. Moreover, if the primary characters of the novel are thus haunted by destruction, some of the secondary characters are even more obviously burned out husks. There is Armitage or Corto, for example: destroyed as a functioning identity in a wartime mission and rebuilt by the artificial intelligence Wintermute as a mere personality shell constantly regressing to the trauma of its fiery destruction. Or there is the consummately decadent Peter Riviera, who is a kind of perverse double of the author himself within the work. Equipped with bio-mech that allows him to project his imagination in holographic form, Riviera is the very artist of look and feel whose elaborate, sadistic holo-dramas give the illusion of depth to a moral emptiness so absolute that it recalls nothing if not the postnuclear rubble rings of Bonn that were the scene of his feral childhood (*Neuromancer,* p. 210).

The distinctive look and feel of *Neuromancer* results from the accumulated weight of such destructivity as it accretes into a palpable patina of everyday decay bearing all the scars of the micro-physics of destruction. In one of many characteristic vignettes, for example, we see this look and feel reflected in the tabletop at the Jarre de Thé teashop:

> The brown laminate of the tabletop was dull with a patina of tiny scratches. With the dex [a drug] mounting through his spine [Case] saw the countless random impacts required to create a surface like that. The Jarre was decorated in a dated, nameless style from the previous century, an uneasy blend of Japanese traditional and pale Milanese plastics, but everything seemed to wear a subtle film, as though the bad nerves of a million customers had somehow attacked the mirrors and the once glossy plastics, leaving each surface fogged with something that could never be wiped away. (*Neuromancer,* p. 9)

Not "city lights, receding," in other words, but bright gloss and surfaces receding, clouded over by a million cosmic ray strikes of destructivity and decay.

Nor is it just the embodied, physical world of the "meat," as Gibson calls it, that is thus affected. While the bright world of cyberspace is ostensibly the opposite of corporeal physicality in *Neuromancer*, the pathos of the novel lies in the way it makes us realize that even cyberspace ultimately reveals behind its polished façade the myriad craquelure that is the interface through which Gibson communicates his infinite history of small tragedies, little losses, measured doses of pain. The emotional core of *Neuromancer*, therefore, is not cyberspace as a circuit-sharp, neon hallucination of the global metropole. Instead, it is the hauntingly rendered virtual beach on which Neuromancer (the other artificial intelligence in the work and the namesake of the novel) at last traps Case with the silicon ghost of his dead girlfriend, Linda Lee. In the distance across the beach, "there seemed to be a city." But strangely, Case can never reach it, no matter how far he walks. The "city lights" recede indeed. Perpetually exiled from the city, the beach resolves for him at last into a smaller, more intimately scaled world—a city of two, village of two, or perhaps even more regressively, cave of two (the ruined concrete bunker where Case shelters with Linda). In this reduced world, the sense of vision usually so dominant in cyberspace ("Lines of light ranged in the nonspace of the mind . . . Like city lights, receding") submits to the ancient, blind wisdom of touch, to the low-information but also somehow higher bandwith, finger-touch digitalism of the body. She loves me; she loves me not . . . : these old binary calculations for what is "real" to human beings are communicated by the one-bit data of cold vs. warm, wet vs. dry, exposure vs. shelter, loneliness vs. embrace. It is the on-and-off love of Case for Linda Lee—enacted in their sexual coupling on that beach—that at last effects "the transmission of the old message" (*Neuromancer,* pp. 233–40).

A perfect, enclosed bubble-world of experience the neuromantic beach thus seems—almost as hermetic as the haunted, regressive "sepulchre there by the sea" in Edgar Allan Poe's poem "Annabel Lee," whose earlier romanticism Gibson's scene by the sea (with "Linda Lee") seems clearly to allude to. Yet look hard enough into the bubble, and even this touching illusion of eternal intimacy—built as it is on the ultimate sand of digital silicon—succumbs to Gibson's distinctive craquelure of destruction: "The music woke him. . . . His vision crawled with ghost hieroglyphs, translucent lines of symbols arranging themselves against the neutral backdrop of the bunker wall. He looked at the backs of his hands, saw faint neon molecules crawling beneath the skin, ordered by the unknowable code. He raised his right hand and moved it experimentally. It left a faint, fading trail of strobed after-images" (*Neuromancer,* p. 241). In the "real" world where Case's body is lying, a companion has hacked into his virtual dream on the beach by playing music in his ears. It is as if the code that underlies the illusion of the beach can only sustain one interface at a time—either that of immersive touch or that of sound. From the viewpoint of the dream, therefore, the result of the interfering music is that the interface stretches too thin and the underlying code shows through as a symptom of decay. Cyberspace itself becomes no more than a "laminate" like that tabletop in the Jarre "dull with a patina of tiny scratches."

Can Gibson's hollowed-out people—his street people, surface people, lichen people—find a way to make destruction and decay survivable, to inhabit destructivity, to burrow within the patina of decay and make a livable space? Can they become like a virus, in other words, that somehow "lives" in and through destructivity, creating a separate, alternate creativity of life? (Case makes his living as a hacker, after all, and the Kuang virus he injects into the corporate computers holding Wintermute and Neuromancer at last helps these two artificial intelligences create a higher, combined consciousness.) That is the essential question of *Neuromancer,* as it is of such later Gibson novels as *Virtual Light,* which introduces Gibson's remarkable, recurring paradigm of the Bay Bridge in San Francisco become a broken mechanism of transport (an allegory for a kind of stopped information superhighway) on which a choked, bustling maze of squatter homes, shops, and what-not scavenged from all of high tech's garbage have accreted into a thick "patchwork" of habitable, if viral, life in the shadow of technological destructivity.[48]

Now we are ready for *Agrippa (A Book of the Dead). Agrippa,* to start with, is a book whose physically distressed form enacts Gibson's distinctive patina of destructivity. As described by Peter Schwenger in a perceptive article, *Agrippa* is an artifact of catastrophe:

> Black box recovered from some unspecified disaster, the massive case opens to reveal the textures of decay and age. Yellowed newspaper, rusty honeycombing, fog-colored cerement enveloping a pale book. . . . The pages are singed at their edges; more fragments of old newspapers are interspersed. And at intervals, engravings by New York artist Dennis Ashbaugh reproduce the commercial subjects of a previous generation, subjects that will later acquire a fuller meaning.[49]

We do not know what "unspecified disaster" left this work behind like the remnants of some half-demolished bomb shelter (though it is suggestive, as Schwenger notes, that Gibson's father worked on the Manhattan Project).[50] But we do know that the work is indeed a testament, compendium, or edition of the generalized destructivity of the twentieth century—the first work, as it were, copy-edited by bomb. Nor is the catastrophe over and done with so that we can expect the trauma to be stabilized. A persistent radioactivity of destructivity continues. We first notice it when Ashbaugh's images (at least as the book was originally conceived) alter in appearance like some picture of Dorian Gray updated to the processes of pixel-rot, gene-splicing, or molecule-creep that are the usual symptoms of cyberpunk's fetishization of digital, biotech, and nanotech fungibility. As Schwenger describes, "Black patches like burns smudge [Ashbaugh's] images. With exposure to light the images gradually fade; the black patches reveal themselves to be the rhythmic chains of the DNA molecule as captured in microphotography."[51] What these auto-destructive yet revelatory images foreshadow is the great auto-da-fé of *Agrippa*, which, true to Gibson's form, is in the end virtual. In a cavity within the last pages of *Agrippa* the book, we find a computer diskette containing the 305 lines of "Agrippa" the poem (the specific contribution of Gibson). The file on the diskette containing the poem can be read on the screen just once, after which a self-encrypting algorithm—the conceptual equivalent of a virus—makes it permanently disappear, leaving behind only the innumerable, hacked copies that now flourish like ghosts on the Internet. If *Agrippa* the book is edited by bomb, then "Agrippa" the poem is edited by a "logic bomb" (a program that, when triggered by a specific condition, runs a destructive routine).[52]

Thus, just as destructivity in Nechvatal's art *becomes* creativity, so destructivity in *Agrippa*—epitomized in its self-canceling poem "Agrippa"—is the ground zero of an alternate creativity. Despite my comparison to the picture of Dorian Gray, however, that alternate creativity is ultimately less *fin-de-siècle* than Romantic, less focused on the creative expression of decay than on a recuperative or re-creative therapy for decay. Not Wilde, in other words, but Wordsworth (the predecessor to Poe in Gibson's Romantic genealogy). Or, put another way, not *The Picture of Dorian Gray* but "Tintern Abbey" with its vision of another kind of picture able to take on a life of its own ("And now, with gleams of half-extinguished thought, / With many recognitions dim and faint, . . . / The picture of the mind revives again"). Indeed, Gibson's "Agrippa" and Wordsworth's "Tintern Abbey" pair up so exactly in their primary topic (the constitution of the self), primary psychological faculty (memory), tone (epiphany balanced against elegy), and genre (what

M. H. Abrams calls the "greater Romantic lyric") that it will repay us to review their similarities in some detail in order to gauge with matching precision what we will see to be the one essential difference of the Gibson poem.[53]

Wordsworth's "Lines Written a Few Miles above Tintern Abbey," conceived one way, asks two interlinked questions: "Who am I?" and "Who am I as a writer?" It is a testament of artistic identity, nested within a witnessing of personal identity. In "Tintern Abbey" (and elsewhere), Wordsworth had answered these tandem questions by equating artistic or "imaginative" sensibility with the foundation of Romantic personal identity: memory, the act of witnessing, conserving, and imaginatively recreating the original creativity of early life amid the destructivity that is the rite of passage into maturity. Destructivity, after all, is everywhere in Wordsworth. Destructivity is familial, as in the death of his father, which prompted the "blasted hawthorn" episode in *The Prelude* (1805; XI.344–88). And it is sociopolitical, as in the trauma of the French Revolution that subliminally haunts "Tintern Abbey" with its fixation on "five years" ago when the poet had just returned from France.[54] But in "Tintern Abbey" and other poems, Wordsworth counters destructivity with a faith in (re)creative memory so deep that it goes beyond the recall of personal origins to the recall of the primeval—of Nature itself as the fount of creation and re-creation. Nature may thus destroy (in the mode of the terroristic "sublime") or make things decay (in the mode of the weathered, textured "picturesque"). But Nature also recreates so that the "picture of the mind revives again." Nature is the spirit of reanimation within destruction. It is Nature's regeneration, or perhaps simply gigantic prosthesis, that supplies all the amputations of existence—all the deaths, revolutions, and other apocalypses major and minor that dull the promise of life and make of it a broken thing called adulthood—with "abundant recompense" (l. 89). In short, Nature is the "organic" continuity that both Wordsworth and Coleridge in the period of "Tintern Abbey" called the One Life:

> Whose dwelling is the light of setting suns,
> And the round ocean, and the living air,
> And the blue sky, and in the mind of man,
> A motion and a spirit, that impels
> All thinking things, all objects of all thought,
> And rolls through all things. . . .
>
> ("Tintern Abbey," ll. 98–103)[55]

In one regard, Gibson's "Agrippa" is fully Romantic because like "Tintern Abbey" it also answers the tandem questions "who am I?" and "who am I as a writer?" by equating artistic identity with a fundamentalism of personal memory. Indeed, Gibson's poem is so much a work of memory that it might be called memory to the second power: it takes as its framing device an antecedent memory-book—a photo album—whose combined appearance of decay and clarity sets the scene for the poem's meditation on destructivity and recuperation. As we learn in the first lines, "Agrippa" is actually the brand name of an old photo album that belonged to Gibson's father:

> A black book:
> ALBUMS
> CA. AGRIPPA
> Order Extra Leaves
> By Letter and Name
>
> (ll. 4–8)

The photo album is a decayed artifact witnessing in its fabric the destructivity—entropic and otherwise—of the twentieth century:

> A Kodak album of time-burned
> black construction paper
>
> The string he tied
> Has been unravelled by years
> and the dry weather of trunks
> Like a lady's shoestring from the First World War
> Its metal ferrules eaten by oxygen
> Until they resemble cigarette-ash
>
> (ll. 9–16)

We recognize here the genealogical source of the blasted look and feel of *Agrippa* and perhaps also of the laminate tabletop at the Jarre or the code-eroded virtual beach in *Neuromancer*. In a sense, the patriarchal photo album is the old testament of loss behind Gibson's poem as well as his cyberpunk fiction of the so-called "near future." The genre of "near future" science fiction, in this regard, is not at all related to prophecy. It is instead a variant of elegy. The compulsion of the "near future" as a fictional genre is that "near" is closer, more intimate, and more deeply sunk into the bone of the "now" than any mere future that can be imagined. There is only one world that can haunt the present so nearly—the past.

Looking into the album, Gibson opens his poem in a kind of fugue memory so deep that it is really ancestral memory. He describes pictures of family history dating from before his own birth or personal memory. (Gibson's father died when he was six, just as Wordsworth's mother died when he was seven and his father when he was thirteen.)[56] Framed by the photo album and its aura of destructivity, what Gibson can see in the pictures is constantly shadowed by a penumbra of loss and erasure. The photo album is a palimpsest of amnesia, of the de-inscription of the past:

> Inside the cover he inscribed something in soft graphite
> Now lost
> Then his name
> W. F. Gibson Jr.
> and something, comma,
> 1924
>
> (ll. 17–22)

Yet for such loss, there follows abundant recompense. Gibson reconstructs from pictures of lost life such vivid memories that we would finally have to call them more "imaginative" than memorial. In the following description inspired by a picture from the album, for example, Gibson reanimates the scene in such a way that we can almost smell the bite of the saw:

> A flat-roofed shack
> Against a mountain ridge
> In the foreground are tumbled boards and offcuts
> He must have smelled the pitch, In August
> The sweet hot reek
> Of the electric saw
> Biting into decades
>
> (ll. 27–33)

It is as if for just a moment the time-arrow of the universe were equipollent in both directions, and entropic destruction (the sharp sawteeth biting, the sawdust flying) were balanced against creation. Destruction vs. creation: in the strange perception of memory there is barely a bit of difference (like 0 versus 1) between the two. What is memory, after all, but a rough-edged blade of mind sawing into decades to release a spray of images, sounds, and—most deeply, as in Proust's madeleine—taste or fragrance of the intense past? Or to change metaphors: stars, astrophysicists tell us, end in certain cases in fiery explosions of destruction that release materials for the creation of new stars and new worlds. So, too, the "sweet hot reek" that forms in Gibson's memory is a little supernova releasing the substance of the deep past as a sudden, sharp perfume of creation.

Once the process of re-creation starts, Gibson in the succeeding portions of his poem can dispense with the mediation of photography to recreate through imaginative memory what lies beyond the last page of his father's album: his own early life. Just as "Tintern Abbey" superimposes two distinct periods of personal memory (separated by "five years," 1793 and 1798) to create its complex diffraction patterns of identity, so "Agrippa" overlays memories of Gibson's childhood with those of his exodus to Canada as a young man to dodge the draft during the Vietnam War (his generation's version of Wordsworth's French Revolution). This is the overall complex of memories that answers the question "who am I?" The more specific question "who am I as a writer?" is then answered by one cluster of memories in particular. These center on an all-night bus station in the southern town of Gibson's youth where—in the heartland of subculture, in the "cool fluorescent cave of dreams" of a magazine rack where a "colored restroom" had once been—he first came under the spell of writing as a vocation:

> There it was that I was marked out as a writer,
> having discovered in that alcove
> copies of certain magazines
> esoteric and precious, and, yes,
> I knew then, knew utterly,
> the deal done in my heart forever,
> though how I knew not,
> nor ever have.
>
> (ll. 247–54)

We might compare a similar moment of vocational dedication in Wordsworth's *Prelude*:

> I made no vows, but vows
> Were then made for me: bond unknown to me
> Was given, that I should be—else sinning greatly—
> A dedicated spirit.
>
> (1805; IV.341–44)

For both Wordsworth and Gibson, memory's recall is prelude to the "calling" of a writer's vocation.

Like Wordsworth's poetry, then, "Agrippa" is a testament to loss that in the end becomes a new testament of recuperation. Gibson's neuromanticism is indeed a disciple of Romanticism.[57] But the payoff for our comparison comes when we see how it sets off Gibson's essential *difference* or heresy from Wordsworth. The One Life in "Tintern Abbey," we remember, is a "motion" and "spirit" that "impels / All thinking things, all objects of all thought, / And rolls through all things." To glimpse Gibson's difference, we need now only substitute the verb *shoots* for *rolls*. Where Wordsworth saw One Life, Gibson sees One Gun. After all, if Wordsworth's most sublime poetry centers on "spots of time" such as the "blasted hawthorn" and "gibbet mast" episodes in *The Prelude,* clearly the equivalent sublime moments in Gibson's "Agrippa" revolve around guns—in particular, one gun he fired accidentally as a child and another he fired more deliberately in later youth (but with equally unexpected results):

> The gun lay on the dusty carpet.
> Returning in utter awe I took it so carefully up
> That the second shot, equally unintended,
>   notched the hardwood bannister and brought
>   a strange bright smell of ancient sap to life
>   in a beam of dusty sunlight.
>   Absolutely alone
>   in awareness of the mechanism.
>
> (ll. 135–42)

> I was seventeen or so but basically I guess
> you just had to be a white boy.
> I'd hike out to a shale pit and run
> ten dollars worth of 9mm
> through it, so worn you hardly
> had to pull the trigger.
> Bored, tried shooting
> down into a distant stream but
> one of them came back at me
> off a round of river rock
> clipping walnut twigs from a branch
> two feet above my head.
> So that I remembered the mechanism.
>
> (ll. 206–18)

We can make sense of the relation between the One Life and One Gun in this way. Both are at base imaginations of the powerful, compulsive agency that underlies authorial identity. What is authorial inspiration? Fundamentally, Wordsworth and Gibson answer, it is obeisance to a terrifying, autonomous agency *without* that is somehow also a poetic agency *within*. We might recall Blake declaring that he wrote his *Milton* "from immediate Dictation . . . without Premeditation & even against my Will."[58] As theorized most strikingly in Harold Bloom's *Anxiety of Influence*, alien agency (equated by Bloom with patriarchal influence) is the signature of a strong writing that comes out of the self but is also *not* of the self or is unconscious to the self.

Gibson's heresy is that he attributes such inspiration to automatic *mechanism* rather than natural organism. In the Enlightenment and early Industrial Revolution, mechanism was the "machine in the ghost" (to reverse the cliché) that had all along haunted Romantic "Nature." We need only exert a slight mental pressure on "Tintern Abbey," for example, to see that there is something uncannily mechanical about a One Life rolling resistlessly through all things like some totalitarian locomotive of the soul. But in "Tintern Abbey," Wordsworth carefully keeps the golem tethered to Nature. Just so, in the "boat stealing" episode of *The Prelude* (1805; I.372–426) Wordsworth tethers the automaton-like "huge and mighty forms that do not live / Like living men" to the autochthonic "huge cliff" that rises overhead. The golems or robots are just metaphor; the natural cliff is reality. In another of his best known poems, "A Slumber Did My Spirit Seal," Wordsworth tethers the automatic motion by which Lucy is "Rolled round in earth's diurnal course / With rocks and stones and trees" to the natural rotation of the planet. As the poet's late-life activism against the railroads in his native Lake District indicates, locomotive-like rollings were for him merely destructive; they became creative only when sanctified by the greater inhumanity he named Nature.[59]

Gibson cuts the tether to nature. Under the ghastly fluorescent lights, by the thrumming bus engines, near "where the long trucks groaned / on the highway" (ll. 261–62), and in hearing of the "timers of the traffic lights" on the streets of his quintessentially urban imagination (l. 257), the inhuman automatism that Wordsworth called Nature reveals itself to be what Gibson fetishistically calls "the mechanism" (where the ritualistic use of the definite article *the* reinforces the sense of alien objectness). The guns are "the mechanism." And so, too, the camera responsible for the photos in his father's album is "the mechanism":

> The mechanism: stamped black tin,
> Leatherette over cardboard, bits of boxwood,

A lens
The shutter falls
Forever
Dividing that from this.

(ll. 98–103)

Even the photo album itself—the crossing point between the old, authoritative technology of the book and new, modern media—is "the mechanism." "The mechanism closes," Gibson says, shutting the album (l.178). Where Romantic Nature was the ghost in the machine that resisted iron mechanism even in the midst of the Enlightenment infatuation with "mechanistic" philosophy, for Gibson "the mechanism" exposes the ghostly gears (and circuits) undergirding contemporary life.

Metaphysics, in short, is for Gibson a matter of engineering. The only "spirit" rolling through the circuitry of all things is the dance of electrons that is the charm of contemporary technology. Out of *this* charm or romance—however destructive it may be to organic Nature—Gibson derives his faith in a new kind of recuperative creativity. How can this be? How can creativity come from mechanical destructivity? The answer inheres in what we understand mechanism to be. As imaged in his koan on the camera (quoted above), Gibson suggests that mechanism is at heart like the shutter "Forever / Dividing that from this." It is a principle of existential discontinuity (of parts "needy for connection," as Donna Haraway says, but barren of genetic or essential affiliation) that during modernity stamped its image on organic life through guns, shutters, and other devices engineered to create discontinuities so radical they cleaved apart limb from limb, and before and after, with a finality as absolute as death.[60] Yet for Gibson, such mechanistic discontinuity and destructivity is also the medium for an alternate mode of continuity and creativity. Like the discontinuous snapshots in his father's album, like the fragmented memories of his own childhood and young manhood, and ultimately like the abruptly juxtaposed scenes of "the mechanism" itself in the poem (camera, album, trucks, traffic light timers, guns, and the final "red lanterns" in Chiyoda-ku "laughing / in the mechanism" [ll. 299–305]), discontinuity *aligns* in "Agrippa" into the montage effect or *picture of discontinuity* that is alone how our epoch knows continuity in the aftermath of the mass destructions of the twentieth century.

Destructivity may be a snip of the scissors that severs the picture of life, in other words, but disassembled vision itself acquires a kind of consistency and integrity. As it were: *I cannot tell you great, heroic stories because too many of the heroes have been sent to the trenches, the beaches, the ovens, or the jungles, and their sagas have been blasted apart by a thousand howitzers and the equally terrifying thud of a million clerical stamps certifying on the passports, birth certificates, and immigration papers of refugees that modern life is fundamentally discontinuous, passing forever* [like Gibson over the Canadian border] *from that to this. But I still have my box of photos and clippings. The very brokenness of these things is the witness of my life.* We are talking, in short, about collage, montage, cut-up, pastiche, assemblage, and so on. We are talking about the entire, spectacular heritage descended from the early twentieth century of auto-destructive works that cut, slice, and otherwise self-fracture themselves in quest of a larger picture of contemporary existence as the survivorship of destructive discontinuity. In Gibson's work, destructivity and discontinuity are virally embedded within life as the possibility of creativity and continuity.

Exhibit 3. With Exhibit 3 we move to a class of transitional digital works that take us beyond the *auto-* in auto-destruction into a limbo where it is undecidable whether the artwork performs only its own destruction or actually *is* destructive beyond its own frame. A well-known example in new media art circles is the work of the Dutch/Belgian duo known as "Jodi" (a name formed by incorporating the first two initials of their given names: Joan Heemskerk and Dirk Paesmans). Jodi's primary medium has been called "browser art," itself a "sub-genre within net art."[61] Browser art may be defined as art that recognizes the authority of the Web browser as the great contemporary art frame but refuses to cede the final word on the look and feel, or even the function, of this frame to the technical, corporate, mass media, and other forces that produce the dominant browser programs. Instead, browser artists either defamiliarize the canvas presented within the dominant browsers (as on Jodi's website at <wwwwwwwww.jodi.org>, where the very domain name spoofs the imperialism of the World Wide Web to the power of three), or create their own alternative browsers (as in the case of Jodi's downloadable browsers named *.com*, *.org*, and so on, after the mainstream Internet domains they spoof).

Jodi's browser art advances an agenda of "retro-avant-gardism," as it might be called, that can be genuinely disturbing because of the "noise" it introduces into information—an effect that Michele White calls an "aesthetic of failure" communicated through "breakdowns, technological confusion, and illegibility."[62] Indeed, the repeated formula of Jodi's art is to use state-of-the-art information technology to create effects of retro-information-technology whose very presence within the data stream of the contemporary appears ipso facto as a corruption of that stream, or "noise." Consider, for example, <wwwwwwwww.jodi.org>, which is one of the entryways into

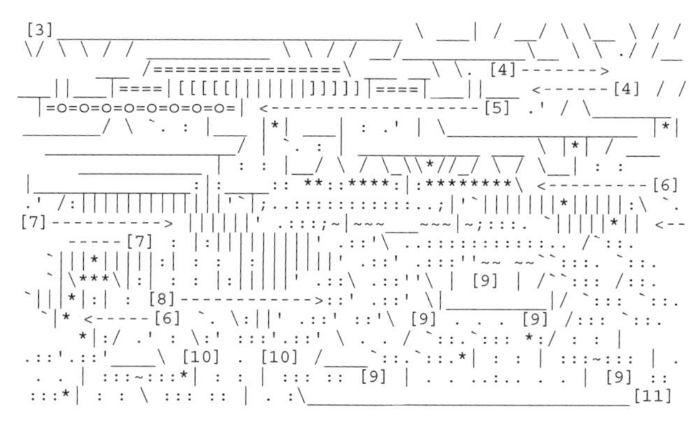

Figure 11.2 Excerpt from <wwwwwwwww.jodi.org>. Line lengths vary depending on the size of the browser window.

Jodi's labyrinth of sites organized (or, perhaps, *dis*-organized) under the jodi.org domain.[63] The top of the <wwwwwwwww.jodi.org> page looks approximately like figure 11.2 when opened on an 800 × 600 resolution screen.

Presented in monochrome green against a black background, these ASCII text characters are "tinged by nostalgia," as Peter Lunenfeld says, because they simulate the look of an early DOS- or CP/M-based personal computer screen in a way that seems to equate that look with mere information noise.[64] A close look at the "noise," however, reveals intricacies of local, self-similar, recursive, or chiasmic patterns hinting at some molecular rather than organic level of order—for example: "|__/ \ / _*//_/ \ / __|" or ":::'.::'\../`::.`:::" It is as if a mirror (or computer screen) were to shatter into a thousand shards that just by accident lined up in inverted patterns or serial progressions akin to the Fibonacci numbers. We sense that some legacy code is at play beneath the contemporary GUI interface that obscures and encrypts that interface.

The impression of retro-styling pursued for the sake of apparent noise is then strengthened if we look at the HTML source code underlying this page (available through the "view source" option in a browser, which opens the code in a text-editing program), as shown in figure 11.3. Readers with even a cursory knowledge of HTML will recognize how radically primitive the code on this page is. All properly formed HTML is opened by the <html> tag and closed by its complement, </html>. Within that pair of tags defining the outer perimeter of a Web page (these tags tell the browser this *is* a Web page), there are two main divisions, the document "header" and "body." Jodi has dispensed almost entirely with the header unit (which would be enclosed between <head> and </head> tags and which includes information about the site as well as script programming for dynamic page behaviors). All that exists of normal header material is the <title> unit. But

```
<html><title> %Location | http://wwwwwwwww.jodi.org </title>
<BODY BGCOLOR="#000000"
TEXT="#00ff00" LINK="#00ff00" VLINK="#00ff00" ALINK="#ffffff">
<font size=5><CENTER><blink><b>
<A HREF="100cc/index.html">
 [3]____________________________________          \___|
               / __/          \          \__ \
              / /            \/           \ \
             / /          _____________          \ \
            / /         __/_____________\__          \ \
           ./ /__ ___    /===================\    ___ __\ \.
    [4]-------> ___||___|====|[[[[[||||||||]]]]]|====|___||___ <------[4]
       / /          |=o=o=o=o=o=o=o=o=| <--------------------[5]
      .' /          \_________ _________/              \ `.
      : |___              |*|               ___| :
      .' | \______________________ |*| ______________________/ | `.
      : | ________________ ____\ |*| /____ ______________ | :
      : |__/         \ /  \_\\*//_/  \ /        \__| :
      : |____________________:|:_____:: **::****:|:*********\ <---------[6]
     .' /:||||||||||||'`|;..:::::::::::..;|'`|||||||*|||||:\  `.
    [7]----------> ||||||' .:::;~|~~~____~~~|~;:::. `|||||*|| <-------[7]
      :  |:|||||||||' .::'\ ..::::::::::::.. /`::. `|||*|||||:|  :
      :  |:||||||||' .::' .:::"~~     ~~``:::. `::. `|\***\|:|  :
      :  |:||||||' .::\ .::"\ |  [9]  | /``::: /::. `|||*|:|  :
    [8]------------>::' .::'   \|______________|/  `::: `::. `|* <-----[6]
       `. \:||' .::' ::'\ [9] .    .    . [9] /::: `::.  *|:/  .'
       :  \:' :::'.::' \ .             . / `::.`::: *:/  :
       :  | .::'.::'_____\   [10] . [10]   /_____`::.`::.*|   :
       :  | :::~:::     |      . . .     |    :::~:::*|   :
       :  | ::: ::  [9] | .   . ..:.. .  . | [9]  :: :::*|   :
       :  \ ::: ::     |      . :\_________________________________[11]

          [.    .    .    .       omitted material      .    .    .]

  </body></html>
```

Figure 11.3 Excerpt of the source code from <wwwwwwwww.jodi.org>, as reproduced in the variable pitch font of this book.

this minimalism is merely a prelude of what is to come. While Jodi respects the need for the body unit (enclosed within <body> and </body> tags), it almost completely dispenses with the subordinate tags within the body that tell a browser how to format information—e. g., tags designating paragraph elements, list structures, table structures, line breaks, horizontal rules, and so on. The only exceptions are the tags "<font size=5> <CENTER><blink><b>," which tell the browser to render the content at a particular font size in boldface, center it, and (in Netscape) make it blink on and off.[65] Everything else is an undifferentiated block of plain text that dispenses even with the enclosing <pre> and </pre> tags that Web authors normally use to tell a browser to format information as primitive "plain text." If the rendered (i.e., manifest) view of the home page is retro-styled noise, then the underlying source code confirms the equation between retro-styling and noise. The source code is so primitive that it might just as well be print on a page lacking any of the advanced media tags that allow a browser intelligently to parse a page into articulated elements.

Indeed, the best way to view this page may be as a printed version of the source code. As others have remarked, the source code and rendered screen have an uncanny way of interpenetrating each other in Jodi's art, reversing or flattening the functions of instrumentality and appearance until they become a single layer of experience.[66] About the version of the <wwwwwwwww.jodi.org> page he accessed in 1997, for example, Lunenfeld remarks, "The source code comes up as a text document, and what is revealed is that there is a whole layer of pictorial, ASCII text art 'below' the surface."[67] Lunenfeld's insight gives us the key to the hidden beauty, and wicked wit, of <wwwwwwwww.jodi.org>. While the normal, GUI view of the page scrambles any visual pattern in the source code because it "wraps" text lines depending on the size of the browser window and compresses multiple white spaces, the source code view reveals that the secret principle of order we noticed earlier amid all the "noise" is based on the careful use of spaces (acting like hard spaces on a printed page of 80 characters' width) to format the code into complex yet balanced minimalist or conceptual visual sculptures that taper, curve, or flare into a memory of Calder-like mobiles (see fig. 11.4). Concealed behind the GUI interface lies a regression not just to "ASCII text art" from earlier in the history of computing (pictures rendered in patterns of text and punctuation) but—at a second level of retroversion—to what at first glance seems to be predigital, modernist art.

But the punchline of Jodi's wit is not fully revealed until we are willing to relinquish all the remaining comforts of a GUI interface and even the graphically pleasing variable pitch font of today's digitally typeset print. If we

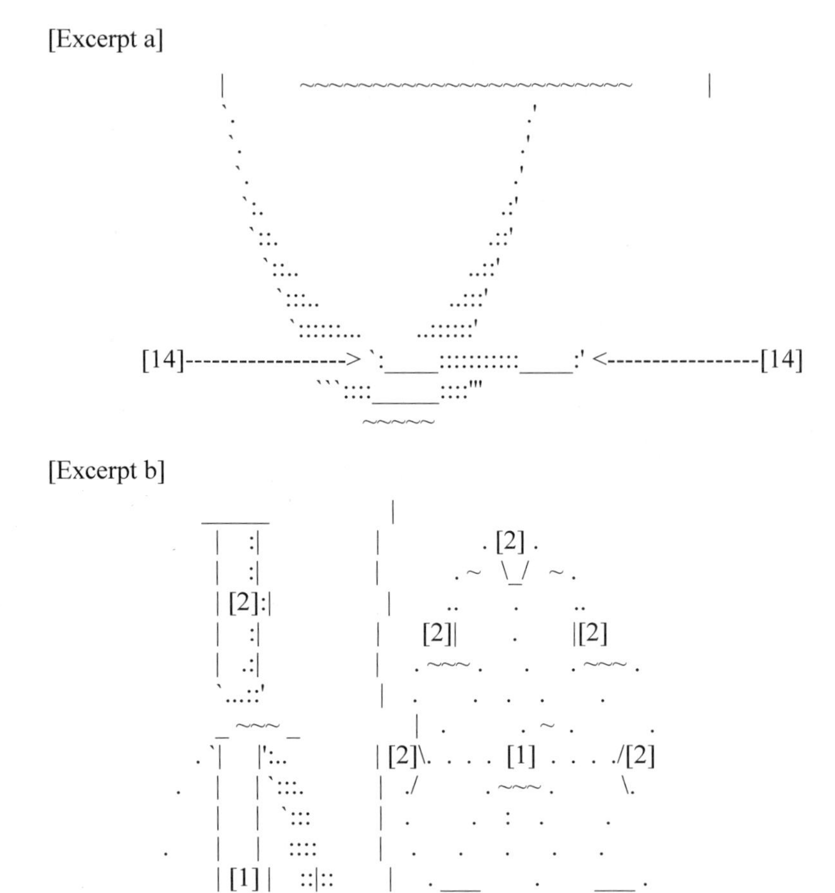

Figure 11.4 Two later excerpts of the source code from <wwwwwwwww.jodi.org>, as reproduced in the variable pitch font of this book.

reset the text-editing program in which we view Jodi's source code to a fixed pitch font like Courier—and so recapitulate the raw, primitive ASCII look of early personal computing—we realize that all the visual patterns we have seen are part of a perfectly symmetrical schematic of the Manhattan Project's Fat Man and Little Boy atom bombs together with a diagram of the fission

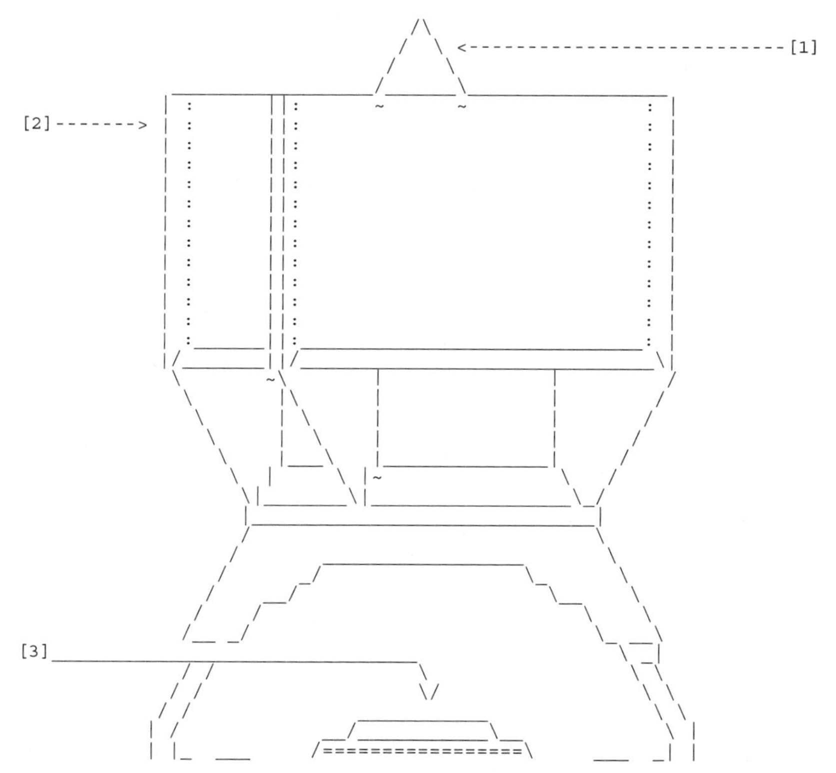

Figure 11.5 Source code for the full schematic of Little Boy, shown in a fixed pitch font (figure continues to page 355).

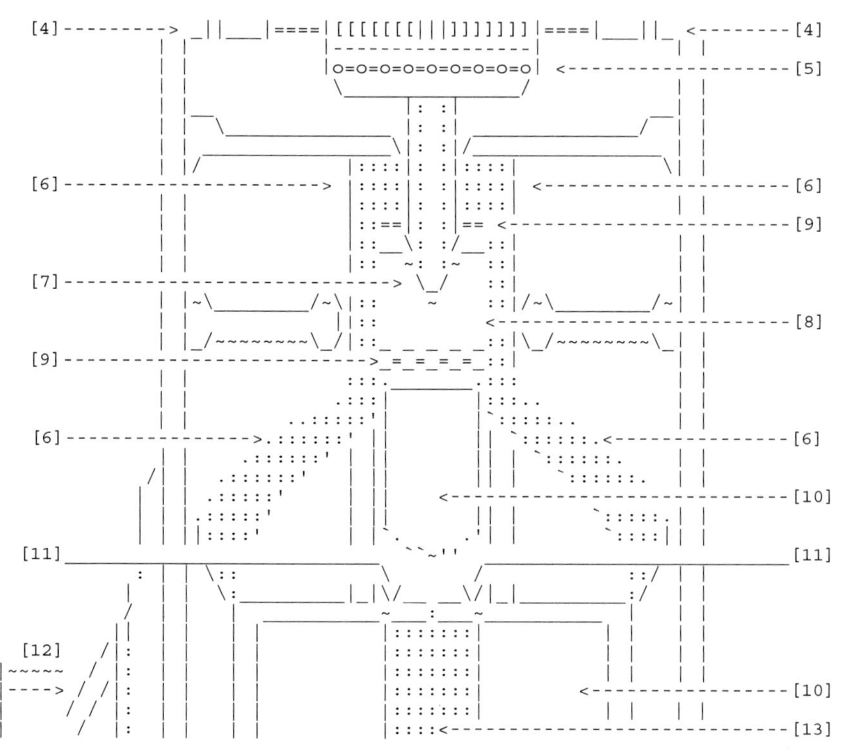

Figure 11.5 continued

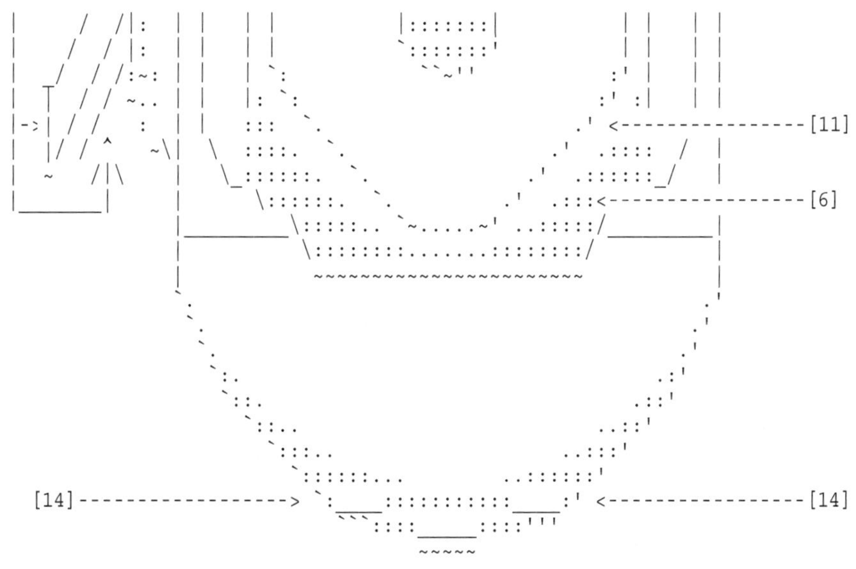

Figure 11.5 continued

pattern of these bombs. Figure 11.5 shows the source code for the full schematic of Little Boy from which the nose cone in figure 11.4a above is taken (seen in a fixed pitch font).[68] The source code of <wwwwwwwww.jodi.org> at last reveals a heritage of the avant-garde that reaches from 1960s-era minimalist geometric forms and conceptualist textual constructs back through Calder's mobiles all the way to Duchamp's Readymades and Marinetti's futurism. The future of computing according to Jodi is a readymade bomb, an icon of the original avant-garde of destructivity. And all of this heritage appears in the manifest view of contemporary GUI computing as just so much scrambled noise.

The noise from Jodi's artistic retro-technology becomes even more striking when we next click on any part of <wwwwwwwww.jodi.org> and jump to a secondary home page in the site <http://wwwwwwwww.jodi.org/100cc/index.html> that manifests with the look and feel of an early radar screen representing terrain features or an isobaric weather map. Superimposed on this radar map are links to incomprehensibly labeled branches of the site, including "BINHEX/message.hqx," "BETALAB," "Reflector/131.24.167," "SURGERY," and "target=."[69] Clicking on these links sends the user to further retro-styled and equally incomprehensible interfaces of information, including jagged bit-mapped screens, wireframe topographies, random assemblages of text and numbers, screens styled after early video games (not only military-themed), and "heads up" targeting displays of the sort projected since the Vietnam War on fighter-jet cockpits. Each of these branches of the site aggressively scrambles the dominant values of "information" (clarity, efficiency) in a noisy manner that recalls such other digital expressions of *détournement* as those created or collected by Irational.org [*sic*] ("an international system for deploying 'irational' information, services and products for the displaced and roaming").[70] For example, clicking on Jodi's "BINHEX/message.hqx" link leads to what appears to be merely interesting, pixelated noise: a background of jaggy streaks in a wide spectrum of colors and a foreground of twice-duplicated bar-graph charts with text rendered as indecipherable, bar code–like inscriptions.

The following excerpt from an interview with Jodi may be taken to be the manifesto of retro-styling for the purpose of noise:

> Do you see electronic media as obscuring communication?
> jodi yes/no
> ill.communication is ok
> ,makes good noise
> ^$%&$%^$%^$^&*&$%$&^(&$^[71]

This is "information designed to resist information," as I have put it, with a vengeance, but with the difference that it is retro- rather than state-of-the-art styling that is cool. The final line here looks like printed binary code from any previous era of information spanning from World War II–period computers (Konrad Zuse's Z1 and Z3 machines) to early personal computers. Or, most subversively, what this latter line and similar lines of apparently random ASCII characters on Jodi's pages clue us to is that the "noise" is not just a matter of old technology obstructing the data stream of new technology. Instead, the perception of noise is inherently part of the substrate of even the most state-of-the-art computing. Jodi's line of random characters resembles an ASCII representation of the binary code of any program or image file—even the most recent—on today's computers.

Put generally, Jodi's subtext is not that old technologies of information are vulnerable to noise and thus introduce nonsense into contemporary, state-of-the-art data. It is that old technologies were all "state of the art" in their own times, and that state-of-the-art technology in each generation has to wrestle anew with the paradox of noise embedded in the modern "mathematical theory of information" that Jodi clearly alludes to in the comment that "ill.communication is ok / ,makes good noise" (a play on Claude Shannon's information theory, which equates information with entropy even as it seeks to isolate the interference of *bad* "noise"). The retro-tech and retro-noise displayed on Jodi's site illustrate how even current state-of-the-art information is in a deep sense nonsense.

But ultimately, "illustrates" is too polite a way to put it. Jodi is not entirely polite. Jodi is also conceptually (though not actually) viral—meaning that its art does not necessarily wait for the viewer of the page to make the mental connection between what is manifest inside the frame of art (the noise of retro-technology) and what can only be pointed to implicitly outside that frame (the noise of current technology). Jodi's work threatens—at least in appearance—to puncture the prophylactic of artistic framing to let noisy code loose on the world. In particular, several pages on the <wwwwwwwww.jodi.org> site pose at least a momentary worry, even for technically savvy users, that on-screen disturbances of nonsense are not just "canned" effects (like art in a frame) but instead active processes running amok on the user's local computer. The bar-graph screen invoked by clicking on Jodi's "BINHEX/message.hqx" link, for example, mimics a kind of disintegration of screen-image integrity that is symptomatic of certain kinds of hardware or software problems. (Zooming in and out repeatedly on an image in a browser such as Internet Explorer 6, for instance, could sometimes multiply and distort images in a manner akin to the twice-duplicated effect of the bar graphs. So, too, running out of graphics memory and related resources on a Windows 95 or 98 computer can create screens that incompletely transform between two states.) Even more disturbing are the streaks of pixelated noise in the background of Jodi's bar-graph screen because they suggest one of the most disabling problems on a computer: a hardware or software glitch involving the graphics adapter and/or monitor that will block even the boot-up process or, sometimes just as bad, the ability to change display settings.[72] Still more disturbing are the animated screens on Jodi's site. Clicking on the "Reflector/131.24.167" link, for example, yields a screen that redirects itself automatically to another page once each second in an endless loop, but with the redirect linking mindlessly to the same page instead of a different page.[73] This "dynamic stasis," as it can be called, simulates the kind of locked or cycling behaviors of a computer about to crash.[74]

The edginess of such misbehaving screens is then honed into fine art in Jodi's *%WRONG Browser* project, whose Web site makes available for download several alternative browser programs that are designed to subvert, invert, or pervert the apparent clarity of information intended by the major, corporate-sponsored browsers. At the time of this writing, there were five such Jodi browsers whose names parody such dominant Internet domains (seen from Jodi's European location) as *.nl, .co.kr, .com.*, and *.org*. As Baumgärtel says, each of these browsers "automatically connects to the net and then proceeds to mangle the web pages by reading their code in the 'wrong' way."

> [The browsers] take over your monitor as soon as they are started. Don't bother trying to use them; they can do that themselves, thank you very much. Each piece connects automatically to random internet addresses in the domain space that their title indicates. You wouldn't know, though, since the programs show the data that they download from these sites as a jumbled mix of digital pulp. You can punch in another URL, but before long the automated net carousel starts to turn again without your interference.
>
> . . . Once started, their . . . browsers mercilessly rip apart HTML data and turn them into a digital collage in the colorful shades of old 8-bit computers. . . . As Dirk Paesmans [of Jodi] puts it, "It looks chaotic, it is not really readable, it is simply not practical. Of course, these are exactly the things we like."[75]

In short, Jodi creates scary browsers. The Jodi browsers are scary because they exhibit three kinds of behaviors simultaneously: they are destructive, they are self-activated, and their activity (as in the case of all "client" programs) occurs *on the user's local computer*. The first two behaviors are no more than the formula for twentieth-century avant-garde, auto-destructive art. The third behavior, however (working in conjunction with the others), stamps Jodi's art with at least some of the symptoms of our contemporary moment of viral aesthetics. Seen in overview, the distinctive genius of Jodi is not so much content or even technique, neither of which is remarkable. Jodi's genius—or in hacker-speak, "exploit"—is the chosen *site* of action. Here it is useful to remember the tradition of the artistic "installation," privileged by minimalism, conceptualism, and other art movements from the 1960s on (including, for example, "land art" or "earthworks"). In this contemporary tradition, an "installation" is art that barely or uneasily fits in formal gallery spaces—art that challenges the legitimacy of such "framed" space (e.g., through installations of rocks or dirt brought into a museum). Perhaps the best way to describe Jodi's art is to call it installation art for the Internet. What is Jodi's installation site? Like many other "Net artists," Jodi does occasionally display in physical exhibition spaces—for example, the Documenta exhibit that the duo was uneasy about.[76] But Jodi's real space is the Internet and, in particular, the seam in the "client/server" system of the Internet where action can seem to occur undecidably both on the server and on the local client machine (e.g., through "client-side" scripts that execute programs on the local machine). The crucial fact about Jodi's *%WRONG Browser* project is thus that users must literally "install" Jodi's perverse browsers on their own machines, where they promptly run berserk. So, too, the most disturbing screens on the <wwwwwwwww.jodi.org> site are those that appear to produce self-activated, destructive, and *local* behaviors on the user's screen. We are well on our way to art as virus, which is to say art that not only performs its own destruction but—exploiting the conditions of the Internet—also contaminates the world outside the frame of art.

Relevant here is a section of Baumgärtel's interview with Jodi that has become canonical in online discussion of the pair:

> [Jodi:] When a viewer looks at our work, we are inside his computer. There is this hacker slogan: "We love your computer." We also get inside people's computers. And we are honored to be in somebody's computer. You are very close to a person when you are on his desktop. I think the computer is a device to get into someone's mind. We replace this mythological notion of a virtual society on the net or whatever with our own work. We put our own personality there.[77]

The site of Jodi's art is not "out there" in gallery space; it is "inside" the user's computer. Or, at least, it is undecidably outside and inside. On the one hand, Jodi expresses reservations about vandalizing the user's computer. The Baumgärtel interview continues:

> There is this rumor that your site causes people's browsers to crash. Is this true?
>
> [Jodi :] No. That is not a challenge. You could shut down anybody's computer with one line of code. That's not interesting.

But on the other hand, Jodi contradicts itself elsewhere in the interview:

> Recently we made this map of the internet, where we took a diagram with all the big back-bones and the names of the major providers. We replaced the names of technical providers with alternative and art sites on the net,

> with links to these sites. We put this piece on the site of the documenta. Every time somebody at the documenta Halle comes to this map and tries to click on one of these links, the computer will crash.

Viral aesthetics either *will* or *will not* make computers crash—an effect of undecidability that is not unlike that of actual computer viruses, whose "payload" varies from purely cosmetic effects (e.g., putting a message on the screen on a particular date) to devastating destruction.

Randall Packer describes Jodi's work as follows:

> Your experience with Jodi might include such misshapen encounters as: a "not found 404" breach in the website that resounds with a warning beep; an interactive user-submission form that strips entries of their vowels; a repository for those filtered vowels funneled from the previous page; a fractured browser interface; fields of numbers endlessly displaying; "good times" complete with corporate icons; more "good times" infected with virus warnings; news of "a 14-year-old sociopath" that "brings my company to its knees"; a virex [Macintosh antivirus program] scan of your computer data; and ASCII characters as pure art. . . . Ultimately, Jodi is Code stripped of all functionality, Code for its aesthetic value, Code as abrasive language, Code as hallucination, Code as theater.

If Jodi is theater, we may say, then its stagecraft transgresses the boundary between stage and audience in the manner not just of experimental theater (Packer's paradigm) but of the infamous "Good Times" virus, a hoax or mere *performance* of a virus in the mid-1990s that uncannily created an effect very much like a real "worm" virus: it prompted users to pass on warnings about the fake virus that in themselves, through their sheer number, created wormlike effects.

Exhibit 4. Other artists or art groups could be exhibited next to Jodi to demonstrate the same edgy, undecidably viral art that makes viewers fear for the integrity of their computer. See, for example, the spectacularly disturbing Web sites of Absurd.org or 0100101110101101.org. Or see *WinGluk Builder* by CooLer, which received an honorable mention in the "software art" prize competition at the 2002 Moscow Read_Me 1.2 software arts festival. *WinGluk Builder* "imitates" viruses and allows users to simulate building viruses.[78] But at this point our fiction of gallerylike "exhibits" crashes, since we are on the verge of breaking out of the exhibit hall into the world at large. My final *Exhibit 4*, therefore, can only be named "exhibit" ironically. It is a twofold exhibit that centers on the conceptual work of the Critical Art Ensemble (CAE) and practical demonstrations by the Electronic Disturbance Theater, 0100101110101101.org, and others.

Founded in 1986 by Steve Kurtz and Steve Barnes, and soon transformed into a "broad based artist and activist collective" of six core members, the Critical Art Ensemble is best known for its series of books arguing for digital activism (sometimes called "hacktivism"), its related publications and projects attacking the biotech industry, and occasional Situationist-style performances, street theater, and other disturbance "art"—all supplemented by appearances on the academic circuit and at Ars Electronica, Documenta, and other international art exhibitions. Technically advanced as well as thoroughly read in postmodern theory (e.g., Deleuze and Guattari, Foucault, Baudrillard, Debord), CAE is now perhaps our most conceptually powerful expositor of destructive creativity and viral aesthetics. It is what the *Processed World* collective, with its "bad attitude," might have become had it progressed from stapling floppy disks to reading pomo theory while cracking code.

As first articulated in 1994 in *The Electronic Disturbance* and subsequently refined in *Electronic Civil Disobedience and Other Unpopular Ideas,* as well as, most recently, *Digital Resistance: Explorations in Tactical Media,* CAE's hacktivist art proceeds on the basis of a primarily Deleuzean critique of the social, economic, political, and military powers of dominance. Or, more accurately, CAE is Deleuze and Guattari with a difference.

Deleuze and Guattari's celebrated *A Thousand Plateaus* (following upon their *Anti-Oedipus* [1972]) is best known for its tour-de-force articulation of "rhizomes," "deterritorialization," "assemblage," "nomadism," "schizophrenia," "bodies without organs," and other delirious paradigms of a generalized philosophy of decentralization. This postmodern philosophy has two distinct personae or tonalities. In what might be called their dominant, "May 1968" persona, Deleuze and Guattari envision a teeming world of experience that refuses to cleave to singular "lines" of identity, much less selves, organisms, states, and other fortresses of bounded identity. Rather, biological, psychological, social, or linguistic life in their vision is constituted through the propagation of zig-zagging, ricochetting, branching, and swarming lines of *de*-identification that, when they intersect with other lines, form up opportunistically and superficially—yet nevertheless vitally—into new, ad hoc "assemblages." That is, precisely because beings are not tied to the same-old, same-old that is their proper, inherent identity, they are free to pursue an instinct for *attachment* ("sticking" to other entities in hybrid entities) as well as *morphosis* ("becoming animal," as Deleuze and Guattari put it) that is much more lively than any life narrowly defined along the monotonous lines of self-identical development (e.g., the tiresome handing down of influence from father to son that Deleuze and Guattari see as the symptom of an oppressive psychoanalytic world-view). Think Bermuda grass, for example, which by rhizomatic propagation touches down superficially, but also ineradicably, on all soils with which it comes in contact. By contrast, what Deleuze and Guattari call the "arborescent" plant with its obsessively linear "taproot" may blow over in a storm.

But just as important, if recessive, is the recognizably post–May 1968 side of Deleuze and Guattari's work: the persona that expresses the particular blend of compromised hope yet belayed disillusion that followed upon the original libertarianism of the May 1968 moment. Thus, for example, we witness the care that *A Thousand Plateaus* takes to acknowledge the inevitability of "knots" or "roots" of "despotic" resistance within the rhizomes of emancipatory decentralization. Centralization and decentralization, for Deleuze and Guattari, are dialectical, but without hope of saving synthesis. Instead of thesis-antithesis-synthesis, there is what amounts to a fractal process by which each pole in the dialectic continually and reciprocally nurtures its opposite within itself. Centralization is so constituted as to produce effects of decentralization as part of its own inner working, and vice versa in the case of decentralization:

> There are knots of arborescence in rhizomes, and rhizomatic offshoots in roots. Moreover, there are despotic formations of immanence and channelization specific to rhizomes, just as there are anarchic deformations in the transcendent system of trees, aerial roots, and subterranean stems. The important point is that the root-tree and canal-rhizome are not two opposed models: the first operates as a transcendent model and tracing, even if it engenders its own escapes; the second operates as an immanent process that overturns the model and outlines a map, even if it constitutes its own hierarchies, even if it gives rise to a despotic channel.[79]

Deleuze and Guattari rest the ethics of their philosophy upon this détente of forces. In the moment of May 1968, we remember, freedom was good and oppression was bad. Ethics was a pure thing. But from the post–May 1968 perspective of *A Thousand Plateaus*, the most ethical position is one that recognizes the constitutive necessity, the instrumentality, of impurity. It is not just necessary, therefore, but *good* that freedom treasures small "despotic formations" of identity within itself. This voice of experience in *A Thousand Plateaus* is heard most plaintively in the recurrent passages in which Deleuze and Guattari counsel that too much freedom is not survivable. The following passage is representative:

> You have to keep enough of the organism for it to reform each dawn; and you have to keep small supplies of signifiance and subjectification, if only to turn them against their own systems when the circumstances demand it, when things, persons, even situations, force you to; and you have to keep small rations of subjectivity in sufficient quantity to enable you to respond to the dominant reality. Mimic the strata. You don't teach the BwO [Body without Organs], and its plane of consistency, by wildly destratifying. . . . If you free it with too violent an action, if you blow apart the strata without taking precautions, then instead of drawing the plane you will be killed, plunged into a black hole, or even dragged toward catastrophe. Staying stratified—organized, signified, subjected—is not the worst that can happen; the worst that can happen is if you throw the strata into demented or suicidal collapse, which brings them back down on us heavier than ever. This is how it should be done: Lodge yourself on a stratum, experiment with the opportunities it offers, find an advantageous place on it, find potential movements of deterritorialization, possible lines of flight, experience them, produce flow conjunctions here and there, . . . have a small plot of new land at all times.[80]

Organized institutions, subjectivities, organisms, and so on cannot simply be blasted apart without nullifying the possibility of living, of retaining some clenched little knot of being or identity from which to make rhizomatic attachments to other knots.

Now we can understand how the Critical Art Ensemble deploys Deleuze and Guattari with a difference. CAE *starts* with the post–May 1968 premise that there is a necessary dialectic between decentralization and centralization, and thus a standing relation between emancipation and "despotic formations." But CAE gives this thesis a unique twist adapted to its predominantly social, political, and economic analysis of the post–May 1968 era—that is, of postindustrialism. For CAE, postindustrialism is when "small" takeovers of decentralization by centralization (in Deleuze and Guattari's words, "small supplies of signifiance and subjectification," "small rations of subjectivity," "a small plot of new land at all times") are subsumed within a wholesale colonization. The great corporate and government despotisms of postindustrialism, that is, simply appropriate rhizomatics as the preferred way of doing business. Fundamental to this analysis is CAE's insight into the central role played in postindustrialism by the new technologies of decentralization, especially electronic and digital communication networks. Where state, economic, social, police, and other forces of repression once fortified themselves in what CAE terms "bunkers" of immobile power, now such forces have taken to the informational circuits to make rhizomatic mobility (e.g., the instantaneous flow of capital in or out of a developing country) the instrument of a new domination. "The archaic model of nomadic power," CAE thus says in its important essay "Nomadic Power and Cultural Resistance," "has evolved into a sustainable means of domination." Put another way: "Elite power, having rid itself of its national and urban bases to wander in absence on the electronic pathways, can no longer be disrupted by strategies predicated upon the contestation of sedentary forces. The architectural monuments of power are hollow and empty, and function now only as bunkers for the complicit and those who acquiesce. They are secure places revealing mere traces of power. . . . These places can be occupied, but to do so will not disrupt the nomadic flow."[81] In short, the protest tactics of the May 1968 generation—for example, occupying university buildings—are now quaint. Real power still has buildings that can be occupied, but they are only decoys. Real power has learned to appropriate the "mob" (i.e., "mobilized") tactics of the street protests of the 1960s and 1970s by taking to the information superhighway. The real street is now the infinitely mobile network where transactions, surveillance, publicity, and other powerful instrumentalities of domination reach a worldwide audience.

Because CAE starts with the premise that centralization and decentralization accommodate each other, it is not interested in Deleuze and Guattari's survivors' ethics of cohabitation or reciprocation between rhizomatics and despotism. CAE instead speaks for a new generation of protest whose ethical and political imperative is to take to the information networks themselves to combat the new world of rhizomatic power. "It is time to turn attention to the electronic resistance," it says.[82] In particular, it is time to invent or adapt methods of viral resistance that obstruct, disrupt, or corrupt the rhizomatic or nomadic data flows upon which contemporary power depends:

> Nomadic power must be resisted in cyberspace rather than in physical space. . . . A small but coordinated group of hackers could introduce electronic viruses, worms, and bombs into the data banks, programs, and networks of authority, possibly bringing the destructive force of inertia into the nomadic realm. Prolonged inertia equals the collapse of nomadic authority on a global level. Such a strategy does not require a unified class action, nor does it require simultaneous action in numerous geographic areas. The

> less nihilistic could resurrect the strategy of occupation by holding data as hostage instead of property. By whatever means electronic authority is disturbed, the key is to totally disrupt command and control. Under such conditions, all dead capital in the military/corporate entwinement becomes an economic drain—material, equipment, and labor power all would be left without a means of deployment. Late capital would collapse under its own excessive weight.[83]

In several passages of *Electronic Civil Disobedience and Other Unpopular Ideas,* CAE refines the concept by thinking in detail about the aims and methods of data trespass, blockage, and "hostage" taking:

> As in CD [civil disobedience], the primary tactics in ECD [electronic civil disobedience] are trespass and blockage. Exits, entrances, conduits, and other key spaces must be occupied by the contestational force in order to bring pressure on legitimized institutions engaged in unethical or criminal actions. Blocking information conduits is analogous to blocking physical locations; however, electronic blockage can cause financial stress that physical blockage cannot, and it can be used beyond the local level. ECD is CD reinvigorated. What CD once was, ECD is now.
>
> For more radical cells ECD is only the first step. Electronic violence, such as data hostages and system crashes, are also an option. Are such strategies and tactics a misguided nihilism? CAE thinks not. Since revolution is not a viable option, the negation of negation is the only realistic course of action. After two centuries of revolution and near-revolution, one historical lesson continually appears—authoritarian structure cannot be smashed; it can only be resisted.
>
> Of prime concern is the development of the tactic of data hostaging, where criminals hold precious research data for ransom. Motivations for such an activity are construed [by state authorities] solely as criminal. . . . But something else of greater interest is beginning to occur. The terror of nomadic power is being exposed. The global elite are having to look into the mirror and see their strategies turned against them—terror reflecting back on itself. The threat is a virtual one. There could be cells of crackers hovering unseen, yet poised for a coordinated attack on the net—not to attack a particular institution, but to attack the net itself (which is to say, the world). A coordinated attack on the routers could bring down the whole electronic

> power apparatus. The vulnerability of the cyber apparatus is known, and now the sign of virtual catastrophe tortures those who created it.[84]

We might think at this point of Deleuze and Guattari's well-known paradigm of the wasp-orchid, according to which two genetically unaffiliated species come into contact with each other and reciprocally adapt their morphology so that the wasp (the pollinator of the orchid) anatomically "traces" the form of the orchid flower, and vice versa—the two organisms thus becoming a hybrid wasp-orchid or orchid-wasp. The vocabulary of "code capture" that Deleuze and Guattari use to describe the resulting assemblage is particularly relevant: "not imitation at all but a capture of code, . . . a becoming-wasp of the orchid and a becoming-orchid of the wasp."[85] The electronic resistance proposed by CAE is a wasp-orchid tactic native to the new codes of postindustrial information where "bugs," not to mention viruses, have an especially potent sting. In CAE's analysis, the dominant power formations of society have captured the code of rhizomatic freedom. Therefore, what more fitting revenge than for the gadfly or viral wasp of "electronic disturbance" to resist the new nomadic dominance by capturing in return the *immobilizing* tactics of traditional repression? Thus we note that CAE's recommended tactics of resistance are *not* guerrilla hit-and-run but a post-guerrilla warfare style that might be called "hit and block." Resistance now forms "knots" of sedentary blockage or "despotic formations" within the electronic rhizomes of a world network dependent on smooth flow. Resistance itself, therefore, organizes in mimicry of Cold War–era military or espionage "cells" ("small but coordinated group of hackers," "cells of crackers") that seem to defy the new postindustrial mutation of "cells" into flexible "teams."

That resistance to dominance seems to require mimicking regressive forms of twentieth-century dominance—precisely the forms of blockage, policing, and unbending discipline that postindustrial dominance itself has abandoned for a fleeter, quicker "chaos management"—might seem ethically troubling. But CAE professes not to be troubled because it can call upon an originally modernist "cadre" concept—avant-gardism—to construe its tactics. For hacktivists to make an effective disturbance in the world-flows of information, CAE believes, they have to be technically skilled. That means that they must be part of the technocratic class and adopt its methods and patterns of work. This in turn means that they need to organize themselves in the spitting image of a crack, technocratic cadre—a kind of elite "tiger team" or "skunk works" able to use the most advanced tools of contemporary power. Such an organization is what CAE calls a "new avant-garde." As it says in its "Addendum: The New Avant Garde,"

> Avant-gardism is grounded in the dangerous notion that there exists an elite class possessing enlightened consciousness. The fear that one tyrant will simply be replaced by another is what makes avant-gardism so suspect among egalitarians, who in turn always return to more inclusive local strategies. While CAE does not want to discourage or disparage the many possible configurations of (democratic) resistance, the only groups that will successfully confront power are those that locate the arena of contestation in cyberspace, and hence an elite force seems to be the best possibility.[86]

Only such a new avant-garde, undecidably artistic and technocratic, will be able to write the code for electronic disturbance.

Is CAE successful in thus legitimating electronic disturbance on the model of a "new avant-garde"? Certainly the group is expansive in its aims for neo-avant-gardism: CAE is ambitious of making neo-avant-gardism a governing aesthetic that can redefine the meaning of "artistic creation." "By appropriating the legitimized authority of 'artistic creation,' and using it as a means to establish a public forum for speculation on a model of resistance within emerging techno-culture, the cultural producer can contribute to the perpetual fight against authoritarianism. . . . The electronic world . . . is by no means fully established, and it is time to take advantage of this fluidity through invention, before we are left with only critique as a weapon."[87] For CAE, "artistic creation" is an aesthetic ideology of destructive creativity that can be positioned in opposition to the dominant contemporary ideology of "creative destruction." Mimicking, in the manner of the wasp-orchid, the official "innovation" mania of the whole corporate world, CAE seeks to "invent" critique anew by capturing the code of creation within a new art and politics of destructivity.

Yet there are signs of strain in CAE's effort to make critical destructivity legitimate in the name of art. We can note here for future reference one dissenting construal of "electronic disturbance" that has periodically plagued the group. This is the charge, expressed well before September 11, 2001, that the views of CAE are "terrorist."[88] After all, if electronic disturbance not only exposes the "terror of nomadic power," but also itself mimics or inhabits the forms and technologies of power, then terrorism would seem to be conceptually part of the tactics that CAE recommends. CAE is sensitive to this issue, attempting to ward it off, for example, by prescribing the moral limits of electronic disturbance, as in the following passages:

> Activists must remember that ECD can easily be abused. The sites for disturbance must be carefully selected. Just as an activist group would not

> block access to a hospital emergency room, electronic activists must avoid blocking access to an electronic site that may have similar humanitarian functions.
>
> Further, if terms are not met, or if there is an attempt to recapture the data, ethical behavior requires that data must not be destroyed or damaged. Finally, no matter how tempting it might be, do not electronically attack individuals (electronic assassination) in the company—not CEOs, not managers, not workers. Don't erase or occupy their bank accounts or destroy their credit. Stick to attacks on the institutions. Attacking individuals only satisfies an urge for revenge without having any effect on corporate or government policy.[89]

So, too, CAE has responded to public accusations of terrorism. CAE's appearance at a conference in London in 1994 led to a charge in the question-and-answer period that CAE was suggesting "not a civil tactic of political contestation at all" but "pure terrorism." CAE's response, which has since become its party line on the issue, is as follows: "CAE found this comment to be very curious because we could not understand who, or more to the point, what this audience member thought was being terrorized. How can terror happen in virtual space, that is, in a space with no people—only information? Have we reached a point in civilization where we are capable of terrorizing digital abstractions? How was it that this intelligent person had come to believe that electronic blockage equaled terror?" And again: "What is frightening to CAE about this scenario is that electronic erasure is perceived as equivalent to being killed in a bomb explosion. Now the perception exists that the absence of electronic recognition equals death."[90]

Never mind that military, police, and other forces of repression do use "virtual" means to repress people in "real" ways, culminating in the impoundment of computers, arrest, and physical incarceration. Never mind that "absence of electronic recognition"—for example, when one's name, face, fingerprints, or passport do not match up with an official database at the myriad checkpoints of contemporary life—leads to the most Kafkaesque scenarios of repression. (How would CAE's views change, we wonder, if even for a moment it took the perspective of illegal or even legal immigrants on what it means to be either included in or excluded from an official database?) And never mind that CAE has argued throughout that the whole point of using digital means of resistance is that "real" power—in all relevant social, political, economic, and other senses—has now migrated into the circuits. Whether or not one agrees with CAE's goals and methods, its punctilious stand on the difference between real people and virtual information *about* people seems remarkably weak. It does not hold up to the same standard of ethical, political, and intellectual rigor (not to mention sophistication) of every other important argument made by the group. If one were a hacker, one would say that this is a network "port" in the CAE system that has been left conspicuously open.

We will come back to this port later. But for now, we can close our exhibit of CAE by giving an abbreviated catalogue raisonné of "electronic disturbances" that indeed reach out beyond the auto-iconoclasm of avant-garde art to block, destroy, disrupt, or corrupt the networks of world power. In this regard, it is best not to concentrate on the practical work of the Critical Art Ensemble itself, which reflects a Situationist preference for small-scale, localized, home-crafted, and primarily parodic *détournement*. For example, consider CAE's "tactical media" installations, including "Artist Books" designed "to insinuate plagiarist texts into cultural institutions," "Radio Bikes" designed for "nomadic broadcast," or toy car tracks set up at public sites to provoke the spectacle of a confrontation with police over playing with toys.[91] In their practical work, the members of CAE are still primarily "performers" in a theater of the absurd or of Situationist spectacle. Thus, in the essay "The Recombinant Theater and the Performative Matrix" in *The Electronic Disturbance*, CAE characteristically speaks of computer hacking as a performative event: "Consider the following scenario: A hacker is placed on stage with a computer and a modem. Working under no fixed time limit, the hacker breaks into data bases, calls up h/er files, and proceeds to erase or manipulate them in accordance with h/er own desires. The performance ends when the computer is shutdown." Or to change paradigms from performance to a spectacle native to the digital world, we can say that CAE creates what are primarily only "demos" of hacktivist art, even if some of its more fully realized demos are packaged as if intended for mass effect.[92]

Let us look instead to others who have practiced electronic disturbance or viral art in ways intended to propagate beyond the theater of spectacle onto the actual networks—the rhizomatic media of TCP/IP, client-server, and peer-to-peer—where CAE itself does not practice. Consider, for example, the Electronic Disturbance Theater (EDT), a group of hacktivists (including Ricardo Dominguez and Stefan Wray) who since 1998 have applied CAE's theory of electronic disturbance in a series of viral political/artistic actions on the Internet. These actions include EDT's instigation of or participation in "swarm" distributed denial-of-service attacks (facilitated by its FloodNet software) against various government and corporate targets—for example, the Mexican government (in support of the Zapatista movement) or eToys.com (in support of the Etoy artist collective whose similarly named Web site was the target of legal action by the company).[93] Or again, consider the European duo named 0100101110101101.org, which collaborated with the epidemiC group at the 2001 Venice Biennale to launch an art-virus named "biennale.py" (a true virus, but with a nondamaging payload and restricted to the Python programming language).[94] These groups are harbingers of the era of virus art. As epidemiC says in describing its 2001 exhibition titled "virii virus viren viry" (at the "d-i-n-a, digital is not analogical" festival in Bologna), "the writing of the source code, that is, the text which programmes a virus to be set loose and do its work can be seen as an aesthetic product." Indeed, such viral programming "has all the credentials of a totally original avant-garde movement, one that is revolutionary and brings innovation both in terms of the generation of new linguistic, psychological, biological and communicational models and on a purely aesthetic level."[95]

We could exhibit more examples of viral aesthetics, but perhaps the above instances are sufficient to suggest the potential of the new aesthetics of "destructive creativity" in the information age. In any case, as I earlier suggested, destructive creativity is only the most sublime of the contenders for a new governing aesthetic responsive to the postindustrial ideology of "creative destruction." Or in more updated terms, destructive creativity is the most "virulent" of the new aesthetics designed to propose an alternate vocabulary of creation/destruction (disturbance, chaos, play, transformation) intended to reconfigure "creative destruction" into a vision not just of global corporate culture but of a flourishing multiplicity of cultures. In a fuller study of information age aesthetics, we would also want to explore what I have called the "new picturesque" of digital art in its relation to the new sublime of destructive creativity. Nor are these two governing ideologies of postindustrial aesthetics necessarily the only contenders. In his "transarchitectures" or "liquid architecture," for instance, Marcos Novak has proposed an aesthetics of "allogenesis" whose goal is the production of the truly other or "alien." The morphed, transient three-dimensional architectures he pulls out of his algorithmically generated fourth dimension of architecture (shapes that shift in time) are reminiscent of the lower dimensional visualizations that string theory physicists use to imagine the postulated ten or eleven spacetime dimensions of the universe (see fig. 11.6). Even in reduced, lower dimensional form, Novak's shapes bear the trace of an unutterable otherness of sensibility—at once entrancing and monstrous—that cannot be assimilated easily to any of the traditional categories of the picturesque, sublime, or (the preferred aesthetics of string theorists) beautiful.[96] All these

Figure 11.6 Marcos Novak, rendering of transarchitectural shape from the *AlloBio* project
Reproduced with permission of Marcos Novak.

struggling aesthetics are attempts to install within the life of knowledge work an "ethos of the unknown" able to mold "cool"—the popular form of such ethos—into more experimental, expansive, diverse, active, and aware forms.

The adequacy of such future aesthetics of the information age, clearly, will henceforth merit reflection, not just in regard to the neo-avant-garde but with respect to the ordinary generations of cool that the neo-avant-garde will mirror, amplify, and help educate.

Chapter 12

Speaking of History

Toward an Alliance of New Humanities and New Arts (With a Prolegomenon on the Future Literary)

I ended the previous chapter by speculating on the "adequacy" of aesthetics in the information age. But what could such adequacy be? Adequate to what, and by what measure? At this point I return to the issue of the relation between the contemporary humanities and the contemporary arts. I have argued that the humanities stand to gain "cool" from the arts, and that the arts might gain in return from the humanities a historical rationale for the new aesthetics (including "viral aesthetics") currently struggling to emerge from the suffocating postindustrial credo of ceaseless, heedless creativity. How this latter aesthetic legitimation is possible is now my concluding topic, for the social, ethical, and aesthetic adequacy of the creative arts in the age of knowledge work seems to me to depend on acquiring such a legitimation. A partnering of the humanities and arts is needed to fulfill the conditions for a future aesthetic.

Remember the "port" we left open in the discussion in chapter 11 of the Critical Art Ensemble's theory of electronic disturbance—that is, the vulnerability of its "new avant-garde" aesthetics to the charge of "terrorism." The validity of such a charge is not the issue here, nor would it be even if most of this book had not been drafted before the terrorist attacks of September 11, 2001. Just because viral aesthetics can be charged with "terrorism" does not necessarily mean that such a charge is discriminate enough—sufficiently sensitive to specific intentions, circumstances, results, and underlying contexts—to be admitted into serious intellectual inquiry, let alone to take precedence in such inquiry. Inversely, just because viral aesthetics practices "virtually" rather than "physically" on the world does not mean—counter to the argument of CAE—that the charge can *never* meaningfully, let alone legally, be made. That would be another kind of indiscriminateness.

Instead, the issue here is the very condition of indiscriminateness that encourages absolutist debates of "art" versus "terrorism." This uncivil condition indicates that while viral aesthetics and other such new aesthetics may be contenders for a governing aesthetic ideology in the age of knowledge work, their ability actually to provide capable governance—that is, to imagine a civil compact not just of the art scene but of its relation to the larger social scene—is blocked by the inability to legitimate the new art. Proponents of viral aesthetics and the "destructively creative" counter-ethos it champions, for example, are especially hard put to explain the cultural value of postcreative art, by which I mean its "ethical" significance in both the broad sense of ethos (the "ethos of the unknown," I called it) and the narrower sense of moral or legal right (the responsibility of its means and ends). The new aesthetics lacks a persuasive rhetoric. After all, such aesthetics follows an avant-garde tradition that prides itself on what can only be described as an antirhetoric of (il)legitimation: stark confrontation, minimalist elision, cryptic allusion, hermetic personal philosophy, difficult modernist or postmodernist theory, or sometimes just astounding inarticulateness. (Think, for example, of Jodi "explaining" its work in interviews or, more characteristically, refusing any explanation on its Web site.) Viral artists thus have no adequate answer to the indiscriminate questions posed by those who either do not discern or do not value their critical art or any critical art: What is socially redeeming about an art that resembles "cyberterrorism," which in turn (as the "viral" metaphor might suggest) resembles "bioterrorism," which itself is the "poor man's version" of nuclear terrorism, and so on up the scale of apocalypse in the contemporary logic of terrorism and counterterrorism?

As demonstrated in the response of the Critical Art Ensemble to the charge of terrorism, viral aesthetics is dumbfounded in the face of such questions. It can offer only such theoretically impoverished responses as CAE's insistence on an essentialist difference between the virtual and physical. Or, inversely but with similar effect, it can answer only with hypertheory, which is unlikely to be found adequate in the court of public opinion. Concepts such as "rhizomatics" and "nomadology" are not going to be understood by the public. And even if they are understood, they would not be persuasive. While knowledge workers may vote for rhizomatic democracy in principle, they also want firewalls for their personal computers; and they kill Bermuda grass on their lawns. Similarly, though less starkly, the other new media aesthetics I instanced previously—complexity art, agent art, transcoding art, gaming art, "alien" art—are also at risk of seeming illegitimate. "Terrorist" may not be the accusation in their case, but instead "frivolous," "perverse," or "just a game." Or, perhaps even more debilitating, the new aesthetics may simply be lumped together with "cool" by the dominant corporate, media, and other institutions of society as part of the undifferentiated bread and circuses of contemporary "entertainment" (the industry that now recruits the "edgiest" and "coolest" musicians, artists, directors, and Web designers).

To explain the legitimacy of the new aesthetic ideologies, I believe, will require precisely the alliance of the arts and contemporary humanities I have posited, beginning in the academy. Humanities education may be all thumbs when it comes to doing cool (relatively few humanities educators can keep up with contemporary information artists, musicians, and electronic literature creative writers in the use of graphics, animation, sound, scripting, and database programming). But the humanities can complement the contemporary information arts by furnishing a paradigm of knowledge that is at once seriously critical of knowledge work and more broadly understandable than "virus." That paradigm is historical knowledge. We can think of it this way: viral aesthetics has improvised a model of "destructive creativity" that undermines the postindustrial ethos of creative destruction from within, reversing the polarity of that ethos to expose the full trauma of destructivity that Schumpeter never underestimated in his original statement of "creative destruction," but which current knowledge work society comfortably obscures behind the façade of boundless "innovation." However, "destructive creativity" is a critique whose absolutism of "destruction" (allegedly "terrorism") vs. "creation" brooks no discourse that can span opposing worldviews to provide a legitimation of art. What the humanities can offer is a long-standing mode of historical interpretation (extending from Enlightenment historiography and nineteenth-century historicism through historical materialism to such recent variants as *Annales* history, structural Marxism, cultural materialism, New Historicism, postcolonial theory, and "cultural criticism") that now perhaps alone has the potential to depict "creative destruction" and "destructive creation" as part of the same process. That process is called *history*.

In the view of the contemporary humanities, history fundamentally transcodes creation as destruction, and vice versa. What is commonly or officially deemed "creative," in this light, destroys, whether deliberately or unwittingly, locally or genocidally. The example offered in Stephen Greenblatt's "Invisible Bullets" essay (one of the hallmarks of the New Historicism) is the creation of the Virginia colony in the sixteenth century, which also had the destructive effect of exposing indigenous populations to European disease. Inversely, what is normally called "destructive" (e.g., subversive, revolutionary, unpatriotic, perverse) creates. As in the case of the original "Terror" of the French Revolution, destruction seeds boundless innovations in the social, economic, military, political, educational, cultural, and other institutions governing modern civil society. In between the poles of creation and destruction, there opens to view the real domain of the humanities in their contemporary form: all the alternative, marginal, minoritarian, subcultural, and residual worlds, at once imperiled and subversive, that are undecidably creative and destructive. In the 1990s the corporate culture of "creative destruction" valued "startups" above all. In that same period, the humanities valued what might be seen as the opposite of startups: minoritarian survivor cultures that challenge the great, dominating forces of postindustrial society less through overt acts of rebellion than through the sheer endurance of their "residual" historical values.

The humanities, in short, have an explanation for the "destructive creation" that undergirds the new arts. That explanation is that the new arts mime the ruthless inexorability of history, which—whether cast in a conservative or liberal light (as "tradition" or "revolution," for instance)—exerts an authority that challenges the much shorter term, market-based authority of postindustrial "creative destruction." What is a thirty-week moving average or even ten-year average in the stock market, for instance, by comparison with the history of the *longue durée* articulated by Fernand Braudel?[1] Whether history is learned intellectually through academic schooling or more viscerally through the ordinary family or social schooling that inculcates what I have called a residual "lag" in work habits, it is difficult today to think of any authority other than history that has the heft to match the ideology of postindustrialism and globalism.

Equally important, the humanities have a rhetoric for their explanation-by-history. We might call this rhetoric a "speech act" of history, a concept that can be elucidated by juxtaposing the New Historicist essay by Greenblatt mentioned above with speech act philosophy. In "Invisible Bullets," Greenblatt recounts the life and works of Thomas Harriot, who, sent by Sir Walter Ralegh "to keep a record of the [Virginia] colony and to compile a description of the resources and inhabitants of the area," produced a publication titled *A Brief and True Report of the New Found Land of Virginia* (1588). Harriot especially took note of how the sleight-of-hand tricks the colonists conjured up with their technology and other knowledge impressed the native peoples as nothing short of supernatural—a thesis that if expressed in the context of European belief would have been deemed (as was often imputed about Harriot) subversively "atheistic" or at best Machiavellian. In Harriot's report, Greenblatt concludes, "we have one of the earliest instances of a significant phenomenon: the testing upon the bodies and minds of non-Europeans or, more generally, the noncivilized, of a hypothesis [in this case the dangerous hypothesis that all religion begins in legerdemain] about the origin and nature of European culture and belief." Greenblatt adds that Harriot's subversive "testing" was complemented by equally subversive acts of "recording" and "explaining," which reserved a space within the register of official Western understanding for alien, nonconformist beliefs. Testing, recording, and explaining together communicated within the very discourse of colonial administration a mental virus of doubt and otherness (doubt of God's truth, of the basis of Western dominion) that was the conceptual equivalent of the "invisible bullets" of disease that decimated the natives. Only, of course, such a conceptual virus was made safe by being quarantined to the "bodies and minds" of non-Europeans. "There is subversion, no end of subversion," Greenblatt famously says (adapting Kafka), "only not for us." Subversion is the thought of the other, and even when introjected within the self is experienced as if projected outward upon others. "Thus the subversiveness that is genuine and radical . . . is at the same time contained by the power it would appear to threaten," Greenblatt argues, where *contain* means both "include" and "restrain." The second half of "Invisible Bullets" then studies Shakespeare to fashion one of the New Historicism's characteristic formulations of art as the premier representation of subversion. Where Harriot as nonartist is a mere journeyman, Shakespeare as creative writer is master of the dangerous art.[2] Whether a Shakespeare play or a Balinese cockfight, Clifford Geertz similarly says in his "Deep Play: Notes on the Balinese Cockfight" (which also concludes upon Shakespeare and may well be read alongside "Invisible Bullets"), artistic representation is a way of "playing with fire."[3]

Subversive aesthetics as theorized in the New Historicism, we can now see with eerie recognition, is none other than contemporary viral aesthetics. But it is such aesthetics articulated in the name of history—in this instance not only because "Invisible Bullets" recounts an episode of literal viral action in the historical past, but because it makes that episode a paradigm of historicity in general as the process of subversion that societies "contain." What is history, the New Historicism asks, but an uncanny replication of the code of otherness within similarity, and vice versa, across societies and ages? Subversion may be the thought of the other, but incorporating the thought of that other—for example, capturing enough of its code through "testing, reporting, and explaining" to understand and be influenced by it—is the hermeneutic of history itself. The viral introjection of otherness is constitutive of the very meaning of history as "change" or becoming-other-than-we-are. Such change, which we remember is also the root issue in Schumpeter's economics, is undecidably creative and destructive.

Now consider what Greenblatt's principles of "testing," "recording," and "explaining" look like from the point of view of speech act philosophy. We can begin by reviewing the evolution of J. L. Austin's thought beyond his initial attempt to contrast "constative" and "performative" speech acts. Austin's uneasiness with this contrast is well known. The effort to rethink the binary of the constative and performative provides the plot of the lectures that became his *How to Do Things with Words*, which eventually introduces the more complicated apparatus of the "locutionary," "illocutionary," and "perlocutionary" to reconfigure the initial terms so that their otherwise unaccountable overlap (constative statements that have a performative aspect) can be governed within a schema making the constative itself an instance of the performative. All statements, that is, perform an action, even if that action is the seemingly neutral one of describing or stating.[4]

A juxtaposition of speech act theory with the New Historicism brings into sharp relief, however, the existence of another, perhaps more fundamental domain of overlap between the constative and performative that Austin, Searle, and others did not characteristically look upon in their fascinated gaze toward what Austin called the "present indicative active" (e.g., "I promise," "I name").[5] Most important, there are the speech acts of remembering, witnessing, testifying, and mourning ("I witness," and so on), to which Greenblatt adds his litany of testing, reporting, and explaining. These "speech acts of history," as they may be called, blur the line between "performing" action in the present and "performing" (in a different sense) actions from the social past. The New Historicism calls the latter, historicist performance acts "dramatization" or "representation." The uncanny blurring or overlap between present indicative active and historicist speech acts may be conceived in the following manner: speech acts of history act in the present for the (impossible) constitutive purpose of (re)constructing or (re)creating past actions that otherwise have no claim to existence either in the present or future—due not just to inadequacies in constative evidence but to what Austin called "unhappiness" (and Lyotard later the "differend") in the conditions of authority or convention that allow a witness of history to appear legitimate in the court of reality.[6] Ultimately, the New Historicism says that the existential status of the past depends on the complex hermeneutic by which the present rehearses the past through such speech acts as "I remember," "I witness," "I mourn," and so on, which in their composite *are* what we know as history. And it says also that the existential status of the past is at least as worthy of thought as the existence-affirming and -making contract between the present and future that brings out the best in Austin and Searle's philosophy of language (as in Searle's work on "promises").[7] When a survivor of the Holocaust says, "I witnessed" (to use Lyotard's fundamental example in *The Differend*), the issue is the relation not between the constative of present reality and the performative that can create a future possibility, but between present reality and a dramatistic performance able to recreate *past* possibility.[8] The constative value of such past possibility, evanescent to the point of extinction, is at stake even before the legal, ethical, economic, and other standing of that value is adjudicated.

Put another way, the issue of historical existence is fundamental, rather than just what Austin calls "perlocutionary" (just a matter of convincing someone that an event in the past happened), because the perlocutionary thesis does not account for all those constative aspects of the performance of history that the New Historicism and cultural criticism in general have discussed under the species name of "representation" and the genus name of cultural "construction." This is because the problem of the constative in historical understanding maps over the problems of both the constative and performative as these are united in the cultural critical notion of "construction." To remember, witness, testify, or mourn some event of history is not just to refer to that event with the aim of having an effect on an audience. It is to construct that event (or agent, action, object, victim) as significantly "real" in the first place amid all the myriad other formulations of events and participants that make a claim for real significance. Moreover, it is to assert that such past events have a reciprocal influence on the construction of present and future reality. "Construction," in other words, looks both ways in time: we construct the past that we believe constructs us. What do speech acts of history "perform," then? They do not perform the "present indicative active"; they perform/construct the historical reality that grounds the very leverage point of the present and indicative in which speech *can* act upon the future. Or, again: speech acts of history overlap unaccountably between the performative and constative because they perform/construct the idea of the constative as a fundamentally historical category. Nothing exists in a way that can be described, referred to, or indicated that does not emerge from constitutive historical experience.

In sum, the constative can seem shallow if exercised only or primarily in contracts between the present and future. This is why some of the signature thought-experiments of original speech act philosophy can seem so thin (abstract, reductive, almost bare of content) despite their "ordinariness." Consider, for instance, some of Austin's first examples of the performative in *How to Do Things with Words*:

> "I do (sc. take this woman to be my lawful wedded wife)"—as uttered in the course of the marriage ceremony.
>
> "I name this ship the *Queen Elizabeth*"—as uttered when smashing the bottle against the stem.
>
> "I give and bequeath my watch to my brother"—as occurring in a will.
>
> "I bet you sixpence it will rain tomorrow."[9]

By contrast, consider the deeper sense of being that may emerge from knowing, for example, not just when, why, how, and in the name of what ideology the good ship *Queen Elizabeth* was commissioned, but also the underlying fact that there indeed *exists* such a relevant history, whose constative status is visible nowhere in the narrow world of the present indicative active. Historical existence is visible only in the performed or constructed universe of the past active.

We can now put Greenblatt's "testing," "recording," and "explaining" in perspective by saying that these eminently administrative procedures are the modern, bureaucratic version of the long tradition of speech acts of history (remembering, witnessing, mourning, and so forth) by which societies perform the history—the conjoined process of creation and destruction—that constructs their reality. Testing, recording, and explaining are the speech acts by which the special performativity of historical knowledge can infiltrate the "performative knowledge" in another sense (the efficient functionality that Lyotard describes) of postindustrialism.[10] We note that Harriot was an intellectual of his time whose *Brief and True Report* was written in his role as what we would today call a bureaucrat (sent by Ralegh "to keep a record of the colony and to compile a description of the resources and inhabitants of the area"). He constructed a subversive history of the origin of Western religion and society by "testing," "recording," and "explaining" it within a report about the New World, thus accommodating radical historicity within administrative discourse. Analogously, Greenblatt is a humanities scholar of our own time whose work here adopts mimetically the bureaucratic, organizational vocabulary of industrialism and postindustrialism to win a place within contemporary knowledge work for an innately "subversive" historical understanding of creative destruction. In other words, given that the age of knowledge work is also the age of the perpetual layoff, perhaps such traditional speech acts of history as "I remember," "I witness," or "I mourn" would by themselves be too dispensable. Legions of the laid-off rehearse those speech acts, after all, and are simply remanded to "outplacement" services for therapy. By contrast, "I test," "I report," and "I explain" are speech acts better suited to sneak historical consciousness into postindustrialism because they seem so integral to the whole suite of surveys, reports, and presentations that are the "ordinary language" of the middle tiers of the corporate world. Yet no test, report, or explanation is valid without a historical baseline. Where postindustrialism extends its baseline back only as far as the last financial quarter or year, the humanities respond by asserting that the real value of knowledge can only be gauged across centuries and millennia.

This, then, is the answer to our question of adequacy about the future aesthetics of the information age: "adequate to what, and by what measure?" The answer that the humanities offer to that question of legitimation is the deep baseline, the profound "moving average," of history. Insofar as the ethos of postindustrial knowledge work has meaning, the humanities say, that meaning can only be found by measuring it not just economically, socially, politically, or even aesthetically, but ultimately ethically against the deep norm of history. This norm is not essentialist (by definition it "changes"). But it does impose a "lag" or "slack" upon the just-in-time norms of postindustrialism, and it is this feeling for a baseline of norms "other" to postindustrialism (yet "contained" within it as the history of its power of creative destruction) that can legitimate the "new avant-garde" and, more fundamentally, the everyday "cool" for which the new arts of information are the outriders.

The humanities thus have an explanation for the new arts of the information age, whose inheritance of a frantic sequence of artistic modernisms, postmodernisms, and post-postmodernisms is otherwise only a displaced encounter with the raw process of historicity. Inversely, the arts offer the humanities serious ways of engaging—both practically and theoretically—with "cool." Together, the humanities and arts might be able to offer a persuasive argument for the humane arts in the age of knowledge work. No doubt this hope is itself too utopian, not sufficiently ironic or skeptical, to be cool. But, in a way, that is my point: the contemporary humanities and arts must not only make contact with the generations of cool but lead them beyond the present limitations of cool. In the final analysis, cool is only a poor beginning, its current "ethos of unknowing" a poor substitute for what could be a much richer "ethos of the unknown." It is the hope of the humanities and arts in the information age to "educate" the cool in something like the root sense of that word—to "lead forth" the generations of cool in search of a genuine unknown residing equipollently in the future, the present, and history.

Finally, then, what about literature, whose "future literary" was the original stake upon which I began this book but whose specific plight I have left in suspense to advance a preliminary inquiry into the general conditions of cool and future aesthetics? What is the unique future of literature in the information age, or does it have a future at all in distinction from the amalgamated data stream of the new media arts in which all images, sounds, text, and so on might seem to be undifferentiated bits? Furthermore, what is the future of "literary history," the specifically literary manifestation of the historical approach of the humanities? How can it contribute to the future literary?

A fuller inquiry into the *differentia specifica* of literature and literary history in the information age, to use the old phrase of the formalists, will have to wait for a future project. (This book was conceived in principle as the first part of a continuing research initiative.) But in a manner that is itself a testament to the value of literary history, we can close with a prolegomenon to such an inquiry by remembering that the problem of the ethos of information—of an ethical understanding of information true *for* as well as *to* life—is a very old one, and that literature above all other knowledges once thought it had the answer.

Classical literature had conceived the literary precisely as the art necessary to value the truth in any mere information (Aristotle: poetry shows more "general truths"; Horace: poetry teaches truth by "informing or delighting"; Sidney: poetry teaches truth by delighting). Only so could information truly be *in*-forming, or what Sidney called *architectonike*, "which stands . . . in the knowledge of a man's self, in the ethic and politic consideration, with the end of well doing and not of well knowing only."[11] In the face of massive new historical and scientific information (skipping ahead several ages), Romantic literature then attempted even more fully to give truth the distinction of the poetic. If factual knowledge had a utility of epochal importance, as Thomas Love Peacock put it in his indictment of uninformed poets (*The Four Ages of Poetry*, a sort of ironic *Workplace 2000* for his time), then utility itself—as Shelley's *Defense of Poetry* answered in its memorable counter-indictment of information—must be accountable to poetry.[12] Only so could information be *architectonike* in Shelley's sense: "We have more moral, political and historical wisdom, than we know how to reduce into practice; we have more scientific and economical knowledge than can be accommodated to the just distribution of the produce which it multiplies. The poetry in these systems of thought, is concealed by the accumulation of facts and calculating processes. . . . We want the creative faculty to imagine that which we know; we want the generous impulse to act that which we imagine; we want the poetry of life."[13]

And so, too, in our own age, the New Critics took their stand against modern industrial and scientific information ("A poem should not mean/ But be") by demanding that knowledge be accountable to an American, pragmatist *architectonike.*[14] Poetry allows us to value, amid the dehumanizing data of modernity, the compound utility + truth that the New Critics named "experience." For Cleanth Brooks, for example, poetry was thus a nonaristocratic Sidneyanism or nontranscendental Shelleyanism, a tonic as much as *architectonike*: "The characteristic unity of a poem . . . lies in the unification of attitudes into a hierarchy subordinated to a total and governing attitude. In the unified poem, the poet has 'come to terms' with his experience."[15] "Experience" was the shadow doxa of technological rationality to which Brooks (above all other New Critics) dedicated poetry as "paradox."[16]

What is the restorative experience, tonic, or ethos of the contemporary knowledge worker? What is the life-informing or governing attitude that literariness—poetic, fictional, hypertextual, or multimedia—must now seek to inform well if it is to help repair the tone of contemporary life? The answer, I have argued, is the "ethos of the unknown," which may be said to be the imagination of the age of knowledge work. This imagination begins with cool. But in the end we need an art *and* literature that are more than cool.

Epilogue

*So now you have spent another decade writing this book, even changing from future to past tense to refer to the new millennium when it became clear that you were not going to make the epochal deadline. Again we wish to ask you the hard questions that pertain to belief—beginning once more with the hardest one: have you believed in your project throughout?**

I was lost in a wilderness of my own making for a time, writing several hundred false pages that sought to define what this book was about. I knew I wanted to write about the fate of the knowledges for which I cared most deeply—historical, literary, artistic, *humane* knowledges—in the new, brazen world of information. But I could not at first perceive what shape to give that topic, and each attempt I made seemed to lead further from the truth. *Un*writing those pages—carving away all the layers to get back to what I cared about—was some of the hardest writing I have ever done. But in the end, when I had carved back to the bone, I found this strange thing left behind that perhaps I could not have seen at all except through the process of unwriting. It was so obvious that it took me years to see it.[1] The strange yet utterly familiar thing I found is what I came to call the "ethos of the unknown." That is what I believe in, and from it the rest of the book grew.

* For the prequel to this dialogue, see the epilogue to my *Wordsworth: The Sense of History*, pp. 500–502.

Yet somehow we do not believe you when you tell such a pure, spare narrative of the journey of your book. Surely, your image of years in the "wilderness" is taken from myth, and an old one at that. Is it not true that the real reason this book took so long is that you found so many idols to worship along the way—above all, technology? Were all those technical projects in which you abandoned yourself really necessary? What about that Voice of the Shuttle *of yours, for example, which has increasingly troubled your conscience because you know you cannot keep it up properly (the links keep breaking and you vainly chase after them in the early hours of the morning as if you were a character in some Borges fantasy of an endless library)? Or what about those three desperate nights until dawn you spent at Christmas one year rebuilding the operating system and services on your department's first Web server as if that were your appointed purpose in life? Was that not an adoration of technology far and away beyond the needs of "research"?*

I wanted to learn enough about the machines and programs to address my topic from the level of code up. That is a testament of belief in its own way. As I have said, the "ethos of the unknown" is not the same as an "ethos of unknowing." As for *VoS*, when I started it, I thought that at one stroke I could further my research and lead other humanists to the Internet. But one stroke led to another.

And it wasn't just technology that was your compulsion, was it? What about your tinkering in business, sociology, anthropology, graphic design, and other fields for which you are also manifestly uncertified? For example, what about those weeks you spent reading Clastres's Chronicle of the Guayaki Indians, *when you already had everything you needed from his chapter on the Guayaki in* Society against the State? *You fell into the* Chronicle *in ways that exceed rational explanation.*

I was moved by that work, and by works in other fields besides. Long projects need to have *some* compulsion to keep them going. But what explanation would satisfy you? A personal or psychological one? Do you wish to know that after immigrating as a child, the first books I read outside school in English were the Tom Swift books? That my father—like every other male member of my family of that generation—won his way to America as an engineer? That even now I turn away from "great literature" at night to read wondrously in science fiction or (lately) explanations of string theory? I suppose my deepest motive is that I wanted this book to bring together the two sides of my intellectual life: my professional work in the humanities and my secret life in technology. Interweaving business, sociology, anthropology, graphic design, and so on in order to build a historical context in which *both* the contemporary humanities and technology participate was my way of making that unification possible. It is like string theory, where you need higher dimensions to see the "supersymmetry" of the fragmentary phenomena of ordinary life. History for me is the only higher dimension in which a grand unification of social life—a unification that somehow also acknowledges the brokenness and contingency of ordinary, lower dimensional life—is possible.*

So now we come to the crux of our inquisition, for it is precisely your desire for a grand unification that we find the most troubling. Is it not true that at heart you wish to know everything, even though rationally you also know that neither you nor any other encyclopedist living today in what you call the New Enlightenment can do more than simulate such knowledge? Is not this book, in other words, just a facsimile of the "global" knowledge that only "learning organizations"—precisely the organizations you criticize—can convincingly simulate? In short, are you not singlehandedly trying to be a knowledge-work corporation?[2]

Would you fault me if that were the case? Would you blame me for wanting to believe that an individual could match the world of corporate knowledge work—at least enough to criticize it? After all, I have no Archimedean fulcrum outside the world on which to rest my lever of critique. I criticize postindustrialism from inside because—here and now, in my place and time—there is no transcendental outside. One must think a little like a corporation to engage with postindustrialism.

How cool of you. In response to that cool student you describe in chapter 10 wearing headphones in the back of your lecture hall, why not just don headphones yourself while lecturing? Isn't critique from inside the corporate knowledge structure just another kind of cool irony, but no more than that?

Irony is part of my project, but it is not my major voice. Shall I instead tell you something very uncool about how this book started?

We have always known that you were not really cool, so we will listen.

I had a vision—or, since I am embarrassed to call it that, let us say only that I had as near a thing to a vision as a scholar is likely to have. It was 1993, and I had stopped to listen to a noon concert on my campus featuring the band Toad the Wet Sprocket, which had started locally before going national. I stood on a slight rise and enjoyed the music. The gentle, smooth sun of Santa Barbara washed down over the upturned faces, tousled hair, and bare shoulders of a few hundred students watching the band—and not just watching, but moving gently, rising up and down in time to that strangely innocent yet also worldly-wise rock that was the signature of this band. It was a moment of delicious pause in the ordinary rhythm of work

* For my fuller notion of history, which is not positivistic, see *Wordsworth: The Sense of History*.

and study, of what ungenerous commentators on Gen X at the time might have called "slack" but that seemed to me to be "recreation" in something near the original, sacral meaning of the word—that is, the experience of a diffuse, saturated re-origination of spiritual meaning in everyday life like that in the Claude paintings of "repose" I once wrote about.[3]

But then, as Shelley said in the "The Triumph of Life," "a Vision on my brain was rolled." It may have been the heavy burden of committee meetings and other administrative duties I carried that year. Or perhaps it was because I was a recent parent and anxious about how to care for my research, my students, and my family simultaneously. In any case, with startlingly bleak, cold vision, I suddenly seemed to see into the future of the students before me. I saw each of those bright, young, free people taken out of the open air and the music, and captured alone in a cubicle. I saw each bathed not in sunlight but in cold office lights and the glow of a computer screen. I saw them, that is, caught not in Weber's "iron cage" but in a "silicon cage" into which not even music would be allowed to flow. (Alas, poor Napster, we now say, as the bard once said, "alas, poor Yorick.")

It was a more muted, even abstract vision, I know, than the visions (and tele-visions, too) that my own generation saw of fellow students coming home in body bags during the years of protest. But it was a real vision, nevertheless, felt in the blood as much as in the mind. The most important concepts I have had come to me that way—as percepts, too, as things that seem to need to be explained in both theoretical and experiential terms. In any case, I recognized the stirrings of something vast but too "slack" to be protest—a kind of slow, sleepy, sad anger awakening not from any single generation but from the deep reservoirs of generations upon generations, from history itself. So I wrote this book of cool sadness, cool anger. Irony is just the tip of that iceberg.

We see that you have learned Socrates' last, best trick. You have escaped the dialectic of argument into myth. Perhaps we were wrong to say that you were borrowing the myth of years in the wilderness. That was the wrong genealogy of belief. Your real myth is that of the cave, into which the generations are forced after seeing the sun. You call that cave the cubicle.

If it assuages your conscience as interrogator, then believe that what I have said is myth.

But is it myth? Or is it fiction—that special refuge from belief that is "literature"? We remind you, at the last, that you are charged with professing literature. However well—or poorly—you have defended yourself on grounds of belief thus far, surely this last charge of infidelity will hold. Your topic was to have been literature in the information age, but—except for a few works discussed along the way—you have instead turned to anything and everything else in "preparation" for that topic. In particular, what about your romance with the digital artists, who have taken the place of honor at the end of this book in place of literary writers?

I do love to see what the artists are doing. But in any case, as I say in my concluding chapter, this book was "conceived in principle as the first part of a continuing research initiative." I think literature will indeed have a place in a new-media world otherwise dominated by the design, visual, and musical arts. But what the eventual nature and position of that literature will be among the convergent data streams of the future is something I do not yet know how to theorize.

Be honest. Are you not romancing the digital arts because in your heart you feel that literature is indeed "dead," as you elegize in your introduction? When was the last time you took pure delight in teaching a poem or novel in the same way you now abandon yourself to the relation between literature and its "contexts"? Or in setting up a server? Have you not become an atheist of literature?

I am a believer at heart. I am searching for a literature capable of that belief—forms and practices of imaginative "authoring" aware of, committed to, humbled by the necessary evolution of literariness amid the tremendous richness (and not a few poverties, too) of new media, with all their new constituencies, institutions, and beliefs crazily mixed up with old ones. Currently, I am learning from the writers, scholars, and archivists of the new "electronic literatures," while also thinking about oral and print literatures that position themselves in relation to information.[4] I seek a literature—and literary history, too—for the cool. But there is much writing—and undoubtedly unwriting—left to do.

Is that a contract?

Well, it is an option.

Critic, remember everything you believe about history. History, including the history of postindustrialism in which this book invests its stock, is too contingent to be "necessary" in the old, universal ways. But it is not optional, either.

I will remember.

Appendixes

Appendix A

Taxonomy of Knowledge Work

While it would be possible to follow John Frow in referring to the totality of mental workers as "intellectuals," in the Gramscian sense of "all of those whose work is socially defined as being based upon the possession and exercise of knowledge, whether that knowledge be prestigious or routine, technical or speculative," I will for clarity use the now more common *knowledge workers* for this catholic purpose.[1] Comparable would be Barbara Ehrenreich's inclusively defined "professional-managerial class" or "professional middle class," though I also include under *knowledge worker,* with necessary qualification, the clerical underclass of this professional-managerial "elite."[2]

In its largest context, *knowledge worker* should thus be read with the encyclopedic amplitude of Fritz Machlup's magisterial accounting in *The Production and Distribution of Knowledge in the United States* (1962) (continued in the extant volumes of his later *Knowledge: Its Creation, Distribution, and Economic Significance*) or of Marc Uri Porat's even broader definition of "information" in the nine-volume report for the U.S. government, *The Information Economy* (1977). Indeed, to follow Paul A. Strassmann's persuasive argument in his *Information Payoff* (1985), the trend since Machlup's and Porat's time to "distribute information work more widely, instead of confining it to typical information occupations" means that *knowledge worker* should now be extended further—even as far as consumers who, for example, use an automated teller machine and thus assume some of the work of clerks and tellers. "I think that perhaps as much as a quarter of the time of *all* 'non-information workers' is devoted to some sort of information work," Strassmann observes.[3] "Knowledge" thus includes in principle all the varieties of scholarship, research, information, advertising, and so on, across the major occupational sectors and on both sides of the "work/leisure" divide. (See the preface to part 2, above, on the contemporary encroachment of work on leisure and home life.)

For the purpose of addressing particular segments of society in their relation to others, however, *knowledge worker* may be subdivided as follows. I conform to Bruce Robbins's *Secular Vocations: Intellectuals, Professionalism, Culture* in reserving *intellectuals* specifically for academic knowledge workers whose current professionalism cannot simply be written off as a "fall" from the high old way of "traditional" or "public intellectuals." *Intellectuals,* that is, names academics whose technical, professional, and managerial function is very much to the point and must be thought positively rather than negatively. By contrast, I follow Alvin Gouldner's *Future of Intellectuals and the Rise of the New Class* in designating corporate, media, government, and military knowledge workers "technical intelligentsia."[4] Or, more fully, in light of my discussion of New Class theory, Gouldner's rubric must be expanded to accommodate several fractions that are no longer necessarily of a piece: technical, professional, and managerial intelligentsia. Finally, there is what I term in this book the "trailing edge" of clerical-level workers within the knowledge-worker world, who are both excluded from the notion of "white collar" in its more elite acceptations (according to the thesis that they constitute a "proletarian" office labor force) and complexly linked to the elite through both willing and unwilling participation in a common ethos of professionalism. (See chapter 1 for a fuller discussion of this matter.)

My taxonomy of knowledge work may thus be summarized:

knowledge workers =
academic intellectuals +
(technical + professional + managerial) intelligentsia +
trailing edge of clerical workers

More complex taxonomies, of course, are possible—ranging, for example, from Daniel Bell's fourfold scheme of scientific, technological, administrative, and cultural estates in *The Coming of Post-Industrial Society* to Erik

Olin Wright's eightfold "expert" or "supervisor" middle classes in *Classes*.[5] In addition, there are salient differentiations *within* the categories of knowledge workers tallied here. In the case of intellectuals alone, for example, the higher education system in the United States includes a wide variety of types and levels of occupations distributed across what Stanley Aronowitz and William DiFazio term different "tiers" of education that often function like separate castes.[6] (Due to my own firsthand familiarity with research-level universities and for reasons of focus, I concentrate in this book on the relation between intellectuals in the higher tiers of the academy and knowledge workers in other social sectors.) The same complexity applies in the other categories of knowledge worker as well. For instance, while I focus on professional-managerial-technical and clerical workers in the private sector, there is also the important matter of similar workers in the public sectors (in government, social agencies, or the military). As Bengt Furåker points out, for example, these latter public sectors are not necessarily analogous to private-sector knowledge work in all essential ways, even if both currently follow common trends of downsizing and information networking.[7]

One important point: as should be clear from the terminology and structure of the taxonomy I sketch here, "knowledge work" is implicitly, but also complexly, a *class* concept. Like Barbara Ehrenreich's *Fear of Falling*, this book is ultimately an inquiry into the ethos of what might be called—too simply—the "*new* new middle class" (i.e., the class that is emerging from the original "new middle class" of salaried white collars who inherited the place of the entrepreneurial and farmer middle classes of the nineteenth century) and of the way this ethos shapes and represents itself in relation to the "others" around it. While for the simplified purposes of this taxonomy I pose knowledge work primarily in terms of occupation and location in society, various parts of my full argument expose it to the underlying intricacy and ambiguity of class theory as it has evolved to treat the historical development of the white-collar middle class.

Appendix B

Chronology of Downsizing (Through the 1990s)

A good grasp of the chronology of downsizing may be pieced together from Robert Tomasko, *Downsizing*; Joseph Boyett and Henry Conn, *Workplace 2000*; and Jeremy Rifkin, *The End of Work;* supplemented by periodic updates in the daily and monthly press.[1]

Downsizing began in the 1980s, but only during its second phase in the 1990s—when middle managers were laid off in unprecedented numbers—did it emerge from the older concept of cyclical "layoffs" (primarily in the blue-collar ranks but also in some professional sectors) as a systemic phenomenon that went to the heart of company identity. Regarding the severity and unprecedented impact of downsizing on middle managers in particular, Michael Hammer (the leading, early "reengineering" advocate) notes that as many as 80 percent of middle managers were vulnerable in companies undergoing reengineering. In the ten years leading up to 1995, 3 million white-collar jobs alone were cut in the United States, and by the mid-1990s white collars accounted for 60 percent of all layoffs.[2]

Downsizing had eliminated 5 million jobs in all in the United States by 1994, and was continuing briskly at the rate of another 2 million jobs annually.[3] Predictions for this phase of downsizing were for an eventual cut of 25 million total jobs (in the private sector) out of a workforce of 90 million.[4] Internationally, the forecast was similar: many European and Far Asian corporations by the mid-1990s were reporting plans to implement American-style restructuring even in the face of enormous social resistance.[5] (In Eastern Europe, of course, the former Soviet bloc underwent a systemic reorganization leading to a major demobilization or deskilling of technical, professional, and bureaucratic knowledge workers—rocket scientists becoming shoe salesmen, for example.) The crash of the Far Asian economies beginning in 1997, accompanied by International Monetary Fund bail-outs that mandated restructuring and other measures, furthered the push.

Evidence that downsizing was indeed structural rather than cyclical lay in the unabated continuation of downsizing even as the U.S. economic recovery and expansion entered its seventh and eighth years in 1997–98 (with overall unemployment diminishing rapidly).[6] Job creation in such boom fields as information technology occurred *concurrently* with downsizing.[7] The trend continued in 1998 when the threat of low-cost imports from the Far East (due to currency devaluation in that region) kept the prices that firms could charge for products down even as per-unit labor costs were going up, thus forcing firms to maintain profitability through the sole remaining tactic of cutting back on units of labor.[8] Indeed, so ingrained was the habit of downsizing by 1998 that even in the face of rising demand for their products or services, some major firms drastically over-downsized and faced what *Business Week* called "a monstrous labor crunch."[9]

Meanwhile, in inverse proportion to downsizing, investment in information technology continued unabated. Rifkin notes that in the 1980s U.S. business invested more than $1 trillion in IT and that by 1992 "virtually every white collar worker in the country had access to $10,000 in information-processing hardware."[10] The surge in low-cost personal computers (in the sub-$1,000 market) from 1997 on also meant that the penetration of IT into the homes of knowledge workers (and others hopeful that their children would one day enter the class) reached unprecedented proportions.

Appendix C

"Ethical Hacking" and Art

The term *hacker* is used antithetically in the literature on computer hacking. In its original sense, *hacker* celebrated the pioneering, ingenious, up-all-hours, no-holds-barred, beyond-geek culture of computer academics and hobbyists—those who explored or jury-rigged systems as part of the normal course of their work because there were as yet no legitimate ways to accomplish a task or because the security stakes were low in a trust community of peers using systems designed to share rather than secure information. The term is still often used in this way by those in the computing community. Tsutomu Shimomura, for example (who came to fame for tracking down the computer criminal Kevin Mitnick), assertively uses *hacker* in this sense to contrast his boon tribe of computer security experts and other "wizards" with the shady computer "underground."[1] But a passage like the following in Shimomura indicates how uncannily the maverick sensibility of legitimate hackers resembles the dark side of the force:

> I've always found it a compelling intellectual challenge—finding the chinks in the armor of a computer or computer network that, unprotected, might enable a digital thief to loot a bank's electronic funds or permit foreign spies to slip into the Pentagon's computers. It's a world that you can't approach just on an academic or a theoretical level. You have to get your hands dirty. The only way you can know for sure the digital locks are strong enough is if you know how to take them apart and completely understand them.[2]

The general truth, perhaps, is that the very mentality of *analysis* (as epitomized in contemporary programmers who "know how to take them apart and completely understand them") verges antithetically into the zone of illegitimacy. This is the zone where *hacker* has become associated with stereotypically young, male, white, middle-class computer "gangs," braggadocio hacker bulletin boards, and solo artistes with cipherlike online names (sometimes descended from the older subculture of telecom hacking or "phreaking").[3]

The term *ethical hacking* is a late variant that mediates between the two senses of *hacker* by envisioning an employable future for illegitimate (and sometimes convicted) members of the computer underground. An article by Sean Donahue in *Business 2.0* titled "New Jobs for the New Economy" defines "Ethical hacker" as follows:

paradigms (e.g., the "nomadism" of Gilles Deleuze and Félix Guattari) to offer persuasive models for an art that might affect the knowledge worker in his or her ordinary cubicle. (See chapters 11 and 12, above, for fuller discussion of both the potential and limitations of ethical hacking, or what I term "destructive creativity," as an artistic paradigm.)

> It's the dream job for a generation of hackers: A company paying you to break into its network. By simulating attacks from the Net or from an internal source, ethical hackers locate weak spots in a company's network. Then they work with clients to prioritize which are the most serious threats, and suggest solutions. . . . Only the "ethical" part of the title is new. The best hackers . . . were illegally penetrating networks long before there was a market for that skill. . . . Salary: $60,000–$140,000.[4]

In a related vein, the *Jargon File* (whose print versions are titled *The New Hacker's Dictionary*) explains the "hacker ethic" antithetically as follows:

> 1. The belief that information-sharing is a powerful positive good, and that it is an ethical duty of hackers to share their expertise by writing open-source code and facilitating access to information and to computing resources wherever possible. 2. The belief that system-cracking for fun and exploration is ethically OK as long as the cracker commits no theft, vandalism, or breach of confidentiality. . . .
>
> Sense 2 is more controversial: some people consider the act of cracking itself to be unethical, like breaking and entering. But the belief that "ethical" cracking excludes destruction at least moderates the behavior of people who see themselves as "benign" crackers (see also *samurai, gray hat*). On this view, it may be one of the highest forms of hackerly courtesy to (a) break into a system, and then (b) explain to the sysop, preferably by email from a superuser account, exactly how it was done and how the hole can be plugged—acting as an unpaid (and unsolicited) tiger team.[5]

It is important to note that the possibility of generalizing the role of socially constructive "ethical hacking" is still primarily a hypothesis. The aesthetic vocabulary in Bruce Sterling's *Hacker Crackdown* is relevant: "Hacking can involve the heartfelt conviction that beauty can be found in computers, that the fine aesthetic in a perfect program can liberate the mind and spirit."[6] I am not entirely comfortable taking up the cause of hacking as a paradigm for a renewed creative arts and cultural criticism, however, because there are few clear models of ethical hackers (in the literal sense defined above) and even fewer analogues in the artistic or cultural critical communities. Such movements as the Critical Art Ensemble and Electronic Disturbance Theater (whose ethos resembles that of the 1980s-era, Bay Area *Processed World* collective) are candidates. But they are perhaps too closely associated with anarchist, Situationist, radical leftist, and/or high-theoretical

Notes

Introduction

1. *The Death of Literature* is the title of Alvin Kernan's 1990 book. Harold Bloom's *Western Canon: The Books and School of the Ages* (1994) begins with "An Elegy for the Canon" and ends with an "Elegiac Conclusion." John Beverley's *Against Literature* (1993), which takes aim at the cultural distinction of "literature" in Latin America, offers one of the most original and strident of many recent arguments against the category of literature. Especially notable is Beverley's discussion of the *testimonio* form as a challenge to traditional literary genres and their encoded social distinctions.

2. Schumpeter, *Capitalism, Socialism and Democracy,* pp. 84, 83.

3. Schumpeter, *Capitalism, Socialism and Democracy,* p. 84; see also p. 32 on the foundational "turmoil" of capitalism.

4. Drucker, *Post-Capitalist Society,* p. 57.

5. Castells, *Information Age,* 1:199.

6. The quote is from "The 21st Century Corporation," the concluding editorial of the *Business Week* special issue of the same title, 28 August 2000, p. 278. A fuller sample of contemporary allusions to Schumpeter would need to include such recent books as McKnight et al., eds., *Creative Destruction;* and Foster and Kaplan, *Creative Destruction;* as well as online political and economic discourse on the topic. For the latter, see, for example, the Progress and Freedom Foundation's manifesto entitled "Cyberspace and the American Dream: A Magna Carta for the Knowledge Age" (discussed further in chapter 8, below). It is axiomatic in the "Magna Carta" that "dynamic competition—the essence of what Austrian economist Joseph Schumpeter called 'creative destruction'—creates winners and losers on a massive scale."

7. The quotation here is from Guillory's *Cultural Capital,* p. 166. For other comments in Guillory that bear on the decline of the social status of literature, see, e.g., pp. x, 81, 153. In his *Death of Literature,* Kernan comments specifically about the decline of literature in the information age—e.g., "An informational model of social change would reveal . . .

without concern for priority, that the photocopier, weak copyright, structuralist poetics, new-style democratic politics, and numerous other social energies are all 'saying' that texts are not so unique and particular as was once thought, or, more positively, that things of all kinds are much the same. In this new informational context, older views of copyright, originality, and artistic creativity would begin to lose plausibility and in time would simply disappear" (p. 127). Miller frames similar issues in the opening pages of his *Black Holes,* pp. 3–5: "Can literary study still be defended as a socially useful part of university research and teaching or is it just a vestigial remnant that will vanish as other media become more dominant in the new global society that is rapidly taking shape?" Miller's work is published on the recto pages of a volume whose verso pages contain Manuel Asensi's *J. Hillis Miller; or, Boustrophedonic Reading.* Citations to Miller's work refer to the recto pages only.

8. *Cultural Capital,* p. 340. In regard to Guillory's concluding imagination in *Cultural Capital* of a universal afterlife of "aesthetics unbound," of aesthetics at the end of the class system, I invoke for comparison Milton's prophetic elegy, "Lycidas." "Lycidas," which mourns the death of Milton's Cambridge friend, is in part also a work about the fate of knowledge workers (species: intellectuals). It closes with these lines:

> Weep no more, woeful Shepherds weep no more,
> For *Lycidas* your sorrow is not dead,
> Sunk though he be beneath the wat'ry floor. . . .
> . . . .
> Henceforth thou [Lycidas] art the Genius of the shore,
> In thy large recompense, and shalt be good
> To all that wander in that perilous flood.

It does not demean Guillory's book, I think, to say that it is ultimately a "Lycidas" or prophetic elegy and that, as with all such elegies (readers not familiar with this genre will also want to read at least Shelley's "Adonais"), the mystery is how one gets from death to the moment of otherworldly release. Guillory (pp. 337–40) fills this gap between deadly, class-based aesthetics and the large recompense of "aestheticism unbound" with Marx and Bourdieu, the latter of whom he constructs as the very precedent for the gap. It is thus Bourdieu himself who utters "surprisingly 'utopian' statements" (p. 339). But such a stop-gap, perhaps, only defers the question, whose undergirding force is heard precisely in Guillory's "surprise." What transitional aesthetics can bridge the rift between class-based and classless aesthetics, between a "distinction" of literature that is now dying and its resurrection in a new body or form? Or, in a less utopian voice, what aesthetics can *represent* itself to itself as transitional in this manner? My argument is that the answer inheres in the avowed aesthetics of contemporary knowledge workers: "cool." "Cool" imagines itself to be what Guillory calls an aesthetics of "universal access" (p. 340), yet what such an aesthetics finally amounts to is the elitist imagination of a universal power to deny access (in the way that a truly cool, hip, slangy, or technically demanding Web page denies as much as it permits transparent access to information; see my definition of "cool" as "information designed to resist information," in the preface to Part III, below). "Cool" is the missing transitional or paradoxical aesthetics in Guillory's scheme because it is a classless imagination of class or, put inversely, class-based imagination of classlessness.

9. By focusing on producers of knowledge (researchers, teachers, professionals, managers) rather than consumers in this book, I mean to give a view of the problem that differs from, yet complements the academic critique of the "culture industry." Part of my argument is that the most influential cultures in contemporary society are those of production rather than of consumption.

I should also mention that in bifurcating knowledge work between, on the one hand, the academy and, on the other, the corporate, media, government, health, military, and other social sectors, I do not mean on principle to elide the trade press and journalistic workers (editors, publishers) who also play an important role in managing literature. In an era when major publishing and journalistic houses are increasingly subject to the management practices of their parent corporate aggregates, such literary managers occupy an uneasy hybrid zone between the academic and corporate realms that would be well worth a study in itself. (On the accelerating impact of postindustrial corporate philosophies on the book trade, including the imposition of "just in time" inventory management, see *Business Week,* "Superstores, Megabooks—And Humongous Headaches," 14 April 1997, pp. 92–94.)

10. For my earlier reflections on cultural criticism, see especially the following works: *Wordsworth: The Sense of History;* "The Power of Formalism: The New Historicism"; "Wordsworth and Subversion, 1793–1804: Trying Cultural Criticism"; and "Local Transcendence: Cultural Criticism, Postmodernism, and the Romanticism of Detail."

11. I am indebted to David Simpson for first steering me to Dirlik. On the relation between postmodernism and postindustrial business, see also essays in organization theory discussed by Dirlik, such as Cooper and Burrell, "Modernism, Postmodernism and Organizational Analysis"; Gergen, "Organization Theory in the Postmodern Era"; and Hearn and Parkin, "Organizations, Multiple Oppressions and Postmodernism." On a similar note, J. Hillis Miller comments, "Theory's opponents lament the falsely supposed suspension of language's referential function in 'deconstruction,' but that suspension actually characterizes the new global economy in all its features" (*Black Holes,* p. 47). (For further reflections on the relation between my argument and Dirlik's, see note 20, below, and chapter 1, note 48, below.)

12. See Kellner, "Intellectuals and New Technologies," p. 440: "In fact, the major research for two of my recent books was done from computer databases and I find that I am spending much more time in databases than in libraries for a variety of research projects. . . . Being an intellectual in the emerging high-tech societies thus requires new skills, the mastery of new technologies and intervention in new public spheres." On the cultural assumptions of operating systems, see Unsworth, "Living Inside the (Operating) System."

13. Greenblatt, *Renaissance Self-Fashioning,* pp. 255–57.

14. See my "Power of Formalism" on the affiliation of the New Historicism with formalism. I use *formalism* here elastically to indicate the whole range of criticisms, from Russian Formalism and New Criticism through deconstruction, that have had a lasting influence on how literary scholars "read closely." The relation between the various close-reading practices and the "technical" is clear even if antithetical. Thus, for example, John Crowe Ransom's 1941 *New Criticism* deploys a master diagram of the relation between poetic meaning and sound that is exaggeratedly scientistic despite its overt purpose of differentiating poetry from determinate scientific meaning (p. 299). The diagram combines the iconography of textbook physics with that of set theory (Venn diagrams) to make it seem that meaning and sound in poetry propagate outward according to the inverse square law. On technicality and the New Criticism in general, see Miller, *Black Holes,* p. 65: "In place of that [the traditional idea that literary study transmits a single culture's permanent values] the New Criticism put, more or less in spite of itself, technical training in the skills of 'close-reading.' Such skills were detached from any fixed cultural values. . . . The New Critics asserted certain universal cultural values while at the same time teaching an ahistorical, technologized form of reading antipathetic to those values."

Similarly, the Russian Formalists approached literature as a matter of "devices" and "techniques." A well-known example is Victor Shklovsky's "Art as Technique." On Shklovsky's "obsession with the machine analogy" and the overall mechanistic emphasis of Russian Formalism, see Steiner, *Russian Formalism,* pp. 44–67.

Finally, on the relation between deconstruction and the technical, see Guillory's discussion of the relation between "the avowed nonscience of rhetorical reading and its transmission in the 'technical' form of a science" (*Cultural Capital,* p. 201). "The refunctioning of rhetoric's *techne* as a kind of technology," he argues, "directly incorporated into the protocols of rhetorical reading a mimesis of the technobureaucratic itself" (p. 262; see also pp. 181, 257).

15. On the corporate "universities," see Micklethwait and Wooldridge, *Witch Doctors,* p. 57; Davidow and Malone, *Virtual Corporation,* pp. 190–95.

16. Davidow and Malone, *Virtual Corporation,* p. 91. Fuller discussion of contemporary corporate philosophy follows in Part I of this book. See also my Web site, *Palinurus: The Academy and the Corporation—Teaching the Humanities in a Restructured Age.*

17. Lyotard,*Postmodern Condition,* especially pp. 31–37.

18. See Dirlik, *Postcolonial Aura,* p. 3: "One of the fundamental contributions of postmodernism—indeed a defining feature of postmodernity—is the questioning of the teleology of the modern, and of other teleologies imbedded in economic, political and cultural narratives that have constituted the idea of the modern; so that it becomes possible once again to conceive the past not merely as a route to the present, but as a source of alternative historical trajectories that had to be suppressed so that the present could become a possibility."

19. See Stephen Greenblatt on Elizabeth as "a ruler without a standing army, without a highly developed bureaucracy, without an extensive police force, a ruler whose power is constituted in theatrical celebrations of royal glory and theatrical violence visited upon the enemies of that glory" (*Shakespearean Negotiations,* p. 64).

20. See Dirlik, *Postcolonial Aura,* pp. ix, 10–11, on "difference." Readers of Dirlik will recognize that my views on the function of history in the age of postindustrial capitalism are broadly consonant with his—but with a contrast related to my specific goal of grasping the role of aesthetics in such a context. One of Dirlik's main arguments is that postmodern, cultural-critical notions of multiculturalism should be replaced by a notion of "multi-historicalism" more aware of the relation between culture and the structural conditions of global capitalism. "In an ideological situation where the future has been all but totally colonized by the ideology of capital," Dirlik writes, "we can ill afford to overlook the critical perspectives afforded by past alternatives that have been suppressed by the history of capital. I take the recovery of these alternatives in memory to be not regressive, but rather as a means to keeping alive alternative visions of society that may yet open up the future in new ways" (pp. 2–3). My difference with Dirlik pertains to his consistent disqualification of postmodern cultural criticism (epitomized in "postcolonialism") as a caretaker of the rain forest of alternative histories—a disqualification that stems from its denial or ignorance of the structural conditions of capitalism in favor of matters of representation and subjectivity. Dirlik's understanding of postmodern cultural criticism does not accommodate the fact—as I have suggested at length in my *Wordsworth: The Sense of History*—that "denial" can itself be a powerful form of historical expression or engagement. (The concept of "denial" is important both to Dirlik, e.g., p. 212, and to my "New Historicist" book on Wordsworth.) Whether manifested as an act of imagination, fantasy, the "bizarre" (Dirlik's most extreme denunciation of postmodern cultural criticism, p. 212), or simple ignorance (which is not necessarily an inert stance in the world of knowledge work), denial may—or may not be—the same as withdrawal or disinterestedness. It depends on the particular contest and context.

The motive for my difference from Dirlik—with whom I am otherwise in profound and multiple agreement—is to save a place at the table for art, which has for some centuries diverged from an exclusive reliance on the aesthetic ideology of mimesis (fidelity to natural or historical reality). Especially since the onset of industrialism, literature and art have been full of different kinds of denials of history. My question in this book concerns how the tradition of aesthetic denial might be brought to bear on the condition of history in the postindustrial age. It is for this reason that I take the especially aggressive form of denial, imagination, fantasy, or bizarre behavior called "hacking" (see below) to be a thought experiment in the art of the future. The idea of the "other" fostered by cultural criticism in the academy, as I see it, might yet be of service in educating the artist of knowledge—whether avant-garde writer and artist or humble Web-page designer in a cubicle—who actively denies the historical necessity of postindustrialism.

21. See Greil Marcus's 1995 *Dustbin of History,* especially pp. 3–28, for a plangent meditation on the obsolescence of the sense of history in contemporary times; see also Marcus's complaint against the "it's history" platitude, p. 22. (Compare Montrose, "Renaissance Literary Studies and the Subject of History," p. 11.) Evan Watkins's 1993 *Throwaways* is an insightful commentary on the culture of innovation and its antithetical other—obsolescence.

22. On the period c. 1900, see Clyde W. Barrow's assessment in his *Universities and the Capitalist State* that the assimilation of the U.S. university system to the corporate form began in the 1890s (especially pp. 31–94). On the period from 1900 to 1930, see Alvin Gouldner's view that in these decades there was a "take-off period" in the demographics of a "New Class" that included both college faculty and the technical-professional-managerial intelligentsia (*Future of Intellectuals and the Rise of the New Class,* p. 15). (Gouldner's discussion is based on occupational statistics studied by Ehrenreich and Ehrenreich, "Professional Managerial Class.") To extend the argument requires distinguishing between the industrial and postindustrial epochs. Fritz Machlup observes that during the overall period from 1900 to 1959 clerical workers (whose numbers increased tenfold) led the rise in knowledge workers. But in the specific years 1940–59 at the end of this period (i.e., the postwar years, commonly accounted the origin of postindustrialism), it was professional-managerial-technical knowledge workers who led the way (*Production and Distribution of Knowledge in the United States,* pp. 387–88) . On the post–World War II era, see also Lyotard, *Postmodern Condition.*

23. J. Hillis Miller's more recent *Black Holes* (1998), which is dedicated to the memory of Readings, extends the lineage of inquiry that runs through these two works.

24. "To put this another way, the appeal to excellence marks the fact that there is no longer any idea of the University, or rather that the idea has now lost all content. As a non-referential unit of value entirely internal to the system, excellence marks nothing more than the moment of technology's self-reflection. All that the system requires is for activity to take place, and the empty notion of excellence refers to nothing other than the optimal input/output ratio in matters of information" (Readings, *University in Ruins,* p. 39).

25. On the millenarianism of recent times, see Castells, *Information Age,* 1:4: "Bewildered by the scale and scope of historical change, culture and thinking in our time often embrace a new millenarism. Prophets of technology preach the new age, extrapolating to social trends and organization the barely understood logic of computers and DNA."

26. Maintaining my academic *Voice of the Shuttle* Web site since its origin in 1994, for example, has often meant justifying the protocols, organization, and vocabulary of academic knowledge to users from other occupational sectors who write to demand conceptual transparency or just plain information. The scholarly or critical works indexed by *Voice of the Shuttle,* that is, stand shoulder to shoulder on the Web with entertainment, commercial, corporate, journalistic, and other resources in an arena where all knowledge is viewed and judged within a single mental horizon called "browsing." As I argue in chapter 4, below, browsing is symptomatic of the kind of attention required by—but also formed in reaction to—knowledge work.

27. My references are to particular controversies. On the uproar over plans by the California State University system (since dropped) to partner with major technology companies, see the resources gathered on my *Palinurus* Web site in the section titled, "Featured Controversy: California State University System 'Technology Infrastructure Initiative': Partnering for Knowledge" <http://palinurus.english.ucsb.edu/CONTROVERSIES-2-calstate.html> On the "Let's pretend we're a corporation" philosophy of the then president of the University of Florida, Gainesville, in 1997, see *Business Week,* "The New U.: A Tough Market Is Reshaping Colleges," 22 December 1997, pp. 96–102. On plans at George Mason University to eliminate programs as part of an effort to restructure for the technological future, see *Washington Post,* "Liberal Arts at GMU Targeted by Degrees: Proposed Cuts Draw Immediate Protests," 15 March 1998, p. B3. On the controversy (with international ramifications) that followed upon the publication in September 1997 of the New Zealand government's Green Paper "A Future Tertiary Education Policy for New Zealand," see, on *Palinurus,* my online article, "The New Zealand 'Green Paper on a Future Tertiary Education' and Its Critics: A New 'Account' of Knowledge": <http:// palinurus.english.ucsb.edu/CONTROVERSIES-1-New-Zealand.html> On the general relation of higher education to corporatization and information technology, see my *Palinurus* Web site, especially the set of bibliographies "Academe and Business" and "Information Tech and the Academy" available through the "Suggested Readings" link in the lefthand navigation frame of the site.

28. Readings, *University in Ruins*; Miller, *Black Holes*; Lauter, *Canons and Context,* especially pp. 175–97, "Retrenchment—What the Managers Are Doing"; Shumar, *College for Sale*; Williams, "Brave New University"; Robins and Webster, *Times of the Technoculture,* pp. 168–218. For a more extensive bibliography of both scholarly and journalistic works dealing with the topic, see the "Academe and Business" section of the "Suggested Readings" on my *Palinurus* Web site.

29. For essays from this point of view, see, for example, Nelson, ed., *Will Teach for Food.* See also *Workplace: The Journal for Academic Labor,* <http://www.workplace-gsc.com/> For a more extensive bibliography on this topic, see "Suggested Readings" on my *Palinurus* Web site, especially the page on "The Economics of Education" <http://palinurus.english.ucsb.edu/BIBLIO-UNIVERSITY-cost.html>

30. Gibson, *Neuromancer,* p. 169: "[McCoy Pauley (a.k.a. Dixie Flatline) to the protagonist, Case:] You ever hear of slow virus before?" "No." "I did, once. Just an idea, back then. But that's what ol' Kuang's all about. This ain't bore and inject, it's more like we interface with the ice so slow, the ice doesn't feel it. The face of the Kuang logics kinda sleazes up to the target and mutates, so it gets to be exactly like the ice fabric. Then we lock on and the main programs cut in, start talking circles 'round the logics in the ice. We go Siamese twin on 'em before they even get restless."

31. I discuss the Critical Art Ensemble and the Electronic Disturbance Theater in chapter 11.

32. In discussing subjectivity in the age of postindustrial production, Dirlik writes: "Only such enriched subjectivities may provide the resources for resistance against a capitalism that, as it generates postmodernity, also produces the minimal and ironic selves that make impossible any significant resistance to its power. The reconstitution of such subjectivity in turn calls for its own autonomous spaces that are immune to the intrusions of capital, or if not immune, which may not be possible, at least seek to provide an 'outside' from which to view, and to deal with, an engineering of society that is the unchanged goal of capitalist modernity" (*Postcolonial Aura,* p. 212).

33. Sidney, *An Apology for Poetry,* p. 156. For a related essay that reads Sidney's *Apology* in relation to information technology, see my "Sidney's Technology."

Part One Preface

1. Lodge, *Nice Work,* pp. 128, 84–85. I have compressed somewhat the scene with the CNC machine. The full scene is complicated by gender and class issues. The "uncanny, almost obscene" movements of the machine (p. 85) thus link up allusively to the "pornographic pin-ups" on the factory walls that Robyn objects to on the previous page. And with regard to class, Robyn's full response to the CNC machine is as follows: "O brave new world . . . where only the managing directors have jobs."

2. Stewart, *Intellectual Capital,* p. x. Such observations on the distinction between head-work and matter-work are now commonplace. See, e.g., Martin Kenney and Richard Florida's *Beyond Mass Production* (1993): "Under past forms of industrial production, including mass-production fordism, much of work was physical. . . . The emergence of digitization increases the importance of abstract intelligence in production and thus requires that workers actively undertake what were previously thought of as intellectual activities. In this new environment, workers are no longer covered with grease and sweat, because the factory increasingly resembles a laboratory for experimentation and technical advance" (p. 54). Similarly, Kevin Kelly writes in his *New Rules for the New Economy* (1998): "The key premise of this book is that the principles governing the world of the soft—the world of intangibles, of media, of software, and of services—will soon command the world of the hard—the world of reality, of atoms, of objects, of steel and oil, and the hard work done by the sweat of brows. Iron and lumber will obey the laws of software, automobiles will follow the rules of networks, smokestacks will comply with the decrees of knowledge. If you want to envision where the future of your industry will be, imagine it as a business built entirely around the soft, even if at this point you see it based in the hard.

"Of course, all the mouse clicks in the world can't move atoms in real space without tapping real energy, so there are limits to how far the soft will infiltrate the hard. But the evidence everywhere indicates that the hard world is irreversibly softening" (p. 2).

See also chapter 1, note 77, below.

3. For Micklethwait and Wooldridge's discussion of *In Search of Excellence,* see *Witch Doctors,* pp. 6, 49, 84–98. *In Search of Excellence* had sold 5 million copies as of 1996.

4. Mattelart, *Mapping World Communication,* pp. 207–8. For the international reach of such business literature and the American mode of management theory it conveys, see Micklethwait and Wooldridge, *Witch Doctors,* pp. 50–53. They note, "In a mammoth pan-Asian survey of business people in 1995, roughly half of the respondents had bought a book by a Western management writer in the previous two years" (p. 53). The survey also found, however, that "nearly the same proportion admitted that they had not finished reading" their purchases. One might conjecture on the basis of such statistics that the new business literature is akin to "myth" in Claude Lévi-Strauss's understanding of the word. It is a kind of discourse whose meaning is not dependent on close attention to its language: "Myth is language, functioning on an especially high level where meaning succeeds practically at 'taking off' from the linguistic ground on which it keeps on rolling" ("Structural Study of Myth," p. 210).

5. Hammer and Champy, *Reengineering the Corporation,* p. 31; Drucker, *Managing in Turbulent Times,* p. 60. See also Drucker's section "Sloughing Off Yesterday" (*Managing in Turbulent Times,* pp. 43–45). On the massive and early influence of Drucker, who almost singlehandedly invented management theory, see Micklethwait and Wooldridge, *Witch Doctors,* chap. 3, "Peter Drucker: The Guru's Guru."

6. I am indebted to discussion with Christopher Newfield (while he was working on his *Ivy and Industry* as well as a subsequent book now in progress on business culture and the academy) for the concept of business prophecy. (For more on business prophecy, see chapter 1, note 47, below.) As regards information technology prophecies in

particular, see my argument in chapter 4 on information technology as an "allegory" of restructuring.

7. Hammer and Champy, *Reengineering the Corporation*, pp. 7, 30.

8. Davidow and Malone, *Virtual Corporation*, pp. 240–41. See also Boyett and Conn, *Workplace 2000*, which inventories "obstacles to self-management" (pp. 240–44) and concludes by tracing them to the recalcitrance of history: "many American workers have resisted . . . because such team systems are so radically different from what most Americans have known before" (p. 241).

9. Boyett and Conn, *Workplace 2000*, pp. 267–68. Similarly, Don Tapscott's *Digital Economy* says to education, in essence, "let my people go." "In the new economy," Tapscott observes, "learning is too important to be left to the schools. Besides as knowledge becomes part of products, production, services, and entertainment, the factory, the office, and the home all become colleges" (p. 37). Thus does learning shift away from "formal schools and universities" to "formal budgeted employee education" on the model of Motorola U., Hewlett-Packard U., Sun U., and so on (pp. 199–200). In general, sustained diatribes against education are so prevalent in the new business literature as to be virtually a rhetorical requirement.

10. On this point, see Ehrenreich, *Fear of Falling*, p. 15. In general, one of the most effective narratives of the transition from the old, entrepreneurial middle classes to the new, salaried, white-collar classes in America remains C. Wright Mills's *White Collar*. I return in the next chapter to the precarious ownership status of "knowledge" capital.

11. "Final Report," pp. 1154–55 (italics mine).

12. Machlup, *Production and Distribution of Knowledge in the United States*; Porat, *Information Economy*.

13. For an impassioned indictment of administrators as the "academic managers" of downsizing in the academy since the 1970s, see the chapter "Retrenchment—What the Managers Are Doing" in Paul Lauter's 1991 work, *Canons and Contexts*, pp. 175–97.

14. Presidential Forum, organized by Elaine Showalter at the 1998 MLA convention, 27 December 1998. The session, titled "Nice Work: Going Public," addressed the present job crisis in literary studies by featuring role models for crossing from the academy to the media industries and to novel writing.

15. Senge, *Fifth Discipline*, p. 13.

16. This irony is especially keen for a scholar like myself, who originally specialized in Wordsworth's poetry (for example, *The Prelude, or Growth of a Poet's Mind*).

17. On the complex mix of intellectual, cultural, and what I term middle-managerial work in a university humanities department, see Watkins, *Work Time*, especially the excellent chapter "Literary Criticism: Work as Evaluation."

18. A version of this argument was presented at the MLA convention in 1998 in a workshop related to the Presidential Forum event "Nice Work: Going Public." It then appeared in the MLA's *Profession* Magazine the next year as "Knowledge in the Age of Knowledge Work." At this point in the original argument, I included the following series of leading questions designed to suggest the scope of the inquiry. Several of these I take up in this book. I quote from the *Profession* article:

> The crucial issues to be advanced at this level [of argument], I propose, include at least the following nine:
>
> 1. What is the difference between, on the one hand, academic poststructuralism, culturalism, and multiculturalism and, on the other hand, the radically distributed, networked model of knowledge now seen in business in the form of flat organizations, information networks, and diversity management?

2. What is the difference between *culture* as the humanities now understands it and *corporate culture?*
3. What do humanities fields that are fundamentally historical in definition (e.g., literary history) have to offer an age of just-in-time and year 2000?
4. Given that the Dearing Report in the United Kingdom, the "White Paper on Tertiary Education in New Zealand," and university corporatization in the United States all use the same language (*accountability, quality* in provocative conjunction with the discourse of *access*), are higher education and business global in the same way? Especially worth study in this regard is the fact that the major contemporary educational reform initiatives are aggressively national (e.g., how New Zealand or United States higher education must evolve to confront an era of global competition).
5. How does information technology as an allegory for the future of the academy differ from information technology as an allegory for the new millennium of business? For example, how does the academy's investment in the library or archive metaphor of knowledge inflect business's investment in the database model? In general, what are the long-range implications of the many recent academic partnerships with information-technology companies that have established channels of ingress for the entire ethos of postindustrial business?
6. What is academic intellectual property or freedom of speech in an age that is devising new means of regulating the ownership and circulation of knowledge work?
7. How does corporatization bear differentially on public, as opposed to private, higher education institutions?
8. Given the paradigm of universal insecurity in the contemporary business world—that is, the flexibility that makes not just permatemp blue-collar and clerical workers but also professional-managerial-technical workers subject to perennial restructuring—can the MLA's present effort to elevate the status of adjunct, part-time, and temporary instructors ("Final Report") be justified without advocating the elimination of the outstanding discrepancy between the academy and contemporary business, namely, tenure (i.e., the exemption from restructuring of a middle-managerial class whose jobs are definitionally different from those of adjunct, part-time, and temporary workers, on whom the burden of restructuring is now artificially concentrated)? And if there is no such credible justification, how might an intellectual safety net—which is not necessarily the same as a job safety net—be instituted such that a redefined freedom of speech and intellectual property in the academy can be protected?
9. Given that contemporary business purports to value critique as the necessary agent of change, how might the kind of critical differences indicated above–embodied in the critical edge of graduate students trained in the humanities–be offered to business as something of distinct value?

19. The preceding two sentences and the last sentence of this paragraph are adapted from my reply to William Pitsenberger's published letter about my 1999 article "Knowledge in the Age of Knowledge Work." Both the letter and my reply appeared in *Profession 2000.* I am indebted to Pitsenberger, who is simultaneously corporate attorney, business executive, and M.A. candidate in English, for personal correspondence suggesting new works to read and fresh ideas.

20. To encourage the exploration of these and other issues, I created in 1998 my World Wide Web resource, *Palinurus: The Academy and the Corporation—Teaching the*

Humanities in a Restructured World: <http://palinurus.english.ucsb.edu>. The site includes a bibliography of print and online resources bearing on the relation between the academy and business, reports on relevant controversies, study questions, and so forth.

Chapter One

1. For convenience I speak of "cultural criticism" in the singular in this chapter, even though it was and is, of course, a constellation of methods and topics that widens into various historicist, multicultural, gender, and other criticisms. At my present level of generalization, I have no ambition to account for the complexity of relationship between the various branches of cultural criticism (e.g., the relationship between American New Historicism and British cultural materialism that Montrose parses ["Renaissance Literary Studies and the Subject of History," pp. 6–7]). My claim for current purposes is only that these kinds draw upon a recognizably common set of assumptions even though the "subject" they deduce (e.g., politicized or not) may not be exactly the same.

2. For a bibliography on "subversion" versus "containment" in cultural criticism, see my "Wordsworth and Subversion," n. 2. Of course, not all 1980s cultural critics avowed a structure that was determinative, as per the Althusserian model, "in the last instance." Yet even so, the ghostly presence of such a structure was always there. In New Historicist criticism of Renaissance literature, for example, one could deconstruct state "power" to show that it was not really there, but "power" was nevertheless something like a sharply defined hole that filled the slot of presence. Its ghostly presence unified and centered the New Historicist argument. As regards the issue of containment versus subversion itself (a particular phrasing of the "structure" versus "agency" problem), I do not imply that all cultural critics were content with an analysis in which active contestation is perpetually contained by overall social inertness. Many critics attempted to think their way out of the impasse. Here I simply point to the existence of the impasse itself.

3. On this point, see Hayden White on Foucault's ruptured "archipelago" history or "chain of epistemic islands" (*Tropics of Discourse,* p. 235).

4. In this regard, cultural criticism took to heart the anti-"functionalist" lesson of its predecessor anthropological structuralisms. From one point of view, for example, Lévi-Strauss's paradigmatic "totemic operator" was the very picture of totalizing structure (*Savage Mind,* p. 152; see figure 8-1 in chapter 8 below). But from another, the totemic operator was nothing if not a structure of imaginary *mediations* of social determination (and of underlying natural determination).

5. Althusser, "Ideology and Ideological State Apparatuses," p. 163. The same model is proposed by Foucault as the paradigm of the civilizing process at the end of *Madness and Civilization* (pp. 241–78), where the direct confinement of madmen becomes internalized as an asylum-structure or conscience in the head.

6. I am using "head" here as a figure for the seat of identity and ideological (un)consciousness. Of course, this seat could be positioned elsewhere—in the genitalia, for example, which then effectively makes that body-part the head. It is important to note that my present diagnosis excludes work on the body that attempts to bypass identity altogether as a foundational category in favor of what Deleuze and Guattari *(Thousand Plateaus)* call the "body without organs."

7. The relation between subject and identity group is complex. By linking them in such formulae as "subject group" or "me and my group," I do not mean to imply that the linkage is monolithic, only that "group" is fundamentally a subject formation to be differentiated from the class formations to which I will turn next.

8. On the phenomenon of general or universal localism in postmodern theory, see my "Local Transcendence"; and Dirlik, "Global in the Local," in his *Postcolonial Aura,* pp. 84–

104. On the "specific intellectual," see Foucault, *Power/Knowledge*, pp. 126–33. See also the discussion of Foucault's "specific intellectual" in Hall's "Answering the Question."

9. Reynolds, *Discourses on Art*, Discourse III.

10. If the cultural right has labeled such politics "tribalism," in other words, that is because groupism is really a purer, more severe Grecian classicism recalling not so much savage nature (the imputation of the Right) but city-state Nature. In its very "rhetoric of blame . . . designed to chastise the target" (as Kenneth Gergen has described it in "Social Construction and the Transformation of Identity Politics"), essentialist identity politics might be said in this context to be as classicist as a partisan of the French Revolution chastising the "aristocrats" while dressed in classical costume.

11. See Gergen's retrospective view of the vexed relation between identity politics and "constructionist" cultural criticism in "Social Construction and the Transformation of Identity Politics."

12. I discuss the universal "detailism" of individual and group identity in cultural criticism in my "Local Transcendence," pp. 93–96.

13. The titles I cite are a retrospective sample of works published from the mid-1980s to the mid-1980s that deal in some central way with the impact of digital culture and/or media on identity, consciousness, subjectivity, self, or the related notion of the "author." For complementary work on identity by communications scholars and sociologists of cyberspace (and on the related issue of virtual "communities"), see, for example, Baym, "Emergence of Community in Computer-Mediated Communication"; Jones, "Understanding Community in the Information Age"; and Lyon, "Cyberspace Sociality: Controversies over Computer-Mediated Relationships." Since the mid-1990s the field of relevant criticism—especially under the two rubrics of "cyberculture" and "new media theory"—has expanded exponentially. See, for example, the following collections of essays: Bell and Kennedy, eds., *Cybercultures Reader;* Herman and Swiss, eds., *World Wide Web and Contemporary Cultural Theory;* Kolko et al., eds., *Race in Cyberspace;* Porter, ed., *Internet Culture;* Trend, ed., *Reading Digital Culture;* and Wardrip-Fruin and Montfort, eds., *New Media Reader.* As regards scholars who think about identity not just in terms of informatics but also in relation to the general field of technology and science, one could add to the example of Haraway, *Simians, Cyborgs, and Women,* such recent books as Hayles, *How We Became Posthuman.* It would also be possible to factor into my argument here the late cultural criticisms of "pragmatics" and "performance," which also conceptualize ungrounded and dynamic identity. Rita Raley's book in progress, with the working title "Transfers: Textuality and the Digital Aesthetic," addresses hypertext and new media in terms of "practice" and "performance." My thanks to Raley for an early manuscript of a chapter, which has since been published in article form as "Reveal Codes: Hypertext and Performance."

An important caveat: there is clearly much variety of method among the scholars I list here, ranging from Turkle's foundation in psychoanalytic theory to the more specifically literary grounding of hypertext theorists. Nevertheless, the shared background of many of these scholars (especially from the mid-1980s to the mid-1990s) in deconstructive and poststructuralist theory—an influence that positions them in the same "after deconstruction" situation as cultural criticism at large—makes it sensible to convene them for my purposes. For the influence of literary and related theory on Turkle, for example, see *Life on the Screen,* pp. 14–15, 17, 272, nn. 6 and 8. For the influence of poststructuralist theory on hypertext theorists, see Landow, *Hypertext;* or Moulthrop, "Rhizome and Resistance." I am indebted on this point to Matthew Kirschenbaum's talk "Hypertext Theory Post-Poststructuralism," presented at the 1997 MLA convention.

14. Poster, *Second Media Age,* p. 93; Landow, *Hypertext,* pp. 71–100; Heim, *Electric*

Language, especially pp. 167–224; Birkerts, *Gutenberg Elegies,* pp. 87–94 ("Paging the Self: Privacies of Reading").

15. Some of the cybercultural works I cite predate the popularization of networking and develop their thought in regard to "hypertext," which, in early, stand-alone information environments like the Storyspace program may be considered an incubator of network consciousness. Here I use the term *networking* in a sense inclusive of the ideals of distributed meaning and reception incubated by early hypertext theory.

16. Turkle, *Life on the Screen,* pp. 170–74, 177–86.

17. Computer-animated "morphing" refers to the technique of rendering an apparently continuous, seamless transition between two images—so that the face of one person, for example, gradually transforms into that of another. Originally created through high-end graphics animation for special effects in movies and videos, "morphing" by the mid-1990s became common enough to appear in cheap, low-end graphics programs that allowed users to select a starting image and an ending image, then automatically generate a series of intervening images for the total animation. For two reflections on the cultural significance of morphing (and of such related phenomena as children's "transformer" toys, body transformation, and so forth), see Turkle, *Life on the Screen,* pp. 170–71, and Dery, *Escape Velocity,* p. 229.

18. Ross, *No Respect,* pp. 209–32; Tuman, *Word Perfect,* pp. 38, 123. See also Simpson, *Academic Postmodern and the Rule of Literature,* pp. 9–11, 20, 118.

19. I borrow the concept from Wacquant, who, in his introduction and review of Bourdieu's work in Bourdieu and Wacquant, *An Invitation to Reflexive Sociology,* refers to the "fuzzy logic of practical sense" (p. 19) and quotes Bourdieu: "habitus is in cahoots with the fuzzy and the vague" (p. 22). See also Bourdieu, *Outline of a Theory of Practice,* on ritual practices as a "fluid, 'fuzzy' abstraction" and "fuzzy systematicity" (pp. 112, 123).

20. Bourdieu and Wacquant, *Invitation to Reflexive Sociology,* p. 119. As instanced in *Distinction,* however, Bourdieu appears to rely less on this fourfold schema of capital than on a generalized notion of "cultural capital" whose components vary (e.g., "educational capital," "political capital").

21. A field is "the system of objective relations within which positions and postures are defined relationally and which governs even those struggles aimed at transforming it" (Bourdieu, *Distinction,* p. 156). The bounding realities of such fields, however, are disputed and tend even in Bourdieu to verge into such metaphors as "playing field" (*Logic of Practice,* pp. 66–67).

22. Bourdieu and Wacquant, *Invitation to Reflexive Sociology,* p. 104. That Bourdieu's theory appears in at least this sense to be a radicalization of poststructuralism is concealed by his adoption of the stance of *not* "theorizing" in direct engagement with the poststructuralist theorists. About the internal logic of fields, Bourdieu writes in *An Invitation to Reflexive Sociology* (p. 103), "the products of a given field may be systematic without being products of a system, and especially of a system characterized by common functions, internal cohesion, and self-regulation." About the relations among fields, he adds, "I believe indeed that there are *no transhistoric laws of the relations between fields,* that we must investigate each historical case separately. Obviously, in advanced capitalist societies, it would be difficult to maintain that the economic field does not exercise especially powerful determinations. But should we for that reason admit the postulate of its (universal) 'determination in the last instance'?" (p. 109).

23. For Bourdieu's quantum physics–like notion of sociology, see, for example, *Outline of a Theory of Practice:* "Even in cases in which the agents' habitus are perfectly harmonized and the interlocking of actions and reactions is totally predictable *from outside,* uncertainty remains as to the outcome of the interaction as long as the sequence has not been completed: the passage from the highest probability to absolute certainty is a qualita-

tive leap which is not proportionate to the numerical gap. This uncertainty, which finds its objective basis in the probabilist logic of social laws, is sufficient to modify not only the experience of practice . . . but practice itself, in giving an objective foundation to strategies aimed at avoiding the most probable outcome" (p. 9).

24. Bourdieu, *Logic of Practice,* p. 68.

25. Bourdieu and Wacquant, *Invitation to Reflexive Sociology,* p. 126.

26. Such class is "the set of agents who are placed in homogeneous conditions of existence imposing homogeneous conditionings and producing homogeneous systems of dispositions capable of generating similar practices"—a "class habitus" (Bourdieu, *Distinction,* p. 101).

27. On the sameness of "difference" in postmodern cultural criticism, see Dirlik, *Postcolonial Aura,* pp. ix, 10–11.

28. Bourdieu and Wacquant, *Invitation to Reflexive Sociology,* p. 108.

29. Bourdieu and Wacquant, *Invitation to Reflexive Sociology,* pp. 107, 133, 201 ([a] and [b] are my notations).

30. Bourdieu, *Distinction,* pp. xi, xii. In relation to my general argument about the lack of essence behind the local/general relation in Bourdieu, see also the following important, if difficult, paragraph, where Bourdieu bonds individual bodies and institutional bodies together on the ground of a "relation between two realizations of historical action": "The proper object of social science, then, is neither the individual, this *ens realissimum* naively crowned as the paramount, rock-bottom reality by all 'methodological individualists,' nor groups as concrete sets of individuals sharing a similar location in social space, but the *relation between two realizations of historical action,* in bodies and in things. It is the double and obscure relation between habitus, i.e., the durable and transposable systems of schemata of perception, appreciation, and action that result from the institution of the social in the body (or in biological individuals), and fields, i.e., systems of objective relations which are the product of the institution of the social in things or in mechanisms that have the quasi reality of physical objects" (*Invitation to Reflexive Sociology,* pp. 126–27). We can decipher the nested logic of "obscurity" in this passage as follows. In the absence of natural essence ("rock-bottom reality"), there is a primary "obscurity" to be observed in the relation between local and general identity *both* as that relation inheres within internal habitus ("schemata of perception . . . that result from the institution of the social [the general] in the body [the local]") *and* as it appears in external social fields ("systems of objective relations which are the product of the institution of the social [the general] in things [the local]"). Then there is also a meta-obscurity—a "double and obscure relation"—between those two primary obscurities. Where the "naive" think there is a continuous natural ground between local and general identity, in other words, Bourdieu sees a doubly empty lack of ground.

31. Bourdieu and Wacquant, *Invitation to Reflexive Sociology,* p. 97.

32. Compare the "space" theory of such late Marxist cultural critics as Soja or Jameson (e.g., the discussions of architecture in Jameson's *Postmodernism*). In such theory, the description of postmodern geographical or architectural space serves as a postmaterialist allegory for social experience. "Space" is what is left when the foundational "matter" has been hollowed out of it.

33. Bourdieu and Wacquant, *Invitation to Reflexive Sociology,* p. 119.

34. See Martin and Szelényi, who compare Bourdieu and Gouldner on the point of their notions of "cultural capital." Gouldner himself cites Bourdieu (Gouldner, *Future of Intellectuals and the Rise of the New Class,* p. 100).

35. Gouldner has surveyed the lineage of New Class critique in the bibliographical appendix to his influential *Future of Intellectuals and the Rise of the New Class;* Frow has

more recently reviewed the field in his *Cultural Studies and Cultural Value* (pp. 89–130). Much of the literature on the subject also surveys at least a portion of the relevant bibliography (e.g., Burris's critical survey of six theorists of the New Class in "Class Structure and Political Ideology"). Besides Gouldner and Frow, I have mainly consulted the following works (not itemized here are related works on "intellectuals" and "professionals"): Bell, *Coming of Post-Industrial Society;* Burris, "Class Structure and Political Ideology"; Carchedi, "Class Politics, Class Consciousness, and the New Middle Class"; Carchedi, *On the Economic Identification of Social Classes;* Dahrendorf, *Class and Class Conflict in Industrial Society;* Ehrenreich and Ehrenreich, "Professional-Managerial Class"; Elliott, "Intellectuals, the 'Information Society' and the Disappearance of the Public Sphere"; Furåker, "Future of the Intelligentsia under Capitalism"; Garnham, "Media and Narratives of the Intellectual"; Martin and Szelényi, "Beyond Cultural Capital"; Poulantzas, *Classes in Contemporary Capitalism;* Schlesinger, "In Search of the Intellectuals"; Touraine, *Post-Industrial Society;* Wright, *Class Structure and Income Determination;* Wright, *Classes;* Wright et al., *Debate on Classes.*

36. A summary view of various schemes of class stratification can be found in Burris's chart (my fig. 1.1). Surveying stratification theory in his *White Collar Working Class,* Sobel distinguishes between theories that center on a broad "new middle class" of knowledge workers and those that focus on the more elite, upper-crust "New Class" of "modern-day philosopher kings, educated technocrats, the brain trust," who are really a "new ruling class" (pp. 14–20). There are strong reasons to keep this distinction in mind. Some labor historians, for example, note that simply to enroll the more proletarianized and feminized clerical ranks in the middle class would obscure the fact that there are really two white-collar classes (see Murolo, "White-Collar Women and the Rationalization of Clerical Work"). Yet there are also valid reasons to suspect any distinction that is too strict. Frow says of the relation between higher and lower tiers of the middle class: "the uncertainties of self-definition and particularly the force of class *aspiration* make this boundary ambivalent" (*Cultural Studies and Cultural Value,* p. 124). What I have called the "trailing edge" effect, by which the lower white-collar ranks must for some purposes be accounted part of the higher, was crucial in the evolution of the new middle class as a whole from the early 1900s on. In the main, the new middle class was created when executive functions at the top of a business were spun off into successively lower and more routine occupations (Zuboff, *In the Age of the Smart Machine,* p. 98). Thus the chores once handled by the nineteenth-century owner-manager entrepreneur gradually devolved upon separate middle-manager and ultimately, at the lowest level, clerical ranks. But such continual differentiation did not at last lead to a clerical "lower" middle class so separate from the upper tiers of that class that it was wholly blue-collar. Instead, a complex relationship of identification developed between strata. From above, the upper tiers exerted a hegemonic pressure that subjected the lower middle class to "professional" standards. From below, reciprocally, the lower middle class identified with, and strove to become, upper middle class in their individual career trajectories, hopes for marriage (after the mass feminization of the clerical ranks), and cross-generation ambitions projected onto sons and daughters. Though not a member of the upper tiers of the new middle class, therefore, the clerical pool secretary nevertheless was expected (and herself expected) to be seen as part of that class as a whole. As Murolo puts it about one of the categories of jobs she studied in the insurance industry that was being "deskilled" by computerization, "the job has been and is being degraded, but it represents a step up for women" (p. 50). Kocka's *White Collar Workers in America, 1890–1940,* is especially useful in thinking about this "trailing edge" effect of class identity, particularly as regards the common need of white-collar ranks to see themselves as different from blue-collars (see pp. 3, 85–86). On the way "professionalism, " especially in the United States (but also with parallels in Germany, Kocka's base of comparison), governed the lower middle class and gave it an

identity in common with its betters, see ibid., pp. 91–92, 147, 263. See also Hochschild, *Managed Heart*, p. 103.

37. On Gouldner's "general theory of capital," see Gouldner, *Future of Intellectuals and the Rise of the New Class*, pp. 21, 27. As regards occupational fields, I do not argue that in New Class critique they are exactly similar to Bourdieu's "fields," only that there is a broad similarity in the two concepts because both differentiate themselves from any totalizing view of social structure.

38. Nor could such contradiction easily be rationalized in Marxist fashion as dialectical (i.e., as a present struggle in structure indicative of an unfolding, unitary future structure). The hidden specter looming over much New Class theory is that at the end of Marxist history lies not any single, integral society but instead a postmodernist "end of history" society filled with middle managers and bureaucrats who are permanently comfortable in their neither-norism.

39. Ehrenreich, *Fear of Falling*, p. 15.

40. The genesis of this now well-established line of thought lies in sustained inquiry into the undecidable politics of the middle class—its "indifference"—initiated in the concluding chapters of C. Wright Mills's *White Collar* (pp. 289–354). See also Ehrenreich, *Fear of Falling*, p. 15, and passim. The New Class seems to be a blob that, when struck by the hammer and sickle of social determination, behaves like mercury: the drops go everywhere, then run back together again in shifting, unpredictable paths.

41. Gouldner, *Future of Intellectuals and the Rise of the New Class*, pp. 27–36.

42. See Wright, *Class Structure and Income Determination*, pp. 26, 39–45. Gouldner makes no bones about declaring that the New Class is not a firmly formed class on the old Marxist model (*Future of Intellectuals and the Rise of the New Class*, p. 8). Frow defines it with even more aggressive fuzziness as a "weakly formed" class (*Cultural Studies and Cultural Value*, pp. 121, 125).

43. See, for example, the national specificity of Gouldner's argument about the fate of the old dominant class relative to the New Class (*Future of Intellectuals and the Rise of the New Class*, pp. 89–92): "The political basis of *détente* in the USSR . . . ," "From the American side . . . ," and so on.

44. The localizing trend is also apparent in such latter-day, post-Enlightenment theories of civilization as Foucault's "micrological" approach to culture. See my "Local Transcendence" on the "detailism" of postmodern cultural criticism.

45. Gouldner, *Future of Intellectuals and the Rise of the New Class*, pp. 7–8, 83. The ambivalence with which Gouldner characterizes the "progressive" potential or vanguardism of the New Class has its precedent in C. Wright Mills's assessment of the middle class in the first half of the twentieth century as "the new little people, the unwilling vanguard of modern society" (*White Collar*, p. xviii).

46. Gouldner, *Future of Intellectuals and the Rise of the New Class*, p. 7. To wrap up the argument at this point by looking from the central exhibit of the New Class to society as a whole then requires hardly any scaling up. Apparently, the New Class is demographically and economically itself already on the scale of the whole—which is to say, of course, of the "*un*whole" or incoherent whole. The New Class is large and ubiquitous enough, first of all, that no one now knows exactly what portions of society it excludes (a point that became clearer in the later twentieth century when even manufacturing-line jobs were being redefined as what Zuboff calls "informated"). Thus, while some of the New Class theories charted by Burris restrict the class to managers and supervisors, others are comprehensive to the point of including routine clerical workers. Second, however, the New Class does not therefore imply social coherence. Rather—and this may be the most

disturbing extension of Wright's thesis—it forecasts a world in which every class, whether capitalist, middle, or laboring, is "contradictory."

47. Tapscott (*Digital Economy*, p. 12) conveniently collects most of these terms from recent business writing. "Fishnet organization" is from Johansen and Swigart, *Upsizing the Individual in the Downsized Organization*, pp. 15–20. "Web" organization is from Helgeson, *Web of Inclusion*.

Corporatism in its postindustrial form has begun to attract the attention of literary and cultural critics, such as Newfield, Farland, and Pitsenberger. I am indebted to talks with Newfield during his early work on *Ivy and Industry* for the concept of business "prophecy." Such prophecy is less an act of prognostication than a permanent mode or genre of discourse (a discourse of "permanent revolution"), as noted by Micklethwait and Wooldridge: "Just as Lenin's *Imperialism, the Highest Stage of Capitalism* (1917) predicted the outbreak of the First World War three years after it had actually happened, management gurus are forever prophesying a future that has already arrived" (*Witch Doctors*, p. 14).

48. See the introduction, above (especially notes 11 and 20), on Dirlik's discussion of the relation between postmodernism and contemporary capitalism. In this book I generally favor the terms *neo-corporate* or *postindustrial* over *postmodern* for reasons that are consonant with Dirlik's work, even though the "postmodern" is the focus of his critique. Extending (and making more severe) the critique of postmodernism initiated most famously in Jameson's *Postmodernism*, Dirlik's *Postcolonial Aura* is a sustained criticism of postmodern academic theory (and, in particular, "postcolonialism" theory) for its concentration on matters of representation, discourse, individual subjectivity, atomistic "localism," and "hybridity" at the expense of any serious consideration of the structural condition of "global capitalism" (which postmodernists, according to Dirlik, dismiss too easily as a totalizing concept). I concentrate on training the same kind of critique upon the extra-academic discourse of neo-corporatism and postindustrialism, which—though it is ostensibly all about the structural conditions of global capitalism—is also in part a free-floating representation. What exactly is "team culture," for example, in relation to historically deep notions of groups or classes? In criticizing the "neo-corporate" and "postindustrial," in other words, I am under no illusions that I am grappling with the Real directly; rather, I am concerned with certain ways of articulating (and also finessing) the real that management theory has made its business. I eschew the term *postmodern*, then, because I wish to focus upon the complex and thick fiction of the "postindustrial." The "postmodern" is a very minor player in this fiction (spoken of in relatively few works of management, organization, or economic theory).

There are other ways in which Dirlik's argument is relevant to mine—e.g., his discussion of how the new corporations rely for control on the management of "culture" and "multiculturalism" (*Postcolonial Aura*, chap. 9, "The Postmodernization of Production and Its Organization"). In general, I have found it useful to think of Castells's trilogy *The Information Age* as a companion in spirit to Dirlik's *Postcolonial Aura*, which does not discuss informationalism. These two texts go far toward establishing the parameters within which to think about the fate of culture amid global postindustrialism.

49. Roszak, *Making of a Counter Culture*, especially pp. 42–83.

50. For some of these euphemisms for downsizing, see Micklethwait and Wooldridge, *Witch Doctors*, p. 9; others I have gathered from a variety of journalistic, business literature, and online resources.

51. Documentation for this phenomenon is available in the "Suggested Readings" bibliography on my Web site *Palinurus: The Academy and the Corporation*. See also introduction, above, note 27. Part of this chapter was first presented at SUNY Buffalo in 1996. At that time, I was coming from the postrecession University of California system to deliver this thesis at the SUNY system, which had just undergone severe budgetary cuts, a

situation that gave the occasion special pathos—perhaps particularly to the graduate students in the audience, who were then suffering under the worst job freeze-out in recent history.

52. On the convergence of the military and business paradigms in the information age, see, for example, Rochlin's chapter "C^{3}I in Cyberspace," in *Trapped in the Net,* pp. 188–209.

53. Vilette, quoted in Mattelart, *Mapping World Communications,* p. 208. Mattelart remarks, "The norms and references of the welfare state, public service, and the constraining play of social forces, all tended to cede power to private interests and the free play of market forces. . . . The corporation and the freedom of the entrepreneur became the center of gravity of society" (ibid., p. 208). Compare Aronowitz and DiFazio, who point out that downsizing is symptomatic of the general rethinking of society called "privatization," by which even "public services are, increasingly, restructured for the needs of private profit" (*Jobless Future,* p. 129). See also Maynard and Mehrtens, *Fourth Wave*: "The other institutions of society—political, educational, religious, social—have a decreasing ability to offer effective leadership: their resources limited, their following fragmented, their legitimacy increasingly questioned, politicians, academics, priests, and proselytizers have neither the resources nor the flexibility to mount an effective response to the manifold challenges we are facing. Business, by default, must begin to assume responsibility for the whole" (pp. 6–7).

54. For CQI in the university, see, for example, the home page of the Penn State Center for Quality and Planning, which "uses organizational change tools to help departments and units develop strategic plans, improve key processes, assess institutional needs, and develop collaborative team environments": <http://www.psu.edu/dept/president/cqi/index.htm> Similar initiatives are in place at many other universities. (I first followed the phenomenon through the HEPROC CQI discussion list, which is now defunct. See the home page of HEPROC, which stands for "Higher Education Processes Network.") For JIT in education, see Hudspeth, "Just in Time Education." See also the various "reengineering" initiatives in higher education—as articulated, e.g., in works by Robert C. Heterick, Jr., James H. Porter, and Herbert Stahlke and James Nyce.

55. Johansen and Swigart, *Upsizing the Individual in the Downsized Organization,* p. xi. The revisionary literature of business has become such a tradition since the early 1990s that many works comfortably recite precisely such shorthand digests or blazons of prevailing business wisdom. Davidow and Malone, for example, write, "Scores of articles and books have been written about such topics as just-in-time supply, work teams, flexible manufacturing, reusable engineering, worker empowerment, organizational streamlining, computer-aided design, total quality, mass customization, and so on" (*Virtual Corporation,* p. 17). See also the pastiche of sound-bites from business literature in Tapscott, *Digital Economy,* p. 12.

56. A 1995 survey found that the average company drew upon 11.8 of 25 leading management techniques in 1993, 12.7 in 1994, and 14.1 in 1995, with dependence on such theories by U.S. firms averaging slightly above that for France and Japan but slightly below that for Britain (Bain & Company, Planning Forum, "Management Tools and Techniques: Survey Results Summary" [London, 1995], cited in Micklethwait and Wooldridge, *Witch Doctors,* p. 15). On the official or semiofficial dissemination of management theory to every desktop in some firms, see the example the technology firm studied by Gideon Kunda, where workers' bookshelves held "mostly technical material, but also a copy of [Peters and Waterman's] *In Search of Excellence,* distributed to all professional and managerial employees" (*Engineering Culture,* p. 51). On the spread of interest in postindustrial management theory from the United States to other countries, see Micklethwait and Wooldridge, *Witch Doctors,* pp. 50–53. Micklethwait and Wooldridge comment in general,

"For better or worse, management theory has played an enormous role in how [macroeconomic] forces have affected people. Although American companies would still have shed jobs without reengineering, the discipline, alongside several other management techniques, probably increased the carnage and certainly affected the way that the cutting was done" (ibid., p. 10).

57. Bourdieu, *Logic of Practice,* p. 68; Bourdieu, *Outline of a Theory of Practice,* p. 164. Guillory, *Cultural Capital,* p. 137, cites this latter passage in launching his chapter on the evolutions of orthodox, heterodox, and "paradox" dogma that produced the New Criticism.

58. Guillory, *Cultural Capital,* pp. 137, 159. An example of a paradox that the New Critics made paradigmatic would be the last lines of Keats's "Ode on a Grecian Urn": "Beauty is truth, truth beauty,—that is all/Ye know on earth, and all ye need to know." See Cleanth Brooks's discussion in *The Well Wrought Urn,* pp. 151–66.

59. See *I'll Take My Stand,* which includes a contribution by John Crowe Ransom. Originally published in 1930, *I'll Take My Stand* represents a kind of chrysalis stage in the metamorphosis of New Criticism from its gestation in the Fugitive Group to later literary critical and academic life stages. See Conkin, *Southern Agrarians;* and Stewart, *Burden of Time, The Fugitives and Agrarians.*

60. See my introduction, note 14, for the way in which the New Criticism also internalized technological rationality in its techniques of close reading.

61. Guillory, *Cultural Capital,* p. 159.

62. Guillory, *Cultural Capital,* p. 175.

63. I am deferring rather than dismissing the stylistic approach of the New Critics, which translated issues of modern sensibility into contrasts between "scientific prose" and "poetic language." As will be clear in Part III, when I address the relation between Web page design and the "work style" of knowledge workers, some variation of formal method is required to understand how industrial and postindustrial cultures are not just "technologically rational" but cool.

64. For example, Marcuse: "the instrumentalistic conception of technological rationality is spreading over almost the whole realm of thought and gives the various intellectual activities a common denominator. They too become a kind of technique" ("Some Social Implications of Modern Technology," p. 153). And also Ellul: "In our technological society, *technique* is the *totality of methods rationally arrived at and having absolute efficiency* . . . in *every* field of human activity. Its characteristics are new; the technique of the present has no common measure with that of the past" (*Technological Society,* p. xxv). Behind Marcuse and Ellul stand the "iron cage" theories of instrumental rationality offered by Weber and other sociologists at the turn of the twentieth century. For a specifically postindustrial formulation of technological rationality oriented toward the new cybernetics, statistical analysis, and systems theory, see Daniel Bell on "intellectual technology," *Coming of Post-Industrial Society,* pp. 27–33.

65. Guillory's description of the New Criticism as "recusant" in *Cultural Capital,* p. 175, occurs in the context of his analysis of the movement's ambivalent relation to mass consumer culture.

66. Castells, *Information Age,* 1:16–17: "The social relationships of production, and thus the mode of production, determine the appropriation and uses of surplus. A separate yet fundamental question is the level of such surplus, determined by the productivity of a particular process of production. . . . Modes of development are the technological arrangements through which labor works on matter to generate the product, ultimately determin-

ing the level and quality of surplus. Each mode of development is defined by the element that is fundamental in fostering productivity in the production process. . . . In the new, informational mode of development the source of productivity lies in the technology of knowledge generation, information processing, and symbol communication."

67. Castells, *Information Age,* 1:14. See also ibid., 1:13–22, for an overall explanation of how the collapse of Soviet and other "statist" modes of production has obscured the fact that the relation between advanced capitalism and "informationalism" is not one of reciprocal necessity.

68. Castells, *Information Age,* 1:18.

69. Johnson, *Interface Culture,* pp. 42–45, 239–42.

70. I take the phrase "data smog" from David Shenk's book by that title.

71. See my "Toward a Theory of Common Sense."

72. In emphasizing the culture of production in the new millennium, my argument may be seen to be a bookend to Jennifer Wicke's brilliant exegesis of the culture of advertising and consumption in "Sublime Lite: Millennial Anticlimax and the American (End of) Century," which she presented at the same 1999 English Institute conference at which I presented parts of this book. The approaches of my discussion and Wicke's essay are equivalent, if staggered, reactions to an earlier epoch of Marxist studies when all things cultural were understood to be ephemera of underlying modes of production. Consumption-side cultural studies have extended the work of the Frankfurt School, Louis Althusser, Michel de Certeau, and others to study the "culture industry" and its reception as "semi-autonomous" of economic production. Production-side cultural studies, I argue, now have the opportunity to apply the lessons of consumption criticism to the realm of production itself. The appropriate object of study in this light is not production per se but semi-autonomous *cultures* of production—extending, for example, from the regulated suppression of affect and ethnicity in Henry Ford's factories through the "service with a smile" era of the service industries to today's corporate culture of "cool" knowledge workers. Dirlik makes a similar point in his chapter entitled "The Postmodernization of Production and Its Organization: Flexible Production, Work and Culture" *Postcolonial Aura,* p. 186). See also du Gay, ed., *Production of Culture/Cultures of Production,* especially the essays by Salaman, "Culturing Production," and du Gay, "Organizing Identity: Making Up People at Work." It can be argued that the culture of production will become increasingly important as the number of hours spent working as opposed to consuming or "leisuring" continues to climb. (See the preface to Part II, below, on this point.)

73. On Zuboff and the vocabulary of vision in computing work, see chapter 3, below.

74. Beniger, *Control Revolution,* p. vi: "From its origins in the last decades of the nineteenth century, the Control Revolution has continued unabated, and recently it has been accelerated by the development of microprocessing technologies. In terms of the magnitude and pervasiveness of its impact upon society, intellectual and cultural no less than material, the Control Revolution already appears to be as important to the history of this century as the Industrial Revolution was to the last." Beniger's general argument is that the recent information revolution is really just part of a longer revolution in the technologies, bureaucracies, and economies of control necessitated by the speed-up of life after the Industrial Revolution.

75. A perusal of the literature of contemporary business IT indicates the frequency of the words *paradox* and *contradiction.* For books on business or information technology that focus on the idea of paradox, see Handy, *Age of Paradox* and Calcutt, *White Noise.* Handy observes sweepingly (pp. 11–12), "We need a new way of thinking about our

problems and our futures. My suggestion is the management of paradox, an idea which is itself a paradox, in that paradox can only be 'managed' in the sense of coping with. . . . I no longer believe in A Theory of Everything, or in the possibility of perfection. Paradox I now see to be inevitable, endemic, and perpetual. The more turbulent the times, the more complex the world, the more paradoxes there are."

The larger ramifications of the "paradoxes" of information technology are clear in the context of what Micklethwait and Wooldridge, in their survey of postindustrial management theory, term the "contradiction" at the heart of the contemporary corporation between the legacy of scientific management (and hence of control in Beniger's sense) and of "humanistic management." The contemporary firm, they say, is the "Contradictory Corporation" (*Witch Doctors,* pp. 16–18).

The following is a small sample of ordinary, rather than thematized, uses of *paradox* in the discourse of contemporary business and information technology: Sproull and Kiesler, *Connections,* p. 101, "It may seem paradoxical that computers, stereotyped as cold and impersonal, can be used to increase personal connections and affiliation"; and Hirschhorn, *Beyond Mechanization,* p. 3, "Paradoxically, just as we are developing technologies characterized by the utmost mathematical abstraction, we must increasingly rely on informal learning and the ability to deal with the unpredictable" (see also p. 169).

76. While postindustrialism is aggressively heterodox in debunking predecessor industrial theory, such heterodoxy is now tantamount to a new orthodoxy. Variation among its proponents does exist, of course. Where such variations are significant in our context of discussion, I will bring them to the fore. In the main, however, my heuristic is to read the new corporatism as a single, broad doctrine in the same way that I previously surveyed academic cultural criticism as a single, if various, evolution. This is because internal variations in neo-corporatism and academism are subordinate to my larger goal of evaluating the relation *between* these two contexts of knowledge.

77. On the matter work versus mind work distinction, see also the preface to Part I, note 2. A revealing synonym for *postindustrial* is *meta-industrial* as used in Harris and Moran, *Managing Cultural Differences,* one of the standard textbooks on business in the age of global competition (pp. 97–99). Another way to gauge the relation between "matter" and "information" in postindustrial economy is by means of econometric analysis. Paul Strassmann, drawing on the work of Charles Jonscher, noted in 1985 that "of the $1087 billion paid by consumers for physical goods, about half was spent on services of the information-handling sector" (*Information Payoff,* p. 188). Even the physical products that one holds in one's hands, in other words, are increasingly phantom objects. They are half information.

78. *PC Magazine* 15, no. 11 (11 June 1996): 46–47, 2–3.

79. Boyett and Conn, *Workplace 2000,* p. 47.

80. Zuboff, *In the Age of the Smart Machine,* p. 98. Attewell writes, "One of the most dramatic recent changes in the managerial work process is the proliferation of computerized management information systems that capture a mass of transactional and production data from the office and factory floor. The fineness of detail of these data is far greater than that available in previous systems. . . . A manager can call up information on aggregate sales orders, production volume, or inventory, and then zoom in to see information on individual transactions or orders" ("Skill and Occupational Changes in U.S. Manufacturing," p. 76).

81. Boyett and Conn, *Workplace 2000,* p. 7.

82. Senge, *Fifth Discipline,* p. 3. See Davidow and Malone: "Everything, is about learning," "the virtual corporation is a learning entity" (*Virtual Corporation,* p. 194).

83. Peters, *Liberation Management,* pp. 108–10.

84. Tapscott, *Digital Economy,* p. 200. *Business Week* noted in 1997 that "the number of corporate 'universities,' formed by business to offer in-house job-related training, has jumped from 1,000 to 1,400 in five years" ("The New U.," p. 100).

85. As Jeremy Rifkin observed in 1995, the 20 percent of Americans on the top and second-to-top tiers of the knowledge sector had accrued more income than the other 80 percent of the population combined, and the trend was toward an even wider spread. Effectively, this has meant that the increased value placed on knowledge work has been sustained by bifurcating the knowledge worker population into those at the top who are the real knowledge workers and those at the bottom (now including everyone from assembly line workers or clerical staff up to fired middle managers) who are simply *called* "knowledge workers" as a badge of their devalued subservience to the whole ideology of knowledge work and its real masters (see Rifkin's chapter "High-Tech Winners and Losers," in *End of Work,* pp. 165–80). Dordick and Wang write a succinct epitaph for the bottom 80 percent: "Despite early predictions that information workers would be riding high in an information economy, they are often the first to lose their jobs as firms consolidate their activities and relocate work, made less valuable by information technology, to low-wage-rate countries. In addition to white-collar and blue-collar workers there are pink-collar workers whose tasks have been downgraded as a result of the application of information technology" (*Information Society,* p. 5).

86. *Business Week* tallied the fate of downsized workers in an article of 13 April 1998, "Downsizing's Painful Effects," based on the latest biennial Labor Department survey of displaced workers at that time. The good news, according to the figures, was that "79% of workers displaced a few years earlier were back at work," though 14 percent and a disproportionate number of workers over fifty-four had dropped out of the labor force. The bad news was the continuing "trend toward downward mobility" noted in previous surveys. Only two-thirds of full-time workers laid off in 1993 and 1994 had full-time jobs again in early 1996; a majority of re-employed workers earned less than previously; and more than one-third had taken a cut of 20 percent or more in pay. See also *Los Angeles Times,* "Missing the Boom-Time Bandwagon," 20 July 1998, pp. A1, A16: "Middle-aged victims of downsizing often find themselves trapped at a lower professional level." See also Rifkin on the "declining middle" (*End of Work,* pp. 170–72). My present focus on the fate of the middle managers and white-collars should not be taken to diminish the significance of blue-collar layoffs (particularly in the 1980s) or clerical layoffs. Part II of this book considers the evolution of the whole white-collar pyramid from the industrial to the postindustrial era.

87. See Mohrman et al., *Designing Team-Based Organizations,* pp. 275–95, on *empowerment* as the "fashionable concept" and "catchword" of the 1990s.

88. On new roles for managers in a team structure, see Hammer and Champy, *Reengineering the Corporation,* p. 77; Boyett and Conn, *Workplace 2000,* pp. 7, 245–49.

89. Boyett and Conn, *Workplace 2000,* p. 7; see also p. 82: "But increased information sharing may be a mixed blessing for American workers. By definition, increased access to information carries with it demands for increased responsibility to use the information for constructive purposes, along with increased accountability for results."

90. The phrase "management by stress" is from Parker and Slaughter, *Choosing Sides,* pp. 16–30. Tomasko's chapter "Demassing: A Blunt Response" in his *Downsizing* is a catalog of the symptoms of middle-management downsizing. See also Micklethwait and Wooldridge, *Witch Doctors,* p. 10; *Los Angeles Times,* "Downsizing Wave Has Reached a Point of Diminishing Returns," 7 July 1996, which quotes UCLA management professor David Lewin on the topic of overwork; and Johansen and Swigart, *Upsizing the Individual in the Downsized Organization,* p. 42: "Reengineering and reinvention leave fewer managerial spans and layers and less continuity. In today's slimmer organizations the middle managers who remain are stretched in ways they never could have imagined. Many are being spanned and layered

beyond their ability to perform. They know that someone, or something, must take over the traditional functions that now go unfilled." Also relevant is their chapter "Love and (Over)Work" (ibid., pp. 63–69); and McArdle et al., "Total Quality Management and Participation: Employee Empowerment, or the Enhancement of Exploitation?" On the ambiguity of the "facilitator" role in management, see Hirschhorn, *Beyond Mechanization,* pp. 136–37. Hirschhorn observes, "Managers often respond to these ambiguities with guile and subterfuge, not because they are dishonest, but because they are confused."

91. On the labor market of the late 1990s, see *Business Week,* "It's the Best of Times—Or Is It?" (12 January 1998) and "Oops, That's Too Much Downsizing" (8 June 1998).

92. Geert Lovink brought <http://www.fuckedcompany.com> and <http://www.internalmemos.com/memos> to my attention in his talk on "The Rise and Fall of the Dot.coms."

93. Davidow and Malone, *Virtual Corporation,* p. 266; Maynard and Mehrtens, *Fourth Wave,* p. 9.

94. On the new fields of business ethics and business and society, see the following page on my *Palinurus* site: <http://palinurus.english.ucsb.edu/BIBLIO-BUSINESS-and-society.html> It was symptomatic that the U.S. press's strenuous effort in 1996 to rationalize Pat Buchanan's early presidential primary wins as a referendum on middle-class job loss, rather than as an expression of sympathy for social extremism, resulted in renewed calls for "corporate responsibility." The *Los Angeles Times* observed in "Downsizing Wave Has Reached a Point of Diminishing Returns" (7 July 1996): "After presidential candidate Patrick J. Buchanan lambasted AT&T for slashing 40,000 jobs—and turned downsizing into a powerful political theme—corporate news releases suddenly took on a new tone. In their merger announcement, California's Pacific Telesis Group, the parent of Pacific Bell, and Texas[-based] SBC Communications trumpeted the potential for actual job growth. And on the cover of June's *Personnel Journal,* a trade magazine for human resources professionals: 'Save Jobs . . . Strategies to Stop the Layoffs.'"

95. The 1995 book by Mohrman et al., *Designing Team-Based Organizations,* may be called a "second generation" work on teams. It assumes that the case for teamworking has already been made and therefore that the relevant problem is how to incorporate teams within the total structure of an enterprise. See Sproull and Kiesler, who in 1991 simply declared as a given: "The fundamental unit of work in the modern organization is the group, not the individual" (*Connections,* p. 25).

96. Lifson, "Innovation and Institutions," p. 315.

97. Maynard and Mehrtens, *Fourth Wave,* pp. 5–6.

98. Hammer and Champy, *Reengineering the Corporation,* pp. 32–33.

99. Mills, *White Collar,* pp. 77–111.

100. Drucker, *Managing in Turbulent Times,* p. 226. Micklethwait and Wooldridge (*Witch Doctors,* p. 77) comment on Drucker's general influence on management theory: "Perhaps Drucker's most insightful observation is that management plays a vital role in all spheres of life, not just in business. It is as important for universities, churches, hospitals, or charities as it is for the manufacturers of soap powder. . . . Management is not only a business concern; it is 'the defining organ of all modern institutions.'"

101. On "postmodern" organizational principles as a means of control, see Dirlik, *Postcolonial Aura,* pp. 202–10; and Hearn and Parkin, "Organizations, Multiple Oppressions and Postmodernism." See my chapter 4, below, for the way centralized forces of control can coopt decentralization in networked information environments.

102. This is not to say, of course, that subject critique in the constructivist mode was not also historical or at least New Historicist in orientation. But historicism in the advanced cultural critical modes is discontinuous rather than continuous with the premise

of identity: the past is appreciated in the intensity of its strangeness. (See my discussion of cultural criticism and history in the introduction, above.) Another way to put this is to say that constructivist historicism is by definition as much a questioning as an endorsement of the foundation of history.

103. On theatricality as an approach to the French Revolution, see Maslan, "Resisting Representation"; and Huet, *Rehearsing the Revolution*.

104. Negroponte is quite succinct on this topic in the section of his *Being Digital* titled "The Nation State as a Mothball": "Nations today are the wrong size. They are not small enough to be local and they are not large enough to be global" (p. 238). A critique of such a thesis would begin with Castells's assertion that states retain an unacknowledged importance: "the globalization thesis ignores the persistence of the nation state and the crucial role of government in influencing the structure and dynamics of the new economy" (*Information Age,* 1:97). If we attend as well to the urgent distinction Castells draws between "nation" and "nation-state" in the information age (ibid., 2:27–52), with the former apparent in resurgent "cultural nationalisms," then it will appear that the notion of a postindustrial global economy radically undervalues the persistence of both "national" and "nation-state" (or cultural and political) units of identity in its imagination of a direct link between the local and global. (With reference to the "imagined communities" thesis of nationhood, it might be said that the local/global nexus is the postindustrial "imagined noncommunity.") In Castells's overarching argument about the resurgence of communitarian identities at the local level (fundamentalisms, nationalisms, and so forth), the local is precisely not networked with the global, and the nation-state concept in the middle is the place to look for the aporia between the two.

105. Discussing working-class cultures and subcultures, John Clarke, Stuart Hall, and others observe, the working class had "its own corporate culture, its own forms of social relationship, its characteristic institutions, values, modes of life" (Clarke et al., "Subcultures, Cultures and Class," p. 41). We can note that such a concept of "customary corporation" should not necessarily be broadened to cover both industrial-age and agrarian-age folk cultures. The Birmingham group's approach to subcultures (about which more later) focused primarily on urban, industrial folk culture.

106. Representative works: Griggs and Louw, eds.,*Valuing Diversity;* Esty, Griffin, and Hirsch,*Workplace Diversity;* Gardenswartz and Rowe, *Managing Diversity* and *Diverse Teams at Work*; Carnevale and Stone, *American Mosaic*; and the sections on multiculturalism in Harris and Moran, *Managing Cultural Differences,* the standard textbook on global diversity. Carnevale and Stone recount the early genesis of diversity management (*American Mosaic,* pp. 91–93). For an excellent study of diversity management from a cultural critical perspective, see Gordon.

Diversity management, of course, has been very U.S.-centered. Carnevale and Stone, for example, discuss multiculturalism in other countries, but conclude that the United States has the clear lead not only in perceiving "diversity" but in seeing it as a management opportunity (*American Mosaic,* pp. 493–501). However, their conclusion is tempered by a recognition that analogous traditions of management in other countries (e.g., European adeptness at managing cross-national workforces) may need only to be readjusted to make the U.S. advantage disappear. On the international dimensions of multiculturalism, see also the chapter "Managing Diversity in the Global Work Culture" in Harris and Moran, *Managing Cultural Differences.*

107. Gordon comments incisively that corporate diversity management seems "progressive" only "because it refuses to inherit historical burdens, only future possibility and progress." It thus suffers "historical amnesia" and "comes to us all fresh and new, unburdened by history, located in the cultural here and now" ("Work of Corporate Culture," pp. 14, 16, 17).

Johnston and Packer's *Workforce 2000* was prepared for the U.S. Department of Labor in 1987. The findings of this report were later reinforced by U.S. Bureau of Labor Statistics forecasts. On the latter, see Carnevale and Stone, *American Mosaic,* pp. 40–44.

108. Thomas, *Beyond Race and Gender,* pp. 9–10.

109. This argument, of course, is necessarily a simplification based primarily on early twentieth-century labor and immigration history. Fully to consider issues of gender, the notion of the line or shift would need to be broadened to include its contrasting complement: the "home" separated from the line/shift. The domestic sphere (whether one's own or another's whose house one tends) was also the locale where the experience of identity groups arose. For convenience, in the present context I will consider gender only as it was fashioned into an identity on the line/shift, in "women's" versus "men's" lines. (One of many vivid memories from my summers during college spent working on various factory lines and in other menial jobs was the rigid distinction in some plants between women's assembly lines, involving the work of fleet fingers, and men's assembly lines, involving the work of mass muscles.)

110. I have previously studied the eighteenth- and early nineteenth-century yeoman household in a British setting in my *Wordsworth,* pp. 236–51.

111. I am not assuming that populations in their original locale prior to migration were in a steady state, or that they did not often experience tensions resulting from long-standing histories of migrations and social collisions. For the purpose of my present argument, however, I simplify by taking the migration from pre-industrial to industrial society as the zero point of origin.

112. I am influenced in these thoughts by Dick Hebdige's *Subculture,* which sketches the intensely localized, working-class youth subcultures of London that emerged in a medium of leisure almost totally quarantined from the culture of work.

113. See Carnevale and Stone, *American Mosaic,* p. 117; Gonzalez and Payne, "Teamwork and Diversity"; Gardenswartz and Rowe, *Diverse Teams at Work.* See also Gardenswartz and Rowe's chapter "Building Multicultural Work Teams" in *Managing Diversity.*

114. On U.S. auto makers and Swedish station-based assembly, see Parker and Slaughter, *Choosing Sides,* p. 13.

115. Thomas, *Beyond Race and Gender,* p. 10.

116. Carnevale and Stone, *American Mosaic,* pp. 116–18. Dirlik observes, "What is of interest here is that transformations in management, required by changes in capitalism that were already under way, were blended with discussions of culture called forth by the need to manage a 'multicultural' work force, and reinforced the importance of culture in management" (*Postcolonial Aura,* p. 194).

117. Gonzalez and Payne, "Teamwork and Diversity," pp. 126–27. The original is a bulleted list; I have inserted numbers for ease of reference.

118. The massive yet precise excision of "roots" I indicate may be read in Thomas's enormous allegory of a "tree" in *Beyond Race and Gender.* The conceit begins: "One way to understand culture is to conceptualize an organization as a tree [here there is a diagram of a tree with roots]. In this organizational tree, the roots are the corporation's culture. These roots, of course, are below the surface, invisible. But they give rise to the trunk, branches, and leaves—the visible parts of the tree. Nothing can take place in the branches and be sustained naturally unless it is congruent with the roots.

"By way of analogy, assume that I live in Georgia and own a grove of oak trees. I like these trees; they provide an enormous amount of pleasure to me. But I would also like to see peaches come into season. So I buy some peach trees and bring them to the grove. Unfortunately, there isn't enough space to plant the new trees without removing some of the oaks, and I don't want to do this.

"I decide instead to graft a healthy peach limb onto one of the oak trees. I begin the process, controlling for moisture, light, temperature, and anything else that might affect the grafting effort. And, for a while, it seems as if I've been successful.

"Then peach season arrives, and the limb produces one very puny peach. . . . The tree knows better. The roots of the oak are sending signals to the peach limb. 'This,' they are saying, 'is an oak tree. . . . Peach limbs don't grow on oaks'" (pp. 13–14).

Striking here is an omission that is at once massive and almost invisible: the entire ethnic/racial sense of "roots" (as in Alex Haley's novel of that name). Inasmuch as the tree of culture survives being cut off from its historical roots, it does so because that entire root system is excavated and replaced by a radically new root system that—and this is the ultimate meaning of Thomas's allegory—*can* allow for trait-by-trait, peaches-into-oaks "grafting." (See also Thomas, pp. 50–59, on "root change" and "growing new roots.")

119. Gonzalez and Payne, "Teamwork and Diversity," p. 127.

120. Gordon, "Work of Corporate Culture," p. 7.

121. Gonzalez and Payne, "Teamwork and Diversity," pp. 126, 128.

122. Harris and Moran, *Managing Cultural Differences,* pp. 10, 124. Compare Kendall's statement that "culture is our way of knowing and doing," p. 82.

123. Davidow and Malone, *Virtual Corporation,* pp. 195–99. For an analogous formulation of cultural diversity as talent diversity, see Griggs, "Valuing Diversity®: Where From . . . Where To?" p. 9: "More and more, organizations can remain competitive only if they can recognize and obtain the best talent, value the diverse perspectives that come with talent born of different cultures, races, and genders, nurture and train that talent, and create an atmosphere that values its workforce." See also Harris and Moran, *Managing Cultural Differences,* p. 109: "The differences of perception that arise from varied academic or training backgrounds, work expertise and experiences, ethnic and national origins can enrich the group's basis for creative problem-solving and achievement."

124. Compare, for example, Louw's use of a "particle" metaphor, complete with quotation from a particle physicist ("No Potential Lost," p. 22). A variety of other tropes depict team members as purely physical systems—e.g., the image of a team as an "energy exchange system" in which "members energize or motivate themselves and one another" in Harris and Moran, *Managing Cultural Differences,* p. 108. Perhaps the most telling of such metaphors is the sustained conceit of cultural diversity as "behavior software," according to which the corporate management of diversity means reprogramming culture at the level of bits (see below).

125. Katzenbach and Smith, *Wisdom of Teams,* p. 18. See also Gonzalez and Payne on team "rituals" ("Teamwork and Diversity," p. 127) and Gardenswartz and Rowe on team "play" (*Diverse Teams at Work,* p. 129). Such team-bonding rituals are not *ab ovo.* Rather, they simulate underlying family, clan, and other rituals in a way that reveals the basic, appropriative nature of "culture" in the new corporatism.

126. As Griggs and Louw put it, "the real value of a diverse workforce is that it nurtures synergistic interactions across difference. It is this synergy that produces unpredictable consequences in terms of breakthrough and results" (*Valuing Diversity,* p. vii). Or again, in the words of Harris and Moran, "For leading-edge organizations, globalism means the *creation of a culture* that embraces diversity to maximize the potential of personnel, especially through cohesive work teams" (*Managing Cultural Differences,* p. 171).

127. Gardenswartz and Rowe, *Managing Diversity,* p. 12 and passim. See also Geert Hoftstede on culture as "mental program" (*Culture's Consequences: International Differences in Work-Related Values;* cited in Dirlik, *Postcolonial Aura,* p. 191).

128. Thomas, *Beyond Race and Gender,* p. 10; see also pp. 43, 172–73.

129. Griggs and Louw, eds., *Valuing Diversity,* p. 6. But see Griggs's reflections on the risk of "taking refuge" in such a broad definition of diversity (ibid., p. 7).

130. Gardenswartz and Rowe, *Managing Diversity,* pp. 20–23.

131. The *Los Angeles Times* noted, "As [California] state affirmative action efforts were dealt another blow by a federal appeals court ruling . . . upholding Proposition 209, corporations are stepping into the breach, aggressively courting women- and minority-owned businesses. . . . Corporate human resources departments remain committed to outreach efforts" ("State May Be in the Minority in Affirmative Action Stance," 10 April 1997).

132. Gardenswartz and Rowe, *Managing Diversity,* pp. 100–101.

133. Harris and Moran, *Managing Cultural Differences,* pp. 105, 160, 10. See Castells, *Information Age,* 1:417: "Furthermore, there is an increasingly homogeneous lifestyle among the information elite that transcends the cultural borders of all societies: . . . the practice of jogging; the mandatory diet of grilled salmon and green salad, with *udon* and *sashimi* providing a Japanese functional equivalent; the 'pale chamois' wall color intended to create the cozy atmosphere of the inner space; the ubiquitous laptop computer; the combination of business suits and sportswear; the unisex dressing style, and so on. All these are symbols of an international culture whose identity is not linked to any specific society but to membership in the managerial circles of the informational economy across a global cultural spectrum."

134. Carnevale and Stone, *American Mosaic,* pp. 125–53.

135. On the creation of lower management and clerical functions from the executive sphere, see Zuboff,*In the Age of the Smart Machine,* p. 98.

136. Thomas and Kochan explain, "Technology, Industrial Relations, and the Problem of Organizational Transformation," p. 216: "Since unions ceded ultimate control over the organization of work to management, the only effective means available for unions to bargain on behalf of their members was rigorously to define and attach prices to the performance of specific jobs. This practice of 'job-control unionism' established a system of industrial jurisprudence centered on what [Philip] Selznick [*Law, Society and Industrial Justice* (New York: Sage, 1969)] termed 'industrial citizenship,' with formalized contracts, detailed job descriptions, and a multistage grievance procedure to adjudicate disputes between rounds of contract negotiations." See also Tolliday and Zeitlin, "Shop Floor Bargaining, Contract Unionism, and Job Control," pp. 232–34, on the history of union seniority and job classification rules.

137. On the history of union seniority and job classifications and the "straw boss" versus "union steward" struggle, see Lichtenstein, "Union's Early Days."

138. Esty, Griffin, and Hirsch, *Workplace Diversity,* pp. 109–11.

139. On "broadbanding," see Greenbaum, *Windows on the Workplace,* p. 93. In their chapter "Hierarchy and Class" Esty, Griffin, and Hirsch conclude a list of "don'ts" with a flattening rule specifically targeted at the team or "task force": "Don't be afraid to mix several levels of employees on task forces or committees. It's good for everyone" (*Workplace Diversity,* p. 119).

140. On the tradition of combining protest with satire in U.S. labor journals, see Carlsson and Leger, *Bad Attitude* (itself a participant in the tradition), p. 16.

141. Parker and Slaughter, *Choosing Sides,* p. 3 (italics in the original).

142. See Parker and Slaughter, *Choosing Sides,* pp. 5, 72, 74–87, 127, for critical analyses of various aspects of declassification and interchangeability. Regarding the physical stress of working on the new just-in-time lines, see their chapter "Management-by-Stress: Management's Ideal Team Concept," pp. 16–30, and the chapter on time-motion studies in the contemporary plant, pp. 88–94. Parker and Slaughter also distinguish between the "initial start-up" of teams and the later reality. During the former phase, teams may well work

according to the corporate vision of worker empowerment and participation. But that is when they contain a higher than usual number of "supervisors, engineers and team leaders" placed there to initiate process designs. Later, once the process has been locked in, it becomes increasingly difficult for teamworkers to change things or participate (ibid., p. 27).

143. Lichtenstein, "Union's Early Days," p. 72.

144. See my discussion of "prosumerism" and its recent extensions in chapter 4, below, for explanations of these concepts.

145. Heckscher, *New Unionism,* pp. 8–9.

146. Boyett and Conn, *Workplace 2000,* p. 263.

147. On the "change in the way unions view cooperation with management" and the effort of some unions to join management in creating "team systems and joint decision-making councils," see *Business Week,* "Look Who's Pushing Productivity," 7 April 1997.

148. *Time* magazine, 19 February 1996. See also the photo of Andreessen in denim jacket and jeans in *Newsweek,* "The Browser War," 29 April 1996, p. 50; and in tweed jacket, tieless shirt, and blue jeans (and holding two bulldogs on his lap) on the contents page of *Business Week,* 13 April 1998, p. 2. The photo on the cover of this same issue of *Business Week,* however, shows him with a tie. David Brooks writes in his *Bobos in Paradise* about the new "bohemian" business upper class: "Corporate America has gone more casual. Microsoft executives appear on the cover of *Fortune* with beanie propeller hats on their heads. Others are photographed looking like mellowing rock stars, wearing expensive collarless linen shirts or multicolored sweaters and rag-wool socks under funky but expensive sandals. Often they'll be shown in jeans, standing proudly in the main hallway of a Rocky Mountain log mansion" (p. 113).

149. On the notion of producer culture, see my discussion above and in note 72 to this chapter. Indeed, it may be argued that the homogenizing effects of mass media and consumption are in all ways secondary to those of production, both because the workplace is the major arena of contemporary experience and because it is precisely the new corporations that produce mass media and mass goods.

150. For an insightful antidote to the thesis of global cultural and economic homogeneity, see Castells's *Information Age,* whose broad emphasis on significant differences in the world—e.g., between U.S. and Japanese styles of service industry growth, or between U.S. and European labor history—makes an antithetical bookend to the Kotkin book *Tribes,* which I discuss below.

151. Kotkin, *Tribes,* pp. 4–5.

152. Rita Raley, who is currently completing a book on global English and the academy, has pointed out to me the perils of the "Anglo-American" conflation in Kotkin's work.

153. Kotkin, *Tribes,* especially pp. 80–84.

154. Kotkin, *Tribes,* p. 89.

155. Kotkin, *Tribes,* pp. 91, 95, 94.

156. On the history and theory of "global English," see Rita Raley, "On Global English and the Transmutation of Postcolonial Studies into 'Literature in English'" and "Machine Translation and Global English." I have also benefited from the manuscript of Raley's book in progress on "Global English and the Academy."

157. Kotkin, *Tribes,* pp. 255–56.

158. Kotkin, *Tribes,* pp. 130–31, 27, 105.

159. Harris and Moran, *Managing Cultural Differences,* p. 87.

160. Morley and Robins comment in their chapter on "Techno-Orientalism: Japan

Panic": "Western stereotypes of the Japanese hold them to be sub-human, as if they have no feelings, no emotions, no humanity" (*Spaces of Identity*, p. 172).

161. My thanks to Stewart Brand for personal correspondence on 9 May 2003 about the origin of the "information wants to be free" phrase. In answer to my query, Brand writes (quoted with permission): "Here's the sequence on 'information wants to be free.' In fall 1984, at the first Hackers' Conference, I said in one discussion session: 'On the one hand information wants to be expensive, because it's so valuable. The right information in the right place just changes your life. On the other hand, information wants to be free, because the cost of getting it out is getting lower and lower all the time. So you have these two fighting against each other.' That was printed in a report/transcript from the conference in the May 1985 *Whole Earth Review*, p. 49. In *The Media Lab* (Viking-Penguin, 1987, still in print) on p. 202 is a section which begins: 'Information Wants To Be Free. Information also wants to be expensive. Information wants to be free because it has become so cheap to distribute, copy, and recombine—too cheap to meter. It wants to be expensive because it can be immeasurably valuable to the recipient. That tension will not go away. It leads to endless wrenching debate about price, copyright, "intellectual property," the moral rightness of casual distribution, because each round of new devices makes the tension worse, not better.' The final iteration for me was in sundry talks I gave in the two years after the *Media Lab* book came out. In those I frequently said, and even put up on an overhead, the following (this one happened to be a national Computer Security conference): 'Information wants to be free (because of the new ease of copying and reshaping and casual distribution), *and* information wants to be expensive (it's the prime economic event in an information age) . . . and technology is constantly making the tension worse. If you cling blindly to the expensive part of the paradox, you miss all the action going on in the free part. The pressure of the paradox forces information to explore incessantly. Smart marketers and inventors quietly follow—and I might add, so do smart computer security people.' Since then I've added nothing to the meme, and it's been living high wide and handsome on its own. I saw in a *Wired*, April [19]97, that Jon Katz opined on p. 186: 'The single dominant ethic in this [digital] community is that information wants to be free.' "

Notable is Brand's use of *free* in a specifically economic context, whereas "information wants to be free" has since also taken on broad political and other overtones.

162. Compare Alain Touraine: "in a post-industrial society, in which cultural services have replaced material goods at the core of production, *it is the defense of the subject, in its personality and in its culture, against the logic of apparatuses and markets, that replaces the idea of class struggle*" (quoted in Castells, *Information Age*, 1:23).

163. I use the expression "pipeline" with ironic allusion to Willam H. Whyte, Jr.'s chapter "The Pipe Line" in his 1956 book *Organization Man*, pp. 109–28. Focused on the easy transition from the college campus to the training programs of the corporation (itself the object of Whyte's critique of "organization" mentality), this chapter stands in bold relief against the Gen X era, where exclusion from the pipeline is the issue. As regards my emphasis on Gen X: evidence from the U.S. Bureau of Labor Statistics indicates that even as the media continue to emphasize the effects of downsizing on baby boomers, Gen X is already the area of critical importance. While the number of boomers crested around 2000, the proportional representation of the group in the workforce had already peaked around1985 at about 55 percent of the total. By 2005, according to the forecast, boomers will increasingly retire from the scene (Carnevale and Stone, *American Mosaic*, p. 40). See also my discussion of Gen X and the notion of "slack" in chapter 9, below.

164. It does not reduce or demean the pathos of Nancy's profound meditation on "community," I believe, to read *Inoperative Community* as in some ways a philosophical allegory of the age of "downsizing." Seen in this context (admittedly partial, yet nonetheless moving), the true community of finitude, extinction, and loss of self that Nancy celebrates

as "inoperative community" is the community of the downsized—i.e., those who are ejected from the false community of the fused "subject" in corporate culture. Translated into Nancy's idiom, downsizing is the "birth" of the community of "death" (see especially chap. 1 of *Inoperative Community*).

165. Rifkin, *End of Work,* p. 197.

166. William Wordsworth, "Tintern Abbey." My identification of Wordsworth as the prophet of knowledge work is anachronistic, of course—but perhaps not too much so. Recent books on the poet by Thomas Pfau *(Wordsworth's Profession)* and Mark Schoenfield *(Professional Wordsworth),* as well as my own discussion of his "vocational imagination" in *Wordsworth,* pp. 332–41, cast him as a "professional" intellectual.

167. Castells, *Information Age,* 1:199 (italics in original). On Japanese *Wa,* see Castells, p. 181; on neo-Confucianism, see Kotkin, *Tribes,* pp. 102, 177–78. Other visions of the information age as ultimately an ethos standing at the end of previous traditions of ethoi can be cited. One of the most extraordinary is that of Yoneji Masuda, author of the 1971 Japanese *Plan for an Information Society: A National Goal toward the Year 2000.* Writing in his 1981 book, *The Information Society as Post-Industrial Society,* Masuda prophesies, "The final goal of Computopia is the *rebirth of theological synergism* of man and the supreme being, or if one prefers it, the *ultimate life force*—expressions that have meaning to both those of religious faith and the irreligious. . . . The ultimate ideal of the global futurization society will be for man's actions to be in harmony with nature in building a synergistic world" (excerpted in Forester, ed., *Information Technology Revolution,* pp. 632–33).

168. Paulson summarizes the central tenets of his 1988 *Noise of Culture* as follows (p. ix): "first, that literature is a noisy transmission channel that assumes its noise so as to become something other than a transmission channel, and second, that literature, so constituted, functions as the noise of culture, as a perturbation or source of variety in the circulation and production of discourses and ideas." Miller comments in his 1999 book *Black Holes,* p. 137: "Works of literature are black holes in the Internet Galaxy. The presence of literature and the literary on the Internet forbids thinking of the Internet as a transparent electronic highway system on which 'information' passes back and forth freely, without interruption, as an open secret." My discussion of the "paradox of information" and "ethos of the unknown" in this book includes the question of the literary within a more massive black hole (like that hypothesized at the center of our galaxy) located at the heart of knowledge work in general.

Paulson's and Miller's theses are important complements to my later definition of "cool" as "information designed to resist information"; see preface to part 3, below. As suggested in my introduction and part 4, however, the "future literary" I meditate orbits around neither creative nor black noise but an edgier paradigm of aesthetically deliberate destruction or "hacking."

Part Two Preface

1. Both Netscape and Microsoft chose names for their Web browsers that connoted the age of sea exploration: "Navigator" and "Internet Explorer," respectively. For a discussion of the relation between exploration metaphors in digital media (specifically, gaming) and colonial-era narratives of the New World, see Fuller and Jenkins, "Nintendo® and New World Travel Writing: A Dialogue." For a broader view of metaphors for the Internet, see Stefik, ed., *Internet Dreams.*

2. On the extensive efforts of Machlup and later Porat (with Rubin) to measure the extent of the "information" and "knowledge" industries in the United States, see below, appendix A, "Taxonomy of Knowledge Work"; and above, the preface to part 1, "Unnice Work." For a survey of the international econometric literature on this topic (including the work of Machlup and Porat), see Dordick and Wang, *Information Society,* pp. 31–58;

and Beniger, *Control Revolution,* p. 22. One vivid, objective confirmation of the common perception that we are now overwhelmed by information ("data smog," as Shenk calls it) is the finding of a 1984 study that from 1960 to 1980 the number of "words" produced in the United States and disseminated through various media grew roughly 2.5 times faster than the number of words consumed (where "words" is a specially defined unit of information also designed to quantify audiovisual and other media) (see Dordick and Wang, *Information Society,* pp. 107–9, 50–52).

3. Just as the Industrial Revolution occurred in two distinct phases (one starting at the end of the eighteenth century and the other at the end of the nineteenth century; see Castells, *Information Age,* 1:34), so the Information Revolution from which the Internet emerged may sensibly be divided into two overlapping periods: the first dominated by electronic analog media and the second by digital media, each with its own sub-chronologies of development. By following Castells in concentrating primarily on the latter—and especially on the convergence of microelectronics, computing, and telecommunications from the 1970s on (*Information Age,* 1:40–46; see also my own variation on Castells's developmental chronology, below)—I am in effect narrowing the relevant spectrum of "information." This makes sense for my purposes because, while "knowledge work" derives from the full social and economic history of twentieth-century work, its present New Class personality coincides with the technological, organizational, and other changes that attended the shift from the broadcast- and leisure-oriented paradigm of analog media to the current, networked and production paradigm of digital media, which is as yet only marginally a leisure category. Of course, digital by no means simply "replaces" analog. Traditional electronic media and entertainment industries, for example, are attempting to move the Internet toward a stronger leisure or "entertainment" orientation in ways reproducing analog-era patterns of dissemination and control, as instanced in the legal victories of the recording industry over MP3.com and Napster in 2000. But the reconfiguration of "information" under the specific influence of the digital paradigm determines the unique ethos of knowledge work I seek to define.

4. As will become clear in the following chapters, understanding the relation of contemporary knowledge workers to cool requires a history of the complex appropriations of cool that related counterculture to subculture in the twentieth century. Two works on the "birth of the cool" are relevant in this regard. One is Greil Marcus's "Birth of the Cool" (1999): "Cool as an idea [as opposed to the "fact of life" of "black irony and reserve"] began in the 1940s, when white artists or would-be artists like William Burroughs and Larry Rivers . . . stepped purposefully into the half-world of heroin, into the oblivion and invisibility of the black America that persisted inside the white nation. Cool began when whites, following such seemingly heroic black addicts as Charlie Parker and, in his mentor's steps, Miles Davis, assumed a superiority, an untouchability, available only to white oblivion seekers, a condescension toward the dominant social reality available only to the addict, to the already dead. Take one step back, and you have cool" (p. 20). (My thanks to Lindsay Waters for pointing me to this essay by Marcus.)

Another work is Lewis MacAdams's *Birth of the Cool: Beat, Bebop, and the American Avant-Garde* (2001), which observes, "The birth of the cool took place in the shadows, among marginal characters, in cold-water flats and furnished basement rooms" (p. 23). As his subtitle indicates, MacAdams goes on to study both the subcultural and countercultural strands of cool.

The "beginning" of cool in twentieth-century Western culture, of course, marks an artificial *terminus a quo* to the story. Robert Farris Thompson's groundbreaking work, "An Aesthetic of the Cool," reminds us that the provenance of the phenomenon may be traced back to West African culture. However, since my focus is specifically on "cool" relations between subculture, mainstream culture, counterculture, and what I will call high-tech "intraculture" toward the close of the twentieth century, I am able to make only light use of Thompson. Even my ability to draw upon the literature on twentieth-century cool is less substantial than it would be if my topic were cool considered in itself rather than as an index of my central concern with contemporary knowledge work.

After some attempts at inclusion, I have also marked as out of bounds Marshall McLuhan's thesis about "cool" versus "hot" media ("Media Hot and Cold," in *Understanding Media*). Though the terminology and context are enticingly close, McLuhan's concept of "cool" at last seemed too indirectly related to cultural cool for me to integrate it without creating an extra layer of bridge concepts or contexts that would have distracted from my argument. Clearly, though, there are perspectives from which McLuhan's "cool" is relevant to such topics I later develop as the style or feeling of cultural cool.

5. "Phyles" is a word I borrow from Neal Stephenson's novel *The Diamond Age,* in which the world is organized into corporate phyles that subsume most of the social, economic, national, and cultural roles of older institutions—a true cradle-to-grave corporate culture.

6. For Mills's fine statement of the "big split" between work and leisure, see *White Collar,* pp. 235–38. Thomas Frank makes a similar point about the excessive preoccupation of critique with consumer culture in his *Conquest of Cool*: "For all of cultural studies' subtle readings and forceful advocacy, its practitioners often tend to limit their inquiries so rigorously to the consumption of culture-products that the equally important process of cultural production is virtually ignored" (pp. 18–19). Frank's highly original book attempts to correct the imbalance by examining the one producer culture most directly and specifically in contact with consumer culture—the advertising industry. On the bias toward analysis of consumption culture in postmodern cultural criticism, see also Dirlik, *Postcolonial Aura,* p. 186.

7. For a digest of *embourgeoisement* theory, see Sobel, *White Collar Working Class,* p. 15. For discussion of the bearing of the *embourgeoisement* thesis on subculture studies, see Clarke et al., "Subcultures, Cultures and Class."

8. Strassmann writes presciently in *Information Payoff* (1985): "In the name of efficiency and convenience, the work now performed by clerical personnel in airlines, libraries, banks, government, insurance companies, and so forth will be done directly by consumers. For instance, many airlines have installed automatic ticketing machines to reduce the number of counter clerks they employ. Since only a few clerks remained (and they are largely devoted to handling inquiries), travellers find themselves lining up behind ticketing machines to do added information work.

"Home computers are enjoying a boom partially because they are picking up increased amounts of work that the consumer cannot afford to have done for him by others. Spreadsheet software and checkbook balancing programs proliferate in the home market as computer owners discover that they can manage their own financial accounting. . . . The increased paperwork burdens of new tax-saving investment and pension schemes have added millions of hours of information work to consumers' ostensibly 'leisure time'" (p. 7). See also ibid., pp. 4–8, 46. In a similar vein, Nicholas Negroponte takes e-mail as his example of the blurred line between office and home: "E-mail is a life-style that impacts the way we work and think. . . . Nine-to-five, five days a week, and two weeks off a year starts to evaporate as the dominant beat to business life. Professional and personal messages start to commingle; Sunday is not so different from Monday" (*Being Digital,* p. 193). Compare Mark Dery: "But the promised Tomorrowland of eternal leisure that was supposed to follow in the wake of these marvels [of computerized and automated communication devices] has faded into history, supplanted by a corporate future where we are always at the beeper's beck and perpetually in motion" (*Escape Velocity,* p. 12). For journalistic reports on similar phenomena, see *USA Today,* "Technology Changing Holiday Habits," 26 May 1998; *New*

York Times, "Talk, Type, Read E-Mail: The Trials of Multitasking," 23 July 1998; and *Los Angeles Times*, "The Endless Electronic Workday," 6 August 1999.

9. Rybczynski, *Waiting for the Weekend*, p. 224.

10. Schor, *Overworked American*, pp. 29, 22: the American worker averaged "an additional 163 hours, or the equivalent of an extra month a year" between 1969 and 1987, accompanied by a nearly 40 percent reduction in leisure time. For the decline in leisure hours, Schor uses a Harris poll of 1988 that covers the period from 1973 on. See also Hochschild, *Time Bind*, pp. 268–69, n. 3, for a survey of the statistical literature on the "time squeeze"; and Dordick and Wang, *Information Society*, p. 116. A Harris poll conducted in 1998 updates the figures to show that Americans work 23 percent more time (an average of 49.9 hours per week in 1998) than in 1973 (*Bloomberg News*, "Americans Spend More Time Working, Survey Says," reprinted in *Los Angeles Times*, 8 July 1998; see also *Los Angeles Times*, "With Labor Day Comes More Labor, Less Pay," 7 September 1998). (But see *Business Week*, "Killer Hours Are a Myth," 3 August 1998, for studies that mitigate or question Schor's statistics.)

Hochschild is particularly shrewd in assessing the increasing impact of work culture not just on the quantity of time left for home life, but also on the *quality* of that time. Parents are effectively "deskilled" at home in the image of assembly-line workers processing "instant mixes, frozen dinners, and take-out meals" (*Time Bind*, p. 209). "Meanwhile a low-grade Tayloresque cult of efficiency has 'jumped the fence' and come home. Home has become the place where people carry out necessary tasks efficiently in the limited amount of time allotted" (ibid., p. 49). Ironically, therefore, the workplace now provides a more expansive sense of culture than the shrunken, desperate family: the "workplace has a large, socially engineered heart while [the] home has gained a newly Taylorized feel" (ibid.).

11. In my allegory here, Steven Spielberg's film *Jurassic Park* is not about the gloriously rendered dinosaurs that erupt anachronistically into normal leisure life (in the venue of a "theme park"). Rather, it is about the computer-equipped control room of the theme park, where technical and managerial knowledge work produces the complete dinosaur experience. True to the *Frankenstein* lineage of such mad-scientist films, knowledge work is the real monster.

12. Hebdige's *Hiding in the Light* is highly pertinent in its analysis of youth subcultures, though we cannot simply expand the application to knowledge-worker "cool" but must also factor in the relation of knowledge workers to both subculture and counterculture. See especially *Hiding in the Light*, p. 35: "For the subcultural *milieu* has been constructed underneath the authorised discourses, in the face of the multiple disciplines of the family, the school and the workplace. Subculture forms up in the space between surveillance and the evasion of surveillance, it translates the fact of being under scrutiny into the pleasure of being watched. It is a hiding in the light.

"The 'subcultural response' is neither simply affirmation nor refusal, neither 'commercial exploitation' nor 'genuine revolt.' It is neither simply resistance against some external order nor straightforward conformity with the parent culture. It is both a declaration of independence, of otherness, of alien intent, a refusal of anonymity, of subordinate status. It is an *in*subordination. And at the same time it is also a confirmation of the fact of powerlessness, a celebration of impotence." In my metaphor of sunglasses worn indoors, I am influenced by the icon of "mirrorshades" in cyberpunk science fiction—e.g., as worn in surgically implanted form by the character Molly in William Gibson's *Neuromancer*. Bruce Sterling writes in the preface to a collection of cyberpunk fiction titled *Mirrorshades:* "Mirrored sunglasses have been a [cyberpunk] Movement totem since the early days of '82. . . . By hiding the eyes, mirrorshades prevent the forces of normalcy from realizing that one is crazed and possibly dangerous" (p. xi).

Chapter Two

1. In Marx and Engels, *Collected Works,* 3:272. Further citations of this work are given parenthetically in the text, by volume and page number. The usual English translation of *fixiert* ("Das Produkt der Arbeit ist die Arbeit, die sich in einem Gegenstand fixiert") is "embodied" or "materialized." However, a variant translation is interesting in the context of my current investigation of "cold": "the product of labour is labour which has been congealed in an object" (Tucker, *Marx-Engels Reader,* p. 71).

2. Archibald's "Using Marx's Theory of Alienation Empirically" reviews later controversy about the confusing relation between these several levels of alienation in Marx's theory.

3. On Taylor's "fight," which became a paradigm in his thought, see Taylor, *Principles of Scientific Management,* pp. 48–53; and Taylor, *Testimony,* pp. 78–85. See also Copley, *Frederick W. Taylor,* 1:157–64. On the Ford Hunger March of 1932, the UAW campaign against Ford later in that decade, and the role of the infamous Ford Service security forces under Harry Bennett, see Sward, pp. *Legend of Henry Ford,* pp. 231–42, 291–429.

4. A fuller study of the rise of the new middle class would need to make it clear that is it was not just Mills's work of 1951 but a succession of similar works in the 1950s—including Riesman, *Lonely Crowd* (1950); Whyte, *Organization Man* (1956); and Packard, *Status Seekers* (1959)—that marked the entrance of the class into American consciousness. (On this moment in the 1950s, see Horowitz's intellectual biography, *C. Wright Mills,* p. 226.) A common theme of this body of works was the recognition that, despite its wholly undramatic and ambiguous nature, the white-collar middle class seemed destined to be the dominant class of the century. Mills writes on his first page: "Yet it is to this white-collar world that one must look for much that is characteristic of twentieth-century existence. . . . By their mass way of life, they have transformed the tang and feel of American experience" (*White Collar,* p. ix). Similarly, on Whyte's first page: "it is their values which will set the American temper" (*Organization Man,* p. 3).

5. See Mills, *White Collar,* pp. 63–67, on the shifting demographics of American white-collar labor relative to wage-workers and the old middle class in the period from 1870 to 1940. By 1940, according to his classification, the old middle class equaled 20 percent of the U.S. workforce (down from 33 percent in 1870); the white-collar "new middle class," 25 percent (up from 6 percent in 1870); and wage-workers, 55 percent (down from 61 percent in 1870). Mills's statistics were based on a variety of government sources (see *White Collar,* pp. 358–59). Kocka, *White Collar Workers in America,* comes up with the comparable statistic that white-collars made up 24 percent of the economically active population by 1940 (p. 19, Table 1.2).

6. Mills, *White Collar,* p. xii. Mills's ironic dramatization (i.e., *un*dramatization) of his "hero" at the beginning of *White Collar,* we may note, is matched by the dramatic frame he puts around this hero at the end: "They are a chorus, too afraid to grumble, too hysterical in their applause. They are rearguarders" (p. 353). Compare Whyte, *Organization Man,* p. 12: "Because his area of maneuver seems so small and because the trapping so mundane, his fight lacks the heroic cast, but it is for all this as tough a fight." A late entry in this tradition of dramatizing the unheroic middle class is Ehrenreich's *Fear of Falling,* which sets the stage for its subject in part as follows: "Nameless, and camouflaged by a culture in which it both stars and writes the scripts, this class plays an overweening role in defining 'America': its moods, political direction, and moral tone" (p. 6).

7. Mills, *White Collar,* p. xviii.

8. *White Collar* precedes Mills's major engagement with Marxism in his late years. But it is already symptomatic of his "plain Marx," or populist-Marx, persona. As he states in his introduction, the book is an attempt to mediate between "liberal"/"pragmatist" and "Marxist"

understandings of society, which both seemed obsolete in the age of white-collar work (*White Collar*, pp. xix–xx). On the influence of American pragmatism (on which he wrote his dissertation), Weber, Marx, and other thinkers on Mills, see Horowitz, *C. Wright Mills*, pp. 117–206, also pp. 28–29, 34, 47. See also Becker, "Professional Sociology."

As regards the broader context of twentieth-century Marxism: a fuller investigation at this point would clearly need to go well beyond Mills, who, despite his eventual reception by the 1960s and 1970s New Left, cannot easily be identified as "Marxist" (and whose thinking about Marxism, in any case, did not become explicit until late in his career). Relevant is the whole diversity of so-called Western Marxism as it included, for example, authors from the Frankfurt School onward. In anticipation of my later mention of Theodore Roszak's work on "counterculture," it would be especially useful to read the aspects of Herbert Marcuse's work that bear on "alienation" and eventually contributed to the countercultural rejection of "technocracy" in the 1960s (at least as Roszak tells it in his chapter on Marcuse and Norman O. Brown in *Making of a Counter Culture*, pp. 84–123).

9. On the introduction of machines into offices on Taylorist principles, see Mills, *White Collar*, pp. 193–95. On Leffingwell's extension of scientific management principles into the white-collar office to accompany the new office machinery, see Zuboff, *In the Age of the Smart Machine*, especially pp. 117–19. Leffingwell and Robinson's *Textbook of Office Management* (originally published in 1932) codifies the way "the principles of scientific management formulated by that famous industrial scientist, Frederick Winslow Taylor," apply "to the conduct of the clerical office" (p. xi).

10. See the chapter "Flow of Work" in Leffingwell and Robinson, *Textbook of Office Management*, pp. 76–82, for the application of assembly-line principles to the office. Their subsequent chapter on mail and messenger services in the office—including the use of pneumatic tubes, conveyor belts, and constantly circulating human messengers—is also relevant. "Flow" in such usage should be distinguished from "continuous-process production" of the sort that Blauner, in his *Alienation and Freedom* (1964) has made paradigmatic of the most advanced stage of contemporary automation. Continuous-process production, as Blauner argues in his chapters on chemical workers, reintroduces elements of flexibility in labor technique. For a discussion of how the "industrial office" was organized to optimize production-line work flow (complete with schematic diagram of the paradigmatic office), see Giuliano, "Mechanization of Office Work" (originally published in 1982), pp. 304–6.

11. The now classic work on deskilling is Braverman, *Labor and Monopoly Capital* (1974), which includes discussion of the clerical and white-collar sectors. On white-collar deskilling and proletarianization, see also Sobel, *White Collar Working Class* (1989), and Zuboff, *In the Age of the Smart Machine* (1984), pp. 113–23. For the effect of white-collar deskilling on the formation of a feminized underclass of clerical workers, see Murolo, "White-Collar Women and the Rationalization of Clerical Work."

12. That even such "nonphysical" white-collar tasks as using a typewriter or calculating machine could generate "body" complaints confirms the general applicability of such analysis for the white-collar realm. In the history of "the white-collar body," especially at its more deskilled levels, Zuboff argues, the routinization of labor is paralleled by dissent over "physical discomfort" (most recently, carpal tunnel syndrome) (*In the Age of the Smart Machine*, p. 141). See Zuboff's chapters "The White-Collar Body in History" and "Office Technology as Exile and Integration" in general for the alienation of the white-collar laborer from the labor process during the twentieth century. Also highly relevant is Greenbaum, *Windows on the Workplace*.

13. See Mitzman, *Sociology and Estrangement*, on the relation between rationalization and alienation theory in turn-of-the-century German sociology.

14. Leffingwell and Robinson, *Textbook of Office Management*, p. 4.

15. Mills, *White Collar,* p. 65.

16. Mills, *White Collar,* p. 64.

17. Twentieth-century theory sometimes even insists that the "later" Marx himself rejected the alienation theory of the "early" Marx. Archibald reviews this problem in the reception of Marx's theory in "Using Marx's Theory of Alienation Empirically," pp. 119–23.

18. Mills, *White Collar,* p. 226. I use *bureaucracy* here in a sense that does not distinguish between official state bureaucracies of the sort that shaped the white-collar middle class in Germany and the looser combination of corporate, educational, and other bureaucracies in the private sphere together with their codes (e.g., "professionalism") that exerted the same influence in the United States. Kocka's unique comparative study of American and German white-collar middle classes is excellent in making use of this distinction to argue that the American white-collar class was historically different from its more literally bureaucratic German counterpart in its relation to capitalism, the lower classes, and ultimately politicization (*White Collar Workers in America,* pp. 141–53). "In America," Kocka writes, "the professions rather than the bureaucracy became a partly illusory model and object of identification for white collar groups for whom the chance for independence and thus also the ideal of the independent businessman had been strongly reduced by the turn of the century" (ibid., p. 147).

19. For the controversy surrounding Blauner's book, see the following: Hull et al., "Effect of Technology on Alienation from Work"; Vallas and Yarrow, "Advanced Technology and Worker Alienation"; and Archibald, "Using Marx's Theory of Alienation Empirically," pp. 128–29. For an overview of the debate on the effect of automation on alienation and "deskilling" from the era of Blauner's thesis on, see Adler, *Technology and the Future of Work,* pp. 6–8.

20. Blauner, *Alienation and Freedom,* p. 17n.

21. Blauner, *Alienation and Freedom,* p. ix.

22. Archibald observes: "Over and above the fact that Marx often was, as Mills aptly remarked, 'brilliantly ambiguous,' one suspects that many of his interpreters have neglected the fact that alienation has several different levels of generality, and confused several different meanings of the terms 'subjective,' 'individual,' and 'psychological'" ("Using Marx's Theory of Alienation Empirically," p. 124). The rupture between the systemic and subjective sides of alienation is the topic of Erickson's "On Work and Alienation" and Seeman's "Alienation and Engagement."

23. Mills writes at the end of the introduction to *White Collar*: "We need to characterize American society of the mid-twentieth century in more psychological terms, for now the problems that concern us most border on the psychiatric. It is one great task of social studies today to describe the larger economic and political situation in terms of its meaning for the inner life and the external career of the individual, and in doing this to take into account how the individual often becomes falsely conscious and blinded. In the welter of the individual's daily experience the framework of modern society must be sought; within that framework the psychology of the little man must be formulated" (p. xx). Again, he writes: "While the modern white-collar worker has no articulate philosophy of work, his feelings about it and his experiences of it influence his satisfactions and frustrations, the whole tone of his life. Whatever the effects of his work, known to him or not, they are the net result of the work as an activity, plus the meanings he brings to it, plus the views that others hold of it" (ibid., p. 215). These passages show Mills at work mixing Marx's theories of alienation and ideology with Freudian psychology and Deweyan pragmatism (especially under the aegis of "experience") to describe modern mentality in terms broader than what each of these influences in his background could individually achieve. The goal was to create a vocabulary capable of addressing collective mental life in a Weberian universe—precisely the goal that later cultural critics

attempted with their structural and discursive theories of collective mentality. (On Mills and Freud, see Horowitz, *C. Wright Mills,* pp. 49, 183.)

24. I make no attempt here to square the differences among these concepts, since my present point concerns only their common feature—the attempt to hold in mind both the structures of social existence and the knowledge/feeling of such existence.

25. Williams, *Marxism and Literature,* p. 132.

26. Some of the arguments I make below were first aired in my postings on 18, 22, and 29 July 1997 to the NASSR-L (North American Society for the Study of Romanticism) discussion list (available online in the list archives). The postings discussed the relation between modern "cool" and Romantic-era sensibility. Thanks to correspondents on the list for helping me to sharpen my views, especially in regard to hermeneutical or interpretive problems in researching the history of the emotions. I am especially grateful to David P. Haney, Alan Richardson, and Hugh Roberts for thought-provoking responses.

27. I draw in particular upon three groups of constructionist emotion research focused on different aspects of twentieth-century experience: social history of the early to mid-twentieth century (and 1960s) in Peter N. Stearns's mode; sociology of the "commercialization" of "emotional labor" in the 1970s and 1980s in Hochschild's influential mode; and sociology of the "informalization" of emotions in the 1970s and 1980s (as in the work of Gerhards, Wouters, and de Swaan). See Peter N. Stearns's introduction to his *American Cool* for an overview of types of emotion history and theory. (See also Stearns and Stearns, "Emotionology.") Wharton, "Affective Consequences of Service Work," reviews the research triggered by Hochschild's thesis of the commercialization of emotional labor in her *Managed Heart.* Hochschild's own appendix "Models of Emotion: From Darwin to Goffman" in *Managed Heart* embeds emotion research within a broad philosophical, psychological, and sociological genealogy. For an overview of the current emotion research field, see Cuthbertson-Johnson et al., *Sociology of Emotions.*

Focusing on constructionist emotionology does not commit us to one side or the other of the nature/culture debate. We need not question the basis of emotional experience in physiological and cognitive structures to see that the relevant issue in our context is the "placement" of such experience within culture. By comparison, we might think of a work that has much influenced emotion research—Elias, *Civilizing Process.* Elias's theory that the civilizing process—as in the history of table manners—has meant learning to manage one's impulses and distancing oneself from bodily and emotional contact implies not so much the extinction of such impulses as the task of learning their "place" at table. Just so, the constructivist emotion research cited above (especially on the commercialization of feeling) is about how modernity has learned the place of the emotions at another kind of table—the boardroom table. (On the relevance of Elias to emotion history, see especially Stearns, *American Cool,* pp. 7–8; and Gerhards, "Changing Culture of Emotions in Modern Society," p. 739.)

28. Stearns, *American Cool,* p. 218; Hochschild, *Managed Heart,* p. 156.

29. My choice of "range and intensity" as the relevant dimensions of emotionality at work is informed by the terms *content, intensity,* and *diversity* in Rafaeli and Sutton, "Expression of Emotion in Organizational Life," pp. 4–7.

30. See the family history chapter of my *Wordsworth,* pp. 225–310, for a fuller survey of the pre-twentieth-century developments I sample here.

31. Gerhards, "Changing Culture of Emotions in Modern Society," p. 741, synonymizes such customary sociality with Habermas's notion of "lifeworld."

32. Hochschild, *Managed Heart,* p. 57. A more systematic statement of the problem, descended from Hochschild's work, may be found in Rafaeli and Sutton, "Expression of Emotion in Organizational Life." The fundamental distinction in the latter work between

"experienced" and "expressed" emotions makes it clear that feeling rules (or "norms," as Rafaeli and Sutton call them, pp. 8–10) negotiate a complex dynamic between emotions one does feel, should feel, and does/should express or display.

33. See Flam, "Emotional Man," which argues that "corporate actors" are organizations "ranging from business firms to charitable foundations" that "produce, apart from everything else . . . *tempered* (restrained, disciplined) but solidified and permanent emotions in place of unpredictable and wavering, often boundless feelings" (p. 225). Flam uses Hochschild's concepts of "feeling rules" and "deep acting" to make her argument.

34. As shown by the alternative contexts I mention here, my study of the genealogy of "cold" emotionality in the context of twentieth-century business could be extended into other institutions of modernity. A pertinent topic of investigation, for example, would be the relation between emotionality and military-political détente in the Cold War.

35. For Beniger's Weberian analysis of bureacracy as "generalized control," see *Control Revolution*, pp. 6, 13–16, 279–80.

36. See Mills, *White Collar*, p. 237: "As the work sphere declines in meaning and gives no inner direction and rhythm to life, so have community and kinship circles declined as ways of 'fixing man into society.' In the old craft model, work sphere and family coincided." For the sake of simplicity, I do not emphasize one further way to relate emotional management in the family and at work. Because the single-family home was the domestic complement to work at the factory or office, it would be possible to correlate the emotional management of domesticity or private life itself (including the sensibility of housework or child raising) with the emotional management of the workplace, whether that correlation is one of complementarity, compensation, or (as we heard above in Stearns's and Hochschild's comparisons of child raising to "management") parallelism. I concentrate in this chapter on emotional management in the nondomestic sphere of production.

37. Taylor, *Testimony*, p. 81.

38. Hochschild, *Managed Heart*, p. 7n. See Barthes, *Elements of Semiology*, pp. 77–78.

39. Taylor, *Testimony*, p. 83; Taylor *Principles of Scientific Management*, pp. 36–37; Taylor, *Testimony*, pp. 40–45.

40. *Shop Management*, pp. 102–4. A more accurate genealogy of the modern notion of management staff would also need to accommodate the fact that Taylor's ideas were not always adopted in detail by the many firms that otherwise swore by "Taylorism" (see Rochlin, *Trapped in the Net*, pp. 54–55). As Beniger points out, factories tended to bypass the schema, if not the principle, of Taylor's rigidly specialized corps of "functional foremen" in favor of the "line-and-staff structure pioneered by the Pennsylvania Railroad in the late 1850s, first adapted to production by the Yale and Towne Lock Company in 1905, and promoted by Harrington Emerson—a former railroad manager himself" (*Control Revolution*, p. 297). A full account of the rise of the concepts of managerial staff and ultimately also middle management would need to consolidate Taylorism within the longer history of the rise of the great modern, "vertically integrated" bureaucratic organizations—a history that, as Alfred D. Chandler, Jr. shows, runs directly through the railroad companies.

41. Taylor, *Shop Management*, p. 131. For *friend* and its variants in Taylor's works, see, for example, *Principles of Scientific Management*, pp. 26–27, 49, 52, 72; *Shop Management*, pp. 185, 196; *Testimony*, pp. 30, 32–33, 59, 66, 83, 85, 128, 145, 204, 280–81, 283.

42. Taylor, *Testimony*, pp. 45, 287.

43. Sward, *Legend of Henry Ford*, p. 33, notes the extent to which Ford's theory of line production complemented Taylor's system. Taylor gave speeches in Detroit in 1909 and 1914 (ibid., pp. 33–34n).

44. Sward, *Legend of Henry Ford*, p. 56; Simonds, *Henry Ford*, p. 137. See Sward,

p. 177, on Ford as the "workingman's best friend." Simonds's biography of Ford thoroughly buys into the Ford-as-workingman's-friend legend.

45. Sward, *Legend of Henry Ford,* pp. 57–59; Batchelor, *Henry Ford,* p. 50.

46. Sward, *Legend of Henry Ford,* p. 312.

47. Sward, *Legend of Henry Ford,* p. 312; Batchelor, *Henry Ford,* p. 53.

48. Cited in Jardim, *First Henry Ford,* pp. 120–21.

49. Batchelor, *Henry Ford,* p. 49. Batchelor draws evidence from Steve Babson et al., *Working Detroit: The Making of a Union Town* (1986).

50. These examples are from Stearns, *American Cool,* p. 55; and Zuboff, *In the Age of the Smart Machine,* p. 34. Zuboff's look at the problem, pp. 31–36, focuses on both nineteenth- and early twentieth-century factories.

51. Charles A. Madison, "My Seven Years of Automotive Servitude," quoted in Batchelor, *Henry Ford,* p. 53.

52. Taylor, *Testimony,* pp. 196–97. Hochschild tellingly concludes the introductory chapter in her *Managed Heart* by noting the rise in popularity of robot jokes (pp. 22–23).

53. Stearns, *American Cool,* p. 4.

54. Leffingwell and Robinson, *Textbook of Office Management,* p. 47.

55. For Leffingwell's research on lighting and other physical aspects of office work, see, for example, his pamphlets *Data on Artificial Lighting* and *Data on Ventilation.* Why it should be common for "girls" to be fainting in the office—whether a function of working conditions, of stereotyping in Leffingwell and Robinson's description, or of both—is another issue worth examining, though I do not take it up here.

56. Cited in Beniger, *Control Revolution,* p. 314. On the apparent contradiction between the philosophies of scientific management and human relations that still seems to bedevil corporations today, see Micklethwait and Wooldridge, *Witch Doctors,* pp. 16–17. For Beniger's discussion of the human relations movement, see *Control Revolution,* pp. 313–15. Also related is Christopher Newfield's discussion of the contradiction between two corporate philosophies, each with its own genealogy: "finance control" and "human relations" ("Corporate Culture Wars").

57. The Hawthorne experiments, which were conducted by Mayo and his industrial research team, are discussed in Mayo's *Human Problems of an Industrial Civilization* (1933) and *Social Problems of an Industrial Civilization* (1945). See also Beniger, *Control Revolution,* p. 314.

58. Stearns, *American Cool,* p. 122. On outbursts of anger among workers, see, for example, Mayo, *Social Problems of an Industrial Civilization,* pp. 61–62.

59. On "balance" and "inner equilibrium," see Mayo, *Human Problems of an Industrial Civilization,* pp. 52, 72. On "emptiness" and "anxieties," see especially Mayo, *Social Problems of an Industrial Civilization,* p. 76. For a notice of "pessimism" and "melancholy," see ibid., p. 61; for a sustained inquiry into "monotony," see the chapter on this topic in Mayo, *Human Problems of an Industrial Civilization.*

60. Mayo, *Social Problems of an Industrial Civilization,* pp. 100, 121–22. The title of Mayo's chapter on "'Patriotism Is Not Enough'" is a quotation from Edith Cavell.

61. Stearns, *American Cool.,* p. 123; also pp. 124–25, 216–17.

62. Whyte, *Organization Man,* p. 36.

63. Smith, *Psychology of Industrial Behavior,* pp. 136–41, 100–101.

64. Burleigh B. Gardner and David G. Moore, *Human Relations in Industry* (1955); and R. A. Sutermeister, "Training Foremen in Human Relations" (1943). Both are quoted in Stearns, *American Cool,* pp. 124, 125.

65. See Whyte, *Organization Man,* pp. 189–91. See, for example, Henry Clay Smith's *Psychology of Industrial Behavior* for the use of tests in "selecting unanxious personnel" and "mature workers" (pp. 107–9, 152–53).

66. Whyte, *Organization Man,* p. 180; also pp. 182–201, 405–10.

67. Stearns, *American Cool,* p. 186; see also p. 219.

68. Both these quotes in Ehrenreich, *Fear of Falling,* p. 27.

69. Stearns, *American Cool,* p. 267; Mills, *White Collar,* p. 236; Rybczynski, *Waiting for the Weekend,* pp. 210–34.

70. Robbins, *Secular Vocations,* pp. 29–56. This movie is set at the turn of the twentieth century in the twilight era of the Old West—i.e., in a cultural segue between the Old West and the clinical professionalism and technological rationality of modernity. Similar in this regard is Sam Peckinpah's *Wild Bunch* (1969), which is set roughly in the same era (just before World War I) and at its infamous climax features the demonic icon of the new age of technology—the machine gun. The massacre at the end is legendary in its violence, but from another point of view it is just about men learning the techniques for running a new machine (one that produces death). The massacre is violence as routine, as assembly line.

71. I have benefited here from the kind of explanation of popular culture that Barbara Ehrenreich suggests in her *Fear of Falling,* which begins with the era of the 1950s and early 1960s. See, for example, her discussion of the nature and role of youth culture within the middle class, of the representation of "juvenile delinquency," and of films and rock and roll (pp. 22–23, 93–95). My thesis about the symbolic displacement by which the middle class at once projected itself into "outsiders" to work and maintained a distance from those outsiders is roughly analogous to Ehrenreich's thesis about the representational function of "the poor" during the middle class's "discovery of poverty" from the Kennedy years on. See, for example, *Fear of Falling,* p. 56, where after taking inventory of condescending mid-century views of the poor as mindless creatures of the present who spend their money without foresight (among other stereotypes), Ehrenreich concludes: "The poor—the invented poor—came to serve as a mirror for the middle class, reflecting its own dread submission to the imperatives of consumption, the tyranny of affluence." In my view, there is ultimately no contradiction between seeing the poor as mirrors of the middle class's own "submission" to consumption and seeing them as representations of outsiders or rebels (i.e., as the "hoods" variously demonized or celebrated in depictions of lower-class youth). Both characterizations paint the poor as "victims" of the system of production. (I defer factoring "the discovery of race" by the middle class into an explanation of cool until I take up the case of counterculture later.)

72. My discussion of cool and subculture in this book is informed by the theory of subcultures in the tradition of the University of Birmingham's Centre for Contemporary Cultural Studies (especially Hebdige's *Subculture* and *Hiding in the Light,* and the essays collected in Hall and Jefferson, eds., *Resistance through Rituals*). Such theory, however, cannot be directly applied to the problem of American cool from the early twentieth century on—not only because of the mismatch between the highly localized context studied by the Birmingham group (e.g., British working-class youth in the 1960s and 1970s) and the broader span of periods, generations, and classes that is my concern, but because in the final analysis the general phenomenon of cool is *not* subcultural, even if the subcultural notion of "style" plays a crucial role in its formation. The question that constantly occurred to me while reading Hebdige's *Subculture* concerned how cool among knowledge workers from the white-collar middle class to the New Class seems at once so exactly like subcultural style (passage after passage in Hebdige could almost be used verbatim to describe the character of information cool) and so unlike. I have also benefited from reading about "cool" in the social history of particular provinces of American and/or Canadian

subculture, including Majors and Billson, *Cool Pose;* and Danesi, *Cool.* Other works that treat the cultural history of cool in relation to subcultures, e.g., from the Jazz Age on, are cited below. The topic of cool and subculture is only one part of the puzzle of how "normal" work culture learns the desire to be cool. The other part of the puzzle, which I turn to in chapter 3, is the relation between cool and counterculture (which itself stands in a complex relation to subculture).

73. I am aware that a fuller argument would not lump together early and mid-twentieth-century relations between mainstream culture and subcultural music. The changing numbers and nature of the "mainstream," let alone of the "subcultures" involved, make a difference that would need to be elaborated. For a study of the social and political history of jazz, see Vincent, *Keep Cool.* Of particular relevance at this point in my argument is Vincent's chapter "The Passing of a Music Revolution," which documents the appropriation of the jazz scene.

74. The essentialism that reads cool as a one-way process of diffusion from authentic street or youth subculture to inauthentic commercial or mainstream culture is well represented in journalistic writings about cool. See, for example, Malcolm Gladwell's piece in the *New Yorker,* "The Coolhunt"; Gary Chapman's column in the *Los Angeles Times* "What Is Cool? Certainly Not a Campaign for It"; the *Los Angeles Times* story "Whatever Happened to Hip?"; *Newsweek*'s article "The Kids Know Cool"; and *Newsweek*'s box story "Will Cigars Stay Hot? How to Track the Trend" (which includes a chart listing four "stages" of cool: Fringe, or Pre-cool; Trendy, or Cool; Mainstream, or Post-cool; and Mutation, or Neo-cool). The most serious and extensive journalistic argument along these lines that I am aware of is the Public Broadcasting Service's *Frontline* broadcast of 27 February 2001, "The Merchants of Cool: A Report on the Creators and Marketers of Popular Culture for Teenagers," which is also represented on the Web in a capacious set of articles and interviews. As it pursues its issues, "Merchants of Cool" touches upon the notion of a dual-directional, "symbiotic relationship" between the media and teens—a more complex notion of the dissemination of cool akin to the dialectical explanation I offer below (see especially the "Themes" page titled "The Symbiotic Relationship between the Media and Teens" that the "Merchants of Cool" Web site devotes to this topic).

75. Barbara Ehrenreich reminds us of the extent to which the *juvenile delinquent* label was used in the 1950s. She comments, "The JD [label] raised the issue of class and provided a comforting distraction from it. If there were other classes, other ways of life than that known by the white middle class, they could be seen as 'deviations'" (*Fear of Falling,* p. 23).

76. See the beginning of Kleinhans's "Cultural Appropriation and Subcultural Expression": "The questions I want to examine are: how do subcultures appropriate from the dominant culture, particularly its mass culture, and how does that dominant mass culture in turn appropriate from subcultures? Does such dual appropriation promote or undermine assimilation and/or identity?" The thesis of "reciprocal appropriation" I propose here also parallels Thomas Frank's approach to counterculture in the 1960s. Frank similarly rejects any linear story of cooptation in favor of exploring what might be called the dialectics of cooptation. Compare, for example, Frank's observation that the counterculture "was triggered at least as much by developments in mass culture . . . as changes at the grass roots" (*Conquest of Cool,* p. 8). In a fuller study we would need to allow for how reciprocal appropriation applies not just to the relation between mainstream culture and either subculture or counterculture but also to the relations of intramural contest/appropriation *within* each of the latter cultures. Hebdige's *Subculture,* for example, drills down to explore the relations between subcultures themselves (especially between black youth subculture in London and various white youth subcultures) and between parent and youth working-class cultures. (See also the formulation of subculture as a "double articulation" to "parent culture" and "dominant culture" in Clarke et al., "Subcultures, Cultures and Class," p. 15.)

77. Unfortunately, I did not learn of Joel Dinerstein's *Swinging the Machine: Modernity, Technology, and African American Vernacular Culture between the World Wars* (2003) in time to draw upon it in my argument. Dinerstein studies the antithetical manner in which African American culture expressed yet resisted its relation to dominant technological society through a "machine aesthetics" of "power, drive, precision, repetition, reproducibility, smoothness" performed in jazz and swing music, dance, storytelling, and folklore (p. 19). "To simplify a bit," he says, "Euro-Americans created the nation's *technology* while African Americans created the nation's *survival technology*. . . . Survival technology consists of public rituals of music, dance, storytelling, and sermonizing that create a forum for existential affirmation through physicality, spirituality, joy, and sexuality—'somebodiness,' as some African American preachers call it—against the dominant society's attempts to eviscerate one's individuality and cultural heritage" (p. 22). Because for African Americans "music is the cultural form that mediates between oppositional and assimilationist trends, between resistance and accommodation," it was African American music and dance in particular that created a survival technology adequate to modern industrialism by "incorporating the dominant society's machines into music" (pp. 105, 119–20). Dinerstein also reflects on the mainstream dissemination of such machine aesthetics. "Swing music and dance constituted significant social practices of cultural resistance that helped Americans regain a sense of their own individual bodies set against assembly line realities," he says; or again, "African American musicians brought the power of machines under artistic control and thus modeled the possibility of individual style within a technological society" (pp. 19, 130). Dinerstein's book provides historical substance for my more anecdotal argument about subcultures and technology here.

This is also the occasion to express regret that much of my book was completed before I had a chance to read the chapters titled "The Hendrix Experience at the Marvel Gym: A Bad Day for the Aesthetically Correct" and "Toy Story: Living the Laws of Gravity" from Lindsay Waters's work in progress, "A Critique of Pure Hipness" (my gratitude to the author for manuscripts of the chapters). Though Waters here focuses on the countercultural moment of the 1960s, which I turn to at the end of my next chapter, these chapters are germane to my present argument about camo-technology because of their analysis of the "marriage of man and machine" witnessed in a 1968 performance by Jimi Hendrix—an undecidably subjective and objective, self-expressive and robotic relation of man to guitar that Waters depicts to frame a subtle meditation on the "mechanical," "artificial," yet nonetheless powerful in art. In Waters's critique of hipness, cultural "hip" is not what I am formulating here as cool mock- or camo-tech. Rather, it is a too aesthetic, cerebral recoil from the "mindless," "mechanical" affect of popular art. "We are too hip to be cool," he says.

78. Hebdige, *Subculture,* pp. 38–39. Compare American hip-hop as Gary Chapman described it in 1996: "The predominant cultural expression of the inner city for the last 15 years, hip-hop, has been a brilliant marriage of street slang and technology. Hip-hop got its start in the 1970s when street DJs started manipulating records on turntables to produce rhythms and surprising collages of sounds and words. Hip-hop musicians have absorbed computers and turned them into bold music machines with sampling, synthesized bass and dense layers of sound and expression" (p. D3).

79. As Hebdige puts it, the commodities that subcultures use as the elements of style are "open to a double inflection: to 'illegitimate' as well as 'legitimate' uses. These 'humble objects' can be magically appropriated; 'stolen' by subordinate groups and made to carry 'secret' meanings: meanings which express, in code, a form of resistance to the order which guarantees their continued subordination" (*Subculture,* p. 18). See also Clarke et al., "Subcultures, Cultures and Class," p. 53. Such arguments contribute to what is by now the well-established thesis of autonomous or contestatory "usage" in cultural studies. As Hebdige says, "It is basically the way in which commodities are *used* in subculture which marks the subculture off from more orthodox cultural formations" (*Subculture,* p. 103).

80. Again, Hebdige's *Subculture* is suggestive. Among the London youth subcultures that Hebdige chronicles were the mods, who exemplify one style of being "outside" of normal working life: "Somewhere on the way home from school or work, the mods went 'missing': they were absorbed into a 'noonday underground' of cellar clubs, discotheques, boutiques and record shops which lay hidden beneath the 'straight world' against which it was ostensibly defined. . . . They lived in between the leaves of the commercial calendar, as it were (hence the Bank Holiday occasions, the week-end events, the 'all-niters'), in the pockets of free time which alone made work meaningful" (*Subculture,* p. 53).

81. Geertz, "Deep Play." If Balinese society rests upon "status," then we might say that modern industrial society rests upon "style." Style is the cockfight of modernity.

Chapter Three

1. Zuboff, *In the Age of the Smart Machine,* pp. 20–21. "Piney Wood" is Zuboff's fictional name for one of the real factories and offices she researched in writing her book. Using pseudonyms for companies is a convention in much business scholarship based on case studies.

2. Zuboff, *In the Age of the Smart Machine,* pp. 124–25.

3. Kraut, "Social Issues and White-Collar Technology," p. 2.

4. The first commercial mainframes appeared in the 1950s; IBM's dominating System/360 computer had its debut in 1964; and DEC's PDP-8 minicomputer arrived in 1965. Mainframe technology was essentially mature by the mid-1970s. See Campbell-Kelly and Aspray, *Computer,* pp. 137–50, 222–26; Castells, *Information Age,* 1:43–44.

5. Feldberg and Glenn, "Technology and the Transformation of Clerical Work," pp. 79–80.

6. Feldberg and Glenn, "Technology and the Transformation of Clerical Work," p. 80; Castells, *Information Age,* 1:246; Bair, "User Needs for Office Systems Solutions," pp. 179–80.

7. See Bikson's table in "Understanding the Implementation of Office Technology," p. 159; see also Zuboff, *In the Age of the Smart Machine,* pp. 415–22. Statistics for the early 1980s and after, however, blur the relevant story here because they also include the use of personal computers in the workplace.

8. As Bair usefully indicates in his analysis of rising curves of technological "awareness" or "share of mind" ("the proportion of people who have a similar conceptualization" of a technology), advanced IT was a "cultural phenomenon" ("User Needs for Office Systems Solutions," pp. 177–79).

9. An excellent example of scholarship on the relationship between technology and mentality, although not production mentality, is Schivelbusch, *Railway Journey,* which studies the ways the railroad changed perceptions of time, space, landscape, and the nature of machinery itself. My thanks to Geoffrey Bowker, who first referred me to Schivelbusch.

10. For an introduction to the "technological determinism" debate, including the critique of "reification," see Daniel Chandler's online essay "Technological or Media Determinism."

11. Zuboff, *In the Age of the Smart Machine,* p. 9. Compare Castells, *Information Age,* 1:32: "What characterizes the current technological revolution is not the centrality of knowledge and information, but the application of such knowledge and information to knowledge generation and information processing/communication devices, in a cumulative feedback loop between innovation and the uses of innovation." See also ibid., 1:61.

12. About the spreadsheet, Rochlin observes: "Once graphics were included, [it] was a nearly perfect expression of what you could do with a computer that you could not do without it; it combined the computer's specialty, memory and calculating power, with the most powerful and most difficult to replicate power of the human mind, the ability to recognize and integrate patterns" (*Trapped in the Net,* p. 28).

13. Sproull and Kiesler, *Connections,* p. 159. The view that computerization is bound up with the act of mental holism or syncretism has wide support in the literature on the history and sociology of information technology. Attewell, for example, notes the extent to which managers and workers in "successfully automated highly integrated plants reported spending more of their time dealing laterally with other departments," scanning "for problems elsewhere that might affect their own department" ("Skill and Occupational Changes in U.S. Manufacturing," p. 79). Hirschhorn describes the "cybernetic framing of problems" as the ability to see "the parts and their valence, function, and performance" in relation "to a whole, in the way they are embedded in a circuit" (*Beyond Mechanization,* p. 39). Such mental framing mixes enhanced awareness on the job with "fringe awareness" (ibid., pp. 92–93). Or as he puts it in an article coauthored with Mokray, "Working with a system of tools ultimately linked together by the computer, the direct personnel must focus increasingly on flow and pattern rather than on the single piece and the particular puzzle it may present" ("Automation and Competency Requirements in Manufacturing," p. 23). Beginning the first chapter of *Life on the Screen* with an image of the simultaneous windows on her computer desktop, Turkle similarly demonstrates the notion of an expanded sensorium: "When I write at the computer, all of these [windows and their contents] are present and my thinking space seems somehow enlarged" (p. 29). Laurel, *Computers as Theatre,* premises her argument about computer interfaces on the need to facilitate a sense of the "whole action" of computing, which she analyzes as a dramatic action. Rochlin documents one of the most extreme forms of holistic information immersion in his study of such operators of critical, zero-fault-tolerance IT as flight controllers, military electronics controllers, nuclear power plant operators, and so forth. In each such occupation, Rochlin finds, there is a state of holistic information awareness—of immersion in the total pattern of information—akin to what the controllers in Navy ship combat operations centers call "having the bubble." "Having the bubble" means being able "to construct and maintain the cognitive map that allows them to integrate such diverse inputs as combat status, information flows from sensors and remote observation, and the real-time status and performance of the various weapons and systems into a single picture of the ship's overall situation and operational status" (*Trapped in the Net,* p. 109).

14. Zuboff, *In the Age of the Smart Machine,* pp. 163, 169, 202. A striking ad campaign in 2003 for IBM's DB2, WebSphere Business Integration, Tivoli Intelligent Management, and Lotus Workplace software illustrates how extreme the "vision" metaphor would become. On a running sequence of several recto pages in such magazines as *Business Week* (see, for example, the issue of 16 June 2003, pp. 95, 97, 99, 101), IBM depicts business people each standing with eyes shut and seemingly alone, yet each also mentally envisioning a world of connected information. The large slogan on each page reads, "Can you see it?" Other copy includes such lines as "See DB2 software connect data, near and far," "See DB2 software create insight, again and again," "See customers connect with partners," "See the problem before it occurs," "See better collaboration."

15. Zuboff, *In the Age of the Smart Machine,* pp. 94, 157.

16. Sproull and Kiesler, *Connections,* p. 81; *Business Week,* 8 August 1983, reprinted in Forester, *Information Technology Revolution,* p. 324; Laurel, *Computers as Theatre,* p. 169; Rochlin, *Trapped in the Net,* p. 126.

17. Zuboff, *In the Age of the Smart Machine,* pp. 58–96, 174–218.

18. Mills, *White Collar,* pp. 189–212.

19. Zuboff, *In the Age of the Smart Machine,* pp. 97–123, 170; see also Murulo, "White-Collar Women and the Rationalization of Clerical Work."

20. Zuboff, *In the Age of the Smart Machine,* p. 6.

21. A banker at Global Bank Brazil tells Zuboff: "With the right data-base technology it becomes cost-effective for us to provide our clients with a continual and accurate picture of their cash position. We can manage their accounts payable and accounts receivable through our system. We can advise them when they need a loan or when they have an excess of funds they should invest" (*In the Age of the Smart Machine,* p. 161).

22. Zuboff, *In the Age of the Smart Machine,* pp. 10, 303. See Castells, *Information Age,* 1: 168–69: the "introduction [of information technology] in the absence of fundamental organizational change in fact aggravated the problems of bureaucratization and rigidity. Computerized controls are even more paralyzing than traditional face-to-face chains of command."

23. Beniger, *Control Revolution,* especially pp. 390–436. For another study of the implications of information technology for organizational and social control, see the chapter "Control and Influence" in Sproull and Kiesler, *Connections,* pp. 103–23.

24. Zuboff, *In the Age of the Smart Machine,* p. 252; Taylor, *Shop Management,* pp. 102–4.

25. Garson, *Electronic Sweatshop,* p. 11. On the challenge that "invisible" information work posed to control and accountability, see Zuboff, *In the Age of the Smart Machine,* pp. 290–96.

26. Greenbaum, *Windows on the Workplace,* p. 44; see also pp. 38–42, 50–67.

27. Garson, *Electronic Sweatshop,* p. 9.

28. Greenbaum, *Windows on the Workplace,* pp. 71–72. As Greenbaum notes, cubicles had been used previously, "but it was not until the 1980s that they were embraced by management strategists" (p. 71). Giuliano, "Office of the Future," discusses the reorganization of the office for information technology after the mid-1970s and provides a schematic floor plan of the paradigmatic cubicle office (pp. 306–7). I am particularly indebted to the comp.human-factors Usenet group in June 1998 for a sustained discussion of life in cubicles compared both to private offices and to the new "break down the walls" architecture of some innovative high-tech firms. (The thread occurred under the subject heading, "offices vs. cubicles.") It was clear from the comments of many of the participants (particularly software designers in this discussion) that workers are well aware that the cubicle is ill suited for knowledge work in at least two, seemingly contradictory ways. On the one hand, a cubicle closes off the worker and inhibits creative collaboration. And on the other hand, it does not close off the worker *enough* to allow for concentrated work free from interruption.

29. For Leffingwell's own thoughts on early, cubiclelike office-partitioning strategies, see the photographs and comments in *Scientific Office Management* (1917), p. 140. One photo of an office partitioned into "booths" is captioned "Making Supervision Easier."

30. Mills, *White Collar,* p. 195.

31. On "idiot-proof" computing, see Greenbaum, *Windows on the Workplace,* pp. 79–80. For other analyses of the fixation on managerial control in early computing, see Zuboff's citations from the work of David Noble, Harley Shaiken, and others (*In the Age of the Smart Machine,* p. 283).

32. Zuboff, *In the Age of the Smart Machine,* pp. 132, 135, 138, 166, 360.

33. In critiquing overly narrow understandings of information technology, Brown and Duguid's *Social Life of Information* uses "tunnel vision" as its leading metaphor (e.g., pp. 1–4, 252).

34. Zuboff, *In the Age of the Smart Machine,* p. 156.

35. Garson,*Electronic Sweatshop,* p. 10.

36. See also Mark Poster's chapters on databases and "panopticism" in *Mode of Information,* pp. 69–98, and *Second Media Age,* pp. 78–94.

37. Zuboff, *In the Age of the Smart Machine,* pp. 316, 326.

38. Though more attuned to the monitoring of consumers than of producers, Poster's chapter "Foucault and Databases" in *Mode of Information* is relevant: "In modern society power is imposed not by the personal presence and brute force of a caste of nobles . . . but by the systematic scribblings in discourses, by the continual monitoring of daily life" (p. 91).

39. Zuboff, *In the Age of the Smart Machine,* p. 333.

40. Source for date of the Post-It note: *Los Angeles Times,* "Looking Back to Office's Future." Brown and Duguid discuss the Post-It note in their chapter on the uses of paper in the information age (*Social Life of Information,* pp. 181–82). See also their discussion of the kinds of "improvisation" that subvert or supplement "forms" in the form-centered world of the modern organization (pp. 108–9).

41. Zuboff, *In the Age of the Smart Machine,* p. 367, 385, 384.

42. Garson, *Electronic Sweatshop,* p. 10.

43. After affecting clerks and switchboard operators, Garson observes, computerization next benumbed secretaries, bank tellers, and service workers, then finally professionals and managers (*Electronic Sweatshop,* p. 10). On the "humbling and humiliating" experience of top-level executives trying to learn the new technology, see *Los Angeles Times,* "Top Execs Slowly Becoming Computer Savvy," 17 January 1998.

44. See the drawings in Zuboff, *In the Age of the Smart Machine,* especially pp. 144–47. One of the "before" drawings I refer to shows the sun hanging in unreal space directly within the office environment; another frames the sun in a landscape or pastoral drawing (complete with tree) hung on the wall.

45. Zuboff, *In the Age of the Smart Machine,* p. 75. The land where cubicles stretched as far as the eye could see (which was only to the next cubicle), we may add, was a land so cold that it is clear why the now ubiquitous office potted philodendron or ficus took root (and, more recently, beautiful orchids and bromeliads in the office lobby). Marking a further evolution in what I termed the mythopoesis of the office landscape, these tropical plants were needed to give compensatory warmth. Everywhere else we see the standard muted textures, colors, and sounds of the cubicle environment (extending to the beige look of computer terminals and the cool phosphors of early monitor screens). But here, around *this* ficus flourishing miraculously under the fluorescent glare, it is as if the sun shone down and we were working in the fields again.

46. I take the phrase "boundary spanning" from Sutton and Rafaeli, "Untangling the Relationship between Displayed Emotions and Organizational Sales," as used on p. 463, for example: "Customers of service organizations often interact with only one or two boundary-spanning employees during a given visit."

47. Roach, "Technology and the Services Sector," p. 118. Roach's detailed study of investment in IT in the service sector provides evidence of the tight coupling of information work to service work.

48. Castells, *Information Age,* 1:210. See also Castells's statistical tables, ibid., 1:282–83, 296. Castells notes, however, that the United States and Canada represent only one paradigm of the service economy. The relatively low proportional rise in service jobs in Japan constitutes what he sees as a different paradigm (ibid., 1:227).

49. For the 1990 study, see V. A. Zeithaml et al., *Delivering Quality Service: Balancing Customer Perceptions and Expectations* (1990), cited in Wharton, "Affective Consequences of

Service Work," p. 205. On BLS projections for 1990–2005, see Castells, *Information Age,* 1: 222–24; see also *Business Week,* "Peering into a New Millennium," 26 January 1998.

50. Wharton, "Affective Consequences of Service Work," pp. 205–6, 210. For other economic definitions of the services, see Mark, "Measuring Productivity in Services Industries," p. 140.

51. For the historical rise of corporate impression and image management, see Marchand, *Creating the Corporate Soul.*

52. Strassmann, *Information Payoff,* p. 4.

53. Mills, *White Collar,* p. xvii; see also pp. 182–88. See Hochschild, *Managed Heart,* p. ix, on the influence upon her work of Mills's chapter "The Great Salesroom."

54. Subsequent work on emotion in the workplace has complemented Hochschild's study of flight attendants with studies of theme park workers, convenience store clerks, bill collectors, and others. For studies of emotional labor management at Disneyland, for example, see Rafaeli and Sutton, "Expression of Emotion as Part of the Work Role," pp. 26–27; and Van Maanen and Kunda, "Real Feelings," pp. 58–70. For studies of convenience store clerks, see Sutton and Rafaeli, "Untangling the Relationship between Displayed Emotions and Organizational Sales"; and for bill collectors, see Sutton, "Maintaining Norms about Expressed Emotions." (Hochschild makes bill collectors her secondary case; see *Managed Heart,* p. 16.)

55. Hochschild, *Managed Heart,* pp. 234–36. For a cross-cultural comparison, see Gerhards's count of the increase in German "emotion workers" from 1925 to 1982, though no systematic correlation between Hochschild's and Gerhards's definitions of occupational groups seems possible ("Changing Culture of Emotions in Modern Society," p. 741).

56. See especially *Managed Heart,* pp. 11, 156–58, 234–42, for Hochschild's statistics and also her identification of the phenomenon as centrally middle class. See also de Swaan on the "organizational middle-class workers" and " 'professional-managerial' middle-class," who in large organizations "no longer performed well-defined routines but engaged in complicated interactions with colleagues, clients, or customers, using their personalities and judgmental capacities as occupational instruments" ("Politics of Agoraphobia," p. 375). On the pseudo-professional status of flight attendants, see Hochschild, *Managed Heart,* p. 103: "Like workers in many other occupations, they call themselves 'professional' [even though they are not in one of the traditional professions] because they have mastered a body of knowledge and want respect for that. Companies also use 'professional' to refer to this knowledge, but they refer to something else as well. For them a 'professional' flight attendant is one who has completely accepted the rules of standardization."

57. Hochschild, *Managed Heart,* pp. 96, 104–5; Rafaeli and Sutton, "Expression of Emotion as Part of the Work Role," p. 27. See also Rafaeli and Sutton, "Expression of Emotion in Organizational Life," pp. 9–10, on the use of "socialization practices" in organizations for "inducing behavior that is consistent with organizational display rules."

58. Van Maanen and Kunda, "Real Feelings," p. 77. See also Kunda, *Engineering Culture.*

59. Van Maanen and Kunda, "Real Feelings," p. 79. See also Rafaeli and Sutton, "Expression of Emotion as Part of the Work Role," p. 27, for other drills in "friendliness."

60. Hochschild, *Managed Heart,* p. 8.

61. Hochschild, *Managed Heart,* pp. 38–42, 110–11, 120.

62. Van Maanen and Kunda, "Real Feelings," pp. 76, 78.

63. Hochschild, *Managed Heart,* pp. 118–19, 90. Compare Rafaeli and Sutton, "Expres-

sion of Emotion in Organizational Life," pp. 13–14. See also their "Expression of Emotion as Part of the Work Role," pp. 32–33, on "emotional dissonance."

64. Hochschild, *Managed Heart,* p. 136; see also p. 21.

65. On the rise of the "corporate culture" concept, see Trice, *Occupational Subcultures in the Workplace,* p. xv; Alvesson, *Cultural Perspectives on Organizations,* p. 5; Newfield, "Corporate Culture Wars," p. 35.

66. Rosen, "Coming to Terms with the Field," pp. 10–11; Trice, *Occupational Subcultures in the Workplace,* pp. 24, 29; Van Maanen and Kunda, "Real Feelings," p. 43.

67. I cite here only authors I consult. On the growth of academic research into corporate culture in general, see Alvesson and Berg, *Corporate Culture and Organizational Symbolism,* pp. 8–18.

68. For an assessment of Martin's approach, see Alvesson, *Cultural Perspectives on Organizations,* pp. 110–18.

69. Peters and Waterman, *In Search of Excellence,* p. 320.

70. In the scholarship on corporate cultures, the standard term for fractional cultures within firms is *subculture.* In the scheme of M. Louis (described in Martin, *Cultures in Organizations,* pp. 89–91), such subcultures can generally be classified as "enhancing subcultures" (e.g., company fanatics who believe fervently in the official company spirit), "countercultures" (dissident subordinate subcultures), and "orthogonal" subcultures (those relatively rare subcultures whose difference does not map onto a contest between dominance and subordination). For reasons of clarity, I will here use the expression *partial* or *fractional cultures* to refer to corporate subcultures. I reserve the term *subculture* in this book for cultures that fall outside the mainstream of white-collar work entirely—that is, "subculture" as studied by the Birmingham cultural studies group.

71. Van Maanen and Kunda, "Real Feelings." See also Kunda's subsequent book, *Engineering Culture.*

72. Van Maanen and Kunda, "Real Feelings," pp. 44–50, 77, 89–90, 44.

73. On "T-groups" in business, see Stearns, *American Cool,* p. 246.

74. Quoted in Campbell-Kelly and Aspray, *Computer,* p. 48.

75. On corporate culture as "control device" and "culture control," see Van Maanen and Kunda, "Real Feelings," pp. 56, 88–89. On management style at the tech firm studied by Van Maanen and Kunda, see p. 72. See de Swaan's general argument that the nature of emotional management has changed as society has passed the responsibility for such management to its large-scale institutions of "production, reproduction, and government" ("Politics of Agoraphobia," p. 369). The new institutions delinearize work, deemphasize management by command, and take up "management through . . . negotiation" (p. 375), with the result that a culture of "controlled decontrol" emerges (p. 378). De Swaan's work bears on the "informalization" problem in the field of emotionology research—i.e., how to interpret the loosening of mores, feeling rules, and standards of interpersonal relations in the latter half of the twentieth century in the West. Is such relaxation symptomatic of freedom, or instead of a rechanneling of control? See the discussions of informalization in Gerhards, "Changing Culture of Emotions in Modern Society"; and in Wouters, "Developments in the Behavioural Codes between the Sexes."

76. Van Maanen and Kunda, "Real Feelings," p. 88.

77. Quoted in Boyett and Conn, *Workplace 2000,* p. 40.

78. Hochschild, *Managed Heart,* pp. 126–36, 114; Van Maanen and Kunda, "Real Feelings," pp. 49–50, 67–68, 82–83.

79. Zuboff, *In the Age of the Smart Machine,* pp. 21–22, 125.

80. See, for example, de Certeau's discussion of the tactic of *la perruque,* which in French refers to ways of doing "the worker's own work disguised as work for his employer" (*Practice of Everyday Life,* p. 25). De Certeau's whole thesis of the tactical "practices" of "everyday life" is relevant to "cool," though my context concerns not so much everyday life as "everyday work." As we will see, the many minor resistances and *perruques* of cool at work—especially when manifested as "bad attitude"—are practices in de Certeau's sense.

81. Roszak, *Making of a Counter Culture,* pp. 1–41, xiv; Gitlin, *Sixties,* p. 31. I use *hip* here with particular allusion to the era of counterculture from the Beats to the hippies. A fuller account of *hip* (and *hep*) would need to relate the term to *cool,* as both arose earlier in the twentieth century.

82. Majors and Billson argue about the cool of inner-city, young black men: "Coolness means poise under pressure and the ability to maintain detachment, even during tense encounters" (*Cool Pose,* p. 2), and again, "Cool pose is a ritualized form of masculinity that entails behaviors, scripts, physical posturing, impression management, and carefully crafted performances. . . . [Young black men] manage the impression they communicate to others through the use of an imposing array of masks, acts, and facades" (p. 4). I am here reading such "cool pose" as dialectically tensed against both the detachment ("Fordization of the face") and the impression management ("service with a smile") of technological rationality. Clearly, such a reading does not exhaust the complexity of the "cool pose," since Majors and Billson's study also looks at the "expressive life-style" and "spontaneous self" of cool (pp. 69–77). There is also the caveat that the resemblance between "cool pose" and technological rationality is reinforced by the discourse of sociological study itself, which imposes its own version of such rationality (as in jargon such as "impression management"). In the following passage from Majors and Billson, for example, it is undecidable where the line is drawn between the mock-technology (the mimicking of technology and technique) practiced by the black youths and men in question and that introduced through specialized discourse and metaphor by their interpreters: "Cool pose is a 'conditioned strength.' What we call 'the problem of selective indiscrimination' occurs because of the conditioned strength the black male has developed through his constant struggle with unyielding systems. This rigid and inflexible strength results in cool behaviors that seem automatic. The posturing of cool pose becomes such a major part of his psyche that even when white males are not present and are of no threat, the black male still operates in a 'high cool gear.' For some men being cool is never switched off" (pp. 41–42).

83. See Frank, *Conquest of Cool,* pp. 8–9, on the need for a serious history and theory of cooptation. Forms of cooptation also affected earlier modes of cool. "Just as the Jazz Age had drowned in amorphous and monotonous 'swing' music," Vincent says in his study of jazz culture, "so R&B was drowned in its rock and roll copies" (*Keep Cool,* p. 5).

84. The political and conservative backlash to counterculture is neatly summarized by Frank, *Conquest of Cool,* pp. 1–4.

85. See Barbara Ehrenreich's incisive discussion of the "discovery of poverty" in *Fear of Falling,* pp. 17–56.

86. Stern and Stern, *Sixties People,* p. 166. Dery cites this quote from Jane and Michael Stern in his discussion of the technological context of counterculture (*Escape Velocity,* p. 25).

87. "The two cultures presented in this book are quite different," Willis says in his introduction, before then drawing resemblances (*Profane Culture,* p. 7).

88. Clarke et al., "Subcultures, Cultures and Class," pp. 57 and following. The difficulty in theoretically formulating (as opposed to describing) counterculture in this essay strongly resembles the difficulty we earlier saw in New Class theory in grasping the "New Class" (see

chapter 1, above). Like the New Class, counterculture is a blurred formation. Whereas working-class subcultures are "clearly articulated, collective structures—often, 'near'- or 'quasi'-gangs," Clarke and colleagues observe, middle-class countercultures are just a "host of variant strands, connections and divergencies within a broadly defined counter-culture *milieu*" (pp. 60–61). And like the New Class as well, counterculture is contradictory in its stance or position within dominant culture: "This 'negating' of a dominant culture, but from *within* that culture, may account for the continual oscillation between two extremes: total critique and—its reverse—substantial incorporation" (p. 62).

89. Roszak, *Making of a Counter Culture,* p. 134.

90. Roszak, *Making of a Counter Culture,* pp. 51–52.

91. Willis, *Profane Culture,* pp. 137–38.

92. Roszak, *Cult of Information,* p. 151. In commenting that "sixties counterculture simultaneously bore the impress of Zbigniew Brzezinski's technetronic age," Dery cites Bruce Sterling's observation: "No counterculture Earth Mother gave us lysergic acid—it came from a Sandoz lab" (*Escape Velocity,* p. 25).

93. Roszak, *Cult of Information,* pp. 150–51.

94. Willis, *Profane Culture,* p. 159. My current description of counterculture is limited to making a particular point. It is insensitive to the internal diversity of styles and options—to the fact, for example, that Pink Floyd can certainly not be crowned the sole avatar of rock style in the era. In Lindsay Waters's book in progress, "A Critique of Pure Hipness," Jimi Hendrix becomes the paradigm for a fundamentally technological, "mechanical," "artificial," and also powerfully affecting experience of popular art that only the too aesthetically pure and censorious "hip," whom Waters distinguishes from mere "hippies" and other groksters, cannot appreciate.

95. Gitlin, *Sixties,* pp. 48–49.

96. For Pynchon's culminating epiphany of the "secular miracle of communication," see *Crying of Lot 49,* p. 149.

97. For Pynchon's mention of the IBM 7094, see *Crying of Lot 49,* p. 93. Farland has explored the relationship between the Beats and information work in her paper "Cyberculture, Business Culture, and the Literary Counterculture, 1950–70," presented at the Modern Language Association convention in Washington, D.C., in 2000. Starting with the post–World War II milieu of counterculture, Farland presents a detailed case for relating the Beats and later countercultural figures to information culture, and also a larger case that literary study needs to situate works of the postwar and postmodern era in light of the paradigm of knowledge work. The following overview statement is congruent with some of the themes of this book: "What is striking in this scene is the centrality of the culture of information to a certain incarnation of the counterculture, and the way in which a certain countercultural strand is linked to the cutting edge of what would become the new knowledge-based economy. But it is also crucial to see that even at this very early moment, post-industrial labor is emphatically linked to leisure and creativity, so that the line between labor and leisure is blurred in a way that will become the hallmark of the post-industrial, information economy" (quoted from the manuscript with permission of author). In regard to my discussion of Ginsberg here, see also Dery, who observes, "Hippiedom inherited the Blakean vision of a return to Eden and the Emersonian notion of a transcendent union with Nature by way of Beat poets such as . . . Allen Ginsberg, whose 'Howl' demonized America as an industrial Moloch 'whose mind is pure machinery'" (*Escape Velocity,* p. 25).

98. There were some famous, but overdetermined, exceptions to the principle of not messing with the equipment, including the guitar-smashing performances of The Who.

99. Frank, *Conquest of Cool,* pp. 95, 31, 111–12; see also p. 9.

100. Brooks, *Bobos in Paradise,* pp. 9–10. The following, especially colorful passage in the book links the discourse of the "bobos" to counterculture and youth subculture: "Even the rules of language have changed. They use short sentences. Nouns become verbs. They eliminate any hint of a prose style and instead tend to talk like 15-year-old joystick junkies. Next year's cost projections? They're insanely great. The product pipeline? Way cool. How'd the IPO go? It cratered. The San Jose conference? Flipped my nuts. Serious mind rub. Real-time life experience. In their conversation and especially in their e-mail, they adopt the linguistic style of Jack Kerouac" (ibid., p. 112).

101. Taylor, *Hiding,* p. 282; Campbell-Kelly and Aspray, *Computer,* p. 245.

102. Rochlin, *Trapped in the Net,* pp. 21–22; Campbell-Kelly and Aspray, *Computer,* p. 238.

103. Brand, "Fanatic Life and Symbolic Death among the Computer Bums," quoted in Dery, *Escape Velocity,* p. 27. Farland also discusses the relationship between counterculture and early hacker and computer culture ("Cyberculture, Business Culture, and the Literary Counterculture, 1950–70"). The term *doping,* which I play upon here, refers to the process of adding selected impurities to nonconductive raw silicon to make it a semiconductor and thus the basis of a transistor.

104. See also Campbell-Kelly and Aspray, *Computer,* pp. 237–40, 244–47; Rochlin, *Trapped in the Net,* pp. 21–22; Castells, *Information Age,* 1:353–54; Dery, *Escape Velocity,* pp. 22–33.

105. Roszak, *Cult of Information,* pp. 138–40, 144; see also Dery, *Escape Velocity,* p. 26.

106. Dery, *Escape Velocity,* pp. 21–41; Rochlin, *Trapped in the Net,* p. 28.

Chapter Four

1. Castells, *Information Age,* 1:151–200. My premise in this chapter—that the proper object of study as regards networking is the *combination* of new information technology and new organizational structure—is reinforced by Peter F. Drucker's argument, in *Innovation and Entrepreneurship,* that the real "new technology" is not IT at all. It is entrepreneurial management (pp. 11–17).

2. In the following sketch of computing history during the era of personal computing and networking, I draw in particular on Castells for a broad understanding of the trends (*Information Age,* especially 1:40–47, 60–65, 246–47); on Campbell-Kelly and Aspray (*Computer,* especially pp. 233–300) and Ceruzzi (*History of Modern Computing,* pp. 207–306) for detailed narratives; and on Polsson's online *Chronology of Events in the History of Microcomputers,* as well as Zakon's *Hobbes' Internet Timeline.* I also draw variously from other accounts or tabulations of computing history in Rochlin, *Trapped in the Net;* Rosenberg, *Social Impact of Computers;* and Straubhaar and LaRose, *Communications Media in the Information Society.* In addition, I have found helpful the now large repository of computing history found in online "computer history" museums and other such resources; for links, see the section "History of Computing and Computer Museums" on the "Cyberculture page" on my *Voice of the Shuttle.*

3. See Castells's discussion of the 1970s as the "technological divide" (*Information Age,* 1:46–47).

4. The early history of the personal computer is now so famous that we need recall here only a few milestones: the primitive Altair 8800 in 1975; the Apple I, 1976; the Apple II (along with the Commodore PET and Radio Shack TRS-80) in 1977; and the early spreadsheet and word-processing programs (including VisiCalc and WordStar) that appeared between 1979 and 1980.

5. *Time* magazine, "Machine of the Year," 3 January 1983; *Business Week*, "Personal Computers Invade Offices," 8 August 1983 (reprinted in Forester, *Information Technology Revolution*, pp. 322–35).

6. Rochlin, *Trapped in the Net*, p. 43.

7. As defined in the *Microsoft Press Computer Dictionary*, client/server architecture is: "An arrangement used on local area networks that makes use of distributed intelligence to treat both the server and the individual workstations as intelligent, programmable devices, thus exploiting the full computing power of each. This is done by splitting the processing of an application between two distinct components: a 'front-end' client and a 'back-end' server. The client component is a complete, stand-alone personal computer (not a 'dumb' terminal). . . . The client portion of the application is typically optimized for user interaction, whereas the server portion provides the centralized, multiuser functionality" (p. 92). The advantages of client/server architecture include greater adaptability of the systems (e.g., programs can be changed on servers and clients independently), faster evolution of new features (software development for personal computers far outpaces that for mainframes), and freedom from whole-system bottlenecks (e.g., waiting for the information services department to run a large job). Such autonomy at the level of the desktop machine has its disadvantages, too, including the ability of individual client users to vary, customize, break, misuse, and even vandalize.

8. *Business Week*, 1 August 1994, p. 14, cited in Greenbaum, *Windows on the Workplace*, p. 109.

9. Straubhaar and LaRose, *Communications Media in the Information Society*, pp. 263–64.

10. Key developments: the 10BASE-T Ethernet specification in 1990 and the evolution of several families of server and networking operating systems.

11. For historical statistics on Internet hosts and domains, see *Internet Domain Survey*.

12. On the "era of Ubiquitous Computing," see the column by Michael J. Miller, editor-in-chief of *PC Magazine*, on "The Fifth Age of Computing" (*PC Magazine*, 26 May 1998).

13. Davidow and Malone, *Virtual Corporation*, p. 140; Rochlin, *Trapped in the Net*, p. 72.

14. Tapscott, *Digital Economy*, p. 100. Compare Davidow and Malone: "The role of management in the business of the future is bound to decline as computers gather and provide the information that was at one time the product of middle management and as employees become better trained and empowered to make decisions. Information and the power it provides will flow to the worker" (*Virtual Corporation*, p. 62).

15. Tapscott, *Digital Economy*, pp. 100, 15–16.

16. Castells, *Information Age*, 1:171; see also 1:151–200 for Castells's larger view of the network enterprise.

17. Networking especially resembles craft because its practices are passed on tacitly from veterans to "newbie" apprentices. Moreover, not just using but administering networks can be compared to craft. This thought occurred to me during one of the sleepless nights I spent in late 1998 installing a series of "service packs," "patches," enhancements, and other upgrades to my English department's Web server—a process that required locating technical documentation scattered unsystematically here and there on the Internet and learning by trial and error (each error requiring a lengthy reboot and sometimes a whole sequence of de- and reinstallation actions) what components had to be upgraded in what order relative to what other components. Only part of this ad hoc process of learning could have been avoided had I been a qualified, "certified" professional because the prac-

tice of maintaining a server requires living in a perpetual limbo of new versions, service packs, patches, and so on. The process of keeping up with security fixes alone on any operating and networking system ensures that a sysadmin operates on a moving front of lore gathered from other sysadmins, FAQ sites, personal experience, custom, or work-around solutions—all difficult to systematize and document.

18. Lifson, "Innovation and Institutions," p. 314.

19. Rochlin, *Trapped in the Net,* pp. 9, 47. Other commentators argue a similar thesis. Strassmann, for example, uses his considerable authority earned from decades of practical consulting work with firms to judge that centralization will be the victor. "Power contests for control between the data-processing centralists and the enthusiastic people who finally wish to be liberated from the heavy hand of computer bureaucrats," he says, are "comparable, on a historical scale, to the feuds between shepherds and farmers." The outcome, he concludes, "is just as predictable. The highly organized computer people will end up running the networks, even though in the process of acquiring this power their role will change. Introducing office automation, then, should be viewed not only as a change in the way individuals work, but also as a process in which centrally directed expertise is used to influence the acceptance of information technology" (*Information Payoff,* p. 59). Similarly, Ralph Carlyle in a 1990 *Datamation* article observes: "As the companies reorganize for the '90s they are also exploding some myths. Chief among these is the notion that decentralization, the theme of the 1980s, will be the favored organizational approach to the emerging global marketplace. . . . Centralized control will increase, new centralized functions and entrepreneurial teams will arise" (quoted in Rosenberg, *Social Impact of Computers,* p. 97). See also Brown and Duguid, *Social Life of Information,* pp. 29–31. A concrete example of such recentralization was the effort of the Hewlett-Packard company in the late 1990s to centralize the creation and maintenance of company intranets, once up to the initiative of individual departments, within its consolidated Electronic Sales Partner intranet system (Cronin, "Bye-Bye, Wild Web.").

20. Davidow and Malone, *Virtual Corporation,* p. 172.

21. Some definitions of terms: *Dynamic packet-filtering firewalls* refers to a kind of hardware and software device that protects security by sitting in between intranets and the public Internet, inspecting and filtering the packets (or individual units of information) in the network stream. *Permission levels* refers to the variety of permissions (none, read, write, change) that can be assigned to different users or groups of users for particular files or directories on a system. *Encrypted login* refers to various means of sending and receiving passwords over networks in protected form (i.e., not "in the clear"). An example of a *log file* would be the record of file accesses kept by every Web server. This record tracks the IP address or domain name of every remote machine that browses the server, the exact pages these machines browse, the time of the visit, and other information. Analysis programs can then be used on the log file to study patterns of usage, the frequency of users from particular nations and domains, and so on. An *IP* (Internet Protocol) address such as 128.111.99.132 specifies the unique identity of each machine on the Internet (or an intranet). IP numbers are the functional addresses underlying URL addresses (whose domain names, e.g., "www.english.ucsb.edu," are translated into IP numbers by a DNS [domain name server]). The above IP number, for example, is the absolute identity of the English Department Web server at the University of California, Santa Barbara. *Spyware programs* monitor or record keystrokes, Web use, and so on.

22. Rochlin, *Trapped in the Net,* p. 8.

23. Quoted in Rosenberg, *Social Impact of Computers,* pp. 104–5.

24. *Network computers* are stripped-down, closed-box, often storageless personal computers that are also sometimes called *thin clients* (meaning either that the bulk of processing and

control is done on the server or that the client must in each instance download a program from the server to be used locally and transiently). *Managed personal computers* are client machines that, together with specialized LAN management software, allow information service staff to manage a far-flung empire of desktop machines by remote control; see the coverage of the managed PC initiative in *PC Magazine,* "Managed PCs," 30 June 1998. On *Internet-use tracking programs,* see *Newsweek,* "bigbrother@the.office.com," 27 April 1998; *C/Net News.com,* "Software to Filter Worker's Access," 4 May 1998; *Los Angeles Times,* "Software Keeps Tabs on Web-Surfing Employees," 20 May 1998, and "Workers Lament Loss of E-Mail Privacy on Job," 11 October 1999. The last mentioned article is especially comprehensive and detailed in its coverage of the issue of monitoring in a networked environment. See also the newer, more aggressive tactics of employee monitoring discussed in the *Los Angeles Times,* "High-Tech Snooping All in Day's Work," 29 October 2000. For related discussion of workplace privacy issues, see chapter 8, below.

25. On the topic of the proliferation of logins and passwords, see *New York Times,* "Forgot a Password? Try 'Way2Many,'" 5 August 1999.

26. Rochlin, *Trapped in the Net,* pp. 47–48.

27. Davidow and Malone, *Virtual Corporation,* pp. 11, 86; Boyett and Conn, *Workplace 2000,* p. 23; Hammer and Champy, *Reengineering the Corporation,* p. 83 (italics in original).

28. My main sources here are the book-length studies of Landauer (*Trouble with Computers*) and Strassmann (*Information Payoff*). On Loveman's work, see Davidow and Malone, *Virtual Corporation,* p. 66; and Landauer, *Trouble with Computers,* pp. 32–33. On Roach and Franke, see Landauer, *Trouble with Computers,* pp. 29, 33–34.

29. See charts showing Roach's and Franke's data in Landauer, *Trouble with Computers,* pp. 31, 34.

30. For commentators who argued for more thorough restructuring to unleash the potential of IT, see Landauer, *Trouble with Computers,* p. 122.

31. On rising productivity in the late 1990s, see *Business Week,* "The Promise of Productivity," 9 March 1998, and "Big Test Ahead for Productivity," 8 June 1998. On the contribution of the Internet to productivity in the late 1990s, see *Los Angeles Times,* "Productivity Jumps with Help from Net," 30 June 1999. Ultimately, however, the New Economy thesis for the rise in productivity in the late 1990s became just as puzzling as the preceding productivity paradox (see *Los Angeles Times,* "Just How Productive Are U.S. Workers?" 9 July 2000). This was especially apparent when U.S. companies boosted productivity in spring 2001, during an economic downturn, by cutting jobs and work hours instead of by increasing their use of IT (the preferred New Economy means) and when in mid-2001 the government revised downward the productivity statistics for the New Economy boom period of 1998 through 2000, making some of the apparent boom a mirage (*Los Angeles Times,* "Job Cuts Boost Efficiency 2.5%," 8 August 2001).

32. Quoted in Birdsall, "Internet and the Ideology of Information Technology."

33. Feldman and March, "Information in Organizations as Signal and Symbol," pp. 177–78. For the specific ways in which information gathering and processing in an organization diverge from rational decision making, see p. 174.

34. Relevant to my thought here is the genealogy of rhetorical speculation that extends from Coleridge on symbol, through the New Critics on "verbal icons," to Paul de Man on allegory versus symbol (in "The Rhetoric of Temporality"). In his *Statesman's Manual,* for example, Coleridge famously defined the difference between allegory and symbol as follows: "Now an allegory is but a translation of abstract notions into a picture-language which is itself nothing but an abstraction from objects of the senses; the principal being more worthless even than its phantom proxy, both alike unsubstantial, and the former shapeless to boot. On the other hand a symbol . . . is characterized by a translu-

cence of the special in the individual or of the general in the especial or of the universal in the general. Above all by the translucence of the eternal through and in the temporal. It always partakes of the reality which it renders intelligible; and while it enunciates the whole, abides itself as a living part in that unity of which it is the representative" (p. 661). The eventual outcome of this genealogy was not only the separation of the orders of the symbolic and allegorical—the former ontologically "full" and the latter a "mere" figure—but, with de Man, a reversal of the Romantic and New Critical privileging of the symbolic over the allegorical. In the de Manian perspective, symbol is at heart an empty allegory. (My discussion here of the confusion between the literal and the figural in the age of networking extends my earlier discussion of a similar literal/figural confusion in the concept of "informating"; see chapter 3, above).

35. The euphemism of such terms as *service packs* is illustrative here. Some notoriously substantial service packs of the 1990s amounted to full-blown reinstallations of the operating system and each of its subordinate programs. The fact that each "official" release of an operating system led to whole dynasties of subsequent service packs (SP1, SP2, SP3, SP4 for the Windows NT 4 server operating system, for example) testifies that the *normal* state of computing systems was Heraclitean flux: you never step in the same operating system twice.

36. Toffler, *Third Wave,* pp. 185, 273–75. Tapscott paints the prosumption scenario as follows: "In the new economy, consumers become involved in the actual production process. They can, for example, enter a new car showroom and configure an automobile on the computer screen from a series of choices. Chrysler can produce special-order vehicles in sixteen days. The customer creates the specs and sets in motion the manufacture of a specific, customized vehicle. In the old economy, viewers watched the evening network news. In the new economy, a television viewer will design a customized news broadcast by highlighting the top ten topics of interest and specifying preferred news sources, editorial commentators, and graphic styles" (*Digital Economy,* p. 62).

37. Davidow and Malone, *Virtual Corporation,* pp. 7, 140. For concrete cases of the blurring of lines between suppliers, designers, assemblers, and others all along the chain of production, see *New York Times,* "Is This the Factory of the Future?" 26 July 1998. Ultimately, the prosumerist principle of co-production can be so extreme that interior elements of individual firms can act as "intrepreneurial" service units competing for customers both within and without the firm (see Boyett and Conn, *Workplace 2000,* pp. 33–34, 37). Everyone, in other words, now "produces" only by flexibly, rapidly, and openly "servicing" someone else.

38. Davidow and Malone, *Virtual Corporation,* pp. 158–59.

39. Bill Gates thus asserts in his *Road Ahead* that "de jure" or "de facto" standards are crucial—preferably, in his view, nongovernmental ones (pp. 50, 66–67, 281). Similarly, *Business Week* quotes Larry Ellison, CEO of Oracle, Inc.: "When you're an e-business, everything is mediated by computers. . . . All the individuality is bled out of the system and replaced by standards. People don't run their own show anymore" ("Oracle: Why It's Cool Again," 8 May 2000, p. 120). Ironically, this is the issue of *Business Week* that shows Ellison on the cover modeling a distinctively Beat look (black turtleneck or pullover sweater, granny sunglasses, goatee-like beard and mustache) under the title, "Despite the Tech Stock Slide . . . Oracle is Cool Again." Transformed into what David Brooks calls Bobo (bohemian bourgeois), Beat is now all about standards *(Bobos in Paradise).*

40. I use *screen* metonymically in the ensuing discussion for the whole circuit of screen, keyboard, mouse, and (recently) voice interaction.

41. Johnson observes (*Interface Culture,* pp. 14–15): "A computer that does nothing but manipulate sequences of zeroes and ones is nothing but an exceptionally inefficient

adding machine. For the magic of the digital revolution to take place, a computer must also *represent itself* to the user, in a language that the user understands.

"In this sense, the term *computer* is something of a misnomer, since the real innovation here is not simply the capacity for numerical calculation. . . . The crucial technological breakthrough lies instead with this idea of the computer as a symbolic system, a machine that traffics in representations or signs. . . . A computer . . . is a symbolic system from the ground up. Those pulses of electricity are symbols that stand in for zeroes and ones, which in turn represent simple mathematical instruction sets, which in turn represent words or images, spreadsheets or e-mail messages. The enormous power of the modern digital computer depends on this capacity for self-representation.

"More often than not, this representation takes the form of a metaphor. . . . These metaphors are the core idiom of the contemporary graphic interface."

Philosophically, of course, it could be said that the material world is quite as fundamentally unintelligible as the informational one: the intricate balance of molecular flux and structural stasis that creates a wooden table, for example, is unknowable except through an elaborate set of physical, optical, cognitive, semantic, and social filters that are de facto an "interface." The difference, it may be suggested, is that informational interfaces are much more shallowly rooted in the whole set of evolved physical, cognitive, and social filters within whose context only the most extraordinary situations could now make it relevant to ponder the difference between the "essence" and "construct" of something as ordinary as a table. In computing, however (e.g., navigating a VRML or three-dimensional interactive display of a table), even the ordinary can be a confusing experience in which the various human filters disconnect from or contradict each other. The *fact* of the "interface" thus obtrudes itself.

"Infinity Imagined" is the title of Johnson's concluding chapter; see also his pp. 42–45.

42. Key steps in the introduction of the graphical user interface include the early development of the concept during the 1960s in Doug Engelbart's Human Factors Research Center at the Stanford Research Institute and David Evans and Ivan Sutherland's Computer Science Laboratory at the University of Utah (Campbell-Kelly and Aspray, *Computer,* pp. 266–67); the further development of the bitmapped screen and GUI during the 1970s in the Xerox Palo Alto Research Center (PARC) (Rochlin, *Trapped in the Net,* p. 24; Ceruzzi, *History of Modern Computing,* pp. 261–62); and finally in the 1980s the debut of Apple's GUI interface followed by that of Microsoft's Windows (and, in the UNIX workstation world, of such interfaces as X-Windows). Much of the evolution of computing hardware and software from the mainframe era on has been driven by the need to satisfy the enormous data-processing and -transfer demands of graphics and multimedia manipulation. For a short history of the development of the GUI interface, see Campbell-Kelly and Aspray, *Computer,* pp. 264–82. On Engelbart and the principle of bitmapping in particular, see Johnson, *Interface Culture,* pp. 11–41.

43. Johnson, *Interface Culture,* pp. 81–82.

44. For "new cultural form" or "metaform," see Johnson, *Interface Culture,* pp. 32, 38. In the chapter titled "The Interface Culture" of his *In the Beginning . . . Was the Command Line,* Stephenson compares GUI computer interfaces to Disneyland and the general need for mediating interfaces: "Disney is a sort of user interface unto itself—and more than just graphical. Let's call it a Sensorial Interface" (p. 52). Stephenson's cyberpunk novels, such as *Snowcrash* and *The Diamond Age,* are highly attuned to the prevalence of corporate cultures in the postindustrial era. It would not be stretching his views too far to say that the "Disney" he makes paradigmatic of the general culture of the interface is ultimately corporate. In the late 1990s, of course, Wall Street and business journalism have identified Disney above all as a corporate power.

45. See, for example, the discussion threads in the comp.human-factors newsgroup in July 1997 and June 1998 on the user-friendliness of Unix (with the subject headers, re-

spectively, "THEORY: Help-systems that really help?" and "Why Is Unix Evil?"). See also the thread in the comp.unix.user-friendly newsgroup in October 1993 with the subject header "unix is user-friendly."

46. Campbell-Kelly and Aspray, *Computer,* p. 267; Johnson, *Interface Culture,* pp. 21–23.

47. E.g., Wichansky and Mohageg, "Usability in 3D," pp. 242–43; James, "American Airlines," pp. 360–61.

48. Wichansky and Mohageg, "Usability in 3D," p. 230.

49. Wiklund, ed., *Usability in Practice,* p. 12.

50. Rosenberg and Friedland, "Usability at Borland," p. 265; Campbell-Kelly and Aspray, *Computer,* p. 256.

51. E.g., Bickel and Jantz, *Bruce & Stan's Guide to God: A User-Friendly Approach;* Easley, *User-Friendly Greek: A Commonsense Approach to the Greek New Testament;* Inkeles and Schencke, *Ergonomic Living: How to Create a User-Friendly Home and Office;* Reimers and Treacher, *Introducing User-Friendly Family Therapy.*

52. Stoll, *Silicon Snake Oil,* pp. 44, 60. *Friend* and its cognates, whether applied to human acquaintances or computers, appear so often in Stoll's book that it is difficult to keep track. Here is a sample: pp. 24, 36, 42, 43, 44, 46, 54, 60, 68, 112, 154, 181, 197, 202.

53. Caplan, "Making Usability a Kodak Product Differentiator," p. 44.

54. Purvis, Czerwinski and Weiler, "The Human Factors Group at Compaq Computer Corporation," pp. 126–29.

55. Laurel, *Computers as Theatre,* p. 102.

56. Laurel, *Computers as Theatre,* pp. 69–70.

57. Dieli et al., "Microsoft Corporation Usability Group," pp. 349–54.

58. From opening screen of Microsoft Excel Help (Microsoft Excel 2000).

59. Laurel, *Computers as Theatre,* p. 7 (italics in original).

60. Laurel, *Computers as Theatre,* p. 144.

61. Laurel, *Computers as Theatre,* p. 147. For Laurel's mention of method acting, see ibid., p. 106.

62. On "Eager," see Laurel, *Computers as Theatre,* p. 108.

63. See Laurel, *Computers as Theatre,* p. 134, on Susanne Bødker's *Through the Interface: A Human Activity Approach to User Interface Design,* which (in Laurel's paraphrase) "asserts that a theory of user interface design must be a subset of a larger theory of human work."

64. Johnson, *Interface Culture,* p. 227. See Johnson's general discussion of this issue, pp. 224–27.

Part Three Preface

1. This page no longer exists in the form described here; see below regarding the later version of the page. The Netscape Communications company was bought by America Online, Inc., in 1999 (which then merged with Time-Warner, Inc., in 2000).

2. Marcel Danesi has studied the teen discourse of "cool" (and similar "pubilect" slang) in his *Cool: The Signs and Meanings of Adolescence.* It would be useful to extend such linguistic and sociological discourse lower in the grade scale, perhaps to as early as kindergarten (when my own daughter began to say "cool" in response to toys that were mechanically and/or stylistically interesting). Lewis MacAdams similarly observes in his *Birth of the Cool,* p. 27, "[In 1995] I heard my three-year-old daughter say 'cool' when she

saw an ad for *Scooby-Doo* reruns on the Cartoon Network." It is likely that the sources of learning about cool at this age are multiple, but certainly one source is commercial. See, for example, the depiction of the Apple Jacks Web site that the Kellogg company put on the back of the cereal box in 1998 as what the box calls "a cool place to hang" and "without a doubt the coolest place for you and your friends to hang out." The Apple Jacks site itself in 1999 offered to "bring the cool new colors of Apple Jacks® cereal straight to your desktop with our two free screen savers!"

3. Labov, "Social and Language Boundaries Among Adolescents," p. 348 (her Table 1). Marcel Danesi, *Cool,* discusses the general phenomenon of contemporary "pubilect" and adolescent culture (drawing in part on Labov).

4. Writing in his weekly newspaper column on digital technology, Gary Chapman observed in 1996: "'Cool' is the holy grail of the cyber-elite. It's the only word that conveys success, that draws crowds in virtual space and that indicates a company 'gets it,' in the cultural vernacular of the times. Cool is completely binary—a company or a Web site is either cool or it's dead, cold" (p. D3).

5. My analogy to tailfins is influenced by Dick Hebdige's discussion of the "streamlining controversy" in mid-century product design. See "Towards a Cartography of Taste, 1935–1962," in his *Hiding in the Light.*

6. My account of information cool draws in part upon my experience canvassing the Web for my *Voice of the Shuttle* Web site. Of particular relevance here was a page on the site titled "Laws of Cool," which served as an informal laboratory in which to collect "cool sites" and to test ideas about cool. This page and its general thesis that "cool" is something that can be thought about was named by Mirsky on his witty *Worst of the Web* site in early 1996, which in turn prompted quite a few anti-academic, formulaic, unfortunately not so witty "you suck" messages in the succeeding weeks, primarily from undergraduate males—messages that I here convert into research material. My thanks to Mirsky for subsequent correspondence about "cool." (Mirsky discontinued his site after November 1996, though another site titled *Worst of the Web* later appeared under different management.)

7. Started by a team that includes Glenn Davis, the originator of the canonical *Cool Site of the Day* site on the Web, Project Cool, Inc. (subsequently part of DevX, Inc.), kept a technically advanced site that for many years has featured a gallery of cool pages, developer resources, forums, and so on, focused on Web authoring. (For information on Glenn Davis's association with both *Cool Site of the Day* and Project Cool, see Gardner, "When It's Your Site That's 'Cool'"; and Narayan, "What Makes a Site Cool? Rules Have Changed since Last Year.") I quote here from a page titled "About the Coolest on the Web," accessed from the site's "Coolest of the Web" frameset in June 1997. Sometime after 1997, "Coolest on the Web" and the "About the Coolest on the Web" page vanished, while the main gallery of "Sightings" of cool pages continued. Though Project Cool's site was folded into the DevX.com site in 2000, the pages I refer to in this book, including "Sightings" and "Previous Sightings," continue to exist at the time of this writing.

8. The language of this "poem" may be lackluster, but its real artistry lay in the graphical design and typography of its presentation. The layout of the page was reminiscent of the New Typography of the Bauhaus era.

9. This is perhaps the time to say something about the apparent truism that cool cannot be known or pinned down without losing its essence. Malcolm Gladwell writes in his "Coolhunt," for example, "The act of discovering what's cool is what causes cool to move on" (p. 78). Or, again, Lewis MacAdams writes in his *Birth of the Cool,* "Anybody trying to define 'cool' quickly comes up against cool's quicksilver nature. As soon as anything is cool, its cool starts to vaporize" (p. 19). This truism is part of the folklore of cool. But it is quite wrong: Cool can certainly be said to be as knowable as (no more, but no less than)

most other complexly dynamic and fugitive physical, social, or other phenomena, such as particle physics or poverty. To say otherwise would be to believe in such an absolute and reduced notion of knowledge—unpragmatically stripped of all such methodologies for constraining the field of uncertainty and the influence of the observer as probability, statistics, psychoanalysis, and, indeed, analysis in general—that *nothing* complex and fugitive could ever be said to be knowable. Fundamentally, I suggest, the too-ready acceptance of the folklore of cool within analytical work on the topic is a category error. Unknowability (a stance that I have called the "ethos of the unknown") is part of cool as an object of study; it is a theme that is internal to cool. Unknowability is not necessarily the fate of the methodology of study.

10. See chapter 1 on paradox and the gesture of information. See also chapter 1, note 168, on the relation between my definition of cool as "information designed to resist information" and William R. Paulson's and J. Hillis Miller's theses of literature as informational "noise" and a "black hole," respectively. Both Paulson's and Miller's paradigms are important complements to mine (though their focus is specifically on literature). But my emphasis in this book is also on the more destructive possibilities of cool—i.e., the way the sabotage recommended by the *Processed World* group (see chapter 8) or "hacking" activates the darker side of "creative destruction." In this light, a comparison with Paulson's emphasis on the "creative," "constructive," or "inventive" potential of noise is especially instructive. For example, "literary language, by its very failure as a system for the communication of preexistent information, becomes a vehicle for the creation of new information" (p. 101). See also Paulson's discussion of "intellectual invention" (*Noise of Culture,* pp. 155–65).

11. See Lévi-Strauss, *Elementary Structures of Kinship.*

12. My thought here is influenced by Pierre Clastres, "The Bow and the Basket," in his *Society against the State.* See my discussion of Clastres in chapter 9, below. Clastres's eloquent, even lyrical essay studies the nighttime, solitary songs of male Guayaki Indians, songs that are precisely designed to be understood by no one other than the individual singer. That is, they are language that does not communicate, that is held back from general social intercourse (and thus analogous to incest as Lévi-Strauss formulates it). "Cool" has this quality of withheld information, of information to be understood incestuously only by those already in the know.

Chapter Five

1. Derrida, "Structure, Sign, and Play in the Discourse of the Human Sciences," especially pp. 288–89.

2. Methodological note: In designing a search strategy, I found especially useful H. Vernon Leighton and Jaideep Srivastava's online publication, "Precision among World Wide Web Search Services (Search Engines): Alta Vista, Excite, Hotbot, Infoseek, Lycos." Based on a set of controlled experiments in early 1997, their study concluded that "Alta Vista, Excite, and Infoseek did a superior job delivering quality relevance." Also useful was Danny Sullivan's *Search Engine Watch* Web site, which includes comparative charts of search engine size, features, and ratings. In July 1998 HotBot ranked highest among the collective reviewers, while AltaVista, Excite, and Infoseek also ranked highly. Another major search engine, Northern Light, was too new to figure in most of the reviews. I did not include among my search tools Lycos, another major search engine, because its results pages did not show total number of matches. I also did not include the search engines of such Web "directories" or "indexes" as Yahoo whose databases were human-filtered and/or categorized (though I consulted the categories for "cool" in Yahoo). None of the search engines, it should be noted, included more than a fraction of the Web in their databases. According to a NEC Research Institute study in 1999, even the largest such databases—those of Northern Light and AltaVista—included no more than about 16 percent of the

Web (*PC Magazine,* 1 September 1999, p. 9). On the limited coverage of search engines, see also *Wall Street Journal,* "Internet Search Engines Trail Web's Growth, Study Shows"; *C/NET News.com,* "Study Finds Web Bigger Than We Think."

Searches on multiple words are especially uncertain. There is little or no documentation of the manner in which individual search engines weigh the following factors in their total matches or relevancy rankings for combinations: exact phrase, words in proximity, words anywhere on page. (The wide variation of results for "cool cool" or the recurrence of the figure, "647,140," in the Excite results, for example, indicates that there are undeclared heuristics at work.) Only some of the engines allow the user to specify an "exact phrase" or "proximity search" option as opposed to "all the words."

For the total number of pages included in each search engine's database, I use the figures from the "Search Engine Features Chart" on the *Search Engine Watch* site. The figures I cite were updated on this source 17 June 1998.

3. On 12 August 2001, for example, Google returned approximately 15.6 million hits for "cool." As reported in their paper at the 2001 ACH/ALLC conference, Rockwell and his group created a system for automatically polling search engines each night over a period of time and compiling statistics on specific cultural topics. Their system, which ran experimentally from September 1999 to January 2000, is a more rigorous approach to tracking terms on the Web than the "snapshot" approach available to me in 1998.

4. Sex sites tend to figure disproportionately in searches because they manipulate the HTML "meta-tag" on Web pages and use other means to force search engines to include them in a wide range of contexts and with a higher ranking order. Meta-tags are certain tags containing information about a page (and other functions) at the top of the HTML source code for a page that is not visible in the browser.

5. A few of the search engines allow searches on a portion of historical Usenet traffic. Especially useful after 2002 is Google Groups.

6. By "encoded page title," I mean the "<title>Page Name</title>" HTML code element that identifies the title of a page to the browser. This title does not display in the browser as part of the page but supplies the text for the browser window title-bar. It also supplies the default title under which a link is stored in a browser's bookmark or favorites list.

7. In his *Electronic Word,* Richard A. Lanham discusses the ability of electronic media to destabilize the relation between looking "at" and "through" a text (e.g., pp. 5, 43). Jay David Bolter and Richard Grusin cite Lanham in their *Remediation* while explicating the double logic of what they call "immediacy" and "hypermediacy." They comment: "A viewer confronting a collage, for example, oscillates between looking at the patches of paper and paint on the surface of the work and looking through to the depicted objects as if they occupied a real space beyond the surface. What characterizes modern art is an insistence that the viewer keep coming back to the surface or, in extreme cases, an attempt to hold the viewer at the surface indefinitely" (p. 41). In a different context, I have myself through the years used the difference between looking at and through a window as one of the main instructional paradigms of my undergraduate course on "Formalism and its Discontents." After discussing Archibald MacLeish's "Ars Poetica" (which became the anthem of the New Criticism), I point to a window in the classroom and ask a student what he or she sees. This opens up a discussion about how to "look" at a poem. As we will see, it is not accidental that the window metaphor puts us in a modernist frame of mind when "looking at" the Web.

8. Johnson observes, "'The principle of the Gothic architecture,' Coleridge once said, 'is infinity made imaginable.' The same could be said of the modern interface" (*Interface Culture,* p. 42). He continues his comparison between interfaces and Gothic cathedrals on pages 42–45.

9. For Coleridge's definition of the symbol, see chapter 4, n. 34.

10. For the point about the adolescent, "marked" pronunciation of "coo-ooool," I am indebted to Valerie Traub, who participated in a conference at the University of Virginia in 1997 where I presented part of my project.

11. Turkle writes: "On the Web, the idiom for constructing a 'home' identity is to assemble a 'home page' of virtual objects that correspond to one's interests. One constructs a home page by composing or 'pasting' on it words, images, and sounds, and by making connections between it and other sites on the Internet or the Web. Like the agents in emergent AI, one's identity emerges from whom one knows, one's associations and connections" (*Life on the Screen,* p. 258). My view differs slightly from Turkle's in this way: the information age identity represented by a home page never actually "emerges" from its networked "associations and connections" because it exists in a relation or stance toward the act of making networked associations and connections perpetually "out there" in the lateral transcendence of information. Cool identity, in other words, emerges not on the page but precisely off it in invisible ways that I seek to clarify below.

12. The unresolved tension between the rational and irrational in cool is also evident in the following statement from an interview with Richard Grimes, one of the staff of *Cool Site of the Day* who in 1996 picked the short list of sites to be nominated for the Cool Site of the Year Award. Coolness, Grimes says, "should be useful, communicate effectively, and be smart enough to appeal to a broad band of audience. . . . I decide what's cool based on a gut feeling. If I go to the site and I am hit in the stomach with the need to explore the whole site, if the site takes my breath away, then it's cool" (Narayan, "What Makes a Site Cool?").

13. *Wall Street Journal,* "Users Are Choosing Information over Entertainment on the Web," 20 July 1998.

14. Critical Art Ensemble, "The Technology of Uselessness," pp. 76–77.

15. These meta-cool sites, which I cull from the annotated links on my *VoS* "Laws of Cool" page, were alive as of 6 June 1997, but many are now defunct. I include references in my list of works cited only for those sites that I discuss substantively. Other sites of the same era, such as the following, do not use the word *cool* but are recognizably cool anthologies: *Der Web des Tages/Web of the Day* (in German), *Dynamite Site of the Nite, Neat Feat of the Moment.*

16. In general, I have been able to be systematic only in reviewing the selections of major, general-purpose anthology sites (especially *Cool Site of the Day,* Project Cool's "Sightings," and Netscape's "What's Cool?") because they keep dated archives. (Netscape's "cool" page and archives have since gone offline.) I have excluded from consideration a category that might be called "business cool" or "institutional cool," by which I mean the corporate, entertainment industry, and even government agency sites selected by the cool anthologies as "cool." Examples include: *Bank of America* (N 7 March 1996); *Eastman Kodak Company* (N 7 March 1996); *HBO* (C 22 May 1996); *Levi Strauss* (N 7 March 1996, P 15 June 1996); *Quicken Financial Network* (N 7 March 1996); *United Parcel Service* (N 7 March 1996); *U.S. Central Intelligence* (C 28 December 1994, N 7 March 1996); *U.S. Internal Revenue Service* (N 7 March 1996); *Wells Fargo Bank* (N 7 March 1996); *Miramax Films* (C 9 January 1996). (Key: C = *Cool Site of the Day;* N = Netscape's *What's Cool?;* P = Project Cool's "Sightings.") Attending to this category would allow us to think about the corporate or institutional appropriation of cool. Cool in the information age, in other words, is just as much a matter of "reciprocal appropriation" as the earlier, twentieth-century kinds of cool I previously studied. However, I have subtracted this category from my discussion because I was not able to determine whether or not some of the corporate appearances on particular cool anthology sites were paid for or otherwise indirectly contingent upon intercorporate relations (even if only at the informal level of employees at a cool anthology taking special note of the corporate Web sites that their firm most frequently deals with). My caution

in this regard is influenced by controversies in the search engine industry, where some search engines allow advertisers to pay for elevating the position of their Web pages in search results (see Danny Sullivan's article "Pay for Placement?" on *Search Engine Watch*).

17. Some of the sites I discuss in this section will undoubtedly be defunct by the time this book is published, while others may have evolved in different directions. As in other instances where I discuss Web sites in this book, I use the analytical present tense unless the past tense is needed to place a site in its specific historical moment in the 1990s or early 2000s. Many old sites can be recovered by using the *Wayback Machine.*

18. See Hebdige's analysis of the "streamlining" effect and the design industry in his *Hiding in the Light,* pp. 58–76.

19. This ironic thought is suggested by one of the most pervasive visual clichés of computer advertising in the late 1990s: the image of a knowledge worker being blasted back into his office chair by the sheer speed of the equipment or software he is operating (hair blown back, tie streaming in the wind, and so on).

20. *Business Week,* "Generation $," 16 August 1999. The picture accompanying the article gathers together all the subcultural fashion motifs I mention. See also the *Los Angeles Times* article, "Smashing the Gen-X Stereotype" (3 September 1999), whose secondary headline reads, "Born between 1961 and 1981, they were supposed to be slackers. But today their successes are as plentiful as Web sites, and the words 'entrepreneur' and 'self-reliant' are more apt descriptions." Perhaps not surprisingly, subcultural motifs also affect the dress of those at the CEO level of technology firms wishing to look cool, as in the Beat-style sunglasses and black pullover that Larry Ellison of Oracle wore on the cover of the 8 May 2000 issue of *Business Week* (under the cover title, "Despite the Tech Stock Slide . . . Oracle Is Cool Again").

Chapter Six

1. On the antithetical assimilation of technological rationality within formalism, see the introduction, note 14. My description here of form as being so deeply inscribed in function that it could no longer be discarded like packaging from the product is inspired by Ransom's discussion of the "containing icon" or "containing body" of poetry (*New Criticism,* p. 291). Ransom's footnote on the same page explicitly rejects the idea that form as a container is merely the throwaway packaging for contents "meant for consumption, so ripe for immediate consumption."

2. The following two passages from Cleanth Brooks are illustrative:

> I have suggested . . . that the poem which meets Eliot's test [of "mature" experience] comes to the same thing as I. A. Richards' "poetry of synthesis"—that is, a poetry which does not leave out what is apparently hostile to its dominant tone, and which, because it is able to fuse the irrelevant and discordant, has come to terms with itself and is invulnerable to irony. . . . The stability is like that of the arch: the very forces which are calculated to drag the stones to the ground actually provide the principle of support—a principle in which thrust and counterthrust become the means of stability. ("Irony as a Principle of Structure," p. 970)

> The structure meant is a structure of meanings, evaluations, and interpretations; and the principle of unity which informs it seems to be one of balancing and harmonizing connotations, attitudes, and meanings. . . . The unity is not a unity of the sort to be achieved by the reduction and simplification appropriate to an algebraic formula. It is a positive unity, not a negative; it represents not a residue but an achieved harmony. (*Well Wrought Urn,* p. 195)

For a review of many of the principles of Russian Formalism, see the retrospective essay by Boris Eichenbaum, "Theory of the 'Formal Method.'"

3. Johanna Drucker notes, "By the end of the nineteenth century, the features of marked typography included: the use of a wide range of type faces, styles, and sizes with mixtures and juxtapositions of these proliferating within a single sheet; the breakup of the page into various zones of activity which received very distinct graphic treatments; the use of circular, shaped, or diagonal elements across the normal horizontal page; the use of vertical elements; and finally, the use of paragonnage—the incorporation of several different typefaces and/or sizes within a single line or word" (*Visible Word,* p. 96).

4. "Asymmetry is the rhythmic expression of functional design," Jan Tschichold declared in the 1928 manifesto, *The New Typography* [*Die neue Typographie*], which helped make asymmetrical composition and starkly unbalanced type sizes and axes the signature features of modernist graphic design (p. 68). More generally, Tschichold observes, the desideratum is "contrast": "The real meaning of form is made clearer by its opposite. We would not recognize day as day if night did not exist. The ways to achieve contrast are endless: the simplest are large/small, light/dark, horizontal/vertical, square/round, smooth/rough, closed/open, coloured/plain; all offer many possibilities of effective design" (p. 70).

5. Tschichold, *New Typography,* p. 70. Tschichold states, "[Asymmetrical movement] must not however degenerate into unrest or chaos. A striving for order can, and must, also be expressed in asymmetrical form. It is the only way to make a better, more natural order possible, as opposed to symmetrical form which does not draw its laws from within itself but from outside" (p. 68).

6. Philip Meggs observes about Herbert Bayer's typography for Bauhaus: "Open composition on an implied grid and a system of sizes for type, rules, and pictorial images brought unity to the designs" (*History of Graphic Design,* p. 294). Similarly, Meggs describes a Tschichold advertisement of 1932 as follows: "Asymmetrical balance, a grid system, and a sequential progression of type weight and size determined by the words' importance to the overall communication are aspects of this design" (p. 300, caption to fig. 19-29). On the grid system, see Kung, "Grid System." The grid may even be seen in the Bauhaus building that Gropius designed for the Dessau phase of the movement (see Julier, *Thames and Hudson Dictionary of 20th-Century Design and Designers,* p. 97).

7. Tschichold states, "In every individual activity we recognize the single way, the goal: **Unity of Life!** So the arbitrary isolation of a part is no longer possible for us—every part belongs to and harmonizes with the whole" (p. 13). The phrase "Unity of Life!" is set flush left and boldface on its own line in the book.

8. Moholy-Nagy formulated the principle of communicational clarity as follows: "Typography is a tool of communication. It must be communication in its most intense form. The emphasis must be on absolute clarity" (quoted in Meggs, *History of Graphic Design,* p. 291). And Tschichold is never more clear than on the topic of clarity:

> The concepts of the New Typography, in use, allow us for the first time to meet the demands of our age for purity, clarity, fitness for purpose, and totality. (p. 7)
>
> [Technical forms], following the laws of nature, are drawn towards greater clarity and purity of appearance. (p. 65)
>
> *The essence of the New Typography is clarity.* This puts it into deliberate opposition to the old typography whose aim was "beauty" and whose clarity did not attain the high level we require today. This utmost clarity is necessary today because of the manifold claims for our attention made by the extraordinary amount of print, which demands the greatest economy of expression. (p. 66)

> The New Typography is distinguished from the old by the fact that its first objective is to develop its visible form out of the functions of the text. It is essential to give pure and direct expression to the contents of whatever is printed; just as in the works of technology and nature, "form" must be created out of function. Only then can we achieve a typography which expresses the spirit of modern man. The function of printed text is communication, emphasis (word value), and the logical sequence of the contents.
>
> Every part of a text relates to every other part by a definite, logical relationship of emphasis and value, predetermined by content. It is up to the typographer to express this relationship clearly and visibly, through type sizes and weight, arrangement of lines, use of colour, photography, etc. (pp. 66–67)
>
> The need for clarity in communication raises the question of how to achieve clear and unambiguous form. (p. 69)

9. Meggs emphasizes the information-centric nature of the eventual International Style: "The visual characteristics of this international style include: a visual unity of design achieved by asymmetrical organization of the design elements on a mathematically constructed grid; objective photography and copy that present visual and verbal information in a clear and factual manner, free from exaggerated claims of much propaganda and commercial advertising; and the use of sans serif typography set in a flush-left and ragged-right margin configuration. The initiators of this style believe that sans serif typography expresses the spirit of a progressive age, and that mathematical grids are the most legible and harmonious means for structuring information" (*History of Graphic Design*, p. 332).

10. See Meggs, *History of Graphic Design*, pp. 359–70, 380–409.

11. On "good design" as a prevalent concept after World War II, see Julier, *Thames and Hudson Dictionary of 20th-Century Design and Designers*, pp. 93–94.

12. Lopez, "Principles of Design," pp. 36, 38.

13. Dondis, *Primer of Visual Literacy*, p. 16; see also pp. 85–103, 112–16. Such is the graphic designer's version, we may say, of that originary work of visual formalism (which influenced the Russian Formalists), Heinrich Wölfflin's *Principles of Art History* (1915) with its analysis of painting into such contrasting "motifs" as the linear versus painterly, planar versus recessional, closed and open form, and so on.

14. Dondis, *Primer of Visual Literacy*, pp. 11, 18; see also p. 146. Other works on design I have consulted include Field et al., eds., *Graphic Arts Manual* (1980); Carter et al., *Typographic Design* (1985); and Heller and Chwast, *Graphic Style* (1988).

15. On the influence of the Container Corporation of America on early to mid-twentieth-century U.S. design, see Meggs, *History of Graphic Design*, pp. 318–25. On design at IBM and the role of Eliot Noyes and Paul Rand, see Julier, *Thames and Hudson Dictionary of 20th-Century Design and Designers*, p. 104; Meggs, *History of Graphic Design*, pp. 385–88; and Rand, *Designer's Art*, pp. 42–45, 135, 219–21. On Müller-Brockman at IBM Europe, see Julier, *Thames and Hudson Dictionary*, p. 137.

16. On the influence of Apple-based software on the design profession, see Julier, *Thames and Hudson Dictionary of 20th-Century Design and Designers*, pp. 20–21.

17. Heller and Chwast, *Graphic Style*, p. 9.

18. Two chapters of Kirschenbaum's dissertation, "A White Paper on Information" and "The Other End of Print: Post-Alphabetic Graphic Design," are especially relevant to issues I take up in this chapter. The quote here is from the former chapter.

19. Bertin, *Semiology of Graphics*, pp. 2, ix–x.

20. Tufte, *Visual Display of Quantitative Information,* p. 13.

21. Tufte, *Visual Display of Quantitative Information,* pp. 91, 105, 107.

22. Unfortunately, I read Manovich's essay (and his other work) after completing the bulk of this book. I would otherwise have liked to interlink my argument more closely with his analysis of the discrete principles and structure of "new media." I have also lately had occasion to read in partial manuscript Manovich's work in progress "Info-Aesthetics: Information as Form," which extends further his comparison of new media to the early twentieth-century avant-garde.

23. Rand, *Designer's Art,* p. 195.

24. Rand, *Designer's Art,* p. 201.

25. Quoted in Sano, *Designing Large-Scale Web Sites,* p. x. Compare Sano's epigraph drawn from Paul-Jacques Grillo's *Form, Function, and Design*: "Interest cannot be created by multiplying various elements into a busy design. By dispersing points of interest, we only create confusion by conflicting effects that, in the end, give no impression at all" (ibid., p. 138).

26. See Sano, *Designing Large-Scale Web Sites,* color pl. 4 and black-and-white illustrations on pp. 73, 187, 191, and passim.

27. Sano, *Designing Large-Scale Web Sites,* p. 190. The orientalism of design is a topic too complex to explore adequately here, especially since it is part of a larger orientalism not just of technology and information technology (as epitomized in the karate fetish of many contemporary video arcade games and films like *The Matrix*) but of postindustrial corporatism in toto, which in the United States could begin, both practically and theoretically, only after the collective formation of a new notion of "Japan, Inc." The view from the other side of the Pacific, one surmises, is just as complex. To what extent is Japanese design sense in media and information technology—the legendary cachet, for example, of the Sony company—an expression not just of deeply rooted Eastern culture but also of selections or mutations of that culture adapted—like the immense eyes of anime heroines—to a faux self-perception or self-exoticization of the East made up to appeal as if to/through Western eyes (and Western consumer tastes)?

On "techno-orientalism," see the chapter titled "Techno-Orientalism: Japan Panic" in Morley and Robins's *Spaces of Identity.* See also Ueno, "Japanimation and Techno-Orientalism." Ueno writes, "I think that the stereotype of the Japanese, which I would like to call 'Japanoid' for not actually Japanese, exists neither inside nor outside Japan. This image functions as the surface or rather the interface controlling the relation between Japan and the other. Techno-Orientalism is a kind of mirror stage or an image machine whose effect influences Japanese as well as other people. This mirror in fact is a semi-transparent or two-way mirror. It is through this mirror stage and its cultural apparatus that Western or other people misunderstand and fail to recognize an always illusory Japanese culture, but it also is the mechanism through which Japanese misunderstand themselves."

A personal anecdote may perhaps be allowed here. I remember the little sleek, curvy, fire-engine-red Sony transistor radio my father bought when we lived in Hong Kong sometime around 1955. It spoke of everything "new" about the world. And, as heard in the plaintive strains of The Platters singing "Smoke Gets in Your Eyes" so often on that radio at that time, "new" meant New World (specifically, a heady mix of the subcultural and mainstream West). Fire-engine red has today given way to matte black or gun-metal gray as the cutting-edge look of technology. But if I had had the English then, I would have said about that magical radio, "cool."

The difference between the affiliation of Western and Japanese design, on the one hand, and the almost total incommensurability of Western and Chinese style (whether ancient or Communist), on the other, is another issue of interest. Chinoiserie in the West must still be given its own room in a house, while neo-Japonesque has broken out into the main areas of living and working where high-tech equipment—TVs, computers, cars, and so on—are part of the ordinary decor. To return full circle in my argument to the origin of International Style, "Euro-design" would also have to be factored in here as a mediating or triangulating aesthetic sensibility between the United States and the East.

28. Waters, *Web Concept & Design*, pp. 78–81.

29. The main *Web Pages That Suck* site is at <http://www.webpagesthatsuck.com/> I refer here to the discussion of techniques gathered on the subpage at <http://www.webpagesthatsuck.com/suckframe.htm> on 14 November 2000. Unfortunately, old pages on the site cannot be recovered through the *WayBack Machine*.

30. I refer to a version of the *Flaunt* site in August 1999. The site has since been redesigned, though the version at the time of this writing also makes marked use of left- and right-centric vertical panels. I borrow the terms *left-* and *right-centric* from the analytical vocabulary of the *Project Cool* site.

31. Compare the Aaron page at <http://www.rockpix.com/> with the example of Storch's design for a two-page *McCall's* spread reproduced in Meggs, *History of Graphic Design*, fig. 22-30. See also the Web sites included in *Critique* magazine's annual Web Crit awards in 2000. Partly because *Critique* ("The Magazine of Graphic Design Thinking") reproduces screen-shots in a horizontal rather than vertical aspect ratio, the award-winners bear a strong resemblance to two-page glossy magazine spreads.

On "typo-photo," see Tschichold, *New Typography*, p. 92: "Since [the New Typography's] aim was to create artistic unity out of contemporary and fundamental forms, the problem of type never actually existed: it had to be sanserif. And since it regarded the photographic block as an equally fundamental means of expression, a synthesis was achieved: photography + sanserif!

"At first sight it seems as if the hard black forms of this typeface could not harmonize with the often soft greys of photos. The two together do not have the same weight of colour: their harmony lies in the contrast of form and colour. But both have two things in common: their objectivity and their impersonal form, which mark them as suiting our age. This harmony is not superficial, as was mistakenly thought previously, nor is it arbitrary: there is only one objective type form—sanserif—and only one objective representation of our times: photography. Hence typo-photo, as the collective form of graphic art, has today taken over from the individualistic form handwriting-drawing.

"By typo-photo we mean any synthesis between typography and photography."

32. See especially Bolter and Grusin's discussion of the "variety of remediations on the World Wide Web," *Remediation*, pp. 197–210.

33. In this discussion I do not extend my argument to the more recent XML or Extensible Markup Language (and the associated XHTML standard for Web presentation). XML is the topic of my recent paper entitled "The Art of Extraction: Toward a Cultural History and Aesthetics of XML and Database-Driven Web Sites." Because it rigorously separates the issues of content and formatting, or logical design and presentation design, XML may be said to be the alter ego of HTML. HTML evolved in a direction that accentuated presentation design at the expense of logical design, but XML evolved precisely to maintain the purity of logical design so that computers can automatically transact documents across different processing, display, and software platforms.

34. Brown and Duguid make this point in a similar way: "In the digital world, moreover, many of the distinctions between designers and users are becoming blurred. We are all, to some extent, designers now" (*Social Life of Information*, p. 4).

35. Julier, *Thames and Hudson Dictionary of 20th-Century Design and Designers*, p. 19. A fuller historical sketch of antidesign in the twentieth century could start with elements of dadaism and futurism. See, for example, Lanham's discussion of Marinetti's typographical work,

SCRABrrRrraaNNG, as a precedent of the way electronic media destabilize the notion of text (*Electronic Word,* pp. 31–34). See also Bolter and Grusin's observation that many Web sites feature "riots of diverse media forms" arranged according to "graphic design principles [that] recall the psychedelic 1960s or dada in the 1910s and 1920s" (*Remediation,* p. 6).

36. See Meggs, *History of Graphic Design,* pp. 446–60; Julier, *Thames and Hudson Dictionary of 20th-Century Design and Designers,* pp. 45–46, 64–65, 127–28.

37. Spalter writes that "this second period featured the increased use of minicomputers, the advent of the personal computer, the commodification of interactive graphics software (requiring no programming by the artist), the widespread adoption of computers by design firms, and the expansion of the computer art community" (*Computer in the Visual Arts,* p. 4). Graphics software after this period became such second nature to designers, for example, that a work like Laurie McCanna's *Creating Great Web Graphics* (1996) simply instructs the user in following certain procedures in Adobe Photoshop and then applying particular effects from Kai's Power Tools or some of the other add-on filters to Photoshop.

38. The effects of what Bolter and Grusin call "remediation," of course, are very clear in graphics software: for example, the insistence of Adobe Photoshop and similar programs on modeling procedures after techniques of manipulating physical art media ("pencils," "air brushes," "layers," "transparency," "masks," and so on). But remediation changes into what might be called premediation—the creation of emergent, as yet inchoate new visual vocabularies—about the time one encounters such digital tools without any clear physical precedent as Photoshop's "Magic Wand" or "History Brush," not to mention such underlying components, protocols, or algorithms of computer imaging as alpha channels, Web-safe or adaptive color swatches, JPEG "lossiness" settings, interlaced or animated GIF images, Gaussian blurs, convolution filters, and so on. For a history of the impact of computer graphics programs on artists, see Spalter, *Computer in the Visual Arts,* pp. 2–36. Spalter's second chapter (ibid., pp. 37–86), on digital painting and raster graphics programs, is especially illuminating in its demonstration of how the computer reconceives visual design. For example, Spalter creates the new term "global touch" (as opposed to "local touch") to describe functions that allow an artist or designer to change multiple or overall aspects of a composition simultaneously. She also explicates how such now common algorithms as Gaussian blurs, convolution filters, and other filtering effects derive from the mathematics of signal processing rather than from traditional media.

39. On intermedia relations early in the twentieth century, for example, see Lanham's discussion of Marinetti's typographical work *SCRABrrRrraaNNG* and the influence of cinema (*Electronic Word,* pp. 32–33). On the challenge to photography by "postphotographic" computer graphics, see Spalter, *Computer in the Visual Arts,* pp. 39–42.

40. However, see Kirschenbaum's shrewd essay "Word as Image in an Age of Digital Reproduction" on the persistent practical differences between texts and images in digital media, which handle ASCII-based characters and bitmapped graphics in fundamentally incommensurable ways (as instanced in the difficulty of making images searchable). "One cannot talk about words as images and images as words," he says, "without taking into account the technologies of representation upon which both forms depend" (p. 141). A fuller statement of my argument here would thus indicate that digitization resulted in the convergence of text and image on certain planes of perception but also in fresh or renewed divergences on other planes of function. The *field* of relation between text and image changed, and consequently the overall structure of similarity and contrast that held modernist typo-photo compositions together was reconfigured in new kinds of similarity and contrast.

41. From the chapter "The Other End of Print: Post-Alphabetic Graphic Design" in Kirschenbaum's online dissertation, *Lines for a Virtual T[y/o]pography.*

42. On the influence of *Ray Gun* on *Wired,* see Pierson, "Welcome to Basementwood," paragraphs 17–18.

43. For a current magazine that retains, and in many ways outdoes, the antidesign sense of the early *Wired,* see *Adbusters,* in which design professionals speak out against the sell-out of the design profession to corporate interests. *Adbusters* is aggressively antidesign in both ideology and practice. Some pages in the 2000 special double issue on "Design Anarchy," for example, parody modernist and International Style design, while others out–*Ray Gun* and out-*Wire* other antidesign magazines in their spectacular, but always professionally produced, illegibility.

44. Spalter, in her chapter on Web graphics, says with some understatement, "Because these markup tags began as a way to indicate document structure rather than nuances of appearance, their capabilities will seem limited to artists and designers familiar with even the most basic design and layout programs" (*Computer in the Visual Arts,* p. 424).

45. In Project Cool's vocabulary for describing Web pages, locking designs into fixed-width tables creates "ice." Project Cool defines *ice* as "a web page that is stiff and frozen to the side of a browser window. It doesn't reflow when you resize the browser and depends on a window being a predefined width to look good" (Project Cool, "Ice"). Ice is the opposite of "liquid" in Project Cool's vocabulary (see note 49, below).

46. Liu, *Lyotard Auto-Differend Page,* "Philosophy of This Page."

47. What comes clear in writing a sentence like this that tries to describe the quick delivery of online artifacts as opposed to the slow or varying rates of their reception is that our vocabulary of temporality is inadequate for mapping the new digital environment. Terms such as *quick, fast, slow,* or *real time* require a qualitative, rather than just quantitative, definition in the new media.

48. David Rodowick discussed the concepts of synchronous and asynchronous digital temporality in his paper, "Cybernetic and Machinic Arrangements" (Modern Language Association convention, 28 December 1997).

49. Project Cool defines *liquid* as follows: "A web page that will reflow to fit whatever size browser window it is poured into. Liquid is the epitome of good web page design and should be emulated whenever possible" (Project Cool, "Liquid").

50. My examples are inspired by the following Web sites: Daniel Chandler, *The Media and Communication Studies Site* (which for some years used the visual metaphor of a spiral-bound notebook); *What Is Miles Watching on TV!* (which depicts a television set); and *Laurie McCanna's Free Art Site* (which in its early version in 1996–98 offered users a menu of links resembling jukebox buttons).

51. Compare the deconstructive idea of "catachresis": the forced, violent use of metaphor to describe things that actually have no proper name.

52. Stephenson, "In the Beginning . . . Was the Command Line," pp. 62–64.

Chapter Seven

1. Shenk, *Data Smog,* pp. 37–38.

2. The number of total "sightings" for this period is approximate because Project Cool retracted some sites after the fact. I also count the use of *fun* on the retrospective, best-of-year page titled "1998: Citing the Sightings."

3. See the Project Cool archive entries for the following dates: 1998—6 September; 1, 12, 19 October; 4 November; 1999—3, 6, 8, 9, 11 January; 14, 19 March; 17 April; 24, 28 May; 29 August. It may be noted that "dreamy," "flavorful," and "wild" also pun on the content of particular sites.

4. *Fairie Queene,* III.xi.54.

5. I benefited from a conversation (and subsequent correspondence) about wit, satire, and cool with Michael Hancher at the University of Minnesota in April 2000 (after a talk in which I presented portions of this book). On the split between passion and control in the picturesque, see my *Wordsworth: The Sense of History,* pp. 61–137, especially 61–65.

6. Relevant here is the emerging field of "emotional" computing, which attempts to endow computers and interfaces with the ability to discern, respond to, and express affect. See Picard, *Affective Computing.*

7. On mirrorshades or mirrored sunglasses as a "totem" of cyberpunk, see Bruce Sterling's preface to *Mirrorshades: The Cyberpunk Anthology,* p. xi. The most famous mirrorshades in cyberpunk fiction are those embedded surgically over the eyes of the Molly character in William Gibson's *Neuromancer.* On constrained or managed emotion, see not only the work of Arlie Hochschild and others discussed previously, but also Stjepan G. Meštrović's *Postemotional Society.*

Chapter Eight

Key to acronyms of organizations cited in this chapter:

ACLU	American Civil Liberties Union
APC	Association for Progressive Communications
CDT	Center for Democracy and Technology
CNOT	Coalition on New Office Technology
CPSR	Computer Professionals for Social Responsibility
DAN	Direct Action Network
D2KLA	(name of Direct Action Network Los Angeles)
EFF	Electronic Frontier Foundation
IGC	Institute for Global Communications
PFF	Progress and Freedom Foundation
SVTC	Silicon Valley Toxics Coalition
UE	United Electrical, Radio and Machine Workers of America

1. David Friedman observes, "Libertarianism is much more important in cyberspace than in real space. . . . Nearly all political discussion online is pro- or antilibertarian. Libertarianism is the central axis" (quoted in McHugh, "Politics for the Really Cool"). Compare Winner, "Cyberlibertarian Myths": "Indeed, there seems to be no coherent, widely shared philosophy of cyberspace that offers much of an alternative [to cyberlibertarianism]. Woven together from available themes and arguments from earlier varieties of social thought, the cyberlibertarian position offers a vision that many middle and upper class professionals find coherent and appealing." Borsook notes, "When I was asked to participate in a survey on the politics of the Net, the questions *presumed* respondents were libertarian, but charitably gave space for outdated contrarian views" ("Cyberselfish").

2. Katz, "Birth of a Digital Nation," p. 1 of 8 successive Web pages. (But see Katz's subsequent "Netizen: The Digital Citizen," which reported his surprise at a survey purporting to show that netizens cared as profoundly about traditional politics as about information "postpolitics.")

3. See Winner, "Cyberlibertarian Myths," paragraphs 5–6, for a similar short list of important cyberlibertarian works and writers.

4. For a fuller list of online "civic organizations," see Doheny-Farina, *Wired Neighborhood,* pp. 189–202. The quotation from the CPSR on its early mission against the military use of computers is from its "History Up to 1994" web page.

5. Winner argues in "Peter Pan in Cyberspace: *Wired Magazine's* Political Vision" that the success of *Wired* inspired established magazines such as *Newsweek,* which started a "Cyberscope" page, and newspapers all over the United States to offer "plugged in" sections.

6. In his "Cyberlibertarian Myths and the Prospects for Community," Winner comments that cyberlibertarian beliefs "can also be found in countless books on cyberspace, the Internet, and interactive media; Nicholas Negroponte's *Being Digital* and George Gilder's *Microcosm* are especially vivid examples. Writers in this strand include Alvin Toffler, Esther Dyson, Stewart Brand, John Perry Barlow, Kevin Kelly, and a host of others that some have called the digerati."

7. Of these works, three are written by authors outside the United States: Barbrook and Cameron (U.K.), Birdsall (Canada), and Verzola (Philippines).

8. "Bill of Information Rights" here imitates the several documents on the Web whose titles or discourse allude to the U.S. Declaration of Independence, Constitution, Bill of Rights, or related documents in the Western history of democratic revolutions (e.g., the French Revolutionary "Declaration of the Rights of Man"). See Barlow, "Declaration of the Independence of Cyberspace"; or Hauben and Hauben, "Proposed Declaration of the Rights of Netizens," part of *Netizens and the Wonderful World of the Net.* A paragraph in the latter declaration reads: "Inspiration from: RFC 3 (1969), Thomas Paine, Declaration of Independence (1776), Declaration of the Rights of Man and of the Citizen (1789), NSF Acceptable Use Policy, Jean Jacques Rousseau, and the current cry for democracy worldwide." "RFC" refers to the foundational technical documents of the Internet, the canon of "requests for comment" (over 3,500 by mid-2003) that proposed the basic ideas, standards, and protocols of networked communication, beginning with the ARPAnet and such early protocols as Telnet or FTP. One of the earliest in the series, RFC 3 on "Documentation Conventions" (later superseded in a series of revisions on the same topic through RFC 30), laid the groundwork for the simultaneously freewheeling and technically precise conventions of discussion that made the RFCs one of the most vigorous and successful forms of technical discourse in the twentieth century—at once wide open to contributions and focused on critical examination and revision. RFC 3 was the free speech act of such technical discourse: "The content of a [Network Working Group] note may be any thought, suggestion, etc. related to the HOST software or other aspect of the network. Notes are encouraged to be timely rather than polished. Philosophical positions without examples or other specifics, specific suggestions or implementation techniques without introductory or background explication, and explicit questions without any attempted answers are all acceptable. . . . There is a tendency to view a written statement as ipso facto authoritative, and we hope to promote the exchange and discussion of considerably less than authoritative ideas."

9. Privacy International "is a human rights group formed in 1990 as a watchdog on surveillance by governments and corporations. PI is based in London, England, and has an office in Washington, D.C. PI has conducted campaigns throughout the world on issues ranging from wiretapping and national security activities, to ID cards, video surveillance, data matching, police information systems, and medical privacy" (Privacy International home page). The Global Internet Liberty Campaign "was formed at the annual meeting of the Internet Society in Montreal. Members of the coalition include the American Civil Liberties Union, the Electronic Privacy Information Center, Human Rights Watch, the Internet Society, Privacy International, the Association des Utilisateurs d'Internet, and other civil liberties and human rights organizations" (Global Internet Liberty

Campaign, "Statement of Principles"). The Electronic Frontier Foundation invokes the U.N. article I mention on its Web page titled, "Preserving Free Expression: Our Fundamental Rights of Freedom of Speech & Press." However, the sentence immediately following the mention of the U.N. in the EFF document returns us to the U.S. provenance by referring to "the U.S. First Amendment holding free speech to be a right."

10. For examples of allusions to Jeffersonian democracy, see Dyson et al., "Cyberspace and the American Dream: A Magna Carta for the Knowledge Age"; Barlow, "Declaration of the Independence of Cyberspace" and "The Economy of Ideas"; and Katz, "Netizen," Web page 3. As Katz puts it, "Far from being distracted by technology, Digital Citizens appear startlingly close to the Jeffersonian ideal—they are informed, outspoken, participatory, passionate about freedom, proud of their culture, and committed to the free nation in which it has evolved." In his "Birth of a Digital Nation," Web page 7, Katz also refers to the U.S. Communications Decency Act of 1996, and the controversy that surrounded it, as "the Net's own Stamp act."

11. It was precisely because he was writing in the wake of CDA and the election season of 1996 that Katz's "Birth of a Digital Nation" connected so forcefully—in its tone as in the response it generated in discussion forums—with the sense of a rising "post"-nation of patriots. In the article, Katz underscores cyberlibertarianism's combined negation of government and religion as follows: "Government doesn't believe that information should be free—witness the fiasco of the Communications Decency Act. Religious organizations, educators, and many parents don't believe information ought to be liberated, either. The realization that children have broken away from many societal constraints and now have access to a vast information universe is one of the most frightening ideas in contemporary America."

12. Winner, "Cyberlibertarian Myths." We note that Alvin Toffler, one of the authors of the PPF's "Magna Carta," was the intellectual confidant of Gingrich. Nuanced statements such as the following from the PPF's "Magna Carta," however, do attempt to find a balance between downsized and activist government: "One further point should be made at the outset: Government should be as strong and as big as it needs to be to accomplish its central functions effectively and efficiently. The reality is that a Third Wave government will be vastly smaller (perhaps by 50 percent or more) than the current one—this is an inevitable implication of the transition from the centralized power structures of the industrial age to the dispersed, decentralized institutions of the Third. But smaller government does not imply weak government; nor does arguing for smaller government require being 'against' government for narrowly ideological reasons.

"Indeed, the transition from the Second Wave to the Third Wave will require a level of government activity not seen since the New Deal."

13. Dyson, Gilder, Keyworth, and Toffler, "Cyberspace and the American Dream: A Magna Carta for the Knowledge Age" (henceforth cited as "Magna Carta").

14. The controversy surrounding the DoubleClick online advertising company, for example, made the list of news items on the home page of the Center for Democracy and Technology on 22 May 2000: "DoubleClick puts hold on tying Personal Info to Online Habits—The Internet advertiser DoubleClick announced on March 2 that it will not move forward on its plans to tie personally identifiable information to Internet users' online surfing habits until government and industry have reached a consensus on privacy rules for the Internet. The move came after CDT and other privacy advocates filed a Statement of Additional Facts and Grounds for Relief with the Federal Trade Commission noting that sensitive information including video titles, salaries, and search terms are being passed to DoubleClick."

15. In his "Economy of Ideas," Barlow describes the music of the Grateful Dead as part of a gift-economy: "In regard to my own soft product, rock 'n' roll songs, there is no question that the band I write them for, the Grateful Dead, has increased its popularity

enormously by giving them away. We have been letting people tape our concerts since the early seventies, but instead of reducing the demand for our product, we are now the largest concert draw in America, a fact that is at least in part attributable to the popularity generated by those tapes." Also relevant is his explanation in this article of the "volunteer" work that helped build the Internet: "And then there are the inexplicable pleasures of information itself, the joys of learning, knowing, and teaching; the strange good feeling of information coming into and out of oneself. Playing with ideas is a recreation which people are willing to pay a lot for, given the market for books and elective seminars. We'd likely spend even more money for such pleasures if we didn't have so many opportunities to pay for ideas with other ideas. This explains much of the collective 'volunteer' work which fills the archives, newsgroups, and databases of the Internet. Its denizens are not working for 'nothing,' as is widely believed. Rather they are getting paid in something besides money. It is an economy which consists almost entirely of information."

16. Center for Democracy and Technology, "Democratic Values for the Digital Age."

17. Barbrook and Cameron, "Californian Ideology." Compare Winner, "Cyberlibertarian Myths": "The first and most central characteristic of cyberlibertarian world view is what amounts to a whole hearted embrace of technological determinism. This is not the generalized determinism of earlier writings on technology and culture, but one specifically tailored to the arrival of the electronic technologies of the late twentieth century. In harmony with the earlier determinist theories, however, the cyberlibertarians hold that we are driven by necessities that emerge from the development of the new technology and from nowhere else."

18. Ayn Rand, "Property Status of the Airwaves," *Objectivist Newsletter,* April 1964, quoted in "Magna Carta."

19. Winner, "Cyberlibertarian Myths and the Prospects for Community." Winner also comments on the reference to Rand in the "Magna Carta."

20. EFF, "Preserving Free Expression."

21. CDT, "Democratic Values for the Digital Age."

22. Winner, "Cyberlibertarian Myths."

23. Borsook, "Cyberselfish." She has since published her book, *Cyberselfish: A Critical Romp through the Terribly Libertarian Culture of High Tech.*

24. My image of bees alludes to the trope of bees in discourse about new media and information technology. Kevin Kelly's *Out of Contol: The New Biology of Machines, Social Systems, and the Economic World,* for example, turns on the metaphors of the hive and swarm (which he also uses in his later *New Rules for the New Economy)*; while Carolyn Guertin's *Queen Bees and the Hum of the Hive: An Overview of Feminist Hypertext's Subversive Honeycombings* employs the metaphor for different purposes. See also Barlow, "Economy of Ideas": "Trying to stop the spread of a really robust piece of information is about as easy as keeping killer bees south of the border."

25. The phrases I quote are from Doheny-Farina, *Wired Neighborhood,,* and Winner, "Cyberlibertarian Myths."

26. Rheingold writes in the introduction of *Virtual Community,* for instance: "I care about these people I met through my computer, and I care deeply about the future of the medium that enables us to assemble. I'm not alone in this emotional attachment to an apparently bloodless technological ritual. Millions of people on every continent also participate in the computer-mediated social groups known as virtual communities, and this population is growing fast. Finding the WELL was like discovering a cozy little world that had been flourishing without me, hidden within the walls of my house; an entire cast of characters welcomed me to the troupe with great merriment as soon as I found the secret door." Also see Rheingold's "Technology, Community, Humanity and the Net."

27. CDT, "Democratic Values for the Digital Age."

28. "Magna Carta." The "Magna Carta" page adds: "The late Phil Salin (in Release 1.0 11/25/91) offered this perspective: 'By 2000, multiple cyberspaces will have emerged, diverse and increasingly rich. Contrary to naive views, these cyberspaces will not all be the same, and they will not all be open to the general public. The global network is a connected "platform" for a collection of diverse communities, but only a loose, heterogeneous community itself. Just as access to homes, offices, churches and department stores is controlled by their owners or managers, most virtual locations will exist as distinct places of private property.'"

29. On the unexpectedly rapid narrowing of the "digital divide" in some areas, see, for example, *Los Angeles Times*. "Univision Adds Site to Bridge the Divide," 29 June 2000, which includes a table of data gathered by the Access Worldwide Cultural Access Group and a telephone survey. The table indicates that Latino households in 2000 were acquiring computers at a significantly faster rate than U.S. households in general (68 percent vs. 43 percent). The divide, however, was still very evident in lower-income households.

30. CDT, "CDT Principles"; EFF, "About EFF."

31. Recall the opposite spin Borsook puts on such an analysis: "Although the technologists I encountered there were the liberals on social issues I would have expected . . . , they were violently lacking in compassion, ravingly anti-government, and tremendously opposed to regulation" ("Cyberselfish").

32. EFF, "About EFF."

33. What is needed here is a detailed analysis, at once philosophical and technical, of online privacy in which encryption is juxtaposed with such related issues as anonymity and pseudonymity. Technically, for example, hard anonymity or pseudonymity of the sort offered by the Zero-Knowledge firm's Freedom software system beginning in 1999 (users could choose "nyms" that identified them only pseudonymously on the Net, even to the Zero-Knowledge firm itself) depends at least in part on strong encryption. (See the Zero-Knowledge Systems page for its Freedom software.) A full study of this issue would also benefit from historical study of the political and criminal use of anonymous speech in the past—for example, in various kinds of blackmail.

34. There are indeed good reasons, for example, having to do with the difference between private individuals and individuals representing institutions of public trust. But my point here is that there are no such reasons integral with the belief structure of cyberlibertarianism, which in practice rides roughshod over niceties of distinction between private and public information. (Thus, if the e-mail of public officials is fair game, then by the logic of cyberlibertarianism so, too, is a video of the private sex life of a TV star and her musician husband, which in at least one infamous instance made the rounds of the Internet.) This is just one example of the thorny questions that can be asked about the credo of privacy and freedom of information in cyberlibertarianism. Fiona Steinkamp is particularly perceptive in her "Ideology of the Internet," which poses a series of questions about anonymity, freedom of information, "freedom to inform," and "freedom to misinform."

35. In using the vocabulary of "boundaries" in this discussion, I am influenced by Corynne McSherry's book on academic intellectual property, which treats the concept of intellectual property as a "boundary object" that is continually transgressed by incommensurable parties (e.g., the university and commercial firms) but nevertheless restabilized in informal or litigated settlements.

36. Such critics as Borsook and Winner single out the conservative elements in cyberlibertarianism, which they see as surprising given what would seem to be the movement's default posture of liberalism. For other critics, such as Paul Treanor, however, it is the

very "hyper-liberalism" of cyberlibertarianism that is the problem (see his "Internet as Hyper-Liberalism").

37. Kole, "Whose Empowerment? NGOs between Grassroots and Netizens," offers a precise definition of *NGO* that differentiates NGOs from each other and from grassroots movements at other levels of organization. NGOs, Kole says, are "non-profit, non-state organizations that represent CBOs [community based organizations], people's initiatives and local issue groups (such as squatters and soup kitchens) in a broader movement for social change."

38. Rheingold's chapter discusses the NGOs and other activist networks in parallel with the WELL and the founding of the EFF. On the NGOs' adaptation to such information-networking systems as Bitnet, see Rheingold's section in the same chapter on "Grassroots and Global: CMC Activists." Kole ("Whose Empowerment?") also points out that such "store and forward" networks as Bitnet and Fidonet were especially instrumental in reaching low-tech areas of the world: "In countries where the Internet is not available, people may still be able to use alternative networks such as FidoNet and CompuServe. Such systems usually offer email, newsgroups, electronic conferencing and file transfer facilities. They are based on store-and-forward technology which means that every connection in the network needs to explicitly be established in order to transport data. Users establish a connection ('dial-up') once or a few times a day, to send and receive a batch of *off-line* prepared messages and files."

39. Fortier, *Civil Society Computer Networks*,. chap. 3. Research on the appropriation of information technology by nongovernmental "civil society" organizations and movements—especially in the disciplines of sociology and political history—is well advanced. The two works I primarily draw upon—Fortier's and Kole's—refer in their turn to other studies.

40. I focus here on networks and organizations devoted to supporting NGO issues through technology. However, other networks and organizations, such as the Global Issues Network, have increasingly sophisticated online presences, with the result that the line blurs between the technology-centered organizations I cite and the NGO movements themselves.

41. Subsequent to the writing of this section of my book, other relevant antiglobalization actions have occurred—for example, in Genoa in July 2001. Genoa was particularly significant in my context not just because of its scale but because it marked the first time the mainstream press emphasized the role played by information technology—the networking of the NGOs—in organizing and staging protest actions.

42. IGC, "About IGC Internet" (my emphasis).

43. APC, "The APC Mission" (my emphasis).

44. From placard at top of D2KLA home page, 6 July 2000.

45. Headline of the Direct Action Network home page, 6 July 2000.

46. Banner at top of Philadelphia Direct Action Group "Call to Action for Domestic and Global Justice" page, 6 July 2000.

47. Privacy International home page; APC, "Online Events," 4 July 2000.

48. The D2KLA home page on 6 July 2000 carried the former motto in a placard at the top and the latter in a headline further down the page.

49. Direct Action Network, "Joint DNC/RNC Call to Action."

50. APC, "Managing Your NGO."

51. IGC home page, 24 May 2000; APC home page, 24 May 2000.

52. SVTC, "Responsible Technology Goes Global!!"

53. On the controversy in environmentalist circles over some unscrupulous "electronics recyclers" who are really "junk exporters" shipping high-tech equipment and toxic

waste to China, India, Pakistan, or the Philippines, see *Chronicle of Higher Education,* "Old Computers Never Die—They Just Cost Colleges Money in New Ways" (14 February 2003). The article cites among other sources the SVTC and another watchdog group, the Seattle-based Basel Action Network.

54. This is an occasion to express regret that much of my book was written before I had a chance to read two expansively conceived books on the nature of "information" that appeared in 1999. My argument would have benefited from close dialogue with these works.

One is N. Katherine Hayles's *How We Became Posthuman: Virtual Bodies in Cybernetics, Literature, and Informatics.* The following passage from the work, for example, is powerful not only in Hayles's argument—about the evolution of a "disembodied" paradigm of information in cybernetics, systems theory, and artificial life research—but in mine about the evolution of information technology in the postindustrial workplace and other quarters of life influenced by the workplace:

> I view the present moment as a critical juncture when interventions might be made to keep disembodiment from being rewritten, once again, into prevailing concepts of subjectivity. I see the deconstruction of the liberal humanist subject as an opportunity to put back into the picture the flesh that continues to be erased in contemporary discussions about cybernetic subjects. Hence my focus on how information lost its body. . . . If my nightmare is a culture inhabited by posthumans who regard their bodies as fashion accessories rather than the ground of being, my dream is a version of the posthuman that embraces the possibilities of information technologies without being seduced by fantasies of unlimited power and disembodied immortality, that recognizes and celebrates finitude as a condition of human being, and that understands human life is embedded in a material world of great complexity, one on which we depend for our continued survival." (p. 5)

The argument I make at present about ergonomics is just the tip of the iceberg—the tip, that is, of what I call the informatic "ice ages—of a much larger argument that could be made about the way embodied experience serves as the crucial witness of the human being as *worker,* not only in Ford and Taylor's day but also in the postindustrial age when it is the new corporations that are "deconstructing" (i.e., reincorporating) the liberal humanist subject through team-based, networked, "flat" restructurings. Knowledge workers in cubicles know in their bones, especially their carpal bones, that information technology does *not* grant "unlimited power and disembodied immortality." Nevertheless, they are "seduced by fantasies" of freedom. The dominant, everyday form of that fantasy is "cool," sustained in part, as I argued in my chapter on "Information is Style," on the ability to treat designer information, a stand-in or avatar for one's own body, precisely as what Hayles calls "fashion accessories." (See also Zuboff's attention to the working body in her *In the Age of the Smart Machine,* especially the sequence of chapters on "The Laboring Body: Suffering and Skill in Production Work," "The Abstraction of Industrial Work," "The White-Collar Body in History," and "Office Technology as Exile and Integration" [which contains a section entitled "The Clerk as a Laboring Body"]. The last-mentioned section addresses specifically the issue of what since the mainframe age of the 1950s and 1960s [but especially in recent decades] has become known as ergonomics: "It is small wonder, then, that so many of these clerks' complaints about the work became complaints about bodily suffering" [p. 141].)

The other book of 1999 I refer to is Albert Borgmann's *Holding On to Reality: The Nature of Information at the Turn of the Millennium.* The full curve of "information" that Borgmann traces along its prehistorical, historical, semiotic, social, cultural, and philosophical coordinates has proven very thought-provoking in my undergraduate and graduate courses on information culture. Especially resonant in my present discussion of the ergonomic or bodily experience of information technology—and well worth being placed in conjunction with Hayles's book—are Borgmann's discussions of the "ancestral environment "of information (chap. 3), virtual experience (chap. 14), and the "ethical" balance between "signs and things" (e.g., his conclusion on "Information and Reality"). One powerful passage from the chapter on virtual experience launched a memorably rich discussion in a graduate seminar I taught in 2000 on "The Culture of Information":

> As for death, Tom Mandel, the *New York Times* tells us, was "one of the first (if not the first) to share on-line, with a wide audience, his own experience of dying." Actually to share a person's mortal illness is to feed, clean, and change that person, to suffer the person's bursts of anger and flights of hallucination. It is to see a person suffer deeply and decay. It is to sleep irregularly and poorly and to feel confined and at times resentful. With all that it can be an occasion of grace and gratitude. In any event it is quite different from checking your e-mail when you are good and ready, to catch up on the progress of the disease, to take in the sentiments of others, to contribute one that reads, "Oh, Tom . . . Damn, damn, damn, damn . . . (Do I get TOS [terms of service violation] for that?) Sweetie . . . I am so sorry and I am so amazed that you can just get on here and blurt it out," and then to log off and go about your daily life. (pp. 191–92; ellipses in original)

One senses behind the deep irony, and deeper wisdom, of this passage the gravity of some profoundly consuming, distracting, various, messy, numbing, soul-killing, and soul-kindling, too, experience of taking care of a dying person compared with which computer "multitasking"—a mere overlay of "windows" upon experience—seems all too clean, tidy, logical, and *not* or *post*-human. If I may add to Borgmann's passage a datum from such a life, or death, experience that much touched me as a youth: one remembers the punched-out hole in the plasterboard of the stairwell in one's house where the poles of the stretcher, carried down by the men in the last extremis, hit the wall by mistake. There it remained for years, a typo on the great blank page of existence that no backspace key of virtuality could erase. Or if it could, there were all too human reasons—a messy blend of the need to witness and the needs of ongoing, distracting life—that deferred the repair. (See also Borgmann and Hayles, "An Interview/Dialogue with Albert Borgmann and N. Katherine Hayles on Humans and Machines," where Borgmann comments, "One thing Katherine and I agree on is that humans are essentially embodied and therefore cannot escape their bodies no matter how or what they think of themselves.")

Of course, thoughts like these may make the pain of carpal tunnel syndrome seem trivial, if not profane. But my book is about the confrontation of human beings not with the cruelties of nature but with those of society—an encounter whose staging is one meaning of "culture." On this representational stage of culture, the tiny, exquisite pains of the nerves tunneling past the carpus bones—just one link up from the "digital" fingertips closer to the central body mass—do very much matter. They matter because of their accumulated weight: a kind of vast, distributed, or store-and-forward network of ache through all the hours, days, and years of everyday life. They matter, too because they *are* representational or symbolic. They are an amalgam of the bodily real and mentally imagined. If the reality of work in the age of the "virtual corporation" is "knowledge," after all, why should not grievances be filed on the grounds of "imagined" pain alone (if the corporate lawyers were to be granted their characterization of it)? Besides, the human truth of pain (as a compadre in graduate school once expressed to me in somewhat different terms) is that any one touch of it is a synapse connected to—and capable of retriggering—others stored in the huge, ungainly, badly wired, constantly short-circuited, bruised memory chip of life. Like striking a gong of hurt that releases pent-up echoes somehow louder than the first sound. It may be that human pain, and the need for communication it motivates, is the archetype of posthuman networking.

55. *Business Week,* "Someone to Watch over You," 10 July 2000. For other examples of the voluminous coverage by the mainstream press of workplace privacy and monitoring, see *Los Angeles Times,* "Staying in Line, Online" (12 October 1997); *Business Week,* "Workers, Surf at Your Own Risk" (12 June 2000).

56. Rosen, *Unwanted Gaze,* p. 80. For statistics on electronic monitoring of the workplace, see ibid., p. 57.

57. Vogel, "Walls Have Eyes," Web pages 1 and 2.

58. For a fuller list of perspectives on workplace privacy, see the *FindLaw Cyberspace Law Center* Web page on "Workplace Privacy." See also the resources created by industry, consulting, or training groups to help corporations manage workplace privacy—for example, the Electronic Messaging Association "Email Privacy Policy Tookit" (available for purchase from the association's home page) or the Fair Measures Corporation's "Internet/Privacy" Web page.

59. On Amy Dean as "an innovative leader of what she calls the 'new labor movement,'" see *San Francisco Chronicle,* "Joe Hill Meets the Microchip," 13 July 1997. On the unions' focus on adapting to the "new economy," see Philip Kelly, "Union Organizing in the New Economy."

60. Bacon, "Organizing Silicon Valley's High Tech Workers." This article appeared on the Web sites of both the Institute for Global Communications (IGC) and *Corporate Watch.*

61. On the "invisible workforce" in Silicon Valley, see Gross, "Silicon Valley," included in the 10 June 1993 issue of the CPSR's *CPU.* See also, in the same issue, Bacon, "Silicon Valley on Strike!" reproduced from a January 1993 issue of the San Francisco *Bay Guardian.* (The article shares some paragraphs and material with Bacon's "Organizing Silicon Valley's High Tech Workers.")

62. On the permatemps and Microsoft, see Philip Kelly, "Union Organizing in the New Economy"; *Seattle Times,* "Microsoft Temps Group Joins Union," 4 June 1999; and the Web site of the Washington Alliance of Technology Workers titled *WashTech: A Voice for the Digital Workforce.* On the general issue of high-tech permatemps, see the PBS *NewsHour* transcript of 10 May 2000 titled "Down in the Valley." For the NGO view of such matters, I consulted in July 2000 the forum on LaborNet titled "Labor.tech: Discussion of Technology Organizing and Issues for Labor Activists."

63. Mander, "Net Loss of the Computer Revolution."

64. CPSR, "Terry Winograd's Thoughts on CPSR's Mission."

65. The exceptions I have in mind include the CPSR working group on Computers in the Workplace, whose Web page "Computers, Work and the Workplace" gathers resources on workplace privacy, labor organization, and workplace health. So, too, there is the CPSR working group on Computers and the Environment.

66. As of the date of this writing, I have not found on the PFF site any discussion of the NGO cyber-political issues. This is not an absolute guarantee of omission (since in Web-based research it is difficult to "read everything"). But it is a strong indication that such issues are not on the PFF front burner.

Winner has written strongly about the procorporate stance of the PFF's "Magna Carta." See, for example, his "Cyberlibertarian Myths": "Characteristic of this way of thinking is a tendency to conflate the activities of freedom seeking individuals with the operations of enormous, profit seeking business firms. In the 'Magna Carta for the Knowledge Age,' concepts of rights, freedoms, access, and ownership justified as appropriate to individuals are marshaled to support the machinations of enormous transnational firms. We must recognize, the manifesto argues, that 'Government does not own cyberspace, the people do.' One might read this as a suggestion that cyberspace is a commons in which people have shared rights and responsibilities. But that is definitely not where the writers carry their reasoning.

"What 'ownership by the people' means, the Magna Carta insists, is simply 'private ownership.' And it eventually becomes clear that the private entities they have in mind are actually large, transnational business firms, especially those in communications."

67. Responding to the 1997 survey initiated by *Wired* in conjunction with the Merrill Lynch Forum (without Katz's advance knowledge), Katz's "Netizen" article observed: "But among the survey's many powerful findings, one in particular caught me by surprise: where I had described them as deeply estranged from mainstream politics, the poll revealed that they are actually highly participatory and view our existing political system positively, even patriotically."

68. That the page by "Lizard" is "blacked out" (the background color is set to black) is part of the virtual street theater in which many other Web pages at the time—a sort of "thousand points of darkness"—also protested the CDA. His page is blacked out, "Lizard" says, "in memory of the First Amendment to the Constitution of the United States. (1789–1996) RIP."

69. De Certeau, *Practice of Everyday Life,* pp. 25–26.

70. Carlsson et al., "Some History of *Processed World.*"

71. Carlsson with Leger, ed., *Bad Attitude,* p. 7. Subsequent citations of this work are given parenthetically in the text.

72. I quote from the reprint of the article in *Bad Attitude,* but the article can also be read online. (See my Works Cited, under "Digit, Gidget.")

73. Floppy disks—pliable, thin-skinned, and lacking a protective shutter over their writing aperture—were the standard storage media for personal computers before hard-shelled diskettes and later media. For a theoretical analysis of the kind of tactics recommended by the *Processed World* collective, see the chapter "Slacker Luddites" in Critical Art Ensemble, *Electronic Civil Disobedience.*

74. See also the interview in *Bad Attitude* titled "CLODO Speaks" (pp. 167–70).

Part Four Preface

1. For examples of psychological or psychoanalytic approaches to information technology, see Turkle, *Life on the Screen;* and Žižek, "From Virtual Reality to the Virtualization of Reality."

2. See, for example, the special issue on "Sexuality and Cyberspace" in *Women and Performance: A Journal of Feminist Theory* 9, issue 17 (1996); Kolko et al., eds., *Race in Cyberspace;* Nakamura, *Cybertypes;* the essays in the section "Performing Identity in Cyberspace" in Trend, ed., *Reading Digital Culture;* and the essays in the sections "Cyberfeminisms," "Cybersexual," and "Cybercolonization" in Bell and Kennedy, eds., *Cybercultures Reader.*

3. See, for example, Borgmann's chapter titled "Ancestral Information."

4. I am thinking in particular of the image of the globe that for many years in the 1990s gave *CNN Headline News* its visual identity.

5. *Information Age,* 1:379.

6. My thanks to Rita Raley, whose studies of the history and theory of "global English" and global information (as well as conversation over the years) have helped educate me about the global dimensions of information culture that lie outside the scope of my project. Particularly relevant are Raley's work in progress titled "Global English and the Academy" and her essays on "Machine Translation and Global English" and "eEmpires."

7. Underlying this way of formulating the question is the premise that there can be—and, indeed, is—more than one dominant idea of the nature of global competition. The U.S. version, on which I standardize here, is thus to be set alongside the European and Asian versions. Relevant is Castells's insight that such notions as postindustrialism and global business tend to flatten out the variety of different permutations of the same phenomena around the world. See, for example, his analysis of Japan and the United States as representing two different paradigms of postindustrialism and informationalism, even though the "predominant model" of postindustrialism is based on the United States (*Information Age,* 1:217–31).

8. Danesi, *Cool: The Signs and Meanings of Adolescence,* p. 37. But see also MacAdams, *Birth of the Cool,* pp. 14–19, for a more detailed, complex tracing of the genesis of "cool" in American usage.

Chapter Nine

1. Clastres, *Society against the State,* p. 11. Subsequent citations of this work are given parenthetically in the text.

2. See also Clastres, *Chronicle of the Guayaki Indians,* pp. 107–8, on nonauthoritarian political "power" among "primitive societies."

3. I speak in the present tense of the Guayaki, among whom Clastres lived in 1963 and 1964. But as his *Chronicle of the Guayaki Indians* makes tragically clear, the Guayaki are now an extinct people (p. 345; see also p. 13).

4. Thompson's study of cool in West Africa is especially suggestive in relation to my current argument, which compares contemporary cool to another tribal society. In a statement that might apply both to the Guayaki men singing by their fire and to knowledge workers producing or using cool information in their cubicles, Thompson writes that cool describes "ordinary lives raised to the level of idealized chieftaincy" ("Aesthetic of the Cool," p. 42).

5. Compare Raymond Williams on the "residual" (*Marxism and Literature,* pp. 121–27). Williams draws a distinction between "archaic" and "residual" (p. 122): "I would call the 'archaic' that which is wholly recognized as an element of the past, to be observed, to be examined, or even on occasion to be consciously 'revived,' in a deliberately specializing way. What I mean by the 'residual' is very different. The residual, by definition, has been effectively formed in the past, but it is still active in the cultural process . . . as an effective element of the present. Thus certain experiences, meanings, and values which cannot be expressed or substantially verified in terms of the dominant culture, are nevertheless lived and practiced on the basis of the residue—cultural as well as social–of some previous social and cultural institution or formation." Since I will not only use *archaic* in an anthropological sense that gives it deeper, prehistorical meaning but also analogize that sense to an oppositional "slack" that is effective in the present, I synonymize *archaic* and *residual.* My argument is that the Guayaki exhibit what may be termed an "archaic residualism" (a prehistorical version of the active residualism Williams defines).

6. The often rapid migration of seasoned programmers and other skilled workers among high-tech firms, for instance, is an extreme illustration that not just layoff- but employee-initiated "turnover" is part of the systemic flexibility and decentralization of the New Economy.

7. In my argument here, both the Guayaki and the cool are "technicians" in a manner that spans the divide Lévi-Strauss hypothesized between primitive "bricoleurs" and modern "engineers" (*Savage Mind,* pp. 16–22).

8. See Clastres's drawings of archery positions or details of the manufacture of bows and arrows in his *Chronicle of the Guayaki Indians* (pp. 34, 280). However, with the excep-

tion of such discussions as that of coati hunting (pp. 198–99), the *Chronicle* is mostly devoid of attention to exact hunting technique, an omission that Clastres accounts for as follows: "the Indians were so agile, so skilled, their gestures so precise and efficient—this was total mastery of the body. But just because of that, and because I could not match them, they were not very eager for my company when they went hunting" (p. 199).

9. In his *Chronicle of the Guayaki Indians,* however, Clastres notes that hunting, no matter how efficient and tightly scripted, is also always more than production: "Is this to say that the men in this society are in some sense victims of economic alienation because they are completely identified with their function as 'producers'? . . . Not at all. Hunting is never considered a burden. Even though it is almost the exclusive occupation of the men, the most important thing they do every day, it is still practiced as a 'sport.' . . . Hunting is always an adventure, sometimes a risky one, but constantly inspiring" (pp. 281–83).

10. See chapter 1, above, p. 39.

11. See chapter 4, above, p. 172. My idiom here of an intimate coupling between technology and technique is influenced by Donna Haraway's discussion of cyborg "connections" and "couplings"—for example, cyborgs "are wary of holism, but needy for connection" (p. 151), and "cyborgs signal disturbingly and pleasurably tight coupling" (p. 152).

12. This is especially evident with the use of a traceroute program that visualizes Internet connections geographically. I use the NeoTrace program from NeoWorx, Inc.

13. See note 5, above, on Williams, the archaic, and the residual. In her response to a portion of this book delivered as a talk at the Doreen B. Townsend Center of the University of California, Berkeley, in 1998, Miryam Sas makes a beautiful, haunting argument about the complex relation between literature and "speed" that is germane to my discussion. Meditating on Baudelaire, Sas writes: "The present moment has two sides: it is both an immediate and effective visible act and the result of a long invisible past, in fact, of repetition or rehearsal in the invisible past. For the writer who has a successful debut, Baudelaire claims, 'every beginning was always preceded and it is the effect of twenty other debuts that others have not known.' Again, he writes, 'In order to write quickly, it is necessary to have already thought a good deal. . . .' The speed of writing reflects back on a long habituation, a constant companionship; the immediacy reveals traces of a silent past. . . . In other words, literary works and old advice books by poets have already had much to say about the kinds of erasures of history that occur within a world of change and information—speed" (p. 37). The eventual question that might be asked within my context is thus: what is the relationship between the peculiar kind of deep slack that literature expresses in its accommodation to industrial or postindustrial speed, and the more popular slack of cool?

14. One might think here of Thompson's identification of cool in West African cultures with the ancestral: "There is more than a symbolic reconciliation of the living with the dead. This is an aesthetic activation, turning ancient objects of thought into fresh sources of guidance and illumination. . . . Black cool [is] antiquity, for as Ralph Ellison has put it, 'We were older than they, in the sense of what it took to live in the world with others'" ("Aesthetic of the Cool," p. 67).

15. A characteristic argument is thus as follows: "Remember, Xers have a different relationship to the mechanisms of the information revolution than older generations. We didn't have to get used to integrating information technology into our work habits. Information technology shapes the way we learn. What looks to many managers like a short attention span is, in fact, a rapid-fire style of information consumption, which makes Xers uniquely suited to the workplace of the future" (Tulgan, *Managing Generation X,* p. 31).

16. Strassmann, *Information Payoff,* pp. 13, 21, 251–52 n. 10. More recently, Brown and Duguid, in *Social Life of Information,* have explored the informal "social" dimension of

information work. Too myopic a focus on information technology itself, they comment, "inevitably pushes aside all the fuzzy stuff that lies around the edges—context, background, history, common knowledge, social resources. But this stuff around the edges is not as irrelevant as it may seem. It provides valuable balance and perspective. It holds alternatives, offers breadth of vision, and indicates choices. It helps clarify purpose and support meaning. Indeed, ultimately it is only with the help of what lies beyond it that any sense can be made of the information that absorbs so much attention" (pp. 1–2). Compare Rochlin, *Trapped in the Net,* on "slack," discussed below.

17. Rochlin, *Trapped in the Net,* pp. 127, 213.

Chapter Ten

1. Axelrod, *Elizabeth I CEO,* p. x. The book encompasses quite a bit of historical context and detail in its effort to present "concise narrative examples of leadership in crisis and triumph" using "the queen's own words whenever possible" (p. 19). The book also includes a timeline of the Elizabethan era and a bibliography of historical and biographical sources.

2. Musashi's book was first translated into English in 1974, just in time to catch the wave of American interest in Japanese management techniques. It became an allegory of the samurai as postindustrial knowledge worker. One adaptation/updating of Musashi, for example, was written by Donald G. Krause under the title *The Book of Five Rings for Executives: Musashi's Classic Book of Competitive Tactics.*

3. I refer to the "desktop themes" titled "Leonardo da Vinci" and "The 60's USA," available for download for Windows 98 from the Microsoft update site.

4. See Baudrillard on history as simulation in "Year 2000 Has Already Happened"; Jameson on history as pastiche, *Postmodernism,* pp. 16–25; and Jencks on historicism and "radical eclecticism" in postmodern architecture, *Architecture Today,* pp. 112–41.

5. I have previously discussed the nature of temporality in relation to historicity in my *Wordsworth: The Sense of History,* pp. 55–59, 160–66. On Napoleon's preoccupation with his watch during battle, see ibid., pp. 402–4. One of the best known statements of the "clockmaker" God thesis was William Paley's *Natural Theology; or, Evidences of the Existence and Attributes of the Deity.*

6. I borrow the concepts of "illud tempus" and "liminality" respectively from Mircea Eliade, *Sacred and the Profane,* and Victor Turner, *Ritual Process.*

7. On the New Historicism, see my "Power of Formalism." In regard to Geertz, my specific allusion is to his essay "Deep Play: Notes on the Balinese Cockfight."

8. Faulkner, "The Bear," in *Go Down, Moses,* pp. 206–8. I am grateful to colleagues and graduate students in my English Department at the University of California, Santa Barbara, who helped me think about Faulkner in this context during an e-mail discussion in September 2000.

9. In the memorable phrase of the Critical Art Ensemble, the cool are "slacker Luddites" who have a complicated hate/love relation with technology: "Quite commonly, a slacker Luddite who hates to slave on h/er computer at work returns home only to sit at the computer again, to desktop publish h/er own magazine. This situation is the opposite of originary Luddism. The slacker Luddite shuns or destroys technology not because of a hatred or fear of it, but because of a hatred for work, while originary Luddites were accustomed to work, but hated and feared the technology. Slacker Luddism is a late capital hybrid, a perfect example of recombinant culture. It synthesizes the tactics of originary Luddism with the zero work ethic of contemporary slackers" (*Electronic Civil Disobedience,* p. 62).

10. See also my discussion of "wit" in chapter 7. On "Fordization of the face," see chapter 2.

11. Again, the chapter "Slacker Luddites" in the Critical Art Ensemble's *Electronic Civil Disobedience* is perceptive: "Implied in the above is another important distinction between Luddites and their apparent descendants: The slacker Luddite is a narcissist. This is not meant in a pejorative way, as they have little choice in the matter. Unlike their predecessors, the slacker Luddites have no sense of everyday life community in the workplace. The dividing of labor into micro-specializations has disrupted this possibility. Electronic salons, though a point of fascination, hardly replace the sedentary and organic interrelationships lost in the economy of late capital. Desirable living conditions are consequently measured by personal pleasure, rather than by contribution to a community" (pp. 62–63).

12. Though much of McSherry's evidence comes from interviews with the academic science community, her issues and conclusions are also germane to the relation of the humanities to private enterprise knowledge work.

13. First published as an essay and later revised as a talk entitled "Some Observations on the Difference between Lay and Professional Reading" at the University of California, Santa Barbara (which I heard in October 2000), Guillory's "Ethical Practice of Modernity" draws on the late Foucault to develop a notion of "ethics" related broadly to my concern with the "ethos" of the cool. Guillory writes (citing from the published version): "let us suppose that conduct can be arrayed along a spectrum extending from morality at one end, the choice between right and wrong, and the aesthetic at the other, the choice among objects of beauty. The ethical then would occupy a terrain in the middle of this spectrum as the choice between goods. This usage would conform more or less to the philosophical problematic descending from Socrates of 'how to live,' which is a question not reducible to adherence to moral law or 'obligation'" (p. 38). Guillory then applies this mediating notion of ethics, or ethos, to the practice of reading: "I would like to propose now that reading belongs to the field of the ethical because it is a practice on the self, and because the motive of pleasure in reading contains within it the potentiality for what was known in the early modern period as 'self-improvement.' . . . Reading is the *principal* ethical practice of modernity, the site where a practice of the self has not been entirely or easily subordinated to the moral code, or rendered solely an instrument of power/knowledge" (p. 39). Reading as ethical practice in this sense, Guillory concludes, is precariously positioned between professional and lay reading practices, both of which tend to turn it to other purposes than the primary issue of ethos: "how to live."

In my argument, it is "cool" that promises—but has yet to realize—what amounts to an ethical practice of "reading." Cool and its included practices (not just reading, but seeing, browsing, and so on) mediate between the dictums of the new morality (corporate culture) and the new aesthetics.

14. A famous example, in light of the usually ascetic nature of his criticism, is Paul de Man's reference to Archie Bunker of TV's *All in the Family* series in "Semiology and Rhetoric," p. 9. My involvement in the University of California, Santa Barbara, English Department's Public Humanities Initiative, for which L. O. Aranye Fradenburg and I organized a conference on "Entertainment Value" (3–4 May 2002), has sensitized me to the general importance of exploring the relationship between the contemporary institutions of education and entertainment. Consider, for example, the prominent role played by the film industry in educating the public about history. Once it was literature that "delighted" (as in Horace's dictum that the aim of poetry is to "inform or delight"). Now, even when the subject is tragic history (e.g., *Braveheart, Amistad, Saving Private Ryan*), it is film that "entertains." For fuller speculations on these issues, see the "Entertainment Value" Web site, which includes an online essay by Fradenburg on "Entertainment Value."

15. How one wishes, by analogy, that the penmanship and typing classes one was taught as a youth might be subsumed within a course on those subjects taught, for example, by Friedrich Kittler.

16. *Secondary orality* is a term I take from Ong, *Orality and Literacy*, pp. 136–38.

17. Gibson, *Burning Chrome*, p. 186.

18. William Gibson and Bruce Sterling's novel *The Difference Engine* is particularly interesting in this context because it is a historical *roman à clef* that tries to consolidate antique technologies with contemporary high tech in a single, vividly imagined alternative Victorian world—one in which steam-driven computers based on Charles Babbage's designs are the dominant industrial "engine" and where our hero, personifying the vitality of the archaic within the contemporary, is a paleontologist.

19. On orientalism in cyberculture and technoculture, see Chun, "Orienting Orientalism, or How to Map Cyberspace"; Ueno, "Japanimation and Techno-Orientalism"; and the chapter on "Techno-Orientalism: Japan Panic" in Morley and Robins's *Spaces of Identity*. See also Warner, "Media Determinism and Media Freedom after the Digital Mutation," on the cyberpunk motif of "kung-fu" hacking. On orientalism in the related context of graphic and techno design, see above, chapter 6, note 27.

20. Suvin, "On Gibson and Cyberpunk SF," p. 353.

21. In his "Media Determinism and Media Freedom after the Digital Mutation," Warner discusses the actions of the hero of *The Matrix* in a way that is germane to my thesis here of the recombinant survival of the archaic within high tech. He writes: "In this film Neo's *hacking of the matrix is accomplished with Kung-fu* and success depends upon learning to have faith in his body. (This translation of a software war into kung-fu has many effects: it means that being tough, hard, and cool—wearing sleek sun glasses and bad black clothes, and having excellent muscle tone—becomes a sign of inner spiritual strength. . . . But I think there is another reason to translate the hacker into an action hero. For humans menaced with a bad engulfment by digitalization, these very physical battles, with the wounds they entail, become a way to reinvest the human body with centrality. However, these bodies are also enhanced with the good effects of digitalization so they are light, fast, and unencumbered by the weight of reality and its laws. This kind of body becomes a testament to a bold new freedom. Looked at within the long history of films that worry the takeover of culture by media, *The Matrix* intensifies both the determinism of media *and* the freedom claimed for the media subject.)"

22. Carl Stahmer, Rita Raley, and Jennifer Jones assisted me in the maintenance and expansion of the original *VoS* site, and other graduate and undergraduate students at my institution have at times assisted. Robert Adlington and Jeremy Douglass redesigned *VoS* as a database-driven site in 2001. See my "Globalizing the Humanities" for further discussion of the history and philosophy of the site.

23. Rita Raley, Carl Stahmer, and Vincent Willoughby assisted in the origin of the *Romantic Chronology*; and Shawn Mummert later assisted in moving resources into a new database-driven site I designed in 1999. Noah Comet also later helped maintain and improve the site. The site has an Editorial Board that has contributed material. My special thanks go to my co-editor, Laura Mandell, who has worked to extend, amend, and recruit editorial content as well as to promote pedagogical use of the chronology even when my own attention was diverted to other technical projects.

24. Contributions to *Palinurus* have been made by Dan Sarel and others.

25. As always in this book, my use of *throwaway* alludes to Watkins's *Throwaways*.

26. Also participating in planning for the project at one point were my colleagues Charles Bazerman and Mark Rose. Since its first three-year development cycle, *Transcriptions* has evolved into a second stage marked by the undergraduate specialization in Literature and the Culture of Information that it oversees in the University of California, Santa Barbara, English Department and its close association with the University of California

Digital Cultures Project, a multi-campus research initiative headquartered at Santa Barbara and directed by William Warner.

27. *Cathedral De Compostela,* Web site created by Laura Fabrick, Geri Ferguson, and Christina Valadez for my undergraduate course in 1999, The Culture of Information: <http://transcriptions.english.ucsb.edu/archive/topics/infoart/cathedral/>

28. This may be the place, however, also to acknowledge that the projects I mention have not always been even in their execution. This was due to a variety of factors in addition to thin or uncertain funding, including the inadequate fit between the way the humanities are conducted in the academy and the way collaborative projects in the sciences or other fields function. Humanities departments, for example, are not characteristically set up to support projects with course relief for supervisors or grant-writing assistance, and grant getting and project participation do not "count" in any regular way in promotion reviews. For a discussion of collaboration in humanities computing consonant with mine, see Martha Nell Smith, "Computing."

29. Christensen, *Romanticism at the End of History,* pp. 188–90.

Chapter Eleven

1. On history versus *la mode rétro,* see Jameson, *Postmodernism,* pp. 16–25.

2. This line of thought has been particularly shaped by my contact in the past few years with such scholars and colleagues as Anna Everett, Lisa Parks, and Constance Penley in the University of California, Santa Barbara, Films Studies Department, and Wendy Chun in Brown University's Modern Culture and Media Department.

3. My argument for an alliance between the humanities and the arts is influenced by my contacts and collaborations over the years with the vibrant community of digital artists in the University of California system and nearby colleges, including Mark Bartlett, Sharon Daniel, George Legrady, Lev Manovich, Robert Nideffer, and Victoria Vesna. Especially important has been the formation of the UC Digital Cultures Project, directed by William Warner; the UC Digital Arts Research Network (DARnet), consisting of a collective of UC digital artists; and the UC Santa Barbara Media Arts and Technology Program Lecture Series, organized by George Legrady. The lecture series has provided an organized framework in which to encounter digital artists not just from California but from around the world. In addition, the move of the Electronic Literature Organization's headquarters to the University of California, Los Angeles, in 2001 had the complementary effect of deepening my familiarity with the scholars and practitioners of digital creative writing (an area of work long fostered by UC Digital Cultures Project member N. Katherine Hayles of UCLA). The following initiatives in digital technology at the University of California, Santa Barbara, have also contributed to my thinking: the Microcosms Project, directed by Mark Meadow and Bruce Robertson, and the Center for Information Technology and Society, directed by Bruce Bimber. I am also grateful to Amr El Abbadi of the UC Santa Barbara Computer Science Dept. and Janet Head of Texas Instruments, Inc., for personal friendship and many leisurely talks about computing (and literature, too) over the years.

4. See for example the 3D renderings of architecture on the *Great Buildings Online* site or the 3D reconstructions of archaeological sites and buildings created by the UCLA Cultural VR Lab. (I am grateful to Jackie Spafford, visual resources curator at the UC Santa Barbara History of Art and Architecture Dept., for help with the latter reference.)

5. A CAVE is an immersive virtual reality room in which stereoscopic digital images are projected on the walls. In the specific sense I invoke here, a *bot* is a program with artificial intelligence attributes that simulates conversation or other interaction with a human being. A classic bot of this sort is ELIZA, which simulated conversation with a psychoanalyst (on ELIZA and other bots, see Turkle, *Life on the Screen,* pp. 77–124).

6. On the Ivanhoe Game, see McGann and Drucker, "Ivanhoe Game." (I have also benefitted from McGann's paper "Narrative, Game, and Performative Poetics," which discusses the game). For the *Villa Diodati* MOO, which uses the originally gamelike attributes of MUDs and MOOs for educational purposes, see the *Romantic Circles* Web site. The *Frankenstein MOO,* created by Broglio and Sonstroem, is also part of *Romantic Circles* (a "Learning Module" in the *Romantic Circles High School* initiative). MUDs were originally text-based, interactive, role-playing, adventure-gaming environments in which players navigated through a fictional landscape that allowed them to perform actions and converse with each other. MOOs are the more socially or educationally oriented descendants of MUDs (there are also technical differences).

7. In my view the issue of using digital technology in instruction is separate from that of distance education. The confusion of these two issues has enabled the sweeping, anti-technology stance of some critics of the instructional use of information technology, perhaps most notably David F. Noble (though his broader perspective on the assimilation of education within postindustrialism is consonant with my own critique). (See, for example, Noble's "Digital Diploma Mills.") Such criticism is often targeted more at the economic and other assumptions of distance education alleged to be the primary motive of administrators and legislators than at the technology itself, which seldom comes in for informed or detailed scrutiny. The kinds of experimental instructional technology I am advocating here—sometimes served remotely to students, but often or even primarily used in the physical classroom with the instructor present—will in the immediate future likely be very expensive in time and cost. They are no boon to those who subscribe to "efficiency," "throughput," or "commodification" in education. But investment in such experimentation will be necessary to bring us to the point where appropriate criticism of instructional IT can begin: with comparative judgments of which technologies do and do not show promise in improving education within the contemporary context of competing media and other social influences. (See also the critical questions and resources gathered on my "Classroom of the Future" Web page.)

8. Thinking of the audience of *Ray Gun* or his other magazines, David Carson observes, "I believe now, if the type is invisible, so is your article, and it's probably not going to get read, because—at least with this audience, and I think it's spreading out more—they're seeing better TV, they're watching video screens. You give somebody a solid page of grey type and say, 'Read this brilliant story,' and a lot of people, they're going to go, 'Doesn't look very interesting. Let's try and find something more interesting'" (quoted in Cloninger, *Fresh Styles for Web Designers,* p. 53). On Carson's design, see chapter 6, above.

9. Horace, "Art of Poetry," p. 72; Sidney, *Apology for Poetry,* p. 146.

10. Moreover, such criticism by text-based scholars in regard to multimedia and browsing is untrue even in regard to "text," whose status as in part a specifically graphical medium was highly charged not just at the time of its emergence from oral culture but again at the beginning of the twentieth century when the "typographical" and "graphic design" arts answered the need to adapt text to the modern graphical and media-oriented world. Mitchell Stephens's work *The Rise of the Image, The Fall of the Word* is relevant to my argument here, as is the work of Walter Ong on the emergence of writing from oral culture and of Johanna Drucker on early twentieth-century typographical art.

11. A nice case study is presented by Steven Johnson in his analysis of the canny prose of the *Suck* online column, which uses linking not to "augment" the reading experience but to alter the rhythm and texture of reading with incisive wit. Johnson comments, "The rest of the Web saw hypertext as an electrified table of contents, or a supply of steroid-addled footnotes. The Sucksters saw it as a way of phrasing a thought" (*Interface Culture,* p. 133; see also pp. 130–37).

12. For images of the Transcriptions studio, see <http://transcriptions.english.ucsb.edu/resources/guides/tech/tech_facilities.asp>

I originally saw the UC Santa Barbara Art Studio Department's electronic studio when it was in the charge of Victoria Vesna, then the campus's digital artist faculty member. I am grateful to Vesna for good talk and collaboration over the years, even after she moved to head up the UCLA Design/Media Arts Department.

13. See, for example, Robert Nideffer's *Proxy* project (agent-based and peer-to-peer); Victoria Vesna's *n 0 time* project (agent-based and peer-to-peer, with screensaver front end); and Sharon Daniel, Mark Bartlett, and Puragra Guhathakurta's *Subtract the Sky* project (collaborative database system).

14. The UC Santa Barbara English Department's Public Humanities Initiative was a series of events in 2000–2002 co-organized by L. O. Aranye Fradenburg and myself that explored the relation between the academic humanities and general society. See <http://www.english.ucsb.edu/initiatives/public-humanities/>

15. See my "Online Resources" page on the Public Humanities Initiative site for listings of initiatives at the intersection of the academic humanities and general society: <http://www.english.ucsb.edu/initiatives/public-humanities/resources/resources-index.asp>

16. This is not to say that the digital creative arts have automatic, adequate, consistent, or unproblematic access to the economically significant part of the "public" (far from it). The point I make here is purely comparative with the humanities.

17. Hayles, *How We Became Posthuman.* Despite Hayles's focus on embodiment, in contrast to my focus on social and cultural history, there is a deep congruence between our two works that may be identified in Hayles's underlying method of study. That method is a historicism so thorough that it not only structures the argument of *How We Became Posthuman* (which proceeds historically through three paradigms of information: as homeostatic system, reflexive or autopoetic system, and virtual emergence), but also figures thematically as the necessary complement to embodiment. As Hayles puts it, her work shows "what had to be elided, suppressed, and forgotten to make information lose its body" (p. 13; see also p. 20). (For a consonant view of history as "denial," see my *Wordsworth: The Sense of History*.) Or again: "By turning the technological determinism of bodiless information, the cyborg, and the posthuman into narratives about the negotiations that took place between particular people at particular times and places, I hope to replace a teleology of disembodiment with historically contingent stories about contests between competing factions, contests whose outcomes were far from obvious" (Hayles, *How We Became Posthuman,* p. 22). The past tense ("became") in Hayles's title is the historical past tense. Gritty, messy, contingent embodiment is the register of historicity, and vice versa.

18. However, it is rapidly becoming clear that some regularized account of the varieties of digital literature and art must soon be given before many of the works become inaccessible due to technological change. I am currently involved in the Electronic Literature Organization's Preservation, Archiving, Dissemination initiative (PAD), part of which involves a technical survey and analysis of the varieties of electronic literature. See the PAD home page and Alan Liu et al., "Excerpt from the Final Technology/Software Report."

19. One index of the popularity of the "patchwork" metaphor in contemporary literary history—as expressed not just in discursive literary histories but also literary anthologies, composition textbooks, and other works that reproduce primary material within a literary-historical framework—is the prevalence of cover illustrations of patchwork quilts (or, less consistently, such conceptual analogues as mosaics, kaleidoscopes, webs, or meshes). See, for example, the covers of the following works: Rico and Mano, eds., *American Mosaic*; Harper and Walton, eds., *Every Shut Eye Ain't Asleep*; Decker and Schwegler, eds., *Decker's*

Patterns of Exposition 13; Scott, ed., *Gender of Modernism*; Perkins and Perkins, eds., *Kaleidoscope;* and Walker, ed., *Graywolf Annual Seven.* Many of these and other recent works of literary history make explicit use of the metaphor of literary history as a "patchwork" in their editorial material as well. Patchwork quilting has also been a notable motif of hypertext fiction, as in such works as Shelley Jackson's *Patchwork Girl* or Deena Larsen's *Samplers.* (In including *creolization* among the terms I use to designate this particular aesthetic effect, I am influenced by Rita Raley, who has used the term to describe some new works of electronic literature; see her "Reveal Codes.")

20. Manovich, *Language of New Media,* especially pp. 218–43, "The Database."

21. As Timothy Allen Jackson put it in "Towards a New Media Aesthetic" (2001): "The time to construct . . . an aesthetics and poetics for new media is now since the ontological shifts of new media are still in progress and no dominant aesthetic model has emerged . . ." (p. 350). Of course, any surmise we can make here of an evolving dominant aesthetic model will need to be tested against future artistic and critical formulations as they appear—including, for example, two important book manuscripts I have recently seen in partial form: Lev Manovich's "Info-Aesthetics: Information and Form" (the sequel to his *Language of New Media*) and Rita Raley's "Transfers: Textuality and the Digital Aesthetic." But in the current *Sturm und Drang* of new media, there is virtue merely in getting an oar into the water to help steer. It is in this spirit that I make my guesses here.

22. The notion of the random and entropic is important in Robinson's *Inquiry into the Picturesque,* which is implicitly a postmodern explanation. See, for example, p. 2: "Both the strength and the vulnerability of the Picturesque stem from its promiscuity. By avoiding a fixed system of rules, picturesque compositions can change and adjust to different conditions. Conversely, by embracing permissive habits in making choices, such compositions run the risk of disappearing into a random background." Or again, on p. 139: "A natural system of compensation as energy flows continuously in search of an entropic conclusion may, at last, underlie the open-ended artifice of the Picturesque."

23. For further discussion of this topic, see my "Art of Extraction: Toward a Cultural History and Aesthetics of XML and Database-Driven Web Sites."

24. In one of its manifestations, for example, Nideffer's *Proxy* project presents itself through a gaming metaphor.

25. In hypothesizing a relation between contemporary digital aesthetics and the older aesthetics of the sublime, I am aided by J. Jennifer Jones's dissertation "Virtual Sublime: Romantic Transcendence and the Transport of the Real," which brings the theory and aesthetics of the sublime into conjunction with the idea of virtual reality. I am grateful to Jones for searching conversations in the past few years regarding the sublime, VR, cyberpunk fiction, Wordsworth, and other topics.

26. Druckrey, with Ars Electronica, eds., *Ars Electronica,* pp. 3–5, 7–8, 11–12.

27. Gamboni, *Destruction of Art,* especially p. 313. Subsequent citations to this work are given parenthetically in the text.

28. As an example of other accounts of iconoclasm in modern art, see the chapter "Anti-Art Gestures in Early Modernism" in Godfrey, *Conceptual Art.*

29. See especially Stiles, "Thresholds of Control," which offers a social and cultural historical explanation of destruction art and cites other male and female destruction artists alongside Metzger. See also Gamboni on Metzger (*Destruction of Art,* pp. 264–65).

30. Stiles and Selz, *Theories and Documents of Contemporary Art,* p. 402.

31. Home, *Assault on Culture.*

32. Quoted in Home, *Assault on Culture.*

33. Stiles, "Selected Comments on Destruction Art."

34. For the purposes of this discussion, my understanding of computer "viruses" is broad. I include *viruses* proper (unauthorized programs that insert themselves via a variety of carriers, including floppy disks, e-mail attachments, and so on into a computer to conduct either innocuous or destructive activity); *worms* (programs whose primary purpose is to use computers to propagate themselves outwards over the net, though they may also have destructive "payloads" on the local computer); *Trojan horses* (destructive programs disguised as something else); *backdoors* (programs that hide on a computer, typically to allow remote access by hackers); and *denial of service attacks* (which do not invade computers per se but instead instigate floods of spurious, automatically generated network traffic intended to block the ability of target computers to serve up information on the Internet). As indicated in my discussion of Joseph Nechvatal's work below, I even include some programs that are not normally considered viral at all but instead derive from cellular automata and artificial life research. For a history and typology of computer viruses, especially from the point of view of the digital art community, see the Web site for the Frankfurt Museum of Applied Arts Digitalcraft exhibition titled *I love you—computer_viruses_hacker_culture.*

35. Blinderman, "Ghost of Electricity." For examples of the drawings from this period, see Nechvatal, *Drawings and Drawing-Based Works from 1980–1987.*

36. Quoted in McCormick, "On the Ecstatic Excess of Joseph Nechvatal."

37. Quoted in Popper, "On Joseph Nechvatal."

38. Joseph Nechvatal, home page. See also his *Ecstasy of Excess,* which reproduces selected works from 1987 to 1991.

39. This description of Nechvatal's working method is based on a personal communication from the artist of 9 June 2003. My thanks to Nechvatal for correspondence about his art.

40. During his residency, Nechvatal worked at the Saline Royale (Fondation Claude-Nicolas Ledoux) computer lab in Arc-et-Senans. For information about Nechvatal's virus projects, see the artist's home page, his online "Biography," and his exhibition catalog, *Joseph Nechvatal.* See also Murphy, "Joseph Nechvatal."

41. See the technical explanation titled "The Model: Notes" by Stéphane Sikora and Joseph Nechvatal on Nechvatal's Web page for Virus Project 2.0. For the code of the cellular automata program in Virus Project 1.0, see Nechvatal, "The Computer Virual Formula." For Nechvatal's theoretical meditation on his "viral" work and his initial cellular automata "virus," see his "Virus Text (We Form a Rhizome with Our Viruses)."

42. *vOluptas 2.0 @ 7.5 min.* was sent to me by the artist in December 2001 as a high-resolution, color photographic reproduction for use in my book. At that time, the work was still conceptually part of Virus Project 2.0. Since then, Nechvatal has gone on to create his new "vOluptuary: an algorithic hermaphornology" series, for which he has written "An Artist's Statement." Created for the exhibit of works from the series at Universal Concepts Unlimited in New York City, 22 May–3 July 2002, the artist's statement explains the visual recombinations performed upon his initial images of human genitalia and intimate body parts in terms of a theory of hermaphroditic recombination inspired by Ovidean myth. Because my discussion of *vOluptas 2.0 @ 7.5 min.* was written earlier, I have not been able fully to integrate the terms of Nechvatal's own interpretation. For example, in retrospect I would have played further variations upon Nechvatal's metaphors of organism and sexuality rather than superimpose my own astronomical metaphors below. However, it is fortuitous that my analysis of the transformations imposed by viral action upon the

pseudo-organic "red balls" in *vOluptas 2.0 @ 7.5 min.* (morphologically related to the bulbous, organic compositional forms in several of the works exhibited at Universal Concepts Unlimited) is essentially homologous with Nechvatal's discussion of sexual transformations.

43. It is unclear from the evidence of the image whether the vertical distortion bands were actually caused by the action of cellular automata or were instead an additional transformational effect (or perhaps part of the original image). Cellular automata programs, in the examples that I myself have run or seen run, either do not produce regular geometries or—when they do produce such geometries—create patterns more symmetrical and/or recursive than that seen in the asymmetrical vertical lines in *vOluptas 2.0 @ 7.5 min.*

44. To quote Nechvatal's "An Artist's Statement" accompanying the "vOluptuary: an algorithic hermaphornology" exhibition at Universal Concepts Unlimited in New York City, these works show "incomprehensible transformation, and, of course, immersive excess." They aim "to depict an imagined realm of political-spiritual chaosmos where new forms of sexual order arise such that any form of order is only temporary and provisional."

45. The publication history of *Agrippa* and its bibliographical status are extraordinarily complicated due both to its experimental physical and digital form (described in my discussion below) and its widespread publicity and partial dissemination in a variety of authorized or unauthorized unofficial versions, including photos of a prerelease mock-up of the art book and many Internet copies of Gibson's text. These para-published versions, as they may be called, have played a predominant role in the work's reception history due not only to the extremely limited run of official copies of *Agrippa* but also to the self-encrypting, disappearing nature of its enclosed digital text, which made it impossible simultaneously to retain the published book in working condition and to read it. (As Kevin Begos said in his letter to me of 26 October 2002, "If collectors/museums want a pure 1st edition, that could only be the unread state"; see the Web copy of this letter I posted with Begos's permission.) The para-versions of the work were thus the only copies that could be read, viewed, or talked about by a wide audience—a limitation that in this case actually amplified the fame of *Agrippa* as more or less a purely conceptual artifact that "fit right in" with "the whole Internet boom . . . in the very early stages" (Begos, ibid). I myself have not had physical access to *Agrippa*. Even Gibson never owned a copy of the finished book, though in his letter to me of 26 August 2002 he recalls that he did at one point own the mock-up seen in the photograph in Peter Schwenger's "*Agrippa*, or The Apocalyptic Book" (p. 621). The definitive account of the publication and bibliographical history of *Agrippa* is Matthew G. Kirschenbaum's discussion in "Awareness of the Mechanism," the introduction to his forthcoming *Mechanisms: New Media and the New Textuality*. (My thanks to Kirschenbaum for this portion of his manuscript.) Kirschenbaum does extensive forensic research in tracing the genesis and online dissemination of *Agrippa*, especially the text by Gibson, as a case study for his larger argument about the complex material histories of new media "virtual" texts.

Because of the intricate bibliographical status of *Agrippa*, it is at times difficult to be certain what aspect of the artifact an interpretive discussion like mine should be addressing. Here I engage primarily Gibson's text in its widely available Internet form (a copy of which I have put online with line numbers for ease of reference) and the concept of the book as a physical artifact. Where I am aware of a significant difference between the concept or mock-up and its realization in the actual book, I indicate that divergence in notes.

46. Gibson, *Neuromancer*, p. 51 (quotation marks and ellipses in the original). Subsequent citations of this work are given parenthetically in the text.

47. Larry McCaffery, for example, notes the influence upon Gibson of "the hard-

boiled writing of Dashiell Hammett," "1940s *film noir,*" and other works ("Interview with William Gibson," p. 264). See also Andrew Ross's chapter "Cyberpunk in Boystown," which relates cyberpunk fiction to the hard-boiled detective and Western genres (*Strange Weather,* pp. 147, 154).

48. "Telepresence had only hinted at the magic and singularity of the thing, and he'd walked slowly forward, into that neon maw and all that patchwork carnival of scavenged surfaces, in perfect awe. Fairyland. Rain-silvered plywood, broken marble from the walls of forgotten banks, corrugated plastic, polished brass, sequins, painted canvas, mirrors, chrome gone dull and peeling in the salt air. So many things, too much for his reeling eye, and he'd known that his journey had not been in vain" (*Virtual Light,* p. 70).

49. Schwenger, "*Agrippa,* or, The Apocalyptic Book," pp. 617–18. Schwenger's excellent reading of *Agrippa* focuses on the poetics of "disappearance" and situates Gibson's poem in the intellectual context of Mallarmé and Blanchot. My reading here, which focuses on "destructivity" and compares the poem to Romantic poetry, varies upon, but complements, Schwenger's. For a photo of the mock-up of *Agrippa* described by Schwenger, see his article, p. 621.

50. Schwenger, "*Agrippa,* or, The Apocalyptic Book," p. 622.

51. Schwenger, "*Agrippa,* or, The Apocalyptic Book," p. 618. Kevin Begos indicated to me in a personal communication on 4 June 2003 that in the final publication Ashbaugh's etchings did not in fact properly alter with exposure to light because the process could not be technically implemented.

52. In a personal communication with me on 4 June 2003, Kevin Begos remembered that the Mac diskette with Gibson's poem was no longer playable within approximately six months because of technical obsolescence (due to rapidly evolving computer technology and software). We thus witness a kind of double disappearance of the work in time: Gibson's text was designed to disappear at the end of reading, but in an unforeseen fashion it soon also disappeared *before* reading.

53. Abrams, "Structure and Style in the Greater Romantic Lyric."

54. On the historical, political, and social background of "Tintern Abbey," which has been controversial in Romanticism studies in recent years, see my discussion and citations of other critics (including Robert Brinkley, Kenneth Johnston, Marjorie Levinson, and Jerome McGann) in *Wordsworth: The Sense of History,* pp. 215–17.

55. On the One Life, see Wordsworth, *The Prelude* (1805), II.430; Coleridge, "The Eolian Harp," l.26.

56. On Gibson's father's death, see Schwenger, "*Agrippa,* or, The Apocalyptic Book," p. 623.

57. For an earlier comparison of "neuromanticism" to Romanticism, see my "Local Transcendence," pp. 75–77.

58. Blake, Letter to Thomas Butts, 25 April 1803 (*Complete Writings,* p. 823).

59. Wordsworth, "Kendal and Windermere Railway: Two Letters Re-Printed from the *Morning Post,*" in *Prose Works,* vol. 3.

60. Haraway, *Simians, Cyborgs, and Women,* p. 151. We might usefully invoke at this point Bob Perelman's contemporary poem "China" (an instance of Language Poetry), which Fredric Jameson has canonized as one of the paradigms of postmodernism (Jameson, *Postmodernism,* pp. 28–30). As in "Agrippa," the framing device of Perelman's poem is an old book of photos. After finding a picture book in Chinatown whose Chinese text he cannot read, Perelman writes a poem composed of radically discontinuous substitute captions—for example:

We live on the third world from the sun. Number three. Nobody tells us
what to do.
The people who taught us to count were being very kind.
It's always time to leave.
If it rains, you either have your umbrella or you don't.

Such is discontinuity so deep, Jameson suggests in his now classic analysis of postmodernism, that it is "schizophrenic."

61. Baumgärtel, "Browsers Bite Back."

62. Michele White, "Aesthetic of Failure," p. 179.

63. For an analysis of the complicated and changing structure of Jodi's various subdomain sites, including <wwwwwwwww.jodi.org>, <oss.jodi.org>, <404.jodi.org>, and others, see Huber, "Only!4!!!!!!!!!!!!!!!!!!!!!!4-for YOUR Private Eyes: An Analysis of the Structure of http://www.jodi.org."

64. Writing about the predecessor to the <wwwwwwwww.jodi.org> page at the address <http://jodi.org> as the latter site existed in 1997, Lunenfeld observes, "The first screen is simple: lines of green characters on a black screen, with a green highlighting function cycling down. Long-term computer users will find their experience tinged by nostalgia: for me, the font, colors, and black background were reminiscent of the first portable computer I ever used, a little Kaypro with a tiny monochrome screen" (*Snap to Grid,* p. 82). The page now at <http://jodi.org> is unrelated.

65. There are also some formatting codes that apply to the whole page (background color, text color, link color, and so forth) that are included as attributes within the body tag itself.

66. On the unconventional relation between source code and rendered appearance in Jodi's art, see Alan Sondheim, "Introduction: Codework"; Michele White, "Aesthetic of Failure"; and Peter Lunenfeld, *Snap to Grid,* pp. 83–84. I am grateful to Rita Raley for recommending many of the secondary readings on Jodi in this part of my chapter, especially on the topic of Jodi's codework.

67. Lunenfeld, *Snap to Grid,* p. 84.

68. The identification of the two bombs and fission pattern depicted in the source code of <wwwwwwwww.jodi.org> is based on my best conjecture. Lunenfeld (*Snap to Grid,* p. 83) reproduces part of the diagram of Fat Man, which is now at the very top of Jodi's page, while I reproduce Little Boy further down. The page has changed since Lunenfeld took his screenshot in 1997, such that the tailfins of Fat Man are no longer there. It is not possible to determine the extent of other changes on the page between 1997 and the time of this writing (August 2003) because Jodi has set the "robots.txt" file on their site to prevent the Wayback Machine and other Web crawlers from archiving their site.

69. More accurately, these labels for links tease technically proficient users with the bare possibilities of meaning. "BINHEX/message.hqx" would seem to refer to a folder containing Macintosh binary files converted into text files; "Reflector/131.24.167" hints at the Internet address (IP number) of a listserv or similar "reflector" program used for managing e-mail discussion forums; and "target=" seems to be the "attribute" part of an HTML tag, used, for example, to make a hypertext link "target" or open a separate window (though the term acquires military connotations in the context of Jodi's depictions of bombs and radar screens). However, even the technical elite would be thwarted in glossing Jodi's link labels because it is never clear from context what is information as opposed to misinformation. Navigating Jodi's pages is akin to using one's mouse as a divining rod, which twitches or dips over hot spots indicating . . . *something.*

70. Irational.org, "irational Curriculum Vitae."

71. Mark Napier, interview with Jodi, 8 April 1997; quoted in Fauconnier, *Web-Specific Art*, "Jodi—<we serve no content>."

72. Try, for example, to reset the display settings when a glitch causes the screen to go to a resolution of 340 by 200 pixels, which means that the control buttons and slider controls for changing the display settings are off screen. My argument about the critical importance of the display apparatus in a computer is based on at least a hundred hours of troubleshooting I have personally done in the past decade.

73. Technically, this is done through a "refresh" attribute in the "meta" tag of the HTML header, but with the URL of the refresh set to the originating page itself.

74. At some point before the time of this writing in August 2003 (when I last revised my discussion of Jodi), Jodi's "aesthetic of failure," as Michele White calls it in her essay of that title, took an extreme form on the page at <http://jodi.org>, which one would ordinarily expect to be the root of the whole Jodi domain with all its subdomains but which has now mutated beyond recognition. Opening this page now produces a completely blacked out screen. After a few seconds, the browser window in which the black screen appears winks out of existence and another, miniature browser window of 100 pixels square opens. This latter window is so abnormal that it cannot be resized and shows nothing other than the minimize, maximize, and close buttons in the window tool bar along with the nonfunctional stubs of other menu bars. The page is a disappearing Cheshire Cat's grin of a page. It is a nonbrowser interpretation of a browser.

75. Baumgärtel, "Browsers Bite Back."

76. See Baumgärtel, "'We Love Your Computer.'" The Documenta exhibitions of contemporary art have been held in Kassel, Germany, approximately every four years since 1955. Jodi exhibited in Documenta 10 in 1997. See the Web site of Documenta 11 for the history of the exhibition: <http://www.documenta.de/documenta_gelb.html>

77. Baumgärtel, "'We Love Your Computer.'"

78. See Read_Me Festival 1.2, "Award Winners." I am grateful to Rita Raley for bringing *WinGluk Builder* to my attention.

79. Deleuze and Guattari, *Thousand Plateaus*, p. 20.

80. Deleuze and Guattari, *Thousand Plateaus*, pp. 160–61.

81. Critical Art Ensemble, *Electronic Disturbance*, pp. 15, 23.

82. Critical Art Ensemble, *Electronic Disturbance*, p. 24.

83. Critical Art Ensemble, *Electronic Disturbance*, p. 25.

84. Critical Art Ensemble, *Electronic Civil Disobedience*, pp. 18, 24, 30.

85. Deleuze and Guattari, *Thousand Plateaus*, p. 10.

86. Critical Art Ensemble, *Electronic Civil Disobedience*, pp. 28–29.

87. Critical Art Ensemble, *Electronic Disturbance*, p. 27.

88. Most of this chapter, and all of the preceding chapters in this book, was written before the terrorist attacks on the United States of September 11, 2001. As I indicate below, however (and anticipated in my introduction), the issue of "terrorism" was built into the context of destructive art and CAE's implementation of such art well before this date. I have thus chosen not to try to exploit, alter, or inflect this issue in retrospect, though no doubt future discussion of this issue will need to be so inflected in light of the various dynamics of global power—including information surveillance and resistance—made more manifest after the events of September 11.

89. Critical Art Ensemble, *Electronic Civil Disobedience*, pp. 18, 19.

90. Critical Art Ensemble, "The Mythology of Terrorism on the Net," in *Digital Resistance,* pp. 30–31, 36–37.

91. The first two examples are from the "Tactical Projects" part of the Critical Art Ensemble's Web site (screens 7 and 9 of this part of the site). The third example is discussed in *Electronic Civil Disobedience,* pp. 52–54.

92. Critical Art Ensemble, *Electronic Disturbance,* p. 62. I am using the term *demo* here not just in its usual sense but in the special sense of the computer art "demo scene," primarily a subcultural phenomenon of young European coders, multimedia artists, hackers, and others. (See, for example, Trixter/Hornet, "PC Demos Explained," and Dave Green, "Demo or Die!" "Demos," perhaps, are the theater of computing. In regard to CAE works that are packaged as if for mass distribution, see, for example, the group's "Child as Audience" project in 2001 (created in collaboration with Creation Is Cruxifixion and the Carbon Defense League). This work is a "package, designed primarily for teenage boys," that "offers a host of radical software, instructions on how to hack a GameBoy, a hard core CD, and a pamphlet on the oppression of youth" ("Tactical Projects," screen 10).

93. *Denial of service attacks* are a species of viral aggression in which technically no invasion of the target computer occurs, though the more potent variant known as *distributed denial of service (DDoS)* can involve intrusions and Trojan horses designed to hijack third-party machines as unwitting agents in an attack. Instead, a flood of automated network traffic, or other means of swamping bandwidth and resources, is directed at the target computer, effectively preventing the target from answering normal requests for Web pages or other services.

On the actions and DDoS tools of the Electronic Disturbance Theater, see the following pages from the organization's Web site: Home page; "Past Actions"; "Electronic Civil Disobedience Archive"; Stalbaum, "The Zapatista Tactical FloodNet"; "Chronology of SWARM"; and Wray, "Electronic Disturbance Theater and Electronic Civil Disobedience." On the participation of EDT in the campaign for eToy vs. eToys.com, which was coordinated by the RTMark group, see CNN.com, "eToys Attacks Show Need for Strong Web Defenses" (21 December 1999); and RTMark, "Press Release: The Brent Spar of E-Commerce: eToys vs. eToy Post-Hearing Press Conference"(25 December 1999). (I thank Christine Lorenz for first bringing to my attention the role of the Electronic Disturbance Theater in the eToys.com action.) It should be noted that EDT's FloodNet tool was designed to allow hacktivists to collaborate willingly in DDoS attacks, whereas many of the DDoS tools that evolved from late 1999 on were true "malware" designed to coopt users' machines unknowingly in an attack.

94. See 0100101110101101.org, "Contagious Paranoia"; and epidemiC, "Biennale.py" and "Press Release." See also Reena Jana's article for *Wired News,* "Want to See Some Really Sick Art?" Also see the Web site for the Frankfurt Museum of Applied Arts Digitalcraft exhibit titled *I love you—computer_viruses_hacker_culture,* which includes a notice of 0100101110101101.org and Biennale.py as well as essays on the history and theory of viruses. I thank Rita Raley for the latter reference.

95. epidemiC, "Beauty of Computer Viruses at the 'D-I-N-A.'"

96. Marcos Novak, talk in the Media Arts and Technology lecture series, University of California, Santa Barbara, 3 March 2002. For some images of Novak's work, see for example his home page and "Liquid~, Trans~, Invisible~: The Ascent and Speciation of the Digital in Architecture. A Story." I am also indebted to correspondence with Novak about his architectural work and ideas after his talk at UC Santa Barbara. My layman's knowledge of string theory, and of the usefulness of lower dimension analogies, is indebted to Greene, *Elegant Universe*; and Kaku, *Beyond Einstein* and *Hyperspace.* Both physicists, but especially Kaku, emphasize the criterion of mathematical "beauty"—an aesthetics of ele-

gant simplicity, symmetry, unity, and seeming inevitability—that drives the search for an ultimate physical theory. (Compare Novak on "beauty" in his "Liquid~, Trans~, Invisible~.") In regard to the comparison I make between Novak's transarchitecture and string theory, a particular point of reference is the physicists' visualizations of higher dimension Calabi-Yau shapes (e.g., Greene, *Elegant Universe*, p. 207).

Chapter Twelve

1. See, for example, Braudel, *On History*, pp. 3–5, 25–54.

2. Greenblatt, "Invisible Bullets," in *Shakespearean Negotiations*, pp. 21–65; the quotes here are from pp. 26, 28, 39, 30.

3. Geertz, "Deep Play," especially p. 444. On the New Historicism and subversion, see my "Power of Formalism: The New Historicism" and "Wordsworth and Subversion, 1793–1804: Trying Cultural Criticism." See also note 2 in chapter 1, above.

4. The crucial portion of *How to Do Things with Words* in this regard is Lecture XI where Austin, for example, observes, "Surely to state is every bit as much to perform an illocutionary act as, say, to warn or to pronounce" (p. 134). See Petrey, *Speech Acts and Literary Theory*, pp. 31–36, on the importance of Austin's Lecture XI. Searle develops the overlap between the constative and performative more complexly in his discussion of reference, predication, and assertion in his *Speech Acts*.

5. The actual examples of the "present indicative active" that Austin gives when he brings up the issue are "I name," "I do," "I bet," and "I give" (*How to Do Things with Words*, p. 56).

6. For the idea of performative "unhappiness" or "infelicity" in Austin (by contrast with constative "wrong" or "incorrect"), see, for example, *How to Do Things with Words*, pp. 15, 25. The section of his book most relevant to the problem I indicate here of statements that perform or (re)enact the constative status of the past is Lecture XI, where Austin invokes the example of a historiographical statement: "Again, in the case of stating truly or falsely, just as much as in the case of advising well or badly, the intents and purposes of the utterance and its context are important; what is judged true in a school book may not be so judged in a work of historical research. Consider the constative, 'Lord Raglan won the battle of Alma,' remembering that Alma was a soldier's battle if ever there was one and that Lord Raglan's orders were never transmitted to some of his subordinates. Did Lord Raglan then win the battle of Alma or did he not? Of course in some contexts, perhaps in a school book, it is perfectly justifiable to say so—it is something of an exaggeration, maybe, and there would be no question of giving Raglan a medal for it. As 'France is hexagonal' is rough, so 'Lord Raglan won the battle of Alma' is exaggerated and suitable to some contexts and not to others; it would be pointless to insist on its truth or falsity" (pp. 143–44). In general, Austin says in this lecture, constative statements—of which historical statements are an instance—can only be verified in a context that is performative or illocutionary: "It is essential to realize that 'true' and 'false,' like 'free' and 'unfree,' do not stand for anything simple at all; but only for a general dimension of being a right or proper thing to say as opposed to a wrong thing, in these circumstances, to this audience, for these purposes and with these intentions" (p. 145).

Perhaps the most relevant aspect of Searle's *Speech Acts* in my present context is his discussion of the "axiom of existence" in reference considered as a speech act (i.e., the axiom that "Whatever is referred to must exist" [p. 77]) and his undergirding notion of "constitutive rules" of illocution (which "do not merely regulate" but "create or define new forms of behavior" [p. 33]). Speech acts of history, as I conceive of them here, are thoroughly performative because in the guise of the constative they both assert the existence of the vanished historical phenomena being referred to and, more fundamentally, construct the rules (or invoke the conventions) of historical validity by which the existence of "existence" is constituted. Statements such as "I remember," "I witness," or "I mourn" are merely the explicit signs of an existential contract with the historical past.

7. On the New Historicism and mourning, see my "New Historicism and the work of Mourning." For Searle on promises, see, for example, *Speech Acts,* pp. 57–62.

8. Very resonant with my argument here is Paul Hernadi's discussion of the "future of the past" at the end of his *Interpreting Events,* whose specific focus (as named in the subtitle of the book) is on "tragicomedies of history on the modern stage." Enveloping speech act theory within a Gadamerian hermeneutical perspective, Hernadi writes suggestively: "historytelling is not a matter of correlating two preexisting spheres of existence in which, respectively, *they were* and *I am.* It is, rather, what speech act theoreticians . . . might well call a 'performative' enactment of that which, thereby, *we will have become*" (p. 219; original emphasis). In general, Hernadi's view of historytelling as a kind of promise speech contract between the present and past is consonant with my discussion here. Hernadi writes: "Whether they live in entirely or predominantly oral, literate, or postliterate (electronic) cultures, the men and women in charge of generating and managing information about the shared past of a particular community have considerable impact on the community's present and future as well" (*Interpreting Events,* pp. 215–16).

9. Austin, *How to Do Things with Words,* p. 5.

10. Lyotard, *Postmodern Condition,* pp. 46–53.

11. Aristotle, *Poetics,* p. 18; Horace, "Art of Poetry," p. 72; Sidney, *Apology for Poetry,* pp. 146, 147.

12. Peacock had blamed Romantic era poetry for ignoring science, history, political economy, and so on to cater only "to that much larger portion of the reading public, whose minds are not awakened to the desire of valuable knowledge" (*Four Ages of Poetry,* p. 514).

13. Shelley, *Defense of Poetry,* p. 526. Comparable is the "What is a Poet?" section that William Wordsworth added in 1802 to his "Preface to *Lyrical Ballads.*" Thinking about the relation of poetry to scientific knowledge, he writes in proto-Shelleyan style, "Poetry is the first and last of all knowledge" (*Prose Works,* vol. 1, p. 141).

14. "A Poem should not mean/But be" from Archibald MacLeish's poem "Ars Poetica" (1926) became a de facto motto of the New Criticism. In battling scientific and historical knowledge, the New Critics simplified what would otherwise have been a bad equation for literature by dropping the issue of utility entirely, as when Cleanth Brooks so elegantly recused the "use of poetry" debate of the times: "If the last sentence seems to take a dangerous turn toward some special 'use of poetry'—some therapeutic value for the sake of which poetry is to be cultivated—I can only say that I have in mind no special ills which poetry is to cure" (*Well Wrought Urn,* p. 209). The difference in theme and tone between this brilliantly insouciant reprise of Arnoldian "disinterestedness," on the one hand, and Shelley's ethical-political interest in the "poetry of life," on the other, is astounding. Indeed, "tone" (so important in New Critical discourse) is in this case the compelling issue. Brooks declines to weigh poetry on the scale of utility, but the cost is the implicit admission that when judged by that standard poetry would at most be useful for "therapeutic value." The poetic *architectonike* has diminished into a tonic, a toning up of the sound mind in a sound body. It is to reinflate the value of such tonic that the New Critics then spoke of poetry as a maturity of "experience" (see below).

15. Brooks, *Well Wrought Urn,* pp. 206–7. See also Brooks's invocation of T. S. Eliot's standard of poetic statements that can be accepted as "coherent, mature, and founded on the facts of experience" ("Irony as a Principle of Structure," p. 970). W. K. Wimsatt envisioned the same experience of *architectonike*: "If it be granted that the 'subject matter' of

poetry is in a broad sense the moral realm, human actions as good or bad, with all their associated feelings, all the thought and imagination that goes with happiness and suffering . . . , then . . . the complexity and unity of the poem, is also its maturity or sophistication or richness or depth, and hence its value" (*Verbal Icon*, p. 82).

16. My account here has presented a severely abbreviated review of literary history and criticism. A fuller review would need to consider the moment between classicism and Romanticism—for example, Dr. Johnson and his standard of moral and palliative usefulness in literature. The moment between Romanticism and the New Critics—including Arnold and Pater, for instance—would also be important.

Epilogue

1. Compare Andy Goldsworthy's remark: "The best of my work, sometimes the result of much struggle when made, appears so obvious that it is incredible I didn't see it before. It was there all the time" ("Introduction").

2. Thanks to my wise and rigorous respondents at the Berkeley Townsend Center for the Humanities in 1998 for a version of this critique. See in particular Ascoli, "Response," pp. 27–28; and the related remarks of Sharon Marcus, "Response," pp. 42–43. Miryam Sas's reflections in the same response session on the way certain literary authors "reinscribe the impulse toward simulation, and the co-optation of history, within a structure that rewrites the meaning of what is useful and reinvents the reductions and necessary compromises involved in moving too fast toward the future" (p. 39) also bear on the issue of the proper simulation of knowledge.

3. For example, Claude Lorrain's *Landscape with the Rest on the Flight into Egypt* (1645); see my *Wordsworth: The Sense of History*, pp. 65–75.

4. I am at present involved in the Electronic Literature Organization's PAD initiative (Preservation/Archiving/Dissemination), which seeks to find ways to retain and migrate historical instances of digital literature within new technological environments.

Appendix A

1. Frow, *Cultural Studies and Cultural Value*, p. 90.

2. See Ehrenreich's debate over terminology in *Fear of Falling*, pp. 5–6.

3. Strassmann, *Information Payoff*, pp. 4–5.

4. Gouldner, *The Future of Intellectuals and the Rise of the New Class*, pp. 48–49.

5. Bell, *Coming of Post-Industrial Society*, p. 375; Wright, *Classes*, p. 88. The new middle class in Wright's table consists of categories 4–11 ("Expert Managers" to "Uncredentialled Supervisors").

6. See the chapter on "A Taxonomy of Teacher Work" in Aronowitz and DiFazio, *Jobless Future*.

7. Furåker, "Future of the Intelligentsia under Capitalism," p. 87.

Appendix B

1. E.g., *Business Week's* report on the latest biennial Labor Department survey of displaced workers in a 13 April 1998 article, "Downsizing's Painful Effects."

2. "A Rage to Re-engineer," *Washington Post*, 25 July 1993, p. H1, cited in Rifkin, *End of Work*, p. 7; Rifkin, *End of Work*, p. 9; *Business Week*, "Downsizing's Painful Effects," 13 April 1998, p. 23.

3. School of Industrial Relations, *The New American Workplace* (Ithaca, N.Y.: Cornell University Press, 1994), cited in Harris and Moran, *Managing Cultural* Differences,

pp. 165–66; “When Will the Layoffs End?” *Fortune,* 20 September 1993, p. 40, cited in Rifkin, *End of Work,* p. 3.

4. Rifkin, *End of Work,* p. 7.

5. For information and statistics on international downsizing, see Rifkin, *End of Work,* pp. 4, 7, 198–207; and Micklethwait and Wooldridge, *Witch Doctors,* p. 32. The severity of the restructuring changes in the instance of Japan was chronicled prominently in a 1996 *Los Angeles Times* article titled “Japanese Jolted by Demands of Future” (14 July 1996, pp. A1, A14–A15). See also the accompanying *Los Angeles Times* story, “A Restructuring Japan Looks to U.S. Models” (14 July 1996, p. D4), and related stories in the paper on 15 July 1996 (pp. A1, A8–A9).

6. *Business Week,* “Big Payoffs from Layoffs,” 24 February 1997, p. 30; “The Downside of Downsizing,” 28 April 1997, p. 26; “It's the Best of Times—Or Is It?” 12 January 1998, pp. 36–38; “Downsizing's Painful Effects,” 13 April 1998, p. 23; “Oops, That's Too Much Downsizing,” 8 June 1998, p. 38.

7. *Business Week,* “An Update on Downsizing,” 25 November 1996, p. 30; *Los Angeles Times,* “Firings and Hirings Shaping Up as the Competing Trends of 1997,” 16 March 1997, p. D5.

8. *Business Week,* “It's the Best of Times—Or Is It?” 12 January 1998, pp. 36–38.

9. “Oops, That's Too Much Downsizing,” 8 June 1998, p. 38.

10. Rifkin, *End of Work,* p. 91.

Appendix C

1. Shimomura, with Markoff, *Takedown.*

2 Shimomura, with Markoff, *Takedown,* p. 13; see also pp. 97, 103.

3. For this use of the term *hacker,* see Parker, *Fighting Computer Crime,* p. 160; and Sterling, *Hacker Crackdown,* part 2, section 2.

4. The definition of “Ethical Hacker” occurs on <http://www.business2.com/articles/mag/0,1640,13064|4,FF.html>

5. This definition of “hacker ethic” occurs in the following section of the *Jargon File:* <http://catb.org/esr/jargon/html/entry/hacker-ethic.html>

6. Sterling, *Hacker Crackdown,* part 2, section 2.

Works Cited

A note regarding Web pages that have moved or become defunct: To find Web sites that have moved to a different URL since this book was written, readers are advised first to try a search engine. To find sites that are now defunct, or to view past versions of continuing sites, readers can use the Internet Archive's *Wayback Machine* <http://www.archive.org/>. Searching for a URL in this archive returns multiple versions of old Web sites, going back to circa fall 1996. Web sites that have vanished and cannot be accessed through the *Wayback Machine* or by other means (e.g., because their content was dynamically generated from databases or because their owners blocked access to automated search engine and indexing "crawlers") are indicated below as "now extinct."

0100101110101101.org. Home page. Retrieved on various dates, 2001–2003. Last retrieved 6 August 2003. <http://0100101110101101.org> Secondary page cited:

"Contagious Paranoia: 0100101110101101.ORG Spreads a New Computer Virus." Retrieved 7 August 2003. <http://0100101110101101.org/home/biennale_py/>

Abrams, M. H. *The Mirror and the Lamp: Romantic Theory and the Critical Tradition.* New York: W. W. Norton, 1958.

———. "Structure and Style in the Greater Romantic Lyric." In *Romanticism and Consciousness: Essays in Criticism.* Ed. Harold Bloom. New York: W. W. Norton, 1970.

Absurd.org. Home page. Last retrieved 14 December 2001. <http://www.absurd.org/>

Adbusters. Special double issue on "Design Anarchy." No. 37 (2001).

Adler, Paul S., ed. *Technology and the Future of Work.* New York: Oxford University Press, 1992.

AltaVista. Search engine. Digital Equipment Corporation (owernership of AltaVista has subsequently changed). Retrieved 6–7 July 1998. <http://altavista.digital.com/>

Althusser, Louis. "Ideology and Ideological State Apparatuses (Notes Towards an Investigation).” In *Lenin and Philosophy and Other Essays.* Trans. Ben Brewster. London: New Left Books, 1971.

Alvesson, Mats. *Cultural Perspectives on Organizations.* Cambridge: Cambridge University Press, 1993.

Alvesson, Mats, and Per Olof Berg. *Corporate Culture and Organizational Symbolism: An Overview.* Berlin: Walter de Gruyter, 1992.

The Amazing Fishcam! [See under Netscape Communications, Inc.]

Ambit Totally Useless: Sixty-Four Exquisite Sites with No Purpose. Ed. Ira Brickman. 29 July 2000. *Ambit: The Web Waystation.* Retrieved 8 November 2000. <http://www.ambitweb.com/useless/useless.html>

American Civil Liberties Union (ACLU) Freedom Network. Home page of ACLU. Retrieved on various dates in 2000; last retrieved 17 November 2000. <http://www.aclu.org/>

Secondary pages cited:

ACLU National Task Force on Civil Liberties in the Workplace. “Workplace Rights: Issue Summary.” Retrieved 6 July 2000. <http://www.aclu.org/issues/worker/iswr.html>

“Workplace Rights: ACLU and Worker’s Rights.” Retrieved 6 July 2000. <http://www.aclu.org/issues/worker/hmwr.html>

Archibald, W. Peter. “Using Marx’s Theory of Alienation Empirically.” *Theory and Society* 6 (1978): 119–32.

Aristotle. *The Poetics.* In *On Poetry and Style: Aristotle.* Trans. G. M. A. Grube. Indianapolis: Bobbs-Merrill, 1958.

Aronowitz, Stanley, and William DiFazio. *The Jobless Future: Sci-Tech and the Dogma of Work.* Minneapolis: University of Minnesota Press, 1994.

Ascoli, Albert Russell. “Response” to Alan Liu, “The Downsizing of Knowledge: Knowledge Work and Literary History.” In *Knowledge Work, Literary History, and the Future of Literary Studies.* Ed. Christina M. Gillis. Doreen B. Townsend Center Occasional Papers, 15. Berkeley: Doreen B. Townsend Center/Regents of the University of California, 1998.

Association for Progressive Communications (APC). Home page. Retrieved on various dates, March–July 2000; last retrieved 4 July 2000. <http://www.apc.org/english/index.htm>

Secondary pages cited:

“The APC Mission.” Retrieved 4 July 2000. <http://www.apc.org/english/about/mission/index.htm>

“Managing Your NGO.” 21 December 1999. Retrieved 10 June 2000. <http://www.apc.org/english/ngos/business/index.htm>

“Online Events.” Retrieved 4 July 2000. <http://www.apc.org/english/ngos/calendar/on_rip.htm>

Attewell, Paul. “Skill and Occupational Changes in U.S. Manufacturing.” In *Technology and the Future of Work.* Ed. Paul S. Adler. New York: Oxford University Press, 1992.

Austin, J. L. *How to Do Things with Words.* 2nd ed. Ed. J. O. Urmson and Marina Sbisà. Cambridge, Mass.: Harvard University Press, 1975.

Axelrod, Alan. *Elizabeth I, CEO: Strategic Lessons from the Leader Who Built an Empire.* Paramus, N.J.: Prentice Hall, 2000.

———. *Patton on Leadership: Strategic Lessons for Corporate Warfare.* Paramus, N.J.: Prentice Hall, 1999.

Bacon, David. "Organizing Silicon Valley's High Tech Workers." *Corporate Watch*. Retrieved 14 July 2000. Sequence of eight Web pages beginning at <http://www.corpwatch.org/trac/internet/globalabor/dbacon1.html>

———. "Organizing Silicon Valley's High Tech Workers." *David Bacon: Stories, Photographs*. Institute for Global Communications (IGC). Retrieved 14 July 2000. <http://www.igc.org/dbacon/Unions/04hitec0.htm>

———. "Silicon Valley on Strike! Immigrants in Electronics Protest Growing Sweatshop Conditions." *Bay Guardian*, 17 January 1993; reprinted in *CPU: Working in the Computer Industry* 3. Computer Professionals for Social Responsibility (CPSR). 10 June 1993. Retrieved 14 July 2000. <http://www.cpsr.org/program/workplace/cpu.003.html>

Bad Attitude: The "Processed World" Anthology. [See under Carlsson, Chris.]

Bair, James H. "User Needs for Office Systems Solutions." In *Technology and the Transformation of White-Collar Work*. Ed. Robert E. Kraut. Hillsdale, N.J.: Lawrence Erlbaum Associates, 1987.

Barbrook, Richard, and Andy Cameron. "The Californian Ideology." Extended mix version. Undated (shorter versions dated 1995–1996). *Hypermedia Research Centre*. School of Communication and Creative Industries, Westminster University, United Kingdom. Retrieved 13 July 2003. Sequence of pages beginning at: <http://www.hrc.wmin.ac.uk/hrc/theory/californianideo/main/t.4.2.html>

Barlow, John Perry. "A Declaration of the Independence of Cyberspace." 8 February 1996. *Electronic Frontier Foundation*. Retrieved 12 September 2001. <http://www.eff.org/~barlow/Declaration-Final.html>

———. "The Economy of Ideas." *Wired* 2.03 (March 1994). *Wired News*. Printer-friendly version of article retrieved 17 June 2000. <http://www.wired.com/wired/archive/2.03/economy.ideas_pr.html>

Barrow, Clyde W. *Universities and the Capitalist State: Corporate Liberalism and the Reconstruction of American Higher Education, 1894–1928*. Madison: University of Wisconsin Press, 1990.

Barthes, Roland. *Elements of Semiology*. Trans. Annette Lavers and Colin Smith. New York: Hill and Wang, 1968.

Batchelor, Ray. *Henry Ford: Mass Production, Modernism and Design*. Manchester: Manchester University Press, 1994.

Baudrillard, Jean. *Selected Writings*. Ed. Mark Poster. Stanford: Stanford University Press, 1988.

———. *Simulations*. Trans. Paul Foss, Paul Patton, and Philip Beitchman. New York: Semiotext(e), 1983.

———. "The Year 2000 Has Already Happened." In *Body Invaders: Panic Sex in America*. Ed. Arthur Kroker and Marilouise Kroker. New York: St. Martin's Press, 1987.

Baumgärtel, Tilman. "Browsers Bite Back." *Eyestorm.com*. 8 March 2001. Retrieved 10 September 2001. <http://www.eyestorm.com/feature/ED2n_article.asp?article_id=234>

———. "'We Love Your Computer': The Aesthetics of Crashing Browsers (Interview with Jodi)." *Telepolis: Magazin der Netzkultur*. 6 October 1997. Retrieved 4 August 2003. <http://www.heise.de/tp/english/special/ku/6187/1.html>

Baym, Nancy K. "The Emergence of Community in Computer-Mediated Communication." In *Cybersociety: Computer-Mediated Communication and Community*. Ed. Steven G. Jones. Thousand Oaks, Calif.: Sage, 1995.

Becker, Howard S. "Professional Sociology: The Case of C. Wright Mills." In Ray C. Rist, ed., *The Democratic Imagination: Dialogues on the Work of Irving Louis Horowitz*. New

Brunswick, NJ: Transaction, 1994. Retrieved online 3 June 2003. <http://home.earthlink.net/~hsbecker/mills.html>

Begos, Kevin, Jr. E-mail to the author on the genesis of *Agrippa (A Book of the Dead)*. 26 October 2002. Posted with permission on the Web, 3 February 2003. Retrieved 24 July 2003. <http://www.english.ucsb.edu/faculty/ayliu/unlocked/begos/letter.html>

Bell, Daniel. *The Coming of Post-Industrial Society: A Venture in Social Forecasting*. New York: Basic, 1973.

Bell, David, and Barbara M. Kennedy, eds. *The Cybercultures Reader*. London: Routledge, 2000.

Beniger, James R. *The Control Revolution: Technological and Economic Origins of the Information Society*. Cambridge, Mass.: Harvard University Press, 1986.

Bertin, Jacques. *Semiology of Graphics: Diagrams, Networks, Maps*. Trans. William J. Berg. Madison: University of Wisconsin Press, 1983. (originally published in French, 1967)

Beverley, John. *Against Literature*. Minneapolis: University of Minnesota Press, 1993.

Bickel, Bruce, and Stan Jantz. *Bruce & Stan's Guide to God: A User-Friendly Approach*. Eugene, Ore.: Harvest House, 1997.

Bikson, Tora K. "Understanding the Implementation of Office Technology." In *Technology and the Transformation of White-Collar Work*. Ed. Robert E. Kraut. Hillsdale, N.J.: Lawrence Erlbaum Associates, 1987.

Birdsall, Willam F. "The Internet and the Ideology of Information Technology." Proceedings of INet Conference, Montreal, 24–28 June 1996. *Internet Society (ISOC)*. Retrieved 2 June 2000. <http://www.isoc.org/isoc/whatis/conferences/inet/96/proceedings/e3/e3_2.htm>

Birkerts, Sven. *The Gutenberg Elegies: The Fate of Reading in an Electronic Age*. Boston: Faber and Faber, 1994.

Blackwell, Lewis, and David Carson. *The End of Print: The Graphic Design of David Carson*. San Francisco: Chronicle Books, 1995.

Blade Runner. Dir. Ridley Scott. Warner Brothers, 1982.

Blake, William. *Complete Writings, with Variant Readings*. Ed. Geoffrey Keynes. London: Oxford University Press, 1974, c. 1966.

Blauner, Robert. *Alienation and Freedom: The Factory Worker and His Industry*. Chicago: University of Chicago Press, 1964.

Blinderman, Barry. "The Ghost of Electricity." In *Joseph Nechvatal: Paintings 1986–1987*, catalogue for Nechvatal exhibition, University Galleries at Illinois State University, Normal, Illinois, 1988. *Joseph Nechvatal Home Page*. Retrieved 30 November 2001. <http://www.eyewithwings.net/nechvatal/blinder.htm>

Bloom, Harold. *The Anxiety of Influence: A Theory of Poetry*. London: Oxford University Press, 1973.

———. *The Western Canon: The Books and School of the Ages*. New York: Riverhead, 1994.

Bødker, Susanne. *Through the Interface: A Human Activity Approach to User Interface Design*. Hillsdale, N.J.: Lawrence Erlbaum, 1991.

Bolter, J. David. *Turing's Man: Western Culture in the Computer Age*. Chapel Hill: University of North Carolina Press, 1984.

Bolter, Jay David, and Richard Grusin. *Remediation: Understanding New Media*. Cambridge, Mass.: MIT Press, 1999.

Borges, Jorge Luis. *Labyrinths: Selected Stories and Other Writings*. Ed. Donald A. Yates and James E. Irby. New York: New Directions, 1964.

Borgmann, Albert. *Holding On to Reality: The Nature of Information at the Turn of the Millennium.* Chicago: University of Chicago Press, 1999.

Borgmann, Albert, and N. Katherine Hayles. "An Interview/Dialogue with Albert Borgmann and N. Katherine Hayles On Humans and Machines." [Anonymous interviewer.] 1999. University of Chicago Press Web site. Retrieved 17 April 2000. <http://www.press.uchicago.edu/Misc/Chicago/borghayl.html>

Borsook, Paulina. "Cyberselfish." *Mother Jones* July–August 1996. *Mojo Wire.* Retrieved 24 May 2000. <http://bsd.mojones.com/mother_jones/JA96/borsook.html>

———. *Cyberselfish: A Critical Romp through the Terribly Libertarian Culture of High Tech.* New York: PublicAffairs, 2000.

Bourdieu, Pierre. *Distinction: A Social Critique of the Judgement of Taste.* Trans. Richard Nice. Cambridge, Mass.: Harvard University Press, 1984.

———. *Homo Academicus.* Trans. Peter Collier. Stanford: Stanford University Press, 1988.

———. *The Logic of Practice.* Trans. Richard Nice. Stanford: Stanford University Press, 1990.

———. *Outline of a Theory of Practice.* Trans. Richard Nice. Cambridge: Cambridge University Press, 1977.

Bourdieu, Pierre, and Loïc J. D. Wacquant. *An Invitation to Reflexive Sociology.* Chicago: University of Chicago Press, 1992.

Bourdieu Forum. Online discussion list. *Spoon Collective.* Accessed on various dates, 1995–96. Current address and archives from 1996 onward available at <http://lists.village.virginia.edu/~spoons/bourdieu/">

Boyett, Joseph H., and Henry P. Conn. *Workplace 2000: The Revolution Reshaping American Business.* New York: Penguin, 1992.

Bradford, Peter, ed. *Information Architects.* Introduction by Richard Saul Wurman. New York: Graphis, 1997.

Brand, Stewart. E-mail to the author. 9 May 2003.

Braudel, Fernand. *On History.* Trans. Sarah Matthews. Chicago: University of Chicago Press, 1980.

Braverman, Harry. *Labor and Monopoly Capital: The Degradation of Work in the Twentieth Century.* New York: Monthly Review Press, 1974.

Broglio, Ron, and Eric Sonstroem. "Frankenstein MOO." [See under *Romantic Circles.*]

Brooks, Cleanth. "Irony as a Principle of Structure." In *Critical Theory since Plato.* Rev. ed. Ed. Hazard Adams. New York: Harcourt Brace Jovanovich, 1992.

———. *The Well Wrought Urn: Studies in the Structure of Poetry.* New York: Harcourt Brace Jovanovich, 1947, 1975.

Brooks, David. *Bobos in Paradise: The New Upper Class and How They Got There.* New York: Simon & Schuster, 2000.

Brown, John Seely, and Paul Duguid. *The Social Life of Information.* Boston: Harvard Business School Press, 2000.

Burris, Val. "Class Structure and Political Ideology." *Insurgent Sociologist* 14, no. 2 (Summer 1987): 5–46.

Business Week

Andreessen, Marc, photos. 13 April 1998, cover and p. 2.

"Big Payoffs from Layoffs." By Gene Koretz. 24 February 1997, p. 30.

"Big Test Ahead for Productivity." By Gene Koretz. 8 June 1998, p. 26.

"The Downside of Downsizing." By Gene Koretz. 28 April 1997, p. 26.

"Downsizing's Painful Effects: Many Workers Don't Bounce Back." By Gene Koretz. 13 April 1998, p. 23.

"Generation $." By Michelle Conlin, with Laura Cohn. 16 August 1999, pp. 34–36.

"It's the Best of Times—Or Is It?" By David Greising, with bureau reports. 12 January 1998, pp. 36–38.

"Killer Hours Are a Myth: Full-timers Aren't Working More." By Dean Foust. 3 August 1998, p. 26.

"Look Who's Pushing Productivity: Labor is Embracing Partnerships to Keep Companies Competitive." By Aaron Bernstein. 7 April 1997, pp. 72–75.

"The New U.: A Tough Market Is Reshaping Colleges." By Keith H. Hammonds and Susan Jackson. 22 December 1997, pp. 96–102.

"Oops, That's Too Much Downsizing." By Aaron Bernstein. 8 June 1998, p. 38.

"Oracle: Why It's Cool Again." By Steve Hamm. 8 May 2000, pp. 115–26. [With photo of Larry Ellison on issue cover under the title, "Despite the Tech Stock Slide . . . Oracle Is Cool Again.]

"Peering into a New Millennium: More Jobs, Trade, and Productivity." By Gene Koretz. 26 January 1998, p. 22.

"The Promise of Productivity." By Christopher Farrell with others. 9 March 1998, pp. 28–30.

"Someone to Watch over You: More Employers Punish Those Who Violate E-mail and Net Rules." By Larry Armstrong. 10 July 2000, pp. 189–90.

"Superstores, Megabooks—And Humongous Headaches. " By Hardy Green. 14 April 1997, pp. 92–94.

"The 21st Century Corporation." 28 August 2000, p. 278.

"An Update on Downsizing." By Gene Koretz. 25 November 1996, p. 30.

"Workers, Surf at Your Own Risk." By Michelle Conlin. 12 June 2000, pp. 105–106.

Calcutt, Andrew. *White Noise: An A–Z of the Contradictions in Cyberculture.* New York: St. Martin's Press, 1999.

Campbell-Kelly, Martin, and William Aspray. *Computer: A History of the Information Machine.* New York: BasicBooks/HarperCollins, 1996.

Caplan, Stanley H. "Making Usability a Kodak Product Differentiator." In *Usability in Practice: How Companies Develop User-Friendly Products.* Ed. Michael E. Wiklund. Boston: Academic Press, 1994.

Carchedi, Guglielmo. "Class Politics, Class Consciousness, and the New Middle Class." *Insurgent Sociologist* 14, no. 3 (Fall 1987): 111–30.

———. *On the Economic Identification of Social Classes.* London: Routledge and Kegan Paul, 1977.

Carlsson, Chris, with Mark Leger, eds. *Bad Attitude: The "Processed World" Anthology.* London: Verso, 1990.

Carlsson, Chris, Adam Cornford, and Greg Williamson. "Some History of *Processed World.*" Written, revised, and published in various places, 1989–1991. *Processed World* Web site. Retrieved 20 July 2003. <http://www.processedworld.com/History/history.html>

Carnevale, Anthony Patrick, and Susan Carol Stone. *The American Mosaic: An In-Depth Report on the Future of Diversity at Work.* New York: McGraw-Hill, 1995.

Carolyn's Diary. Carolyn L. Burke. 31 October 2000. Retrieved 8 November 2000. <http://carolyn.org/Diary.html>

Carter, Rob, et al. *Typographic Design: Form and Communication.* New York: Van Nostrand Reinhold, 1985.

Castells, Manuel. *The Information Age: Economy, Society and Culture.* 3 vols. Malden, Mass.: Blackwell, 1996–98.

Center for Democracy & Technology (CDT). Home page. Retrieved on various dates 1999–2000; last retrieved 6 June 2000. <http://www.cdt.org/>

Secondary pages cited:

"CDT Mission." 22 May 2000. Retrieved 22 May 2000. <http://www.cdt.org/mission/>

"CDT Principles." 6 June 2000. Retrieved 6 June 2000. <http://www.cdt.org/mission/principles.shtml>

"Democratic Values for the Digital Age: Summary of CDT Activities 1999—Work Plan 2000." January 2000. Retrieved 6 June 2000. <http://www.cdt.org/mission/activities2000.shtml>

Center for Information Technology and Society. Dir. Bruce Bimber. University of California, Santa Barbara. Home page. Retrieved 30 July 2003. <http://www.cits.ucsb.edu/>

Ceruzzi, Paul E. *A History of Modern Computing.* Cambridge, Mass.: MIT Press, 1998.

Chamot, Dennis. "Electronic Work and the White-Collar Employee." In *Technology and the Transformation of White-Collar Work.* Ed. Robert E. Kraut. Hillsdale, N.J.: Lawrence Erlbaum Associates, 1987.

Chandler, Alfred D., Jr. *The Visible Hand: The Managerial Revolution in American Business.* Cambridge, Mass.: Harvard University Press, 1977.

Chandler, Daniel. [See under *The Media and Communication Studies Site.*]

Chapman, Gary. "What Is Cool? Certainly Not a Campaign for It." *Los Angeles Times,* 2 December 1996, pp. D1, D3.

Christensen, Jerome. *Romanticism at the End of History.* Baltimore: Johns Hopkins University Press, 2000.

The Chronicle of Higher Education. "Old Computers Never Die—They Just Cost Colleges Money in New Ways." By Scott Carlson. 14 February 2003: A33. *Chronicle.com.* Retrieved 18 July 2003.

Chun, Wendy. "Orienting Orientalism, or How to Map Cyberspace." In *Asian America.net: Ethnicity, Nationalism, and Cyberculture.* Ed. Rachel C. Lee and Sau-ling Cynthia Wong. New York: Routledge, 2003.

Clarke, John, et al. "Subcultures, Cultures and Class." In *Resistance through Rituals: Youth Subcultures in Post-War Britain.* Ed. Stuart Hall and Tony Jefferson. London: Hutchinson and Centre for Contemporary Cultural Studies, University of Birmingham, 1976.

Clastres, Pierre. *Chronicle of the Guayaki Indians.* Trans. Paul Auster. New York: Zone, 2000.

———. *Society against the State: Essays in Political Anthropology.* Trans. Robert Hurley, with Abe Stein. New York: Zone, 1989.

Cloninger, Curt. *Fresh Styles for Web Designers: Eye Candy from the Underground.* Indianapolis: New Riders, 2002.

C/NET News.com. CNET Networks, Inc. Retrieved on various dates, 1998–2003; last retrieved 13 July 2003. <http://www.news.com/>

Particular *C/NET News* stories:

"Software to Filter Workers' Access." By Beth Lipton. 4 May 1998. Retrieved 1 November 2000. <http://news.cnet.com/news/0-1004-200-328964.html?st.ne.fd.mdh>

"Study Finds Web Bigger Than We Think." By Associated Press, special to *C/NET News.com*. 26 July 2000;. Retrieved 27 July 2000. <http://news.cnet.com/news/0-1005-200-2356979.html?tag=st.ne.1002.thed.ni>

CNN.com. Home page. Cable News Network LP, LLLP, an AOL Time Warner Company. Retrieved various dates. <http://www.cnn.com/>

Secondary page cited:

"eToys Attacks Show Need for Strong Web Defenses." By Ellen Messmer. 21 December 1999. <http://www.cnn.com/1999/TECH/computing/12/21/etoys.attack.idg/>

Coalition on New Office Technology (CNOT). Home page. Originally retrieved 6 July 2000 at <http://www.rsi.deas.harvard.edu/CNOT/> Last retrieved 19 July 2003 at <http://www.eecs.harvard.edu/cnot/>

Coleridge, Samuel Taylor. *Samuel Taylor Coleridge*. Ed. H. J. Jackson. Oxford: Oxford University Press, 1985.

comp.human-factors. Usenet newsgroup. <news:comp.human-factors>

Various online postings in the thread on "THEORY: Help-systems that really help?" (on the "user friendliness" of Unix). July 1997. Retrieved 19 June 2003 in archive available through *Google Groups* advanced search page. <http://www.google.com/advanced_group_search?hl=en>

Various online postings in the thread on "Why Is Unix Evil?" (on the "user friendliness" of Unix). June 1998. Retrieved 19 June 2003 in archive available through *Google Groups* advanced search page. <http://www.google.com/advanced_group_search?hl=en>

Various online postings in the thread on "offices vs. cubicles." June 1998. Retrieved on various dates in June 1998.

comp.unix.user-friendly. Usenet newsgroup. Various online postings in the thread on "unix is user-friendly." October 1993. <news:comp.unix.user-friendly> Retrieved 19 June 2003 in archive available through *Google Groups* advanced search page. <http://www.google.com/advanced_group_search?hl=en>

Computer Professionals for Social Responsibility (CPSR). Home page. Retrieved on various dates 1999–2000; last retrieved 17 November 2000. <http://www.cpsr.org/>

Secondary pages cited:

CPSR Computers in the Workplace Working Group. "Computers, Work and the Workplace." 5 March 1999. Retrieved 17 June 2000. <http://www.cpsr.org/program/workplace/workplace-home.html>

CPSR Cyber-Rights Working Group. "Cyber-Rights Home Page." Retrieved 22 May 2000. <http://www.cpsr.org/cpsr/nii/cyber-rights/cyber-rights.html>

CPU: Working in the Computer Industry 3. 10 June 1993. Retrieved 14 July 2000. <http://www.cpsr.org/program/workplace/cpu.003.html>

"History Up to 1994." Retrieved 26 May 2000. <http://www.cpsr.org/cpsr/history.html>

Winograd, Terry. "Terry Winograd's Thoughts on CPSR's Mission." Text dated 25 October 1996; Web page dated 30 October 1996. Retrieved 26 May 2000. <http://www.cpsr.org/cpsr/winnog.html>

Conkin, Paul K. *The Southern Agrarians*. Knoxville: University of Tennessee Press, 1988.

CoolLinks.com. MkMedia, Inc. Retrieved 1 November 2000. <http://www.coollinks.com/>

Cool Site of the Day. Originated by Glenn Davis. Ed. Richard Grimes, et al. *InfiNet*. Retrieved 19 December 1999. <http://cool.infi.net/>

Secondary pages cited:

"How to Be Cool" (also titled "What is Coolium?"). Retrieved 16 July 1998. <http://cool.infi.net/coolium.html>

Still Cool Archive. Various pages of "Previous Sightings" retrieved through cgi-scripted links on this page, 16 July 1998. <http://www.coolsiteoftheday.com/cgi-bin/stillcool.pl>

Cooper, Robert, and Gibson Burrell. "Modernism, Postmodernism and Organizational Analysis: An Introduction." *Organization Studies* 9, no. 1 (1988): 91–112.

Copley, Frank Barkley. *Frederick W. Taylor: Father of Scientific Management.* Vol. 1. 1923; rpt., New York: Augustus M. Kelley, 1969.

Corporate Watch. (Renamed *CorpWatch* in March 2001.) Home page. Transnational Resource and Action Center. Retrieved on various dates, 1997–2001; last retrieved 4 August 2001. <http://www.corpwatch.org/>

Secondary pages cited:

"The Dark Side of High Tech Development." [See under Smith, Ted.]

"Feature #2: The High Cost of High Tech." 10 February 1997. Retrieved 24 May 2000. <http://www.corpwatch.org/trac/feature/hitech/index.html>

"The Net Loss of the Computer Revolution." By Jerry Mander. [See under Mander, Jerry.]

Crash Site. Big Gun Project. Retrieved in various versions 1997–1998; last retrieved 7 February 1998. <http://www.crashsite.com/>

Critical Art Ensemble (CAE). Home page. Retrieved 14 September 2001. <http://www.critical-art.net/> [Note: Navigation on the Critical Art Ensemble site occurs through a Flash interface that sometimes prevents the citation of specific links.]

Secondary pages cited:

Digital Resistance: Explorations in Tactical Media. Brooklyn, NY: Autonomedia. 2001. Retrieved 14 September 2001 as .pdf files through links under the category "Book Projects" on the Critical Art Ensemble home page.

Electronic Civil Disobedience and Other Unpopular Ideas. Brooklyn, NY: Autonomedia. 1996. Retrieved 14 September 2001 as .pdf files through links under the category "Book Projects" on the Critical Art Ensemble home page.

The Electronic Disturbance. Brooklyn, NY: Autonomedia. 1994. Retrieved 14 September 2001 as .pdf files through links under the category "Book Projects" on the Critical Art Ensemble home page.

"Tactical Projects" ("Tactical Media"). Branch of the Critical Art Ensemble's home page devoted to art installations of "tactical media." Retrieved 14 December 2001. <http://www.critical-art.net/tactical_media/index.html>

"The Technology of Uselessness. " In *Electronic Civil Disobedience and Other Unpopular Ideas.* Brooklyn, NY: Autonomedia. 1996. Retrieved 14 September 2001 as .pdf files through links under the category "Book Projects" on the Critical Art Ensemble home page.

"Critical Art Ensemble Timeline." [No author identified.] *Drama Review,* 44, no. 4 (Winter 2000). Retrieved online as .pdf file, 9 December 2001. <http://mitpress.mit.edu/journals/DRAM/44-4/pdf/cae_timeline.pdf>

Critique: The Magazine of Graphic Design Thinking. Web Crit 2000 Awards issue. No. 18 (2000).

Cronin, Mary J. "Bye-Bye, Wild Web." *Fortune,* 27 October 1997. *Fortune.com.* 27 October 1997. Retrieved 21 October 1997. <http://pathfinder.com/fortune/digitalwatch/1027dig3.html>

Cuthbertson-Johnson, Beverley, et al. *The Sociology of Emotions: An Annotated Bibliography.* New York: Garland, 1994.

D2KLA. Home page. Retrieved 6 July 2000. <http://d2kla.org/>

Dahrendorf, Ralf. *Class and Class Conflict in Industrial Society.* Stanford: Stanford University Press, 1959.

Danesi, Marcel. *Cool: The Signs and Meanings of Adolescence.* Toronto: University of Toronto Press, 1994.

Daniel, Sharon, Mark Bartlett, and Puraga Guhathakurta. *Subtract the Sky.* Last retrieved 19 July 2002. <http://arts.ucsc.edu/sdaniel/new/subtract.html>

DARnet. [See UC DARnet.]

Davidow, William H., and Michael S. Malone. *The Virtual Corporation: Structuring and Revitalizing the Corporation for the 21st Century.* New York: HarperCollins, 1992.

DC Independent Media Center (Washington, D.C.). Home page. Retrieved 6 July 2000. <http://dc2.indymedia.org/>

Deal, Terence E., and Allan A. Kennedy. *Corporate Cultures: The Rites and Rituals of Corporate Life.* Reading, Mass.: Addison-Wesley, 1982.

Deb and Jen's Land O' Useless Facts. Ed. Deborah Henigson and Jennifer Godwin. 13 July 1998. Retrieved 16 July 1998. <http://www-leland.stanford.edu/~jenkg/useless.html>

Debord, Guy. *The Society of the Spectacle.* Trans. Donald Nicholson-Smith. New York: Zone, 1995.

de Certeau, Michel. *The Practice of Everyday Life.* Trans. Steven Rendall. Berkeley: University of California Press, 1984.

Decker, Randall E., and Robert A. Schwegler, eds. *Decker's Patterns of Exposition 13 (Instructor's Edition).* 13th ed. New York: HarperCollins, 1992.

Deleuze, Gilles, and Félix Guattari. *A Thousand Plateaus: Capitalism and Schizophrenia.* Trans. Brian Massumi. Minneapolis: University of Minnesota Press, 1987.

de Man, Paul. "The Rhetoric of Temporality." In *Blindness and Insight: Essays in the Rhetoric of Contemporary Criticism.* 2nd ed. rev. Minneapolis: University of Minnesota Press, 1983.

———. "Semiology and Rhetoric." In *Allegories of Reading: Figural Language in Rousseau, Nietzsche, Rilke, and Proust.* New Haven: Yale University Press, 1979.

———. "Shelley Disfigured." In Harold Bloom et al., *Deconstruction and Criticism.* New York: Seabury, 1979.

Denham, John. "Cooper's Hill." Ed. Jack Lynch. Rutgers University, Newark. Retrieved 12 July 2003. <http://newark.rutgers.edu/~jlynch/Texts/cooper.html>

Derrida, Jacques. "Structure, Sign and Play in the Discourse of the Human Sciences." In *Writing and Difference.* Trans. Alan Bass. Chicago: University of Chicago Press, 1978.

Dertouzos, Michael L. *What Will Be: How the New World of Information Will Change Our Lives.* New York: HarperCollins, 1998.

Dery, Mark. *Escape Velocity: Cyberculture at the End of the Century.* New York: Grove, 1996.

de Swaan, Abram. "The Politics of Agoraphobia: On Changes in Emotional and Relational Management." *Theory and Society* 10 (1981): 359–87.

Dieli, Mary, et al. "The Microsoft Corporation Usability Group." In *Usability in Practice: How Companies Develop User-Friendly Products.* Ed. Michael E. Wiklund. Boston: Academic Press, 1994.

"Digit, Gidget." (Pseudonym of Stephanie Klein.) "Gidget Gets Fired: Introduction to 'Sabotage: The Ultimate Video Game.'" *Processed World,* no. 5 (Summer 1982). Reprinted in Chris Carlsson with Mark Leger, eds. *Bad Attitude: The "Processed World" Anthology*. London: Verso, 1990, 60–61. Online version also available on *Processed World* Web site. Retrieved 20 July 2003. <http://www.processedworld.com/Issues/issue05/05gidget.htm>

———. "Sabotage: The Ultimate Video Game." *Processed World*, no. 5 (Summer 1982). Reprinted in Chris Carlsson with Mark Leger, eds. *Bad Attitude: The "Processed World" Anthology*. London: Verso, 1990. Online version also available on *Processed World* Web site. Retrieved 20 July 2003. <http://www.processedworld.com/Issues/issue05/05sabotage.htm>

The Digital Cultures Project. [See UC Digital Cultures Project.]

Dinerstein, Joel. *Swinging the Machine: Modernity, Technology, and African American Culture between the World Wars*. Amherst: University of Massachusetts Press, 2003.

Direct Action Network (DAN). Home page. D2KLA. Retrieved 6 July 2000. <http://d2kla.org/dan.html>

Secondary pages cited:

"Joint DNC/RNC Call to Action: Challenge the Democratic and Republican Parties This Summer in Philadelphia and Los Angeles!" Retrieved 6 July 2000. <http://d2kla.org/dan_dncrnc.html>

Dirlik, Arif. *The Postcolonial Aura: Third World Criticism in the Age of Global Capitalism*. Boulder: Westview, 1997.

Disgruntled: The Business Magazine for People Who Work for a Living. Counterpoint Press, Inc. Retrieved on various dates, 1997–2000; last retrieved 3 March 2000. <http://www.disgruntled.com/>

Documenta 11. Home page for exhibition. Retrieved 23 July 2002. <http://www.documenta.de/documenta_gelb.html>

Doheny-Farina, Stephen. *The Wired Neighborhood*. New Haven: Yale University Press, 1996.

Donahue, Sean. "New Jobs for the New Economy." *Business 2.0*. July 1999. Retrieved from *Business 2.0* online site, 17 December 2001. <http://www.business2.com/articles/mag/0,,13064,FF.html> [Definition of "ethical hacker" at <http://www.business2.com/articles/mag/0,1640,13064|4,FF.html>]

Dondis, Donis A. *A Primer of Visual Literacy*. Cambridge, Mass.: MIT Press, 1973.

Dordick, Herbert S., and Georgette Wang. *The Information Society: A Retrospective View*. Newbury Park, Calif.: Sage, 1993.

Drucker, Johanna. *The Visible Word: Experimental Typography and Modern Art, 1909–1923*. Chicago: University of Chicago Press, 1994.

Drucker, Peter F. *Innovation and Entrepreneurship: Practice and Principles*. New York: Harper & Row, 1985.

———. *Managing in Turbulent Times*. New York: Harper & Row, 1980.

———. *Post-Capitalist Society*. New York: HarperCollins, 1993.

Druckrey, Timothy, with Ars Electronica, eds. *Ars Electronica: Facing the Future: A Survey of Two Decades*. Cambridge, Mass.: MIT Press, 1999.

du Gay, Paul. "Organizing Identity: Making Up People at Work." In *Production of Culture/Cultures of Production*. Ed. Paul du Gay. London: SAGE, in association with Open University, 1997.

du Gay, Paul, ed. *Production of Culture/Cultures of Production*. London: SAGE, in association with Open University, 1997.

Dyson, Esther, George Gilder, George Keyworth, and Alvin Toffler. "Cyberspace and the American Dream: A Magna Carta for the Knowledge Age." [See under Progress & Freedom Foundation.]

Easley, Kendell H. *User-Friendly Greek: A Commonsense Approach to the Greek New Testament.* Nashville, Tenn.: Broadman & Holman, 1994.

EDUCAUSE. Home page. Last retrieved 19 July 2002. <http://www.educause.edu/>

Ehrenreich, Barbara. *Fear of Falling: The Inner Life of the Middle Class.* 1989; rpt., New York: HarperCollins, 1990.

Ehrenreich, Barbara, and John Ehrenreich. "The Professional-Managerial Class." In *Between Labor and Capital.* Ed. Pat Walker. Boston: South End, 1979.

Eichenbaum, Boris. "The Theory of the 'Formal Method.'" In *Russian Formalist Criticism: Four Essays.* Trans. Lee T. Lemon and Marion J. Reis. Lincoln: University of Nebraska Press, 1965.

Electronic Disturbance Theater (EDT). Home page (*Electronic Civil Disobedience*). Retrieved 14 December 2001. <http://www.thing.net/~rdom/ecd/ecd.html>

Secondary pages cited:

"Chronology of SWARM." 10 September 1998. Retrieved 7 August 2003. <http://www.thing.net/~rdom/ecd/CHRON.html>

"Electronic Civil Disobedience Archive." Ed. Stefan Wray. Retrieved 16 November 1999. <http://www.nyu.edu/projects/wray/ecd.html>

"Past Actions." Retrieved 7 August 2003. <http://www.thing.net/~rdom/ecd/pastactions.html>

Stalbaum, Brett. "The Zapatista Tactical FloodNet: A Collaborative, Activist and Conceptual Art Work of the Net." Retrieved 7 August 2003. <http://www.thing.net/%7Erdom/ecd/ZapTact.html>

Wray, Stefan. "The Electronic Disturbance Theater and Electronic Civil Disobedience." 17 June 1998. Retrieved on various dates, 1999–2003; last retrieved 7 August 2003. <http://www.thing.net/~rdom/ecd/EDTECD.html>

———. "On Electronic Civil Disobedience." 20–22 March 1998. Retrieved on various dates, 1999–2003; last retrieved 7 August 2003. <http://www.thing.net/~rdom/ecd/oecd.html>

Electronic Frontier Foundation (EFF). Home page. Retrieved on various dates 1999–2000; last retrieved 17 November 2000. <http://www.eff.org/>

Secondary pages cited:

"About EFF." Retrieved 22 May 2000. <http://www.eff.org/abouteff.html>

Barlow, John Perry. "A Declaration of the Independence of Cyberspace." [See under Barlow.]

"Defining Digital Identity: How Do You Define Yourself—And Protect Your Privacy?" Retrieved 12 June 2000. <http://www.eff.org/identity.html>

"EFF 'Censorship & Free Expression' Archive." 18 April 2000. Retrieved 22 May 2000. <http://www.eff.org/pub/Censorship/>

"EFF 'Privacy, Security, Crypto, & Surveillance' Archive." 15 November 2000; retrieved 27 November 2000. <http://www.eff.org/pub/Privacy/>

"EFF 'Privacy—Surveillance' Archive." Subarchive in "EFF 'Privacy, Security, Crypto, & Surveillance' Archive." 9 June 2000. Retrieved 12 June 2000. <http://www.eff.org/pub/Privacy/Surveillance/>

"EFF 'Privacy—Workplace Monitoring & Employer/Employee Privacy Conflicts' Archive." Subarchive in "EFF 'Privacy, Security, Crypto, & Surveillance' Archive." 18 April 2000. Retrieved 24 May 2000. <http://www.eff.org/pub/Privacy/Workplace/>

"Preserving Free Expression: Our Fundamental Rights of Freedom of Speech & Press." Retrieved 22 May 2000. <http://www.eff.org/freespeech.html>

Electronic Literature Organization. Home page. Last retrieved 30 July 2003. <http://www.eliterature.org/>

Secondary pages cited:

Electronic Literature Directory. Database dir., Robert Kendall; programmer, Nick Traenkner. Last retrieved 30 July 2003. <http://directory.eliterature.org/>

Preservation, Archiving, Dissemination Initiative (PAD). Home page. Retrieved 31 July 2003. <http://www.eliterature.org/pad/>

Alan Liu et al. "Excerpt from the Final Technology/Software Report" (PAD Initiative). April 2003. Retrieved 31 July 2003. <http://www.eliterature.org/pad/content/excerpts_tech_report.php>

Electronic Messaging Association (EMA). Home page. Retrieved 28 November 2000. <http://www.ema.org/>

Eliade, Mircea. *The Sacred and the Profane: The Nature of Religion*. Trans. Willard R. Trask. New York: Harcourt, Brace, 1959.

Elias, Norbert. *The Civilizing Process*. Vol. 1: *The History of Manners*. Trans. Edmund Jephcott. New York: Pantheon, 1978.

Elliott, Philip. "Intellectuals, the 'Information Society' and the Disappearance of the Public Sphere." In *Media, Culture and Society: A Critical Reader*. Ed. Richard Collins et al. London: Sage, 1986.

Ellul, Jacques. *The Technological Society*. Trans. John Wilkinson. New York: Alfred A. Knopf, 1967.

epidemiC. Home page. Retrieved 15 December 2001. <http://www.epidemic.ws/>

Secondary pages cited:

"The Beauty of Computer Viruses at the 'D-I-N-A.'" 23 April 2001. Retrieved 16 September 2001. <http://www.epidemic.ws/print_e.html>

"Biennale.py." Retrieved 7 August 2003. <http://www.epidemic.ws/biannual.html>

"Press Release: A Virus in the Venice Biennale." 1 June 2001. Retrieved 15 December 2001. <http://www.epidemic.ws/prelease.txt>

Erickson, Kai. "On Work and Alienation." *American Sociological Review* 51 (1986): 1–8.

Esty, Katharine, Richard Griffin, and Marcie Schorr Hirsch. *Workplace Diversity*. Holbrook, Mass.: Adams, 1995.

European Commission. "Creating a User-Friendly Information Society: Working Document on the Information Society Technologies (IST) Programme (Thematic Programme II, Fifth Framework Programme)." 5 November 1997. *European IT Conference & Exhibition*. EITC 97 Secretariat. Retrieved 29 Oct. 2000. <http://www.cordis.lu/esprit/src/istwork.htm>

Excite. Search engine. Excite, Inc. Retrieved 6–7 July 1998. <http://search.excite.com/>

Fabrick, Laura, Geri Ferguson, and Christina Valadez. *Cathedral de Compostela*. 1999. The Culture of Information. Course site. Instructor, Alan Liu, Department of English, University of California, Santa Barbara. Retrieved 17 July 2002. <http://transcriptions.english.ucsb.edu/archive/topics/infoart/cathedral/>

Fair Measures Corporation. "Internet/Privacy." Retrieved 6 July 2000. <http://www.fairmeasures.com/privacy.html>

Farland, Maria Magdalena. "Cyberculture, Business Culture, and the Literary Counterculture, 1950–70." Paper presented at session on "Brain Work: Representations of Postindustrial Labor in American Literature," MLA convention, Washington, D.C., 28 December 2000. Manuscript courtesy of the author.

Fauconnier, Sandra. *Web-Specific Art: Het World Wide Web als artistiek Medium.* Master's thesis, Ghent University, Belgium, 1997. Published online. Last retrieved 6 August 2003. <http://www.spinster.be/web-specific-art/>

Secondary pages cited:

"jodi—<we serve no content>." Retrieved 6 August 2003. <http://www.spinster.be/web-specific-art/hfdst3/E3.html>

Faulkner, William. *Go Down, Moses, and Other Stories.* New York: Vintage, 1973.

Feldberg, Roslyn L., and Evelyn Nakano Glenn. "Technology and the Transformation of Clerical Work." In *Technology and the Transformation of White-Collar Work.* Ed. Robert E. Kraut. Hillsdale, N.J.: Lawrence Erlbaum Associates, 1987.

Feldman, Martha S., and James G. March. "Information in Organizations as Signal and Symbol." *Administrative Science Quarterly* 26 (1981): 171–86.

Field, Janet N., et al., eds. *Graphic Arts Manual.* New York: Arno, 1980.

FindLaw Cyberspace Law Center. "Workplace Privacy." FindLaw. Retrieved 24 May 2000. <http://cyber.findlaw.com/privacy/workplace.html>

Flam, Helena. "'Emotional Man': II. Corporate Actors as Emotion-Motivated Emotion Managers." *International Sociology* 5 (1990): 225–34.

Flanagin, Andrew J. "Social Pressures on Organizational Website Adoption." *Human Communication Research* 26 (2000): 618–46.

Flanagin, Andrew J., and Miriam J. Metzger. "Internet Use in the Contemporary Media Environment." *Human Communication Research* 27 (2001): 153–81.

Flanders, Vincent. *Web Pages That Suck: Learn Good Design by Looking at Bad Design.* Web site. Retrieved 16 July 1998. <http://www.webpagesthatsuck.com>

Secondary pages cited:

"Cool Sites." Retrieved 16 July 1998. <http://www.webpagesthatsuck.com/cool.html>

[Untitled page]. Retrieved 14 November 2000. <http://www.webpagesthatsuck.com/suckframe.htm>

Flanders, Vincent, and Michael Willis. *Web Pages That Suck: Learn Good Design by Looking at Bad Design.* Book and CD-ROM. San Francisco: Sybex, 1998.

Flaunt. [See under Wright, Shauna.]

Forester, Tom, ed. *The Information Technology Revolution.* Cambridge, Mass.: MIT Press, 1985.

Fortier, François. *Civil Society Computer Networks: The Perilous Road of Cyber-politics.* Diss. York University, Toronto, 1996. *CRIT-ICT*/Distributed Knowledge Project, York University, Toronto. Retrieved 2 June 2000. <http://www.yorku.ca/research/dkproj/fortier/>

Foster, Richard N., and Sarah Kaplan. *Creative Destruction: Why Companies That Are Built to Last Underperform the Market—and How to Successfully Transform Them.* New York: Currency/Doubleday, 2001.

Foucault, Michel. *Madness and Civilization: A History of Insanity in the Age of Reason.* Trans. Richard Howard. New York: Vintage, 1965.

———. *Power/Knowledge: Selected Interviews and Other Writings, 1972–1977.* Ed. Colin Gordon. Trans. Colin Gordon et al. New York: Pantheon, 1980.

Fradenburg, L. O. Aranye, "Entertainment Value." 2002. *Public Humanities Initiative.* Department of English, University of California, Santa Barbara. Retrieved 25 May 2002. <http://www.english.ucsb.edu/initiatives/public-humanities/resources/fradenburg-on-entertainment-value.html>

Frank, Thomas. *The Conquest of Cool: Business Culture, Counterculture, and the Rise of Hip Consumerism.* Chicago: University of Chicago Press, 1997.

Freiberger, Paul, and Michael Swaine. *Fire in the Valley: The Making of the Personal Computer.* 2nd ed. New York: McGraw-Hill, 2000.

Frow, John. *Cultural Studies and Cultural Value.* Oxford: Oxford University Press, 1995.

FuckedCompany.com. Retrieved 7 May 2003. <http://www.fuckedcompany.com/>

Fukuyama, Francis. *The End of History and the Last Man.* New York: Free Press, 1992.

Fuller, Mary, and Henry Jenkins. "Nintendo® and New World Travel Writing: A Dialogue." In *Cybersociety: Computer-Mediated Communication and Community.* Ed. Steven G. Jones. Thousand Oaks, Calif.: Sage, 1995.

Furåker, Bengt. "The Future of the Intelligentsia under Capitalism." In *Intellectuals, Universities, and the State in Western Modern Societies.* Ed. Ron Eyerman et al. Berkeley: University of California Press, 1987.

Gamboni, Dario. *The Destruction of Art: Iconoclasm and Vandalism since the French Revolution.* New Haven: Yale University Press, 1997.

Gardenswartz, Lee, and Anita Rowe. *Diverse Teams at Work: Capitalizing on the Power of Diversity.* Chicago: Irwin, 1994.

———. *Managing Diversity: A Complete Desk Reference and Planning Guide.* Burr Ridge, Ill.: Irwin/San Diego: Pfeiffer, 1993.

Gardner, Elizabeth. "When It's Your Site That's 'Cool.'" *Web Week* 1, no. 5 (September 1995). Republished in *Internet World Daily.* 19 December 1999. Retrieved 15 July 1998. <http://www.internetworld.com/print/1995/09/01/undercon/1-5coolsite.html>

Garnham, Nicholas. "The Media and Narratives of the Intellectual." *Media, Culture and Society* 17 (1995): 359–84.

Garson, Barbara. *The Electronic Sweatshop: How Computers Are Transforming the Office of the Future into the Factory of the Past.* 1988; rpt., New York: Penguin, 1989.

Gates, Bill. "The Case for Microsoft: Why Windows and Microsoft Office Should Stay under One Roof." *Time* magazine, 15 May 2000, p. 57.

Gates, Bill, with Nathan Myhrvold and Peter Rinearson. *The Road Ahead.* Rev. ed. New York: Penguin, 1996.

Geertz, Clifford. "Deep Play: Notes on the Balinese Cockfight." In *The Interpretation of Cultures: Selected Essays.* New York: Basic Books, 1973.

Gergen, Kenneth J. "Organization Theory in the Postmodern Era." In *Rethinking Organization: New Directions in Organization Theory and Analysis.* Ed. Michael Reed and Michael Hughes. London: Sage, 1992.

———. "Social Construction and the Transformation of Identity Politics." 1995. Retrieved 6 July 2001. < http://www.swarthmore.edu/SocSci/kgergen1/text8.html>

Gerhards, Jürgen. "The Changing Culture of Emotions in Modern Society." *Social Science Information* 28 (1989): 737–54.

Gibson, William. "Agrippa (A Book of the Dead)." Text of poem in *Agrippa (A Book of the Dead),* with text by William Gibson and etchings by Dennis Ashbaugh. New York: Kevin Begos, 1992. [Widely available on the Internet; see note under Gibson and Ashbaugh, *Agrippa (A Book of the Dead).* Version cited here has line numbers for reference.] 3 February 2003. Retrieved 24 July 2003. <http://www.english.ucsb.edu/faculty/ayliu/unlocked/gibson/agrippa.html>

———. *Burning Chrome.* New York: Berkley, Ace Books, 1987.

———. E-mail to the author. 26 August 2002.

———. *Neuromancer.* New York: Ace Books, 1984.

———. *Virtual Light.* New York: Bantam, 1993.

Gibson, William, and Dennis Ashbaugh. *Agrippa (A Book of the Dead).* Text by Gibson; etchings by Ashbaugh. New York: Kevin Begos, 1992. [Limited-edition art book whose text by Gibson is commonly read on the World Wide Web in many transcriptions. Because some elements of the work—including the poem by Gibson—were designed to decay or disappear after reading, the work is usually accessible only in descriptions, mock-ups, photographs, and transcriptions in varying degrees of fidelity.]

Gibson, William, and Bruce Sterling. *The Difference Engine.* New York: Bantam, 1991.

Gilder, George. *Wealth and Poverty.* New York: Basic, 1981.

Ginsberg, Allen. *Howl, and Other Poems.* San Francisco: City Lights, 2002, c. 1959.

Gitlin, Todd. *The Sixties: Years of Hope, Days of Rage.* Rev. ed. New York: Bantam, 1993.

Giuliano, Vincent E. "The Mechanization of Office Work." In *The Information Technology Revolution.* Ed. Tom Forester. Cambridge, Mass.: MIT Press, 1985.

Gladwell, Malcolm. "The Coolhunt." *New Yorker,* 17 March 1997, pp. 78–88.

Global Internet Liberty Campaign. Home page. 21 April 2000. Retrieved 22 May 2000. <http://www.gilc.org/>

Secondary pages cited:

"Regardless of Frontiers: Protecting the Human Right to Freedom of Expression on the Global Internet." Retrieved 15 August 2000. <http://www.gilc.org/speech/report/>

"Statement of Principles." Retrieved 19 November 2000. <http://www.gilc.org/about/principles.html>

Global Issues Network. Home page. Retrieved 6 July 2000. <http://www.globalissues.net/>

Godfrey, Tony. *Conceptual Art.* London: Phaidon, 1998.

Goldsworthy, Andy. "Introduction." In *Andy Goldsworthy: A Collaboration with Nature.* New York: Harry N. Abrams, 1990.

Gonzalez, Rafael, and Tamara Payne. "Teamwork and Diversity." In *Valuing Diversity: New Tools for a New Reality.* Ed. Lewis Brown Griggs and Lente-Louise Louw. New York: McGraw-Hill, 1995.

Google. Search engine. Google, Inc. Retrieved on various dates; last retrieved 14 July 2003. <http://www.google.com>

Secondary page cited:

"Google Search Solutions." Retrieved 3 June 2002. <http://www.google.com/services/>

Gordon, Avery. "The Work of Corporate Culture: Diversity Management." *Social Text* 44, vol. 13, no. 3 (Fall–Winter 1995): 3–30.

Gouldner, Alvin W. *The Future of Intellectuals and the Rise of the New Class: A Frame of Reference, Theses, Conjectures, Arguments, and an Historical Perspective on the Role of Intellectuals and Intelligentsia in the International Class Contest of the Modern Era.* New York: Seabury, 1979.

The Graduate. Dir. Mike Nichols. United Artists, 1967.

Great Buildings Online. Ed. Kevin Matthews. Artifice, Inc. Retrieved 19 July 2002. <http://www.greatbuildings.com/>

Green, Dave. "Demo or Die! *Wired* 3.07 (July 1995). *Wired News*. Retrieved 7 August 2003. Sequence of five web pages beginning at <http://www.wired.com/wired/archive/3.07/democoders.html?topic=%3Cdroplink>

Greenbaum, Joan. *Windows on the Workplace: Computers, Jobs, and the Organization of Office Work in the Late Twentieth Century*. New York: Monthly Review Press, 1995.

Greenblatt, Stephen. *Renaissance Self-Fashioning: From More to Shakespeare*. Chicago: University of Chicago Press, 1980.

———. *Shakespearean Negotiations: The Circulation of Social Energy in Renaissance England*. Berkeley: University of California Press, 1988.

Greene, Brian. *The Elegant Universe: Superstrings, Hidden Dimensions, and the Quest for the Ultimate Theory*. 1999; rpt. New York: Vintage, 2000.

Griggs, Lewis Brown. "Valuing Diversity®: Where From . . . Where To?" In Lewis Brown Griggs and Lente-Louise Louw, ed., *Valuing Diversity: New Tools for a New Reality*. New York: McGraw-Hill, 1995.

Griggs, Lewis Brown, and Lente-Louise Louw, eds. *Valuing Diversity: New Tools for a New Reality*. New York: McGraw-Hill, 1995.

Gross, Andrew. "Silicon Valley: A Divided Workforce." *CPU: Working in the Computer Industry* 3. Computer Professionals for Social Responsibility (CPSR). 10 June 1993. Retrieved 14 July 2000. <http://www.cpsr.org/program/workplace/cpu.003.html>

Guertin, Carolyn. *Queen Bees and the Hum of the Hive: An Overview of Feminist Hypertext's Subversive Honeycombings. BeeHive* 1, no. 2 (July 1998). Retrieved 26 September 2001 through javascript link on the following page: <http://beehive.temporalimage.com/archive/12arc.html>

Guillory, John. *Cultural Capital: The Problem of Literary Canon Formation*. Chicago: University of Chicago Press, 1993.

———. "The Ethical Practice of Modernity: The Example of Reading." In *The Turn to Ethics*. Ed. Marjorie Garber, et al. New York: Routledge, 2000.

———. "Some Observations on the Difference Between Lay and Professional Reading." Paper presented at Public Humanities Initiative forum, University of California, Santa Barbara, 5 October 2000. [Expanded and revised version of the published essay "Ethical Practice of Modernity" listed above.]

Hall, Gary. "Answering the Question: What Is an Intellectual." *Surfaces* 6 (1996). *Surfaces Electronic Journal*. 22 December 1996. Retrieved online 6 July 2001. <http://pum12.pum.umontreal.ca/revues/surfaces/vol6/hall.html>

Hall, Stuart, and Tony Jefferson, ed. *Resistance Through Rituals: Youth Subcultures in Post-War Britain*. London: Hutchinson and Centre for Contemporary Cultural Studies, University of Birmingham, 1976.

Hammer, Michael, and James Champy. *Reengineering the Corporation: A Manifesto for Business Revolution*. New York: HarperCollins, 1993.

Handy, Charles. *The Age of Paradox*. Boston: Harvard Business School Press, 1994.

Haraway, Donna J. *Simians, Cyborgs, and Women: The Reinvention of Nature*. New York: Routledge, 1991.

Harper, Michael S., and Anthony Walton, ed. *Every Shut Eye Ain't Asleep: An Anthology of Poetry by African Americans since 1945*. Boston: Little, Brown and Company, 1994.

Harris, Philip R., and Robert T. Moran. *Managing Cultural Difference*. 4th ed. Houston: Gulf Publishing, 1996.

Hauben, Ronda, and Michael Hauben. "Proposed Declaration of the Rights of Netizens." In Ronda Hauben and Michael Hauben, *The Netizens and the Wonderful World of the Net: An Anthology*. Retrieved 14 June 2000. <http://studentweb.tulane.edu/~rwoods/netbook/ch.d13_rights.html>

Hayles, N. Katherine. *How We Became Posthuman: Virtual Bodies in Cybernetics, Literature, and Informatics*. Chicago: University of Chicago Press, 1999.

Hayles, N. Katherine, and Albert Borgmann. "An Interview/Dialogue with Albert Borgmann and N. Katherine Hayles." [See under Borgmann, Albert, and N. Katherine Hayles.]

Hearn, Jeff, and Wendy Parkin. "Organizations, Multiple Oppressions and Postmodernism." In *Postmodernism and Organizations*. Ed. John Hassard and Martin Parker. London: Sage, 1993.

Hebdige, Dick. *Hiding in the Light: On Images and Things*. London: Routledge, 1988.

———. *Subculture: The Meaning of Style*. London: Methuen, 1979.

Heckscher, Charles C. *The New Unionism: Employee Involvement in the Changing Corporation*. Ithaca, N.Y.: ILR/Cornell University Press, 1996.

Heim, Michael. *Electric Language: A Philosophical Study of Word Processing*. 2nd. ed. New Haven: Yale University Press, 1999.

Helgesen, Sally. *The Web of Inclusion: A New Architecture for Building Great Organizations*. New York: Doubleday, 1995.

Heller, Steven, and Seymour Chwast. *Graphic Style: From Victorian to Post-Modern*. New York: Harry N. Abrams, 1988.

HEPROC (Higher Education Processes Network). R&R Publishers, Inc. Retrieved 11 July 2001. <http://heproc.org/>

Herman, Andrew, and Thomas Swiss, eds. *The World Wide Web and Contemporary Cultural Theory*. New York: Routledge, 2000.

Hernadi, Paul. *Interpreting Events: Tragicomedies of History on the Modern Stage*. Ithaca, N.Y.: Cornell University Press, 1985.

Hertz, Garnet. *The Stimulator*. 1997. ConceptLAB. Retrieved 8 November 2000. <http://www.conceptlab.com/simulator/index.html>

Heterick, Robert C., Jr., ed. *Reengineering Teaching and Learning in Higher Education: Sheltered Groves, Camelot, Windmills, and Malls*. CAUSE Professional Paper Series, 10. *Educause*. Retrieved 11 July 2001. <http://www.educause.edu/ir/library/text/pub3010.txt>

Hirschhorn, Larry. *Beyond Mechanization: Work and Technology in a Postindustrial Age*. Cambridge, Mass.: MIT Press, 1984.

Hirschhorn, Larry, and John Mokray. "Automation and Competency Requirements in Manufacturing: A Case Study." In *Technology and the Future of Work*. Ed. Paul S. Adler. New York: Oxford University Press, 1992.

Hobart, Michael E., and Zachary S. Schiffman. *Information Ages: Literacy, Numeracy, and the Computer Revolution*. Baltimore: Johns Hopkins University Press, 1998.

Hochschild, Arlie Russell. *The Managed Heart: Commercialization of Human Feeling*. Berkeley: University of California Press, 1983.

———. *The Time Bind: When Work Becomes Home and Home Becomes Work*. New York: Henry Holt, 1997.

Home, Stewart. *The Assault on Culture: Utopian Currents from Lettrisme to Class War*. London: Aporia/Unpopular, 1988. Excerpt from chap. 11 retrieved online, 17 January 2001. <http://www.entartetekunst.org/Metzger/metzg01_en.html>

Horace (Quintus Horatius Flaccus). "Art of Poetry." In *Critical Theory since Plato*. Rev. ed. Ed. Hazard Adams. New York: Harcourt Brace Jovanovich, 1992.

Horowitz, Irving Louis. *C. Wright Mills, an American Utopian*. New York: Free Press, 1983.

HotBot. Search engine. Wired Digital, Inc. Retrieved 6–7 July 1998. <http://www.hotbot.com/>

Huber, Hans Dieter. "Only!4!!!!!!!!!!!!!!!!!!!!!!4-for YOUR Private Eyes: An Analysis of the Structure of http://www.jodi.org." *art.net.dortmund.de*. Retrieved 27 July 2003. <http://www.art.net.dortmund.de/eng/per/hub_jodi/jod_hu_fr.html>

Hudspeth, D. R. "Just In Time Education." *Educational Technology* 32, no. 6 (June 1992): 7–11.

Huet, Marie-Hélène. *Rehearsing the Revolution: The Staging of Marat's Death, 1793–1797*. Trans. Robert Hurley. Berkeley: University of California Press, 1982.

Hull, Frank M., et al. "The Effect of Technology on Alienation from Work: Testing Blauner's Inverted U-Curve Hypothesis for 110 Industrial Organizations and 245 Retrained Printers." *Work and Occupations* 9 (1982): 31–57.

I love you—computer_viruses_hacker_culture. Web site for Digitalcraft exhibition at Frankfurt Museum of Applied Arts, May–June 2002. Retrieved 24 July 2002. <http://www.digitalcraft.org/index.php?artikel_id=283&PHPSESSID=112e369a525daf034546e1cb567385c7>

I'll Take My Stand: The South and the Agrarian Tradition, by Twelve Southerners. 1930; reprinted with introduction and biographical essays, Baton Rouge: Louisiana State University Press, 1977.

Independent Media Centers. [See under city location of specific indymedia centers: e.g., Los Angeles, Seattle, and "DC" (Washington, D.C.) independent media centers.]

Information Infrastructure Taskforce (IITF) Committee on Applications and Technology (National Information Infrastructure [NII] Initiative). "What It Takes to Make It Happen: Key Issues for Applications of the National Information Infrastructure." 25 January 1994. *ibiblio*. Retrieved 20 June 2003. <http://www.ibiblio.org/pub/archives/whitehouse-papers/1994/Jan/1994-01-25-National-Information-Infrastructure:-Key-Issues>

Information Supercollider. Electrical Engineering and Computer Science Department, Harvard University. Retrieved in various versions 1995–2000; last retrieved 10 November 2000. <http://www.eecs.harvard.edu/collider.html>

Infoseek. Search engine. Infoseek Corporation. Retrieved 6–7 July 1998. <http://www.infoseek.com/>

Inkeles, Gordon, and Iris Schencke. *Ergonomic Living: How to Create a User-Friendly Home and Office*. New York: Simon & Schuster, 1994.

Institute for Global Communications (IGC). Home page. *IGC Internet*. Retrieved various dates, March–July 2000; last retrieved 4 July 2000. <http://www.igc.org/igc/gateway/index.html>

Secondary page cited:

"About IGC Internet." Retrieved 4 July 2000. <http://www.igc.org/igc/gateway/about.html>

InternalMemos.com. RCK Group, Inc. Retrieved 7 May 2003. <http://www.internalmemos.com/memos/>

Internet Domain Survey. 2003. Internet Software Consortium. Retrieved 17 June 2003. <http://www.isc.org/ds/>

Irational.org. Home page. Retrieved 6 December 2001. <http://www.irational.org/cgi-bin/front/front.pl>

Secondary page cited:

"irational Curriculum Vitae." Retrieved 6 December 2001. <http://www.irational.org/cgi-bin/cv/cv.pl>

Jackson, Shelley. *Patchwork Girl by Mary/Shelley and herself.* CD-ROM. Watertown, Mass.: Eastgate Systems, 1995.

Jackson, Timothy Allen. "Towards a New Media Aesthetic." In *Reading Digital Culture.* Ed. David Trend. Malden, Mass.: Blackwell, 2001.

James, Janice S. "American Airlines." In *Usability in Practice: How Companies Develop User-Friendly Products.* Ed. Michael E. Wiklund. Boston: Academic Press, 1994.

Jameson, Fredric. *Postmodernism, or the Cultural Logic of Late Capitalism.* Durham, N.C.: Duke University Press, 1991.

Jamieson, David, and Julie O'Mara. *Managing Workforce 2000: Gaining the Diversity Advantage.* San Francisco: Jossey-Bass, 1991.

Jana, Reena. "Want to See Some Really Sick Art?" *Wired News,* 27 June 2001. Retrieved 15 December 2001. <http://www.wired.com/news/culture/0,1284,44728,00.html>

Jardim, Anne. *The First Henry Ford: A Study in Personality and Business Leadership.* Cambridge, Mass.: MIT Press, 1970.

The Jargon File, Version 4.3.3. Current editor: Eric S. Raymond. 20 September 2002. Retrieved 5 May 2003. <http://catb.org/esr/jargon/html/> (Note: versions of *The Jargon File* have also appeared in print under the title *The New Hacker's Dictionary.)*

Jencks, Charles. *Architecture Today.* Rev. and enlarged ed. New York: Harry N. Abrams, 1988.

Jodi [Joan Heemskerk and Dirk Paesmans]. *wwwwwwwww.jodi.org.* Retrieved on various dates, 2001–2003; last retrieved 5 August 2003. <http://wwwwwwwww.jodi.org/>

———. *%WRONG Browser.* Retrieved on various dates, 2001–2003; last retrieved 5 August 2003. <http://www.wrongbrowser.com/>

Johansen, Robert, and Rob Swigart. *Upsizing the Individual in the Downsized Organization: Managing in the Wake of Reengineering, Globalization, and Overwhelming Technological Change.* Reading, Mass.: Addison-Wesley, 1994.

Johnson, Steven. *Interface Culture: How New Technology Transforms the Way We Create and Communicate.* San Francisco: HarperEdge/HarperCollins, 1997.

Johnston, William B., and Arnold E. Packer. *Workforce 2000: Work and Workers for the Twenty-first Century.* Prepared for the U. S. Department of Labor. Indianapolis: Hudson Institute, 1987.

Jones, J. Jennifer. "Virtual Sublime: Romantic Transcendence and the Transport of the Real." Diss., University of California, Santa Barbara, 2002.

Jones, Steven G. "Understanding Community in the Information Age." In *Cybersociety: Computer-Mediated Communication and Community.* Ed. Steven G. Jones. Thousand Oaks, Calif.: Sage, 1995.

Julier, Guy. *The Thames and Hudson Dictionary of 20th-Century Design and Designers.* London: Thames and Hudson, 1993.

Jurassic Park. Dir. Steven Spielberg. Amblin/Universal, 1993.

Kaku, Michio. *Hyperspace: A Scientific Odyssey through Parallel Universes, Time Warps, and the Tenth Dimension.* 1994; rpt. New York: Anchor/Doubleday, 1995.

Kaku, Michio, and Jennifer Thompson. *Beyond Einstein: The Cosmic Quest for the Theory of the Universe.* Rev. ed. New York: Anchor/Random House, 1995.

Katz, Jon. "Birth of a Digital Nation." *Wired* 5.04 (April 1997). *Wired News.* Retrieved 19 May 2000. Sequence of eight Web pages beginning at <http://www.wired.com/wired/5.04/netizen.html> Single-file printer version available at <http://www.wired.com/wired/5.04/netizen_pr.html>

———. "Netizen: The Digital Citizen." *Wired* 5.12 (December 1997). *Hotwired.* Retrieved 13 July 2003. Sequence of eight web pages beginning at <http://hotwired.wired.com/special/citizen/>

Katzenbach, Jon R., and Douglas K. Smith. *The Wisdom of Teams: Creating the High-Performance Organization.* 1993; rpt. New York: HarperBusiness/HarperCollins, 1994.

Kellner, Douglas. "Intellectuals and New Technologies." *Media, Culture and Society* 17 (1995): 427–48.

Kellogg Company. Apple Jacks® cereal box. 1998.

———. *Kellogg's Apple Jacks* Web site. 27 November 1999. Kellogg, Inc. Retrieved 19 December 1999. <http://www.applejacks.com/>

Kelly, Kevin. *New Rules for the New Economy: 10 Radical Strategies for a Connected World.* New York: Viking, 1998.

———. *Out of Contol: The New Biology of Machines, Social Systems, and the Economic World.* Reading, Mass.: Addison-Wesley, 1994.

Kelly, Philip F., Jr. "Union Organizing in the New Economy: Can New Efforts Reverse Labor's Long Goodbye?" *Labor Watch,* March 2000. Capital Research Center. Retrieved 14 July 2000. <http://www.capitalresearch.org/LaborWatch/lw-0300.htm>

Kendall, Frances E. "Diversity Issues in the Workplace." In *Valuing Diversity: New Tools for a New Reality.* Ed. Lewis Brown Griggs and Lente-Louise Louw. New York: McGraw-Hill, 1995.

Kenney, Martin, and Richard Florida. *Beyond Mass Production: The Japanese System and Its Transfer to the U.S.* New York: Oxford University Press, 1993.

Kernan, Alvin. *The Death of Literature.* New Haven: Yale University Press, 1990.

Kirschenbaum, Matthew G. "Hypertext Theory Post-Poststructuralism." Paper presented at MLA convention, Toronto, 28 December 1997.

———. "The Awareness of the Mechanism." Introduction to "Mechanisms: New Media and the New Textuality." Book in progress. [Manuscript courtesy of the author.]

———. *Lines for a Virtual T[y/o]pography: Electronic Essays on Artifice and Information.* Diss. University of Virginia, 1999. Published online; password for most of the material may be requested of the author. Last retrieved 12 July 2003. <http://www.iath.virginia.edu/~mgk3k/dissertation/>

———. "The Word as Image in an Age of Digital Reproduction." In *Eloquent Images: Word and Image in the Age of New Media.* Eds. Mary E. Hocks and Michelle R. Kendricks. Cambridge, Mass: MIT Press, 2003

Kittler, Friedrich A. *Discourse Networks, 1800/1900.* Trans. Michael Metteer with Chris Cullens. Stanford: Stanford University Press, 1990.

Klein, Stephanie. [See under Digit, Gidget.]

Kleinhans, Chuck. "Cultural Appropriation and Subcultural Expression: The Dialectics of Cooptation and Resistance." Department of Radio/Television/Film, Northwestern University. 14 November 1994. Retrieved 26 July 2001. <http://www.rtvf.nwu.edu/people/kleinhans/cult_and_subcult.html>

Kocka, Jürgen. *White Collar Workers in America, 1890–1940: A Social-Political History in International Perspective.* Trans. Maura Kealey. London: Sage, 1980.

Kole, Ellen S. "Whose Empowerment? NGOs between Grassroots and Netizens (1st Draft)." 16 May 1998. *CRIT-ICT*/Distributed Knowledge Project, York University, Toronto. Retrieved 2 June 2000. <http://www.yorku.ca/research/dkproj/crit-ict/ek1.htm>

Kolko, Beth E., et al., eds. *Race in Cyberspace.* New York: Routledge, 2000.

Kotkin, Joel. *Tribes: How Race, Religion, and Identity Determine Success in the New Global Economy.* New York: Random House, 1992.

Krause, Donald G. *The Book of Five Rings For Executives: Musashi's Classic Book of Competitive Tactics.* London: Nicholas Brealey, 1999.

Kraut, Robert E. "Social Issues and White-Collar Technology: An Overview." In *Technology and the Transformation of White-Collar Work.* Ed. Robert E. Kraut. Hillsdale, N.J.: Lawrence Erlbaum Associates, 1987.

Kunda, Gideon. *Engineering Culture: Control and Commitment in a High-Tech Corporation.* Philadelphia: Temple University Press, 1992.

Kung, Hans. "The Grid System." In *Graphic Arts Manual.* Ed. Janet M. Field et al. New York: Arno, 1980.

LaborNet. "Labor.tech: Discussion of Technology Organizing and Issues for Labor Activists." Discussion forum. Retrieved through "guest" log-on to the *LaborNet Forums,* 15 July 2000. Discussion forum gateway page: <http://www.labornet.org/forums.htm>

Labov, Teresa. "Social and Language Boundaries among Adolescents." *American Speech* 67 (1992): 339–66.

Landauer, Thomas K. *The Trouble with Computers: Usefulness, Usability, and Productivity.* Cambridge, Mass.: MIT Press, 1995.

Landow, George P. *Hypertext: The Convergence of Contemporary Critical Theory and Technology.* Baltimore: Johns Hopkins University Press, 1992.

Landow, George P., ed. *Hyper/Text/Theory.* Baltimore: Johns Hopkins University Press, 1994.

Lanham, Richard A. *The Electronic Word: Democracy, Technology, and the Arts.* Chicago: University of Chicago Press, 1993.

Larsen, Deena. *Samplers: Nine Vicious Little Hypertexts.* CD-ROM. Watertown, Mass.: Eastgate Systems, 1997.

Laurel, Brenda. *Computers as Theatre.* Reading, Mass.: Addison-Wesley, 1991, 1993.

Lauter, Paul. *Canons and Contexts.* New York: Oxford University Press, 1991.

Leffingwell, William Henry. *Data on Artificial Lighting: Supplementary to Section II of "Scientific Office Management."* Chicago: A. W. Shaw, 1917.

———. *Data on Ventilation: Supplementary to Section II of "Scientific Office Management."* Chicago: A. W. Shaw, 1917.

———. *Scientific Office Management.* Chicago: A. W. Shaw, 1917.

Leffingwell, William Henry, and Edwin Marshall Robinson. *Textbook of Office Management.* 2nd ed.. New York: McGraw-Hill, 1943.

Legrady, George. *Pockets Full of Memories.* Installation, Centre Pompidou, Paris, 2001. Online representation retrieved 1 January 2002. <http://legrady.mat.ucsb.edu/pfom_lang.html>

Leighton, H. Vernon, and Jaideep Srivastava. "Precision among World Wide Web Search Services (Search Engines): Alta Vista, Excite, Hotbot, Infoseek, Lycos." 29 August 1997. Winona State University Library. Retrieved 20 December 1999. <http://www.winona.msus.edu/library/webind2/webind2.htm>

Leopoldseder, Hannes. "Foreword." In *Ars Electronica: Facing the Future: A Survey of Two Decades.* Ed. Timothy Druckrey, with Ars Electronica. Cambridge, Mass.: MIT Press, 1999.

Lévi-Strauss, Claude. *The Elementary Structures of Kinship.* Rev. ed. Trans. James Harle Bell, John Richard von Sturmer, and Rodney Needham. Ed. Rodney Needham. Boston: Beacon, 1969.

———. *The Savage Mind.* Chicago: University of Chicago Press, 1966.

———. "The Structural Study of Myth." Trans. Claire Jacobson and Brooke Grundfest Schoepf. In *Structural Anthropology.* New York: Basic, 1963.

Levy, Steven. *Hackers: Heroes of the Computer Revolution.* Garden City, N.Y.: Anchor/Doubleday, 1984.

Lichtenstein, Nelson. "The Union's Early Days: Shop Stewards and Seniority Rights." In *Choosing Sides: Unions and the Team Concept.* Ed. Mike Parker and Jane Slaughter. Boston: South End Press, 1988.

Lifson, Thomas B. "Innovation and Institutions: Notes on the Japanese Paradigm." In *Technology and the Future of Work.* Ed. Paul S. Adler. New York: Oxford University Press, 1992.

Lipnack, Jessica, and Jeffrey Stamps. *The Age of the Network: Organizing Principles for the 21st Century.* Essex Junction, Vt.: Oliver Wight, 1994.

Liu, Alan. "The Art of Extraction: Toward a Cultural History and Aesthetics of XML and Database-Driven Web Sites." Paper presented at "Interfacing Knowledge" conference, University of California, Santa Barbara. 10 March 2002. Audio recording of talk: <http://dc-mrg.english.ucsb.edu/conference/2002/voice/Sun_1513_AlanLiu.mp3> Revised version presented at Modern Language Association convention, New York. 29 December 2002.

———. "The Downsizing of Knowledge: Knowledge Work and Literary History." Abridged version (abridged for publication by Randolf Starn) of paper originally delivered at Doreen B. Townsend Center, University of California, Berkeley. In *Knowledge Work, Literary History, and the Future of Literary Studies.* Ed. Christina M. Gillis. Doreen B. Townsend Center Occasional Papers, 15. Berkeley: Doreen B. Townsend Center/Regents of the University of California, 1998.

———. "Globalizing the Humanities: 'Voice of the Shuttle: Web Page for Humanities Research.'" *Humanities Collections* 1, no. 1 (1998): 41–56.

———. "Knowledge in the Age of Knowledge Work." *Profession* (1999): 113–24.

———. "Local Transcendence: Cultural Criticism, Postmodernism, and the Romanticism of Detail." *Representations* 32 (Fall 1990): 75–113.

———. *The Lyotard Auto-Differend Page.* 1995. University of California, Santa Barbara. Retrieved 11 July 2003. <http://www.english.ucsb.edu/faculty/ayliu/research/auto/lyotgate.htm>

Secondary page cited:

"Philosophy of this Page." 3 August 1995. Retrieved 11 July 2003. <http://www.english.ucsb.edu/faculty/ayliu/research/auto/whypage.htm>

———. NASSR-L Postings (archived postings to Listserv of the North American Society for the Study of Romanticism):

"Formal vs. Historical Value, plus 'Cool.' " 29 July 1997. Retrieved 4 June 2003. <http://listserv.wvu.edu/cgi-bin/wa?A2=ind9707&L=nassr-l&T=0&O=A&P=11604>

"Further Thoughts on the Hemans Thread" [on cool and sentimentality]. 22 July 1997. Retrieved 4 June 2003. <http://listserv.wvu.edu/cgi-bin/wa?A2=ind9707&L=nassr-l&T=0&O=A&P=5757>

"Re: Hemans" [on cool and sentimentality]. 18 July 1997. Retrieved 4 June 2003. <http://listserv.wvu.edu/cgi-bin/wa?A2=ind9707&L=nassr-l&T=0&O=A&P=4126>

———. "The New Historicism and the Work of Mourning." *Studies in Romanticism* 35 (1996): 553–62.

———. "The Power of Formalism: The New Historicism." *ELH* 56 (1989): 721–71.

———. Reply [to William Pitsenberger's Letter to the Editor in regard to Alan Liu's "Knowledge in the Age of Knowledge Work"]. *Profession* (2000): 186–88.

———. "Sidney's Technology: A Critique by Technology of Literary History." In *Acts of Narrative*. Ed. Carol Jacobs and Henry Sussman. Stanford: Stanford University Press, 2003.

———. "Toward a Theory of Common Sense: Beckford's *Vathek* and Johnson's *Rasselas*." *Texas Studies in Literature and Language* 26 (1984): 183–217.

———. *Wordsworth: The Sense of History*. Stanford: Stanford University Press, 1989.

———. "Wordsworth and Subversion, 1793–1804: Trying Cultural Criticism" *Yale Journal of Criticism* 2, no. 2 (Spring 1989): 55–100.

Liu, Alan, et al. "Excerpt from the Final Technology/Software Report" (PAD Initiative). [See Electronic Literature Organization.]

Liu, Alan, ed. "Classroom of the Future: Open Planning Forum for a Digital Cultures Casebook." 3 November 2000; Web site created 2 November 2000. The Digital Cultures Project, University of California. Retrieved 19 July 2002. <http://dc-mrg.english.ucsb.edu/conference/2000/PANELS/ALiu/classroom-of-future.html>

———. *Palinurus: The Academy and the Corporation (Teaching the Humanities in a Restructured World)*. Home page. March 1998. University of California, Santa Barbara. Retrieved 19 October 2000. <http://palinurus.english.ucsb.edu>

Secondary pages cited:

"Business and Society." Retrieved 11 July 2001. <http://palinurus.english.ucsb.edu/BIBLIO-BUSINESS-and-society.html>

"The Economics of Education." Retrieved 21 October 2000. <http://palinurus.english.ucsb.edu/BIBLIO-UNIVERSITY-cost.html>

"Featured Controversy: California State University System 'Technology Infrastructure Initiative': Partnering for Knowledge." Retrieved 15 December 1999. <http://palinurus.english.ucsb.edu/CONTROVERSIES-2-calstate.html>

"The New Zealand 'Green Paper on a Future Tertiary Education' and Its Critics: A New 'Account' of Knowledge." Retrieved 15 December 1999. <http://palinurus.english.ucsb.edu/CONTROVERSIES-1-New-Zealand.html>

"*Palinurus* Rationale." Retrieved 30 July 2003. <http://palinurus.english.ucsb.edu/RATIONALE-main.html>

———. *Voice of the Shuttle: Web Site for Humanities Research*. Created 1994. University of California, Santa Barbara. Last retrieved 16 June 2003. <http://vos.ucsb.edu>

Secondary pages cited:

"Cyberculture." <http://vos.ucsb.edu/browse.asp?id=2710>

"The Laws of Cool." <http://vos.ucsb.edu/browse.asp?id=2715>

Liu, Alan, and L. O. Aranye Fradenburg, eds. *Public Humanities Initiative.* Web site of Public Humanities Initiative, Department of English, University of California, Santa Barbara. 2000–2002. Retrieved 19 July 2002. <http://www.english.ucsb.edu/initiatives/public-humanities/>

Secondary pages cited:

Entertainment Value. Web site for "Entertainment Value" conference. University of California, Santa Barbara, 3–4 May 2002. Retrieved 25 May 2002. <http://www.english.ucsb.edu/entertainment/>

"Online Resources." Last retrieved 31 July 2003. <http://www.english.ucsb.edu/initiatives/public-humanities/resources/resources-index.asp>

Liu, Alan, and Laura Mandell, eds. *Romantic Chronology.* [See under Mandell, Laura, and Alan Liu]

Liu, Alan, et al., ed. *Transcriptions: Literary History and the Culture of Information.* Home page of Transcriptions Project. Department of English, University of California, Santa Barbara. Initiated 1998. Retrieved 30 July 2003. <http://transcriptions.english.ucsb.edu>

Secondary pages cited:

"About Transcriptions." Last revised 26 September 2002. Retrieved 30 July 2003. <http://transcriptions.english.ucsb.edu/about/index.asp>

"Transcriptions Technology Facilities." Last rev. 5 December 2002. Retrieved 31 July 2003. <http://transcriptions.english.ucsb.edu/resources/guides/tech/tech_facilities.asp>

"Lizard's All-Purpose, Multi-Functional, Free Speech: Civil Disobedience, Enemies List, and Survival Guide." *Ask Mr. Lizard, Global Village Grouch.* Retrieved 22 May 2000. <http://www.dnai.com/~lizard/civdis.htm>

Loden, Marilyn, and Judy B. Rosener. *Workforce America! Managing Employee Diversity as a Vital Resource.* Homewood, Ill.: Business One Irwin, 1991.

Lodge, David. *Nice Work.* New York: Penguin, 1988.

Long, Karawynn. *The Rest of Me.* Retrieved in various versions, 1997–2000; last retrieved 10 November 2000. <http://www.karawynn.net/>

Lopez, Jon. "Principles of Design." In *Graphic Arts Manual.* Ed. Janet M. Field et al. New York: Arno, 1980.

Los Angeles Independent Media Center. Home page. Retrieved 6 July 2000. <http://la.indymedia.org/>

Los Angeles Times

"Americans Spend More Time Working, Survey Says." From Bloomberg News. 8 July 1998, p. D3.

"Downsizing Wave Has Reached a Point of Diminishing Returns." By Martha Groves. 7 July 1996, p. D12.

"The Endless Electronic Workday." By Nancy Rivera Brooks. 6 August 1999, pp. C1, C10.

"Firings and Hirings Shaping Up as the Competing Trends of 1997." By Martha Groves. 16 March 1997, p. D5.

"High-Tech Snooping All in Day's Work." By Greg Miller. 29 October 2000, pp. A1, A26, A27.

"Japanese Jolted by Demands of Future." By Teresa Watanabe and David Holley. 14 July 1996, pp. A1, A14, A15.

"Job Cuts Boost Efficiency 2.5%." By Peter S. Gosselin. 8 August 2001, pp. A1, A16.

"Just How Productive Are U. S. Workers?" By David Friedman. 9 July 2000, pp. M1, M6.

"Looking Back to Office's Future." By Marla Dickerson. 19 January 1998, Special Section on Careers, pp. 32–34.

"Missing the Boom-Time Bandwagon." By Stuart Silverstein. 20 July 1998, pp. A1, A16.

"Productivity Jumps with Help from Net." By Leslie Helm. 30 June 1999, pp. A1, A2, A13.

"A Restructuring Japan Looks to U. S. Models." By Hilary E. MacGregor. 14 July 1996, p. D4.

"Smashing the Gen-X Stereotype." By Martin Miller. 3 September 1999, pp. A1, A28.

"Software Keeps Tabs on Web-Surfing Employees." By Lawrence J. Magid. 20 May 1998, p. D8.

"State May Be in the Minority in Affirmative Action Stance." By Vicki Torres. 10 April 1997, p. D1.

"Staying in Line, Online." By Karen Kaplan. 12 October 1997, pp. D1, D12.

"Top Execs Slowly Becoming Computer Savvy." By Jennifer Pendelton. 17 January 1998, pp. D1, D3.

"Univision Adds Site to Bridge the Divide." By Lee Romney. 29 June 2000, pp. C1, C10.

"Whatever Happened to Hip?" By Mary McNamara. 22 February 2000, pp. E1, E4.

"With Labor Day Comes More Labor, Less Pay." By Jodi Wilgoren. 7 September 1998, pp. A1, A12.

"Workers Lament Loss of E-Mail Privacy on Job." By Jube Shiver, Jr. 11 October 1999, pp. A1, A10.

[See also Chapman, Gary.]

Louw, Lente-Louise. "No Potential Lost: The Valuing Diversity® Journey-An Integrated Approach to Systemic Change." In Lewis Brown Griggs and Lente-Louise Louw, ed., *Valuing Diversity: New Tools for a New Reality*. New York: McGraw-Hill, 1995.

Lovink, Geert. "The Rise and Fall of the Dot.coms." Paper presented at University of California Digital Cultures Project and University of California, Santa Barbara, Transcriptions Project, University of California, Santa Barbara, 14 April 2003.

Lunenfeld, Peter. *Snap to Grid: A User's Guide to Digital Arts, Media, and Cultures*. Cambridge, Mass.: MIT Press, 2000.

Lyon, David. "Cyberspace Sociality: Controversies over Computer-Mediated Relationships." In *The Governance of Cyberspace: Politics, Technology and Global Restructuring*. Ed. Brian D. Loader. London: Routledge, 1997.

Lyotard, Jean-François. *The Differend: Phrases in Dispute*. Trans. Georges Van Den Abbeele. Minneapolis: University of Minnesota Press, 1988.

———. *The Postmodern Condition: A Report on Knowledge*. Trans. Geoff Bennington and Brian Massumi. Minneapolis: University of Minnesota Press, 1984.

MacAdams, Lewis. *Birth of the Cool: Beat, Bebop, and the American Avant-Garde*. New York: Free Press, 2001.

MacGregor, Douglas. *The Human Side of Enterprise*. New York: McGraw-Hill, 1960.

Machlup, Fritz. *Knowledge: Its Creation, Distribution, and Economic Significance.* Vol. 1: *Knowledge and Knowledge Production.* Princeton: Princeton University Press, 1980.

———. *The Production and Distribution of Knowledge in the United States.* Princeton: Princeton University Press, 1962.

MagicURL Mystery Trip. Ryan Scott. Retrieved on various dates 1996–99; last retrieved 8 November 1999. <http://www.netcreations.com/magicurl/>

Majors, Richard, and Janet Mancini Billson. *Cool Pose: The Dilemmas of Black Manhood in America.* New York: Simon & Schuster, 1992.

Mandell, Laura, and Alan Liu, eds. *Romantic Chronology.* Created 1995–96. Miami University, Ohio/University of California, Santa Barbara. Retrieved 14 November 2000. <http://english.ucsb.edu:591/rchrono/>

Mander, Jerry. "The Net Loss of the Computer Revolution." Excerpt from Jerry Mander and Edward Goldsmith, eds., *The Case Against the Global Economy, and for a Turn Toward the Local.* San Francisco: Sierra Club Books, 1996. Excerpt in *Corporation Watch* feature on "The Battle for the Future of the Internet: Corporate Cybermall or Global Town Hall?" Retrieved 24 May 2000. <http://www.corpwatch.org/trac/feature/feature1/mander.html>

Manovich, Lev. "Avant-Garde as Software." In *Ostranenie.* Ed. Stephen Kovats. Frankfurt: Campus Verlag, 1999. Retrieved online from Manovich's Web site, 21 January 2002. <http://www.manovich.net/docs/avantgarde_as_software.doc>

———. *The Language of New Media.* Cambridge, Mass.: MIT Press, 2001.

Marchand, Roland. *Creating the Corporate Soul: The Rise of Public Relations and Corporate Imagery in American Big Business.* Berkeley: University of California Press, 1998.

Marcus, Greil. "Birth of the Cool," *Speak* (Fall 1999): 16–25.

———. *The Dustbin of History.* Cambridge, Mass.: Harvard University Press, 1995.

Marcus, Sharon. "Response" to Alan Liu's "Downsizing of Knowledge: Knowledge Work and Literary History." In *Knowledge Work, Literary History, and the Future of Literary Studies.* Ed. Christina M. Gillis. Doreen B. Townsend Center Occasional Papers, 15. Berkeley: Doreen B. Townsend Center/Regents of the University of California, 1998.

Marcuse, Herbert. "Some Social Implications of Modern Technology." In *The Essential Frankfurt School Reader.* Ed. Andrew Arato and Eike Gebhardt. New York: Continuum, 1988.

Mark, Jerome A. "Measuring Productivity in Services Industries." In *Technology in Services: Policies for Growth, Trade, and Employment.* Ed. Bruce R. Guile and James Brian Quinn. Washington, D.C.: National Academy Press, 1988.

Martin, Bill, and Ivan Szelényi. "Beyond Cultural Capital: Toward a Theory of Symbolic Domination." In *Intellectuals, Universities, and the State in Western Modern Societies.* Ed. Ron Eyerman et al. Berkeley: University of California Press, 1987.

Martin, Joanne. *Cultures in Organizations: Three Perspectives.* New York: Oxford University Press, 1992.

Marx, Karl, and Frederick Engels. *Collected Works.* Vol. 3. New York: International, 1975.

Marxhausen, Paul, ed. *Computer Related Repetitive Strain Injury Page.* 1996. Department of Electrical Engineering, University of Nebraska-Lincoln *Engineering Electronics Shop Pages.* Retrieved 18 July 2000. <http://www.engr.unl.edu/eeshop/rsi.html>

Maslan, Susan. "Resisting Representation: Theater and Democracy in Revolutionary France." *Representations* 52 (Fall, 1995): 27–51.

The Matrix. Dir. Andy Wachowski and Larry Wachowski. Warner Bros., 1999.

Mattelart, Armand. *Mapping World Communication: War, Progress, Culture*. Trans. Susan Emanuel and James A. Cohen. Minneapolis: University of Minnesota Press, 1994.

Maynard, Herman Bryant, Jr., and Susan E. Mehrtens. *The Fourth Wave: Business in the 21st Century*. San Francisco: Berrett-Koehler, 1993.

Mayo, Elton. *The Human Problems of an Industrial Civilization*. 1933; rpt., New York: Viking, 1960.

———. *The Social Problems of an Industrial Civilization*. Boston: Division of Research, Graduate School of Business Administration, Harvard University, 1945.

McArdle, Louise, et al. "Total Quality Management and Participation: Employee Empowerment, or the Enhancement of Exploitation?" In *Making Quality Critical: New Perspectives on Organizational Change*. Ed. Adrian Wilkinson and Hugh Willmott. London: Routledge, 1995.

McCaffery, Larry. "An Interview with William Gibson." In *Storming the Reality Studio*. Ed. Larry McCaffery. Durham, N.C.: Duke University Press, 1991.

McCanna, Laurie. *Creating Great Web Graphics*. New York: MIS Press, 1996.

———. *Laurie McCanna's Free Art Site*. Laurie McCanna. Retrieved on various dates, 1996–2003; last retrieved 8 July 2003. <http://www.mccannas.com/>

McCormick, Carlo. "On the Ecstatic Excess of Joseph Nechvatal." 1987. *Joseph Nechvatal Home Page*. Retrieved 30 November 2001. <http://www.eyewithwings.net/nechvatal/mccor.htm>

McGann, Jerome. "Narrative, Game, and Performative Poetics." Paper presented at "Narrative at the Outer Limits" conference, Interdisciplinary Humanities Center, University of California, Santa Barbara, 4 May 2001.

McGann, Jerome, and Johanna Drucker. "The Ivanhoe Game." Institute for Advanced Technology in the Humanities, University of Virginia, Charlottesville. Retrieved 19 July 2002. <http://www.iath.virginia.edu/~jjm2f/IGamehtm.html>

McHugh, Josh. "Politics for the Really Cool." *Forbes* magazine, 8 September 1997. *Forbes.com*. Retrieved 23 May 2000. <http://www.forbes.com/forbes/97/0908/6005172a.htm>

McKnight, Lee W., et al., ed. *Creative Destruction: Business Survival Strategies in the Global Internet Economy*. Cambridge, Mass.: MIT Press, 2001.

McLuhan, Marshall. *The Gutenberg Galaxy: The Making of Typographic Man*. Toronto: University of Toronto Press, 1962.

———. *Understanding Media: The Extensions of Man*. Cambridge, Mass.: MIT Press, 1994.

McSherry, Corynne. *Who Owns Academic Work? Battling for Control of Intellectual Property*. Cambridge, Mass.: Harvard University Press, 2001.

The Media and Communication Studies Site. Ed. Daniel Chandler. University of Wales, Aberystwyth. Last retrieved 11 September 2001. <http://www.aber.ac.uk/media/Functions/mcs.html>

Secondary page cited:

"Technological or Media Determinism." 1995. Retrieved 25 October 2000. <http://www.aber.ac.uk/media/Documents/tecdet/tecdet.html>

Mediocre Site of the Day. Ed. Jensen Harris. Now extinct. Retrieved 2 June 1997. <http://pantheon.cis.yale.edu/~jharris/mediocre.html>

Meggs, Philip B. *A History of Graphic Design*. 2nd ed. New York: Van Nostrand Reinhold, 1992.

Meštrović, Stjepan G. *Postemotional Society*. London: Sage, 1997.

Michi-Web Random Link to a Random Page. [Now extinct.] Retrieved on various dates, 1996. <http://www.cris.com/~jmeddaug/randrand.shtml>

Micklethwait, John, and Adrian Wooldridge. *The Witch Doctors: Making Sense of the Management Gurus.* New York: Times Books/Random House, 1996.

Microcosms: Objects of Knowledge. Dir. Mark Meadow and Bruce Robertson. Project home page. Ed. Jackie Spafford. University of California, Santa Barbara. Retrieved 30 July 2003. <http://www.microcosms.ihc.ucsb.edu/>

Microsoft Corporation

Microsoft Excel 2000. Release 9.0.3821 SR-1. Copyright 1995–99.

Usability Research. Home page of Microsoft Usability Group. Retrieved 9 June 1998. <http://www.microsoft.com/usability/>

Microsoft Press Computer Dictionary. 3rd ed. Redmond, Wash.: Microsoft Press, 1997.

Mike and Anthony's Wired Room. Retrieved on various dates, 1997. <http://saturn.dsu.edu/room/>

Miller, J. Hillis. *Black Holes.* Stanford: Stanford University Press, 1999. [Published on the recto pages in a volume that contains on the verso Manuel Asensi, *J. Hillis Miller; or, Boustrophedonic Reading.*]

Miller, Michael J. "The Fifth Age of Computing." *PC Magazine,* 26 May 1998, p. 4.

Miller, Rand, and Robyn Miller. *Myst.* CD-ROM. Cyan, Inc./Broderbund Software, Inc., 1994.

———. *Riven.* CD-ROM. Cyan, Inc./Broderbund Software, Inc., 1997.

Mills, C. Wright. *White Collar: The American Middle Classes.* New York: Oxford University Press, 1956.

Milton, John, "Lycidas." *Complete Poems and Major Prose.* Ed. Merritt Y. Hughes. Indianapolis: Bobbs-Merrill, 1957.

Mitzman, Arthur. *Sociology and Estrangement: Three Sociologists of Imperial Germany.* New York: Alfred A. Knopf, 1973.

Mkzdk. Stephen Miller. Retrieved in various versions, 1996–2000; last retrieved 10 November 2000. <http://www.mkzdk.org/>

Modern Language Association Committee on Professional Employment. "Final Report of the MLA Committee on Professional Employment." *PMLA* 113 (1998): 1154–77.

Mohrman, Susan Albers, Susan G. Cohen, and Allan M. Mohrman, Jr. *Designing Team-Based Organizations: New Forms for Knowledge Work.* San Francisco: Jossey-Bass, 1995.

Montrose, Louis. "Renaissance Literary Studies and the Subject of History." *English Literary Renaissance* 16 (1986): 5–12.

Morley, David, and Kevin Robins. *Spaces of Identity: Global Media, Electronic Landscapes, and Cultural Boundaries.* London: Routledge. 1995.

Motel Americana: Exploring Classic Roadside Architecture since 1995. Ed. Jenny Wood and Andy Wood. 2000. Retrieved 8 November 2000. <http://www.sjsu.edu/faculty/wooda/motel.html>

Moulthrop, Stuart. "Rhizome and Resistance: Hypertext and the Dreams of a New Culture." In *Hyper/Text/Theory.* Ed. George P. Landow. Baltimore: Johns Hopkins University Press, 1994.

Murolo, Priscilla. "White-Collar Women and the Rationalization of Clerical Work." In *Technology and the Transformation of White-Collar Work.* Ed. Robert E. Kraut. Hillsdale, N.J.: Lawrence Erlbaum Associates, 1987.

Murphy, Jay. "Joseph Nechvatal." *Galleries Magazine.* 1995. *Joseph Nechvatal Home Page.* Retrieved 30 November 2001. <http://www.eyewithwings.net/nechvatal/murphy.htm>

Murray, Janet H. *Hamlet on the Holodeck: The Future of Narrative in Cyberspace.* Cambridge, Mass.: MIT Press, 1997.

Nakamura, Lisa. *Cybertypes: Race, Ethnicity, and Identity on the Internet.* New York: Routledge, 2002.

Nancy, Jean-Luc. *The Inoperative Community.* Ed. Peter Connor. Trans. Peter Connor et al. Minneapolis: University of Minnesota Press, 1991.

Narayan, Shoba. "What Makes a Site Cool? Rules Have Changed since Last Year." *Web Week* 2 no.12 (19 August 1996). Republished in *Internet World Daily.* 19 August 1996. Retrieved 15 May 2000. <http://www.internetworld.com/print/1996/08/19/undercon/rules.html>

Nechvatal, Joseph. *An Ecstasy of Excess, with an Essay by Noemi Smolik.* 4050 Mönchengladbach 1, Germany: Juni-Verlag, n.d.

———. E-mail to the author. 9 June 2003.

———. Home page. Retrieved on various dates 2001–2003; last retrieved 2 August 2003. <http://www.nechvatal.net/> or <http://www.eyewithwings.net/nechvatal/>

Secondary pages cited (pages authored by Nechvatal only; writings on this Web site about Nechvatal by others are cited under individual author names):

"An Artist's Statement by Joseph Nechvatal for His May Exhibit at Universal Concepts Unlimited." Retrieved 24 July 2003. <http://www.eyewithwings.net/nechvatal/voluptuary/text.html>

"Biography of Joseph Nechvatal." Retrieved 2 August 2003. <http://www.eyewithwings.net/nechvatal/biography.html>

"The Computer Virual Formula." Retrieved 30 November 2001. <http://www.eyewithwings.net/nechvatal/virusfor.html>

Computer Virus Project 2.0. Retrieved 24 July 2003. <http://www.eyewithwings.net/nechvatal/virus2/virus20.html>

Drawings and Drawing-Based Works from 1980–1987. Retrieved 2 August 2003. <http://www.eyewithwings.net/nechvatal/early/early.html>

"Virus Text (We Form a Rhizome with Our Viruses)." Retrieved 3 August 2003. <http://www.eyewithwings.net/nechvatal/virustxt.html>

vOluptuary: an algorithic hermaphornology. Web page for Nechvatal's exhibit of the "vOluptuary: an algorithic hermaphornology" series at Universal Concepts Unlimited, New York City, 22 May to 3 July 2002. Retrieved 24 July 2003. <http://www.eyewithwings.net/nechvatal/algorithic.html>

———. *Joseph Nechvatal.* Catalog of "Computer Virus Exhibition," Saline Royale d'Arcet-Senans, 10 September–1 November 1993. Arc-et-Senans: Fondation Claude-Nicolas Ledoux/Fonds Régional d'Art Contemporain de Franche-Comté, 1993.

Negroponte, Nicholas. *Being Digital.* New York: Vintage, 1995.

Nelson, Cary, ed. *Will Teach for Food: American Labor in Crisis.* Minneapolis: University of Minnesota Press, 1997.

The Net. Dir. Irwin Winkler. Columbia Pictures. 1995.

Netscape Communications, Inc. [Later part of AOL-TimeWarner, Inc.] *The Amazing Fishcam!* 1997. Retrieved 28 June 1997. <http://home.netscape.com/fishcam/fishcam.html>

———. "What's Cool?" 7 March 1996. Retrieved 28 June 1997. <http://www3.netscape.com/

escapes/whats_cool.html> [Title shown on page is "What's Cool," but the link from the Netscape home page added the question mark.]

———. "What's Cool." 2 July 1998. Retrieved 2 July 1998. <http://home.netscape.com/netcenter/cool.html> [Link to page from the Netscape home page is also "What's Cool" without a question mark.]

———. "What's Cool: Editorial Policy." 2 July 1998. Retrieved 7 July 1998. <http://home.netscape.com/netcenter/cool/editorial.html>

The New Hacker's Dictionary. [See under *The Jargon File*.]

Newfield, Christopher. "Corporate Culture Wars." In *Corporate Futures: The Diffusion of the Culturally Sensitive Corporate Form*. Ed. George E. Marcus. Chicago: University of Chicago Press, 1998.

———. *Ivy and Industry: Business and the Making of the American University, 1880–1980*. Durham, N.C.: Duke University Press, 2003.

Newman, John Henry. *The Idea of a University*. Garden City, N.Y.: Image Books, 1959.

Newsweek

"bigbrother@the.office.com." By Deborah Branscum. 27 April 1998, p. 78.

"The Browser War: Microsoft and Netscape Fight It Out for Control of the Internet." By Steven Levy. 29 April 1996: 47–50.

"The Kids Know Cool." By Gregory Beals and Leslie Kaufman. 31 March 1997, pp. 48–49.

"Will Cigars Stay Hot? How to Track the Trend." 21 July 1997, p. 59.

New York Times

"Forgot a Password? Try 'Way2Many.' " By Jennifer 8. Lee. 5 August 1999. *New York Times on the Web*. Retrieved 8 August 2001. <http://query.nytimes.com/search/advanced>

"Is This the Factory of the Future?" By Saul Hansell. 26 July 1998. *New York Times on the Web*. Retrieved 7 August 2001. <http://query.nytimes.com/search/advanced>

"Talk, Type, Read E-Mail: The Trials of Multitasking." By Amy Harmon. 23 July 1998. *New York Times on the Web*. Retrieved 21 July 2002. <http://query.nytimes.com/search/advanced>

Nideffer, Robert. *Proxy*. University of California, Irvine. Retrieved 19 July 2002. <http://proxy.arts.uci.edu/>

Noble, David F. "Digital Diploma Mills: The Automation of Higher Education." *First Monday* 3, no. 1 (5 January 1998). Retrieved 19 July 2002. <http://www.firstmonday.dk/issues/issue3_1/noble/index.html>

Northern Light. Search engine. Northern Light Technology, LLC. Retrieved 6–7 July 1998. <http://www.northernlight.com/>

Novak, Marcos. E-mail to the author, 7 March 2002.

———. Home page. Retrieved 24 July 2002. <http://www.centrifuge.org/marcos/>

———. "Liquid~, Trans~, Invisible~: The Ascent and Speciation of the Digital in Architecture. A Story." Web site for Digital/Real—Blobmeister, First Built Projects exhibition. Deutsches Architektur Museum, Frankfurt, 30 May–5 August, 2001. Retrieved 24 July 2002. <http://www.a-matter.de/digital-real/eng/mainframe.asp?sel=17>

———. Talk in the Media Arts and Technology lecture series, University of California, Santa Barbara, 3 March 2002.

Ode to Lynx. Poem by "Zeigen." Image capture by Mike Batchelor. Retrieved 9 July 2003. <http://www.batch.com/ode-to-lynx.html>

Ong, Walter J. *Orality and Literacy: The Technologizing of the Word.* London: Methuen, 1982.

Online Cameras. AvEdis. Retrieved on various dates, 1996–1997; last retrieved 13 May 1997. <http://www.rdrop.com/users/avedis/htcamera.htm>

Ouchi, William G. *Theory Z: How American Business Can Meet the Japanese Challenge.* Reading, Mass.: Addison-Wesley, 1981.

Packard, Vance. *The Status Seekers: An Exploration of Class Behavior in America and the Hidden Barriers That Affect You, Your Community, Your Future.* New York: David McKay, 1959.

Packer, Randall. "Net Art as Theater of the Senses: A HyperTour of Jodi and Grammatron." *Beyond Interface* online exhibition. Curator, Steve Dietz. Museums and the Web conference, Toronto, 22–25 April 1998. Retrieved 6 August 2003. <http://www.archimuse.com/mw98/beyondinterface/bi_frpacker.html>

Paley, William. *Natural Theology; or, Evidences of the Existence and Attributes of the Deity.* 12th ed. London: J. Faulder, 1809. University of Michigan Humanities Text Initiative. 1998. Retrieved 15 July 2002. <http://www.hti.umich.edu/cgi/p/pd-modeng/pd-modeng-idx?type=HTML&rgn=TEI.2&byte=53049319>

Parker, Donn B. *Fighting Computer Crime: A New Framework for Protecting Information.* New York: John Wiley & Sons, 1998.

Parker, Mike, and Jane Slaughter. *Choosing Sides: Unions and the Team Concept.* Boston: South End Press, 1988.

Paul's a Computer Geek. [See under Schrank, Paul.]

Paul's (Extra) Refrigerator. Paul Hass. Retrieved 26 September 2000. <http://www.hamjudo.com/cgi-bin/refrigerator>

Paulson, William R. *The Noise of Culture: Literary Texts in a World of Information.* Ithaca, N.Y.: Cornell University Press, 1988.

PC Magazine

"The Fifth Age of Computing." By Michael J. Miller. 26 May 1998, p. 4.

"Managed PCs: Order and Flexibility in One Package." By Cade Metz. 30 June 1998, pp. 126–89.

"Search Engines' Tiny Bite." 1 September 1999, p. 9.

Peacock, Thomas Love. *The Four Ages of Poetry.* Excerpted in *Critical Theory since Plato.* Rev. ed. Ed. Hazard Adams. New York: Harcourt Brace Jovanovich, 1992.

Penn State Center for Quality and Planning. (Name later changed to Office of Planning and Institutional Assessment.) Home Page. 27 June 2001. Pennsylvania State University. Retrieved 11 July 2001. <http://www.psu.edu/dept/president/cqi/index.htm>

Perkins, Barbara, and George Perkins, eds. *Kaleidoscope: Stories of the American Experience.* New York: Oxford University Press, 1993.

Peters, Tom. *Liberation Management: Necessary Disorganization for the Nanosecond Nineties.* New York: Fawcett Columbine, 1992.

Peters, Thomas J. [Tom], and Robert H. Waterman, Jr. *In Search of Excellence: Lessons from America's Best Run Companies.* New York: Warner, 1982.

Petrey, Sandy. *Speech Acts and Literary Theory.* New York: Routledge, 1990.

Pfau, Thomas. *Wordsworth's Profession: Form, Class, and the Logic of Early Romantic Cultural Production.* Stanford: Stanford University Press, 1997.

Philadelphia Direct Action Group. Home page (*The Party's Over*). Retrieved 6 July 2000. <http://www.thepartysover.org/>

Secondary page cited:

"Call to Action for Domestic and Global Justice." Retrieved 6 July 2000. <http://www.thepartysover.org/calltoaction.html>

Picard, Rosalind W. *Affective Computing*. Cambridge, Mass.: MIT Press, 1997.

Pierson, Michele. "Welcome to Basementwood: Computer Generated Special Effects and *Wired* Magazine." *Postmodern Culture* 8, no. 3 (May 1998). Retrieved 15 November 2000. <http://jefferson.village.virginia.edu/pmc/text-only/issue.598/8.3pierson.txt>

Pitsenberger, William. Letter to the Editor, [Response to Alan Liu's "Knowledge in the Age of Knowledge Work."] *Profession* (2000): 185–86.

Poe, Edgar Allan. *Collected Works*. Vol. 1. Ed. Thomas Ollive Mabbott. Cambridge, Mass.: Harvard University Press, 1969.

Polsson, Ken. *Chronology of Events in the History of Microcomputers*. 11 December 1997. Retrieved 19 January 1998. <http://www.islandnet.com/~kpolsson/comphist.htm>

Pope, Alexander. *The Poems of Alexander Pope: A One-Volume Edition of the Twickenham Text with Selected Annotations*. Ed. John Butt. Rpt.; New Haven.: Yale University Press. 1963

Popper, Frank. "On Joseph Nechvatal." 2001. *Joseph Nechvatal Home Page*. Retrieved 30 November 2001. <http://www.eyewithwings.net/nechvatal/popper.html>

Porat, Marc Uri, with Michael R. Rubin. *The Information Economy*. 9 vols. Washington, D.C.: U. S. Department of Commerce, Office of Telecommunications, 1977.

Porter, David, ed. *Internet Culture*. New York: Routledge, 1996.

Porter, James H. "Business Reengineering in Higher Education: Promise and Reality." *CAUSE/EFFECT* 16, no. 4 (1993). *Educause*. Retrieved 11 July 2001. <http://www.educause.edu/ir/library/text/cem934a.txt>

Poster, Mark. *The Mode of Information: Poststructuralism and Social Context*. Chicago: University of Chicago Press, 1990.

———. *The Second Media Age*. Cambridge: Polity, 1995.

Poulantzas, Nicos. *Classes in Contemporary Capitalism*. Trans. David Fernbach. London: Verso, 1978.

Privacy International. Home page. *Privacy.org*. Retrieved 22 May 2000. <http://www.privacy.org/pi/>

Privacy Rights Clearinghouse. Home page. 20 May 2000. Retrieved 24 May 2000. <http://www.privacyrights.org/>

Secondary page cited:

"Fact Sheet #7: Employee Monitoring: Is There Privacy in the Workplace?" March 1993; rev. August 1997. Retrieved 24 May 2000. <http://privacyrights.org/FS/fs7-work.htm>

The Professionals. Dir. Richard Brooks. Columbia, 1966.

Progress & Freedom Foundation (PFF). Home page. Retrieved on various dates, 1999–2000; last retrieved 17 September 2001. <http://www.pff.org/>

Secondary pages cited:

Dyson, Esther, George Gilder, George Keyworth, and Alvin Toffler. "Cyberspace and the American Dream: A Magna Carta for the Knowledge Age." Release 1.2. 22 August 1994. Retrieved 22 May 2000. <http://www.pff.org/position.html>

"Mission Statement." Retrieved 22 May 2000. <http://www.pff.org/what_we_do.htm>

Project Cool. Home page. Project Cool, Inc. Retrieved 19 December 1999. <http://www.projectcool.com/> [Project Cool was later acquired by DevX, Inc. and its pages folded into the DevX.com site. However, versions of many of the pages I discuss, including "Sightings" and the archive of "Previous Sightings," continue to exist and may be accessed at the new address: <http://archive.devx.com/projectcool/sightings/previous.html>]

Secondary pages cited:

"1998: Citing the Sightings." Retrieved 8 September 1999. <http://www.projectcool.-com/sightings/1998citings.html>

"About the Coolest on the Web." Retrieved 12 June 1997. [Accessed through the *Project Cool* site's "Coolest of the Web" frameset: <http://www.projectcool.com/coolest/> under the link for "About the Coolest."]

"Ice." Retrieved 8 September 1999. <http://www.projectcool.com/sightings/ice.html>

"Liquid." Retrieved 8 September 1999. <http://www.projectcool.com/sightings/liquid.html>

"Sightings" (and "Previous Sightings"). Various pages of archived "sightings" of "cool" Web sites made available through scripted navigation links accessed by the "Previous Sightings" link on this page. Retrieved 16 July–8 September 1999. <http://www.projectcool.com/sightings/>

Public Broadcasting Service. *The Merchants of Cool: A Report on the Creators & Marketers of Popular Culture for Teenagers.* Web site supporting the *Frontline* television feature of this title of 27 February 2001. *Frontline* Web site, February 2001. Retrieved 1 March 2001. <http://www.pbs.org/wgbh/pages/frontline/shows/cool/>

Secondary page cited:

"The Symbiotic Relationship between the Media and Teens." Retrieved 1 March 2001. <http://www.pbs.org/wgbh/pages/frontline/shows/cool/themes/symbiotic.html>

———. "Down in the Valley." Transcript of PBS *NewsHour with Jim Lehrer* show. 10 May 2000. *PBS Online NewsHour.* Retrieved 28 November 2000. <http://www.pbs.org/newshour/bb/cyberspace/jan-june00/silicon_valley_5-10.htm>

Purvis, Cynthia J. Roe, Mary Czerwinski, and Paul Weiler. "The Human Factors Group at Compaq Computer Corporation." In *Usability in Practice: How Companies Develop User-Friendly Products.* Ed. Michael E. Wiklund. Boston: Academic Press, 1994.

Pynchon, Thomas. *The Crying of Lot 49.* New York: Perennial Classics/HarperCollins, 1999.

Rafaeli, Anat, and Robert I. Sutton. "Expression of Emotion as Part of the Work Role." *Academy of Management Review* 12 (1987): 23–37.

———. "The Expression of Emotion in Organizational Life." *Research in Organizational Behavior* 11 (1989): 1<n.42. [See also under Sutton and Rafaeli.]

Raley, Rita. "eEmpires." *Cultural Critique* (forthcoming).

———. "Global English and the Academy." Unpublished book manuscript, courtesy of the author.

———. "On Global English and the Transmutation of Postcolonial Studies into 'Literature in English.'" *Diaspora* 8:1 (1999): 51–80.

———. "Machine Translation and Global English." *Yale Journal of Criticism* 16, no. 2 (2003): 291–313.

———. "Reveal Codes: Hypertext and Performance." *Postmodern Culture* 12, no. 1 (September 2001). Retrieved 30 July 2002. <http://muse.jhu.edu/journals/pmc/v012/12.1raley.html> Text-only version: <http://www.iath.virginia.edu/pmc/text-only/issue.901/12.1raley.txt>

Rand, Paul. *A Designer's Art.* New Haven: Yale University Press, 1985.

Ransom, John Crowe, *The New Criticism.* Norfolk, Conn.: New Directions, 1941.

Read_Me Festival 1.2. "Award Winners." Web site for software arts festival. DOM Cultural Center, Moscow, 18–19 May 2002. Retrieved 24 July 2002. <http://www.macros-center.ru/read_me/adden.htm>

Readings, Bill. *The University in Ruins.* Cambridge, Mass.: Harvard University Press, 1996.

Reimers, Sigurd, and Andy Treacher, with Carolyn White. *Introducing User-Friendly Family Therapy.* New York: Routledge, 1995.

Remembering Nagasaki. San Francisco Exploratorium. Retrieved 15 November 2000. <http://www.exploratorium.edu/nagasaki/>

The Rest of Me. [See under Long, Karawynn.]

Reynolds, Sir Joshua. *Discourses on Art.* Ed. Robert R. Wark. 1959; rpt., New Haven: Yale University Press, 1975.

"RFC 3: Documentation Conventions." By Stephen D. Crocker. 9 April 1969. *RFC Editor Homepage.* Retrieved 14 July 2003. <ftp://ftp.rfc-editor.org/in-notes/rfc3.txt>

"RFC 30: Documentation Conventions." By Stephen D. Crocker. 4 February 1970. *RFC Editor Homepage.* Retrieved 14 July 2003. <ftp://ftp.rfc-editor.org/in-notes/rfc30.txt>

Rheingold, Howard. "Technology, Community, Humanity and the Net." 29 April 1999. *Intellectual Capital.com.* Retrieved 27 November 2000. <http://www.intellectualcapital.com/issues/issue225/item4242.asp>

———. *The Virtual Community: Homesteading on the Electronic Frontier.* New York: HarperCollins, 1994. *Howard Rheingold Virtual Community Services.* Retrieved 26 October 2000. <http://www.rheingold.com/vc/book/>

Rico, Barbara Roche, and Sandra Mano, eds. *American Mosaic: Multicultural Readings in Context.* Boston: Houghton Mifflin, 1991.

Riesman, David, with Reuel Denney and Nathan Glazer. *The Lonely Crowd: A Study of the Changing American Character.* New Haven: Yale University Press, 1950.

Rifkin, Jeremy. *The End of Work: The Decline of the Global Labor Force and the Dawn of the Post-Market Era.* New York: G. P. Putnam, 1995.

Roach, Stephen S. "Technology and the Services Sector: America's Hidden Competitive Challenge." In *Technology in Services: Policies for Growth, Trade, and Employment.* Ed. Bruce R. Guile and James Brian Quinn. Washington, D.C.: National Academy Press, 1988.

Robbins, Bruce. *Secular Vocations: Intellectuals, Professionalism, Culture.* New York: Verso, 1993.

Robins, Kevin, and Frank Webster. *Times of the Technoculture: From the Information Society to the Virtual Life.* London: Routledge, 1999.

Robinson, Sidney K. *Inquiry into the Picturesque.* Chicago: University of Chicago Press, 1991.

Rochlin, Gene I. *Trapped in the Net: The Unanticipated Consequences of Computerization.* Princeton: Princeton University Press, 1997.

Rockett's New School. CD-ROM. Mountain View, Calif.: Purple Moon Media, 1997.

The Rock-n-Roll Gallery Collection of Richard E. Aaron. Retrieved 22 August 1999. <http://www.rockpix.com/>

Rockwell, Geoffrey, et al. "Tracking Culture on the Web: An Experiment." Paper presented at panel on "Digital Culture," 2001 Joint International Conference of the Association for Computers and the Humanities and Association for Literary and Linguistic

Computing, New York University, 15 June 2001. (See also online abstract, retrieved from conference site 7 July 2003. <http://www.nyu.edu/its/humanities/ach_allc2001/papers/rockwell/index.html>)

Rodowick, David N. "Cybernetic and Machinic Arrangements." Paper presented at MLA convention, Toronto, 28 December 1997.

Rogers, Everett M., and Judith K. Larsen. *Silicon Valley Fever: Growth of High-Technology Culture.* New York: Basic, 1984.

Romantic Circles. General editors, Neil Fraistat, Steven E. Jones, Carl Stahmer. University of Maryland. Last retrieved 19 July 2002. <http://www.rc.umd.edu/>

Secondary pages cited:

Frankenstein MOO. Created by Ron Broglio and Eric Sonstroem. "Learning Module" of *Romantic Circles High School.* May 2001. Retrieved 19 July 2002. <http://www.lcc.gatech.edu/~broglio/rc/frankenstein/>

Villa Diodati. Web-MOOspace. Retrieved 19 July 2002. <http://www.rc.umd.edu:7000/>

Rosen, Jeffrey . *The Unwanted Gaze: The Destruction of Privacy in America.* New York: Random House, 2000.

Rosen, Michael. "Coming to Terms with the Field: Understanding and Doing Organizational Ethnography." *Journal of Management Studies* 28 (1991): 1–24.

Rosenberg, Daniel, and Liam Friedland. "Usability at Borland: Building Best of Breed Products." In *Usability in Practice: How Companies Develop User-Friendly Products.* Ed. Michael E. Wiklund. Boston: Academic Press, 1994.

Rosenberg, Richard S. *The Social Impact of Computers,* 2nd ed. San Diego: Academic Press, 1997.

Ross, Andrew. *No Respect: Intellectuals and Popular Culture.* New York: Routledge, 1989.

———. *Strange Weather: Culture, Science, and Technology in the Age of Limits.* London: Verso, 1991.

Roszak, Theodore. *The Cult of Information: A Neo-Luddite Treatise on High Tech, Artificial Intelligence, and the True Art of Thinking.* Rev. ed. Berkeley: University of California Press, 1994.

———. *The Making of a Counter Culture: Reflections on the Technocratic Society and Its Youthful Opposition.* 1969; rpt. with new introduction, Berkeley: University of California Press, 1995.

RTMark. "Press Release: The Brent Spar of E-Commerce: eToys vs. eToy Post-Hearing Press Conference." 25 December 1999. *RTMark.* Retrieved 7 August 2003. <http://www.rtmark.com/etoyprxmas.html>

Rybczynski, Witold. *Waiting for the Weekend.* New York: Penguin, 1991.

Salaman, Graeme. "Culturing Production." In *Production of Culture/Cultures of Production.* Ed. Paul du Gay. London: SAGE, in association with Open University, 1997.

San Francisco Chronicle. "Sunday Interview—Joe Hill Meets the Microchip: Amy Dean, Chief Executive Director of the South Bay AFL-CIO Central Labor Council, Thinks It's Time for Labor to Make a Hard Drive in Silicon Valley." By Mark Simon. 13 July 2000. *SFGate.com.* Retrieved 14 July 2000. <http://www.sfgate.com/cgi-bin/article.cgi?file=/chronicle/archive/1997/07/13/SC24580.DTL>

Sano, Darrell. *Designing Large-Scale Web Sites: A Visual Design Methodology.* New York: John Wiley & Sons, 1996.

Sas, Miryam. "Response" to Alan Liu's "Downsizing of Knowledge: Knowledge Work and Literary History." In *Knowledge Work, Literary History, and the Future of Literary Stud-*

ies. Ed. Christina M. Gillis. Doreen B. Townsend Center Occasional Papers, 15. Berkeley: Doreen B. Townsend Center/Regents of the University of California, 1998.

Schivelbusch, Wolfgang. *The Railway Journey: The Industrialization of Time and Space in the 19th Century.* Berkeley: University of California Press, 1986.

Schlesinger, Philip. "In Search of the Intellectuals: Some Comments on Recent Theory." In *Media, Culture and Society: A Critical Reader.* Ed. Richard Collins et al. London: Sage, 1986.

Schoenfield, Mark. *The Professional Wordsworth: Law, Labor, and the Poet's Contract.* Athens: University of Georgia Press, 1996.

Schor, Juliet B. *The Overworked American: The Unexpected Decline of Leisure.* New York: Basic, 1992.

Schrank, Paul. *Paul's a Computer Geek.* Retrieved in various versions, 1997–99; last retrieved 3 December 2000. <http://pgeek.com>

Secondary page cited:

Design. Retrieved in various versions, 1997–99; last retrieved 3 December 2000. <http://pgeek.com/design.htm>

Schumpeter, Joseph A. *Capitalism, Socialism and Democracy.* New York: Harper and Row, 1975.

Schwenger, Peter. "*Agrippa,* or, The Apocalyptic Book." *South Atlantic Quarterly* 92 (1993): 617–26.

Scott, Bonnie Kime, ed. *The Gender of Modernism: A Critical Anthology.* Bloomington: Indiana University Press, 1990.

Search Engine Watch. Ed. Danny Sullivan. Internet.com/Mecklermedia, Inc. Last retrieved 12 September 2001. <http://searchenginewatch.com/>
Secondary pages cited:

"The Major Search Engines." Retrieved 6 July 1998. <http://searchenginewatch.internet.com/facts/major.html>

"Search Engine Features Chart." 17 June 1998. Retrieved 7 July 1998. <http://searchenginewatch.internet.com/webmasters/features.html>

"Search Engine Reviews." Retrieved 6 July 1998. <http://searchenginewatch.internet.com/resources/reviews.html>

"Search Engine Reviews Chart." Retrieved 6 July 1998. <http://searchenginewatch.internet.com/reports/reviewchart.html>

Sullivan, Danny. "Pay for Placement?" 4 September 2001. Retrieved 12 September 2001. <http://www.searchenginewatch.com/resources/paid-listings.html>

Searle, John R. *Speech Acts: An Essay in the Philosophy of Language.* Cambridge: Cambridge University Press, 1969.

Seattle Independent Media Center. Home page. Retrieved 6 July 2000. <http://seattle.indymedia.org/>

Seattle Times. "Microsoft Temps Group Joins Union." By Keith Ervin. 4 June 1999. *SeattleTimes.com.* Retrieved 28 November 2000. <http://seattletimes.nwsource.com/news/business/html98/temp_19990604.html>

Seeman, Melvin. "Alienation and Engagement." In *The Human Meaning of Social Change.* Ed. Angus Campbell and Philip E. Converse. New York: Russell Sage Foundation, 1972.

The Semi-Existence of Bryon. [Now extinct.] Retrieved on various dates, 1996–1999; last retrieved 5 September 1999. <http://www.geocities.com/~semi_bryon/>

Senge, Peter M. *The Fifth Discipline: The Art and Practice of the Learning Organization.* New York: Doubleday, 1990.

Shelley, Percy. *A Defense of Poetry.* In *Critical Theory since Plato.* Rev. ed. Ed. Hazard Adams. New York: Harcourt Brace Jovanovich, 1992.

———. "The Triumph of Life." In *The Norton Anthology of English Literature.* Vol. 2. 6th ed. M. H. Abrams, general editor. New York: W. W. Norton, 1993.

Shenk, David. *Data Smog: Surviving the Information Glut.* New York: HarperCollins, 1997.

Shimomura, Tsutomu, with John Markoff. *Takedown: The Pursuit and Capture of Kevin Mitnick, America's Most Wanted Computer Outlaw—By the Man Who Did It.* New York: Hyperion, 1996.

Shklovsky, Victor. "Art as Technique." In *Russian Formalist Criticism: Four Essays.* Trans. Lee T. Lemon and Marion J. Reis. Lincoln: University of Nebraska Press, 1965.

Shumar, Wesley. *College for Sale: A Critique of the Commodification of Higher Education.* London: Falmer, 1997.

Sidney, Sir Philip. *An Apology for Poetry.* In *Critical Theory since Plato.* Rev. ed. Ed. Hazard Adams. New York: Harcourt Brace Jovanovich, 1992.

Sikora, Stéphane, and Joseph Nechvatal. "The Model: Notes." On Web page for Joseph Nechvatal's Virus Project 2.0. Retrieved 24 July 2003. <http://www.eyewithwings.net/nechvatal/virus2/virus20.html>

Silicon Valley Toxics Coalition (SVTC). Home page. Retrieved 11 July 2000. <http://www.svtc.org/>

Secondary page cited:

"Responsible Technology Goes Global!!" 11 August 1998. Retrieved 18 July 2000. <http://www.svtc.org/icrt.htm>

Simon, Mark. "Sunday Interview—Joe Hill Meets the Microchip: Amy Dean, Chief Executive Director of the South Bay AFL-CIO Central Labor Council, Thinks It's Time for Labor to Make a Hard Drive in Silicon Valley." [See under *San Francisco Chronicle.*]

Simonds, William Adams. *Henry Ford: His Life, His Work, His Genius.* Indianapolis: Bobbs-Merrill, 1943.

Simpson, David. *The Academic Postmodern and the Rule of Literature: A Report on Half-Knowledge.* Chicago: University of Chicago Press, 1995.

The Simulator. Garnet Hertz. 1997. ConceptLAB. Retrieved 8 November 2000. <http://www.conceptlab.com/simulator/index.html>

Smith, Henry Clay. *Psychology of Industrial Behavior.* New York: McGraw-Hill, 1955.

Smith, Martha Nell. "Computing: What Has American Literary Studies to Do with It?" *American Literature* 74, no. 4 (2002): 833–57.

Smith, Ted. "The Dark Side of High Tech Development." *Corporate Watch.* Transnational Resource and Action Center. 10 February 1997. Retrieved 11 July 2000. <http://www.corpwatch.org/trac/feature/hitech/overview.html>

Snow, C. P. *The Two Cultures and the Scientific Revolution.* New York: Cambridge University Press, 1959.

Sobel, Richard. *The White Collar Working Class: From Structure to Politics.* New York: Praeger, 1989.

Soja, Edward W. *Postmodern Geographies: The Reassertion of Space in Critical Social Theory.* London: Verso, 1989.

Sondheim, Alan. "Introduction: Codework." *American Book Review,* 22, no. 6 (September/

October 2001). Retrieved 27 July 2003. <http://www.litline.org/ABR/issues/Volume22/Issue6/sondheim.pdf>

Spalter, Anne Morgan. *The Computer in the Visual Arts.* Reading, Mass.: Addison-Wesley, 1999.

Spenser, Edmund. *Edmund Spenser's Poetry: Authoritative Texts, Criticism.* Ed. Hugh Maclean. New York: W. W. Norton, 1968.

Sperte, Sean. Home page. Retrieved 22 August 1999. <http://members.xoom.com/_XOOM/sperte/index.html>

Sproull, Lee, and Sara Kiesler. *Connections: New Ways of Working in the Networked Organization.* Cambridge, Mass.: MIT Press, 1991.

Stahlke, Herbert F. W., and James M. Nyce. "Reengineering Higher Education: Reinventing Teaching and Learning." *CAUSE/EFFECT* 19, no. 4 (1996): 44–51. *Educause.* Retrieved 11 July 2001. <http://www.educause.edu/ir/library/html/cem9649.txt>

Stalbaum, Brett. "The Zapatista Tactical FloodNet: A Collaborative, Activist and Conceptual Art Work of the Net." [see Electronic Disturbance Theater.]

Stearns, Peter N. *American Cool: Constructing a Twentieth-Century Emotional Style.* New York: New York University Press, 1994.

Stearns, Peter N., with Carol Z. Stearns. "Emotionology: Clarifying the History of Emotions and Emotional Standards." *American Historical Review* 90 (1985): 813–36.

Stefik, Mark, ed. *Internet Dreams: Archetypes, Myths, and Metaphors.* Cambridge, Mass.: MIT Press, 1996.

Steiner, Peter. *Russian Formalism: A Metapoetics.* Ithaca, N.Y.: Cornell University Press, 1984.

Steinkamp, Fiona. "The Ideology of the Internet." Retrieved 19 May 2000. <http://www.univie.ac.at/philosophie/bureau/steinkamp.html>

Stephens, Mitchell. *The Rise of the Image, the Fall of the Word.* New York: Oxford University Press, 1998.

Stephenson, Neal. *Cryptonomicon.* New York: Avon, 1999.

———. *The Diamond Age, or, A Young Lady's Illustrated Primer.* New York: Bantam, 1995.

———. *In the Beginning . . . Was the Command Line.* New York: Avon, 1999.

———. *In the Beginning Was the Command Line.* 1999. File "command.txt" downloaded in compressed form on 9 August 1999 from *Crytonomicon* Web site. <http://www.cryptonomicon.com/beginning.html>

———. *Snow Crash.* New York: Bantam, 1992.

Sterling, Bruce. *The Hacker Crackdown: Law and Disorder on the Electronic Frontier.* New York: Bantam, 1992. [Also available online from numerous sources, including versions with preface and epilogue added in 1994.] Retrieved 27 December 1999 from <http://www.lysator.liu.se/etexts/hacker/>

Sterling, Bruce, ed. *Mirrorshades: The Cyberpunk Anthology.* 1986; rpt., New York: Ace Books, 1988.

Stern, Jane, and Michael Stern. *Sixties People.* New York: Alfred A. Knopf, 1990.

Stewart, John L. *The Burden of Time, The Fugitives and Agrarians: The Nashville Groups of the 1920's and 1930's, and the Writing of John Crowe Ransom, Allen Tate, and Robert Penn Warren.* Princeton: Princeton University Press, 1965.

Stewart, Thomas A. *Intellectual Capital: The New Wealth of Organizations.* New York: Doubleday, 1997.

Stiles, Kristine. "Selected Comments on Destruction Art." In *Book for the Unstable Media.*

Ed. Alex Adriaansens et al. 1992. *Book for the Unstable Media* Web site. V2_Organisation, Institute for the Unstable Media. Retrieved 25 November 2001. <http://www.v2.nl/publicaties/unst_media/stiles.html>

———. "Thresholds of Control: Destruction Art and Terminal Culture." In *Ars Electronica: Facing the Future: A Survey of Two Decades.* Ed. Timothy Druckrey, with Ars Electronica. Cambridge, Mass.: MIT Press, 1999.

Stiles, Kristine, and Peter Selz, eds. *Theories and Documents of Contemporary Art: A Sourcebook of Artists' Writings.* Berkeley: University of California Press, 1996.

Stillbirth and Neonatal Death Support (SANDS). [Later named SIDS Western Australia.] Home page. Ed. Tim Law. Retrieved on various dates, 1996–2003; last retrieved 12 July 2003. <http://www.sandswa.org.au/>

Stoll, Clifford. *Silicon Snake Oil: Second Thoughts on the Information Highway.* New York: Anchor/Doubleday, 1995.

Stone, Allucquère Rosanne. *The War of Desire and Technology at the Close of the Mechanical Age.* Cambridge, Mass.: MIT Press, 1996.

Stone, Lawrence. *The Family, Sex and Marriage in England 1500–1800.* New York: Harper and Row, 1977.

Strassmann, Paul A. *Information Payoff: The Transformation of Work in the Electronic Age.* New York: Free Press/Macmillan, 1985.

Straubhaar, Joseph, and Robert LaRose. *Communications Media in the Information Society.* Belmont, Calif.: Wadsworth, 1996.

Sullivan, Danny. [See under *Search Engine Watch.*]

Sutton, Robert I. "Maintaining Norms about Expressed Emotions: The Case of Bill Collectors." *Administrative Science Quarterly* 36 (1991): 245–68.

Sutton, Robert I., and Anat Rafaeli. "Untangling the Relationship between Displayed Emotions and Organizational Sales: The Case of Convenience Stores." *Academy of Management Journal* 31 (1988): 461–87. [See also under Rafaeli and Sutton]

Suvin, Darko. "On Gibson and Cyberpunk SF." In *Storming the Reality Studio.* Ed. Larry McCaffery. Durham, N.C.: Duke University Press, 1991.

Sward, Keith. *The Legend of Henry Ford.* 1948; rpt., New York: Russell & Russell, 1968.

Tabitha's Days at Work (Mpeg). Ed. Martin Richard Friedman. 1994. Retrieved 8 November 2000. <http://www-white.media.mit.edu/~martin/snaps/>

Tapscott, Don. *The Digital Economy: Promise and Peril in the Age of Networked Intelligence.* New York: McGraw-Hill, 1996.

Taylor, Frederick Winslow. *The Principles of Scientific Management.* New York: Harper & Brothers, 1911.

———. *Shop Management.* In *Scientific Management, Comprising "Shop Management," "The Principles of Scientific Management," "Testimony before the Special House Committee."* 1947; rpt., Westport, Conn.: Greenwood, 1972.

———. *Testimony before the Special House Committee.* In *Scientific Management, Comprising "Shop Management," "The Principles of Scientific Management," "Testimony before the Special House Committee."* 1947; rpt., Westport, Conn.: Greenwood, 1972.

Taylor, Mark C. *Hiding.* Chicago: University of Chicago Press, 1997.

The Tele-Garden. Co-directors, Ken Goldberg and Joseph Santarromana. 1995–97. University of Southern California. Retrieved 9 November 2000. <http://www.usc.edu/dept/garden/>

Thomas, R. Roosevelt, Jr. *Beyond Race and Gender: Unleashing the Power of Your Total*

Work Force by Managing Diversity. New York: AMACOM/American Management Association, 1991.

Thomas, Robert J., and Thomas A. Kochan. "Technology, Industrial Relations, and the Problem of Organizational Transformation." In *Technology and the Future of Work*. Ed. Paul S. Adler. New York: Oxford University Press, 1992.

Thompson, Robert Farris. "An Aesthetic of the Cool." *African Arts* 7, no. 1 (Autumn 1973): 41–43, 64–67, 89.

Time magazine

Cover with picture of Marc Andreessen titled "The Golden Geeks." Cover story titled "High Stakes Winners." By James Collins. 19 February 1996, pp. 43–47.

"Machine of the Year: The Computer Moves In." By Otto Friedrich. 3 January 1983, pp. 14–24.

Toffler, Alvin. *The Third Wave*. New York: Bantam, 1981.

Tolliday, Steven, and Jonathan Zeitlin. "Shop Floor Bargaining, Contract Unionism, and Job Control: An Anglo-American Comparison." In *On the Line: Essays in the History of Auto Work*. Ed. Nelson Lichtenstein and Stephen Meyer. Urbana: University of Illinois Press, 1989.

Tomasko, Robert M. *Downsizing: Reshaping the Corporation for the Future*. Rev. ed. New York: AMACOM/American Management Association, 1990.

Tönnies, Ferdinand. *Community and Society (Gemeinschaft und Gesellschaft)*. Trans. and ed., Charles P. Loomis. East Lansing: Michigan State University Press, 1957.

Touraine, Alain. *The Post-Industrial Society—Tomorrow's Social History: Classes, Conflicts and Culture in the Programmed Society*. Trans. Leonard F. X. Mayhew. New York: Random House, 1971.

Transcriptions Project. [See under Liu, Alan, et al., ed.]

Treanor, Paul. "Internet as Hyper-Liberalism." *Telepolis* 1996. Republished on Paul Treanor's untitled Web site on "Liberalism, market, ethics; nationalism, geopolitics, the state; the future of Europe; urban theory and planning." Retrieved 16 July 2003. <http://web.inter.nl.net/users/Paul.Treanor/net.hyperliberal.html>

Trend, David, ed. *Reading Digital Culture*. Malden, Mass.: Blackwell, 2001.

Trice, Harrison M. *Occupational Subcultures in the Workplace*. Ithaca, N.Y.: ILR Press, 1993.

Trixter/Hornet [pseud.]. *PC Demos Explained*. "Final update" version. 21 September 1998. *Hornet Demogroup/The Oldskool PC*. Retrieved 14 December 2001. <http://www.oldskool.org/demos/explained/>

Tschichold, Jan. *The New Typography: A Handbook for Modern Designers*. Trans. Ruari McLean. Berkeley: University of California Press, 1995.

Tucker, Robert C., ed. *The Marx-Engels Reader*. 2nd ed.. New York: W. W. Norton, 1978.

Tufte, Edward, R. *The Visual Display of Quantitative Information*. Cheshire, Conn.: Graphics Press, 1983.

Tulgan, Bruce. *Managing Generation X: How To Bring Out the Best in Young Talent*. Santa Monica, Calif.: Merritt, 1995.

Tuman, Myron C. *Word Perfect: Literacy in the Computer Age*. Pittsburgh: University of Pittsburgh Press, 1992.

Turkle, Sherry. *Life on the Screen: Identity in the Age of the Internet*. New York: Simon & Schuster, 1995.

Turner, Victor W. *The Ritual Process: Structure and Anti-Structure.* Ithaca, N.Y.: Cornell University Press, 1977.

UC DARnet (University of California Digital Arts Research Network). University of California Multicampus Research Group. Home page. Retrieved 30 July 2003. <http://www.ucdarnet.org/>

UC Digital Cultures Project. Dir. William Warner. University of California Multicampus Research Group. Home page. Last rev. 25 June 2003. Retrieved 30 July 2003. <http://dc-mrg.english.ucsb.edu/>

UCLA Cultural VR Lab. Home page. University of California, Los Angeles. Retrieved 19 July 2002. <http://www.cvrlab.org/index.html>

Ueno, Toshiya. "Japanimation and Techno-Orientalism." Retrieved 23 July 2003. <http://www.t0.or.at/ueno/japan.htm>

Unsworth, John. "Living Inside the (Operating) System: Community in Virtual Reality." In *Computer Networking and Scholarly Communication in the Twenty-First-Century University.* Ed. Teresa M. Harrison and Timothy Stephen. Albany: SUNY Press, 1996. "Draft" version of 29 May 2001 retrieved from *Postmodern Culture* Web site, 27 June 2001. <http://www.iath.virginia.edu/pmc/Virtual.Community.html>

URouLette. Jill M. Sheehan. Retrieved in various versions, 1996–2000; last retrieved 10 November 2000. <http://www.uroulette.com/>

USA Today. "Technology Changing Holiday Habits." By Stephanie Armour. 26 May 1998. *USA Today.com.* Retrieved 27 May 1998. <http://www.usatoday.com/life/cyber/tech/ctc804.htm?st.ne.fd.mnaw>

The Useless WWW Pages. Ed. John Gephart IV. *Go2Net.* Retrieved 8 November 2000. <http://www.go2net.com/useless/>

Vallas, Steven Peter, and Michael Yarrow. "Advanced Technology and Worker Alienation: Comments on the Blauner/Marxism Debate." *Work and Occupations* 14 (1987): 126–42.

Van Maanen, John, and Gideon Kunda. "'Real Feelings': Emotional Expression and Organizational Culture." *Research in Organizational Behavior* 11 (1989): 43–103.

Verzola, Roberto. "Towards a Political Economy of Information." March 1998. CRIT-ICT/Distributed Knowledge Project, York University, Toronto. Retrieved 2 June 2000. <http://www.yorku.ca/research/dkproj/crit-ict/rv3.htm>

Vesna, Victoria. *n 0 time.* Design/Media Arts Department, University of California, Los Angeles. Last retrieved 19 July 2002. <http://notime.arts.ucla.edu/notime3/>

Vincent, Ted. *Keep Cool: The Black Activists Who Built the Jazz Age.* London: Pluto, 1995.

Vogel, Jennifer. "The Walls Have Eyes: The Many Ways Bosses Spy on Employees." *Working Stiff* site, *PBS Online.* Retrieved 24 May 2000. Sequence of five web pages beginning at <http://www.pbs.org/weblab/workingstiff/features/ionu.html>

Walker, Scott, ed. *The Graywolf Annual Seven: Stories from the American Mosaic.* St. Paul, Minn.: Graywolf, 1990.

Wall Street Journal

"Internet Search Engines Trail Web's Growth, Study Shows." 8 July 1999. By Associated Press. *Wall Street Journal Online.* Retrieved 8 July 1999. <http://interactive.wsj.com/articles/SB931389114102902503.htm>

"Users Are Choosing Information Over Entertainment on the Web." By Jared Sandberg. 20 July 1998. *Wall Street Journal Interactive Edition.* Last retrieved 3 August 2001 under the title "Mundane Matters: It Isn't Entertainment That Makes the

Web Shine; It's Dull Data" from Dow Jones News/Retrieval Publications Library. <http://interactive.wsj.com/nrw-docs/transfer.html>

Wardrip-Fruin, Noah, and Nick Montfort, ed. *The New Media Reader.* Cambridge, Mass.: MIT Press, 2003.

Warner, William. "Media Determinism and Media Freedom after the Digital Mutation: The Internet, *The Matrix* and Napster." Paper presented at Participatory Design Conference 2000, Graduate Center, City University of New York, 1 December 2000. *The Digital Cultures Project,* University of California. Retrieved 23 August 2001. <http://dc-mrg.english.ucsb.edu/committee/warner/MediaDeterminism.html>

Washington Alliance of Technology Workers (WashTech). Home page (*WashTech: A Voice for the Digital Workforce*). 29 June 2000. Retrieved 14 July 2000. <http://www.washtech.org/index.php3>

Washington Post. "Liberal Arts at GMU Targeted by Degrees: Proposed Cuts Draw Immediate Protests." By Victoria Benning. 15 March 1998, p. B3.

Waters, Crystal. *Web Concept & Design: A Comprehensive Guide for Creating Effective Web Sites.* Indianapolis: New Riders, 1996.

Waters, Lindsay. "A Critique of Pure Hipness." Unpublished manuscript, 1999.

Watkins, Evan. *Throwaways: Work Culture and Consumer Education.* Stanford: Stanford University Press, 1993.

———. *Work Time: English Departments and the Circulation of Cultural Value.* Stanford: Stanford University Press, 1989.

Wayback Machine. Internet Archive. Last retrieved 8 July 2003. <http://www.archive.org/web/web.php>

Web Pages That Suck: Learn Good Design by Looking at Bad Design [Web site; see under Flanders, Vincent.]

Weber, Max. *The Protestant Ethic and the Spirit of Capitalism.* Trans. Talcott Parsons. 1930; rpt. London: Routledge, 1992.

Wharton, Amy S. "The Affective Consequences of Service Work: Managing Emotions on the Job." *Work and Occupations* 20 (1993): 205–32.

What's Inside Jeremy's Wallet? Jeremy Wilson. 5 October 1999. Retrieved 8 November 2000. <http://www.inforamp.net/~xeno/wallet/>

What Is Miles Watching on TV! Miles Michelson. 21 December 1995. Retrieved 8 November 2000. <http://www.csua.berkeley.edu/~milesm/ontv.html>

White, Hayden. *Tropics of Discourse: Essays in Cultural Criticism.* Baltimore: Johns Hopkins University Press, 1978.

White, Michele. "The Aesthetic of Failure: Net Art Gone Wrong." *Angelaki* 7, no. 1 (2002): 173–94.

Whyte, David. *The Heart Aroused: Poetry and the Preservation of the Soul in Corporate America.* New York: Currency/Doubleday, 1994.

Whyte, Willam H., Jr. *The Organization Man.* New York: Simon and Schuster, 1956.

Wichansky, Anna M., and Michael F. Mohageg. "Usability in 3D: Silicon Graphics, Inc." In *Usability in Practice: How Companies Develop User-Friendly Products.* Ed. Michael E. Wiklund. Boston: Academic Press, 1994.

Wicke, Jennifer. "Sublime Lite: Millennial Anticlimax and the American (End of) Century." Paper presented at the English Institute, Cambridge, Mass., 2 October 1999.

Wiklund, Michael E., ed. *Usability in Practice: How Companies Develop User-Friendly Products.* Boston: Academic Press, 1994.

The Wild Bunch. Dir. Sam Peckinpah. Warner Bros., 1969.

Williams, Jeffrey. "Brave New University." *College English* 61 (1999): 742–51.

Williams, Raymond. *Marxism and Literature.* Oxford: Oxford University Press, 1977.

Willis, Paul E. *Profane Culture.* London: Routledge & Kegan Paul, 1978.

Wimsatt, W. K., Jr. *The Verbal Icon: Studies in the Meaning of Poetry.* With two preliminary essays written in collaboration with Monroe C. Beardsley. Lexington: University of Kentucky Press, 1954.

Winner, Langdon. "Cyberlibertarian Myths and the Prospects for Community (Draft for Comment)." 1997. *Langdon Winner's Home Page.* Retrieved 14 July 2003. <http://www.rpi.edu/~winner/Cyberlib.html>

———. "Do Artifacts Have Politics?" *Daedalus* 109, no. 1 (Winter 1980): 121–36.

———. "Mythinformation." In Langdon Winner, *The Whale and the Reactor: A Search for Limits in an Age of High Technology.* Chicago: University of Chicago Press, 1986.

———. "Peter Pan in Cyberspace: *Wired* Magazine's Political Vision." *Educom Review* 30, no. 3 (May–June 1995). *Educause.* Retrieved 24 May 2000. <http://www.educause.edu/pub/er/review/reviewArticles/30318.html>

———. "Technē and Politeia." In Langdon Winner, *The Whale and the Reactor: A Search for Limits in an Age of High Technology.* Chicago: University of Chicago Press, 1986.

Winograd, Terry. "Terry Winograd's Thoughts on CPSR's Mission." [See under Computer Professionals for Social Responsibility.]

Wired magazine. Condé Nast Publications, Inc.

Wölfflin, Heinrich. *Principles of Art History: The Problem of the Development of Style in Later Art.* Trans. M. D. Hottinger. 1932; rpt., New York: Dover, 1950.

Women & Performance: A Journal of Feminist Theory. Special issue on "Sexuality and Cyberspace." Issue 17 (vol. 9, no. 1, 1996).

Wordsworth, William. *The Prelude: 1799, 1805, 1850: Authoritative Texts, Context and Reception, Recent Critical Essays.* Norton Critical Edition. Ed. Jonathan Wordsworth, M. H. Abrams, and Stephen Gill. New York: W. W. Norton, 1979.

———. *The Prose Works of William Wordsworth.* Ed. W. J. B. Owen and Jane Worthington Smyser. 3 vols. Oxford: Oxford University Press, 1974.

———. *William Wordsworth.* Ed. Stephen Gill. Oxford: Oxford University Press, 1984. [Shorter poems of Wordsworth are cited from this edition.]

Workforce 2000: Work and Workers for the 21st Century. Prepared by William B. Johnston and Arnold E. Packer for the U. S. Department of Labor. Indianapolis: Hudson Institute, 1987.

Workplace: The Journal for Academic Labor. Retrieved 27 June 2001. <http://www.workplace-gsc.com/>

Worst of the Web. Mirsky. [Site extinct in August 1997, after which URL is used for "The NEW Mirsky.com".] Mirsky-Style Productions. Retrieved on various dates, 1996–97. <http://www.mirsky.com/wow/>

The Worthless Page. Retrieved 8 November 2000. <http://www.nktelco.net/worthless/>

Wouters, Cas. "Developments in the Behavioural Codes between the Sexes: The Formalization of Informalization in the Netherlands, 1930–85." *Theory, Culture & Society* 4 (1987): 405–27.

———. "On Status Competition and Emotion Management." *Journal of Social History* 24 (1991): 699–717.

Wray, Stefan. [See Electronic Disturbance Theater.]

Wright, Erik Olin. *Classes.* London: Verso, 1985.

———. *Class Structure and Income Determination.* New York: Academic, 1979.

Wright, Erik Olin, et. al. *The Debate on Classes.* London: Verso, 1989.

Wright, Shauna. *Flaunt.* Retrieved 22 August 1999. <http://www.flaunt.net/>

Yahoo! Yahoo!, Inc. Retrieved 6–7 July 1998. <http://www.yahoo.com/>

Zakon, Robert H'obbes'. *Hobbes' Internet Timeline.* Version 6.0. 5 February 2003. Retrieved on various dates, 2001–2003; last retrieved 16 June 2003. <http://www.zakon.org/robert/internet/timeline/>

ZDNet News. CNET Networks, Inc. Retrieved on various dates, 1999–2000; last retrieved 19 November 2000. <http://www.zdnet.com/zdnn/>

Zero-Knowledge Systems, Inc. *Freedom* (home page for Freedom® software.) Retrieved 4 March 2000. <http://www.freedom.net/>

Žižek, Slavoj. "From Virtual Reality to the Virtualization of Reality." In *Electronic Culture: Technology and Visual Representation.* Ed. Tim Druckrey. New York: Aperture, 1996.

Zuboff, Shoshana. *In the Age of the Smart Machine: The Future of Work and Power.* New York: Basic Books, 1988.

Index